Astrophysics and Space Science Library

Volume 405

More information about this series at http://www.springer.com/series/5664

Leonty Miroshnichenko

Solar Cosmic Rays

Fundamentals and Applications

Second Edition

Springer

Leonty Miroshnichenko
IZMIRAN
Moscow, Russia

ISSN 0067-0057 ISSN 2214-7985 (electronic)
ISBN 978-3-319-09428-1 ISBN 978-3-319-09429-8 (eBook)
DOI 10.1007/978-3-319-09429-8
Springer Cham Heidelberg New York Dordrecht London

Library of Congress Control Number: 2014946075

Cover illustration: Artist illustration of a coronal mass ejection. Image credit: NASA/MSFC

Printed on acid-free paper

Springer is part of Springer Science+Business Media (www.springer.com)

To those who will explore the Moon, Mars and other planets…

Preface

Sol Lucet Omnibus

It turned out to be really a rare and happy occasion to know exactly when and how a new branch of space physics was born, namely, a physics of solar cosmic rays. It happened on 28 February and 7 March 1942 when for the first time two "cosmic ray bursts" were recorded on the Earth. But only several years after, when two similar events occurred on 25 July 1946 and, particularly, on 19 November 1949, the Sun was unambiguously identified as the source of high-velocity particles with energies up to $\geq 10^{10}$ eV. Due to such a high energy, these relativistic particles have been called "solar cosmic rays" (SCRs), in distinction from the "true" cosmic rays of galactic origin, or galactic cosmic rays (GCRs), discovered in 1912. Between 1942 and the beginning of the space era in 1957, only extremely high-energy solar particle events could be occasionally recorded by cosmic ray ground-level detectors and balloon-borne sensors.

Relativistic SCR events are those where solar protons contain enough energy to initiate nuclear cascades in the atmosphere that can penetrate the surface of the Earth. At the beginning of the 1970s (e.g., Duggal and Pomerantz 1971; Duggal et al. 1971), these events acquired a special international name – "Ground Level Events," or "Ground Level Enhancements" (GLEs) – because they produce sudden increases in the intensity of the secondary cosmic radiation measured by surface cosmic ray detectors (see also Simpson 1990; Cliver 2009). Solar cosmic ray events were initially identified by Forbush (1946) who made the first association between solar flare activity and GLEs. The early GLEs were observed by cosmic ray ionization chambers whose primary response was to the secondary muons generated by the interaction of the incident high-energy particles (protons >4 GeV) with the atmosphere. Since then, the detection techniques varied considerably, and the study of SCRs became an essential part of solar and solar-terrestrial physics. At present (the middle of 2014), a total of 71 GLEs have been recorded by ground-based detectors.

During the last 50 years, the physics of the Sun has been developing very extensively in many directions. Alongside the traditional branches of investigation (solar activity and cyclicity, solar flares, solar wind, energetic solar particles and their influences on the Earth's environment), some new possibilities appeared necessitating further study of the Sun (solar neutrinos, helioseismology, etc). A number of new phenomena have been discovered, for instance, coronal mass ejections (CMEs) and coronal holes (CHs); also, high-energy neutrons and gamma rays from solar flares were observed for the first time. As a result, enormous amounts of diverse data were obtained from different techniques (ground-based telescopes and satellite detectors) in different "channels" of observation (solar energetic particles (SEPs), solar wind, electromagnetic waves, coronal and/or interplanetary shocks, etc.).

However, even though our understanding of the solar processes has evolved dramatically during the past 30–40 years, there are still many unanswered questions to be solved. It especially concerns the physics of particle acceleration at the Sun, or, in other words, the production of SCRs. It should be emphasized that, in spite of the very impressive achievements in other areas, SEPs of different energies have been and still serve to be one of the most generous sources of data about the Sun.

From the astrophysical point of view, the Sun represents a unique stellar laboratory where we can directly observe cosmic ray generation, i.e., acceleration of charged particles (ions and electrons) to very high energies. Solar cosmic rays produced in solar flares (and probably in some other high-energy solar processes) are one of the most important manifestations of solar activity (SA) and one of the main agents in solar-terrestrial relationships (STRs). The astrophysical aspects of solar cosmic ray physics (magnetic structure and plasma dynamics in the sources of accelerated particles, their maximum number and energy, occurrence rate of registration, production of neutrons, high-energy gamma rays, and neutrinos in flares, etc.) are of enormous interest.

Now, after 72 years of observation, we can define three basic lines of fundamental space research related to solar cosmic rays: (1) physics of the Sun (eruptive processes – solar flares, coronal mass ejections, and related phenomena; structure and dynamics of magnetic fields in the solar atmosphere; mechanisms of particle acceleration); (2) physics of the heliosphere (structure, dynamics, and turbulence of the interplanetary magnetic field, or IMF; models of particle propagation); and (3) geophysics (interaction of SCRs with the terrestrial envelopes – magnetosphere, ionosphere, and neutral atmosphere) and mechanisms of solar-terrestrial relations (STRs).

Among applied aspects, I would point out, first of all, to the prediction of SCR flux for the needs of practical astronautics (cosmonautics), i.e., provision of radiation safety for the crew and spacecraft equipment. The problem acquires specific importance in the context of the development of some very ambitious projects to set up space power stations at geosynchronous orbits with the term of operation of up to 30 years, as well as in connection with an increased duration and distance of multipurpose spacecraft flights in circumterrestrial and interplanetary orbits.

From the very beginning, it should be noted that it was not my intention to present a comprehensive analysis of the problem. Nevertheless, I have tried to give an up-to-date summary of my knowledge of SCR generation and propagation. The present monograph differs from the reviews published earlier in three main aspects: (1) it presents the problem in a self-contained form, in all its aspects – from a historical outline to the present state of the problem, from the main concepts and hypotheses to modern models, from astrophysical aspects to geophysical and astronautical applications; (2) it includes a large amount of new data which have not yet been described in the review literature; and (3) it contains an extensive bibliography which gives a fair idea about the historical development of the problem and covers impartially the main achievements and failures in this field. The book is implied not only to be one of the many reviews in solar physics but will also serve as a useful manual (guide) in this rapidly developing field of space research. I would be happy if his efforts stimulate a new interest to the problem, especially from the new generation of investigators. The book may be of relevance to a few graduate courses and will be useful, hopefully, at the postgraduate level as well.

Space constraints do not allow me to explain every time the solar-terrestrial nomenclature used in the current English language literature. To make clear the jungle of terms, I recommend to the readers a list of standard terms described in detail in the *Illustrated Glossary for Solar and Solar-Terrestrial Physics* (Eds.: A. Bruzek and C.J. Durant, 1977). The Glossary is designed to be a technical dictionary that will provide solar and geophysical workers with concise information on the nature and properties of the phenomena of solar and solar-terrestrial physics. Some terms are updated in Appendixes 1 and 2. The monograph contains 12 chapters, and their contents cover five principal "blocks":

1. Production of SCRs: solar flares, coronal mass ejections, other high-energy phenomena at the Sun; particle acceleration and release (Chaps. 1, 2, 3, and 4).
2. Energy spectrum of SCRs in their sources; interaction and transport processes in the corona; production of relativistic particles in the extended coronal structures (including CME-driven shocks) (Chaps. 5, 6, and 7).
3. Interplanetary propagation: physical and sounding aspects of SCR studies (Chap. 8). Arising heliospheric aspects are also considered here.
4. Energy spectrum of SCRs in the geosphere and its geophysical effects (Chaps. 9 and 10).
5. Solar energetic particles and radiation hazard in space (Chap. 11).
6. Chapter 12 contains the summary and conclusions. I consider existing problems and arising matters, outline some promising prospects in this field, and try to compile a list of research tasks for future studies.

In two separate Appendixes (in the Backmatter), I present the main acronyms and Author's Index. At the end of the book, the relevant References are listed (in alphabetical order). I have tried to present the most significant observational results concerned with particle acceleration at the Sun, energetic particles in interplanetary space, and geophysical effects of SCRs known up to the beginning of 2014. Some recent review papers published since 2001 (year of publication of the first edition of this book) are also included in the monograph. Nevertheless, more than a few interesting problems which do deserve attention probably remain unmentioned, and I apologize most sincerely to those of my colleagues whose work I have been unable to discuss properly in this book, for one reason or another.

The book was thought in the atmosphere of "information burst" in space research that occurred during the last three decades, and in Chap. 12 I have made an attempt to summarize the most interesting theoretical concepts, models, and ideas which do deserve attention in the context of solar cosmic ray investigations. I hope that the book will be helpful for a wide enough circle of space physicists and geophysicists. Some results may be of interest for those whose fields are theoretical physics or plasma physics.

I acknowledge heartily joint works and fruitful discussions, severe criticisms, valuable advices, and innumerable helpful comments from several tens of my colleagues in the former Soviet Union, Russia, United States of America, Mexico, China, Czechoslovakia, Germany, Switzerland, Spain, Italy, Canada, Finland, Australia, and other countries. My special acknowledgments and warmest thanks go to my numerous Mexican colleagues and friends from Instituto de Geofísica UNAM (Mexico City) and to my long-standing colleagues from IZMIRAN (Troitsk) and especially from my "Alma Mater," Moscow State University (Moscow), for their generous support, cordial help, and constant cooperation and for providing favorable conditions for scientific work.

I am extremely indebted and cordially grateful also to my wife Nina for her invaluable help, indispensable support, and incomparable patience over several decades of my research work.

IZMIRAN, Troitsk, Moscow, Russia Leonty I. Miroshnichenko
7 November 2013–1 May 2014

Contents

1990, 2003a; Miroshnichenko and Petrov 1985). During "space era" years, a new class of observations has revealed two distinct populations of solar energetic particles (SEP), with completely different origins, based upon the abundances, ionization states and time profiles of the particles as well as the longitude distribution and the radio, optical, X-ray and gamma-ray associations of the events (Reames 1995a, b, c, 1996, 1999). These observations gave rise to a new classification of SEP events, namely, to their separation on two distinct groups (impulsive and gradual ones), though there are also hybrid events in which both impulsive and gradual phenomena occur (Cliver 1996, 2009).

In addition, several new methods were proposed to determine energy spectra of SCR near the Earth by the data of observations inside magnetosphere, at ionospheric heights, and upon data of ground-based observations at isotropic and anisotropic phases of GLE (see, *e.g.*, Miroshnichenko 1990, 2001; Miroshnichenko and Petrov 1985 and references therein). First data were obtained on flare gamma radiation (Ramaty et al. 1975) and solar neutrons (Chupp et al. 1982; Kocharov 1983; Chupp 1996). It was found a certain association of large SPEs with fast (>400 km s^{-1}) CMEs, ejection profiles of solar protons (>10 MeV) being correlated with CME heights in some events (Kahler 1994). Several attempts were undertaken to confirm and substantiate the existence of upper energy limit of SCR spectrum (for detail see Miroshnichenko 1990, 1994, 1996, 2001).

It is even more important to note that in recent years a number of new acceleration models for ions and electrons were proposed based on different initial grounds (see Chap. 5). One group of these models relies directly on reconnection theory of solar flares (*e.g.*, Litvinenko and Somov 1995; Somov 1996, 2012), second one involves stochastic acceleration by plasma turbulence (*e.g.*, Miller 1991) and/or acceleration by shock waves (*e.g.*, Kallenrode and Wibberenz 1997; Berezhko and Taneev 2003) provided the mechanism of particle return to shock front does really exist (for more detail see an excellent monograph of Berezinsky et al. (1990) and several comprehensive reviews (Forman et al. 1986; de Jager 1986; Scholer 1988; Vlahos et al. 1989; Chupp 1996; Miller et al. 1997; Priest and Forbes 2000). It seems clear, however, that all three basic acceleration mechanisms could be simultaneously involved in some flares.

On the other hand, the largest and most energetic particle events at the Earth's orbit seem to be associated with shock waves driven out into interplanetary space by CMEs (*e.g.*, Reames 1996, 1999). Evidently, serious modeling efforts are required to reconcile different approaches to the problem and to eliminate obvious controversies in interpretation of recent observational data. In the whole, the problem of SCR spectrum formation remains unresolved: if in the range of low or moderate energies the spectrum and composition are determined by intimate local plasma processes (*e.g.*, Miroshnichenko 1987, 1995; Vlahos 1989; Miller et al. 1997) deeply inside the solar atmosphere, the spectrum formation in relativistic range is very likely to govern by large-scale, extended magnetic structures high in the corona (Perez-Peraza et al. 1992; Chertok 1995; Miroshnichenko 1997, 2001; Miroshnichenko et al. 1996, 2000).

1.2 Contribution to Solar-Terrestrial Relationships

Besides those aspects, the study of solar energetic particles is very important for some applied and geophysical problems, such as radiation hazard in space, radio wave propagation in high latitude regions, possible meteorological effects of SCR and other phenomena affecting man technology at the Earth's surface and in the nearest terrestrial environment. The detrimental effects of solar particles on terrestrial systems are well documented. These range from radiation damage of spacecraft electronic and solar arrays (*e.g.*, Kreinin and Grigorieva 1979; Adams and Gelman 1984; Miroshnichenko and Petrov 1985; Smart and Shea 1989b) after powerful SPEs, to production of induced voltages on telephone and power cables and corrosion on pipelines during severe geomagnetic storms (*e.g.*, Lanzerotti et al. 1991, 1995).

As it was stated by numerous studies, an occurrence rate, or a frequency of SPE registration near the Earth's orbit is determined not only by the chosen energy threshold of the measuring device (Dodson et al. 1975; Akinyan et al. 1983; Bazilevskaya et al. 1986, 1990a; Sladkova et al. 1998; Logachev et al. 2014), but to a great extent depends on the conditions of forming the observed SCR spectrum. Spectral features of SCR have a decisive importance for evaluating radiation dose and its dynamics (Miroshnichenko and Petrov 1985; Gussenhoven et al. 1988; Miroshnichenko 2003a, b). It is quite obvious a connection between SCR spectrum research and the tasks of geophysical and applied character (*e.g.*, Shea and Smart 1993b; Miroshnichenko 1992b, 2003a, b, 2008).

For over 55 years by now the effect has been studied of additional ionization of lower ionosphere in polar regions with the intrusion of solar protons (polar cap absorption of radio waves, or PCA effect). In recent years there have been obtained theoretical proofs and observational evidences (see Chap. 10) of an important role of SCR in depletion and general dynamics of terrestrial ozone layer, in changes of conductivity in global circuit of atmosphere electricity, in some other aeronomical and meteorological processes (see, *e.g.*, Roble 1985; Pudovkin and Raspopov 1992; Miroshnichenko 2008 and references therein).

The question of possible participation of SCR in meteorological processes is not studied in full yet (*e.g.*, Loginov and Sazonov 1978; Migulin et al. 1987; Tinsley and Deen 1991; Miroshnichenko 2008, 2011). More than 20 years ago, Pudovkin and Raspopov (1992) have suggested a physical mechanism concept of solar activity influence on the lower atmosphere and climate based on the experimental data and estimations carried out. Their main idea is a variability of atmosphere transparency (the change of "meteorological" solar constant, S_m) and, therefore, the change of solar energy flux penetrated in lower atmosphere due to the variations of SCR flux and intensity of galactic cosmic rays (GCR) modulated by solar activity (see Chap. 10). When estimating the above mentioned geophysical effects, precise quantitative information on absolute intensity and exponent of SCR spectrum is needed. Thus, the approach accepted by the author (Fig. 1.2), at least in general features, corresponds to the present state of SCR problem, including its fundamental and applied aspects. Of course, we should not overestimate relative role of SCR. However, their contribution may represent significant part of the physical mechanism of solar-terrestrial relationships, alongside with CMEs and other solar-interplanetary drivers of geophysical disturbances. At any rate, it is out of doubts now that the appearance of considerable flux of SEPs near the Earth is not isolated, exotic phenomena, but one of important manifestations of significant disturbance in the Sun-Earth space.

As to applied aspects, in 1985 an attempt was undertaken (Bengin et al. 1985; Miroshnichenko and Petrov 1985) to develop a new approach to the problem of diagnostics and prediction radiation conditions in space taking into account the dynamics of SCR radiation characteristics, stochastic nature of solar phenomena and requirements of necessary wholeness and precision of prediction (see Chap. 11). Since then a series of new prediction models was suggested, for example, an interplanetary proton fluence model JPL 1991 (Feynman et al. 1993; Getselev et al. 1996a; Feynman 1997). Although most major SPEs seem to be associated with CMEs, the solar flare process is the most commonly assumed source of solar protons. As it emphasized by Shea and Smart (1993b), we do not yet understand how the Sun accelerates ions to relativistic energies, nor how to predict the fluence from an individual flare, but we have assembled enough data to be able to place some preliminary, but nevertheless realistic, limits on the extent and severity of SPEs.

1.3 Pivot of the Problem

Previous reviews and monographs are mainly concentrated on the SCR dynamics at the Sun, in the interplanetary space, and in the Earth's environment, or have not been actualized with the present specialized publications on the most keen problems of solar flare and solar-terrestrial physics (STP). In contrast to many of modern researchers, the author (Miroshnichenko 2001) continue to proceed from the suggestion (conviction) that a source function (energy spectrum, chemical composition, and time profile) may serve as a pivot of SCR problem, and the formation of source function, its evolution and measurements are key questions of the problem under consideration (see Fig. 1.2). In other words, a consistent treatment of SCR dynamics must rest upon the analysis of their energy-charge distributions at different stages of SPE. Such an approach includes equally both acceleration models for treatment of spectrum formation at/near the Sun and propagation models for description of SCR intensity-time profiles near the Earth. In particular, this concept involves two fundamental process of general physical interest in astrophysics and space sciences, namely, acceleration of charged particles in space plasmas and their interaction with space magnetic fields, waves and matter.

We consider this approach to be physically justified and methodically convenient, as it allows for a comprehensive analysis of variety of physical processes associated with the build-up and development of SPE and the processes involved in

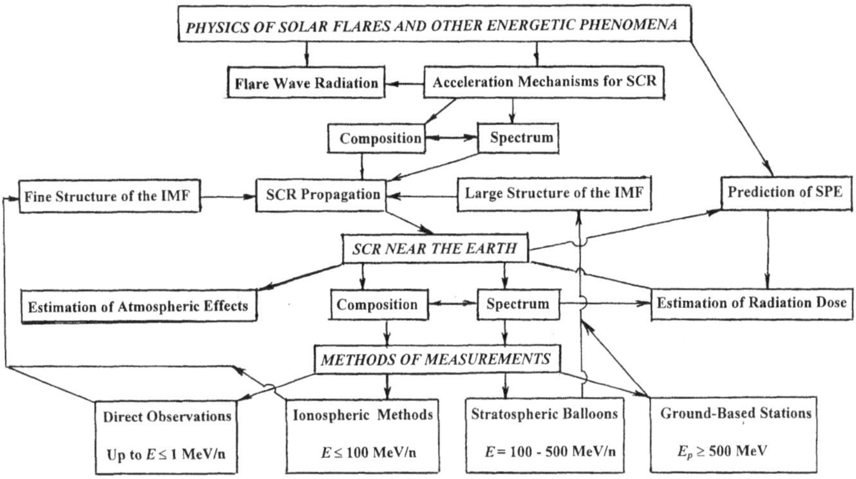

Fig. 1.3 General scheme of research of solar cosmic rays and the system of their "feedbacks" with other problems of solar-terrestrial physics (Adapted from Miroshnichenko and Petrov 1985). In the *lower part* of the scheme are given measurement methods and possible applications of SCR study

the formation of particle composition, energy spectrum and their evolution. On the other hand, it may also help to estimate the role of SCR in the dynamics and energetics of the flare phenomena. We believe that the proposed approach must in the end to be useful for any attempt to develop a self-consistent solar flare model. Therefore, it is made emphasis on theoretical models of SCR acceleration within the context of flare build-up models, and on models of SCR transport in the corona and interplanetary space. Besides, it enables to improve the techniques of SCR flux prediction, to emphasize existing difficulties and to single out problems unresolved.

At such approach, main present problems of SCR study and the system of their "feedbacks" with other problems of STP may be shown in the form of block-scheme (Fig. 1.3). Its key blocks are the processes of formation, evolution and observation of SCR spectrum. Upper parts of the scheme ("Source physics" and "SCR near the Earth") reflect, basically, the physical processes in which take part accelerated particles in the source, in the solar corona, in interplanetary medium and in the near-terrestrial space. In the lower part of the scheme are given measurement methods and possible applications of SCR research for sounding of physical conditions in different areas (ranging from the source to the Earth), as well as for predicting purposes.

From the scheme it is seen, in particular, that SCR spectrum is determined through the mechanism of acceleration and determines, in its turn, a number of the flare effects – SCR energetics, generation of neutral and electromagnetic radiation, etc. After acceleration the spectrum of escaping particles undergoes the primary deformation in corona (energy losses of different nature for protons and electrons, drift, diffusion and other effects). The influence of coronal magnetic field is displayed in a temporary delay of accelerated particles which depends upon their energy.

Further evolution of ejection spectrum occurs through the processes of particle transport in interplanetary space due to the influence of large-scale structure of interplanetary magnetic field (IMF) and its small-scale inhomogeneities (fluctuations). Particles are led by lines of force and experience drifts as well as scattering over IMF inhomogeneities (*e.g.*, Toptygin 1985), these effects are dependent on particle energy, too. Such a dependence causes the delay of particles coming to the Earth relatively to the moment of their release from the corona and determines the form of intensity-time profile of SCR near the Earth and the degree of deformation of observed spectrum if compared with ejection one. Besides, particles of low energies may undergo convection, adiabatic deceleration and additional acceleration in the interplanetary medium (*e.g.*, Miroshnichenko 1992b).

There are three main empirical approximations that are commonly used to describe the observed spectrum. They are a power-law in energy

$$D(E) = D_0 E^{-\gamma} \tag{1.6}$$

a power law in rigidity,

$$D(R) = D_0 R^{-\gamma} \tag{1.7}$$

and an exponential in rigidity,

$$D(R) = D_0 \exp(-R/R_0), \tag{1.8}$$

where D_0 is the normalization coefficient, and the parameters γ and R_0 are energy (rigidity) dependent and may change with time during a SPE. Moreover, observed SCR spectrum during a large SPE near the Earth may cover of 4–5 orders of the energy value (from >1 MeV to >10 GeV), and difference in the intensity of particles (protons) at the ends of the spectrum (due to its great steepness in the range of high energies) may amount to 6–8 orders (Miroshnichenko 1994, 1996, 2001). This causes certain methodical difficulties in measuring SCR near the Earth (on the GCR background) and interpretation of the data obtained. To give an adequate fitting of the SCR spectrum in a wide energy interval one has to align, so to say, the results of a few kinds of measurements (onboard the satellites, in the stratosphere, at terrestrial surface, *etc.*), thus introducing additional errors (uncertainties) into the spectral characteristics. The same is true when estimating the spectrum of accelerated particles in the source directly by gamma-ray, neutron, or other solar flare data (Ramaty and Murphy 1987).

In its turn, the reconstruction of spectra near the Sun ("in the source") based on the observation data near the Earth is associated with the use of models which so far do not give an adequate idea of specific features of interplanetary transport of accelerated particles in different energy ranges (Miroshnichenko et al. 1999). A certain contribution into the process of formation of the observed SCR spectrum is introduced by possible temporal trapping in the corona, prolonged ejection and considerable anisotropy of ejected particles. Judging by observations of different kind of emission of flares and taking into account the results of theoretical simulation of acceleration it may be stated that formation of the SCR spectrum and charge distribution in the source has its own spatial structure and time dynamics (*e.g.*, Miroshnichenko 1993). To separate the effects of long duration acceleration, temporal containment and extended propagation of SEPs is a rather difficult research task.

This general scheme may be undergone to considerable modification, if we take into account some other sources of acceleration, for example, CME-driven shock (*e.g.*, Zank et al. 2000, 2007) or two converging shock waves (Kallenrode 2003). Some observational evidence of this latter possibility seems to obtain in the analysis of cosmic ray variations in July 1959 (Chirkov and Filippov 1977). Many details of such a picture should be anew investigated, but the main role of these additional sources reduces to the deformation of original intensity profiles and spectra produced by basic (flare) source. In Chap. 3 we discuss existing dilemma "Flare-CME", or "Solar Flare Myth". To our opinion, this dilemma may be resolved if we accept a general concept of drastic eruption in the solar atmosphere as a common phenomena for generation a pair "Flare-CME" implying close topological and/or physical links between both.

In this context, if we accept the source function as a pivot for the study and treatment of different aspects of SCR problem, the energy (rigidity) spectra of SEPs should be considered as a whole, without any division on relativistic and non-relativistic parts.

1.4 General Characteristics of Solar Particle Events

Although the solar flare process is the most commonly assumed source of energetic solar particles, recent research indicate that the coronal mass ejection (CME) may be the phenomenon that is associated with the release of solar protons into the interplanetary medium (Kahler et al. 1984; Kahler 1994, 1996; Reames 1999). Since most major flares are associated with solar mass ejections, it is still customary to refer to solar proton events as emanating from solar flares, and we will continue to use this nomenclature through the book. As it will be shown in other Chapters, many research efforts are still required to separate these two energetic solar phenomena and to make clear their basic physics and underlying mechanisms.

During a solar flare, electromagnetic radiation such as X-ray and radio emission is generated by the hot plasma and travels at the speed of light through interplanetary space. This type of radiation takes ~8.33 min to reach the Earth, and it is usually the first indication that a major flare has occurred. The onset of an increase in solar X-ray emission detected by sensors on the Earth-orbiting satellites of GOES type is approximately simultaneous with the visual observations of a solar flare usually made in the $H\alpha$ wavelength.

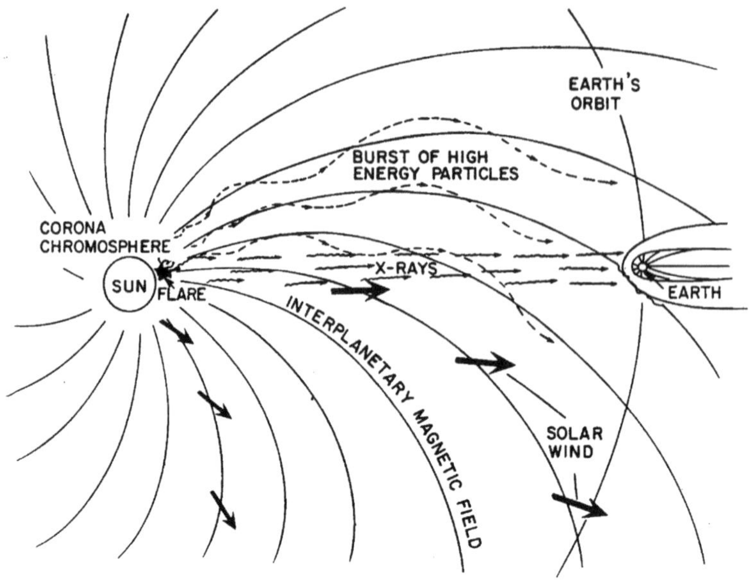

Fig. 1.4 Characteristics of the idealized structure of the interplanetary medium (Smart and Shea 1989a, b, 1993)

Unlike solar electromagnetic radiation, both the onset time and maximum intensity of SEP flux in a given (detection) point in space is dependent, first of all, upon the energy of the particle. The location (heliolongitude) of the flare with respect to detection point is also very important. Under idealized circumstances, from "well-connected" solar flares (50°W–70°W), relativistic solar protons can reach the Earth ($R_E = 1.0$ AU) within 10–15 min of the onset of the flare; 10 MeV protons take approximately 80–90 min to reach the same distance. This directionality results because solar electrons and ions, being charged particles, spiral along the interplanetary magnetic field lines. The IMF topology, in its turn, is determined by the solar wind outflow and the rotation of the Sun. During "quiet" conditions this topology can be approximated by an Archimedean spiral as illustrated in Fig. 1.4 (Smart and Shea 1989a, b, 1993).

As shown by numerous observations (*e.g.*, Miroshnichenko and Petrov 1985), SCR diffusion in interplanetary space proceeds mainly along the IMF lines (the parallel diffusion coefficient considerably exceeds the transverse one). Hence, the shape of the field line being known, one can localize the escape site of accelerated particles from the corona.

The method for estimating heliolongitude of the connection line between the Sun and the observation point (Nolte and Roelof 1973) proceeds from the assumption that solar wind propagates quasi-radially (see Fig. 1.4), its velocity not changing from the moment it leaves the corona until it reached the observational point. A slight solar wind acceleration that actually takes place with distance is partially

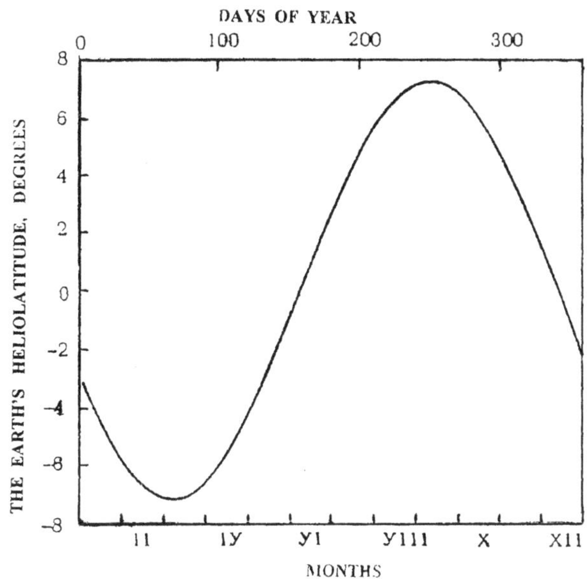

Fig. 1.5 Variations of the heliolatitude projection of the Earth during a year (Bazilevskaya et al. 1986)

compensated by the effect of plasma corotation with the Sun at a distance $r_A \leq r$, where r_A is the critical Alfvén point.

The connection longitudes are estimated with an accuracy of 10 %, provided the solar wind velocity at the observation point does not change by more than 50 % a day. The Carrington longitude (see Bruzek and Durrant 1977) of Sun-Earth connection, Φ_c, is given by the expression:

$$\Phi_c = \Phi_0 + (R_E/U)\,\Omega \tag{1.9}$$

where Φ_0 is the Carrington longitude of the central meridian, U is the solar wind velocity at the observation moment, $R_E = 1$ AU is the distance from the Earth to the Sun, and $\Omega = 13.3°$ per day is the angular rotation velocity of the Sun.

According to (1.9), the connection longitude changes continuously owing to both the rotation of the Sun and variations of the solar wind velocity. If the latter did not change, the connection longitude would uniformly shift eastwards at a rate of 13.3° a day owing to the Sun rotation. When the Earth gets into a high-speed stream of solar wind, the connection point sharply (in a jump) shifts eastwards due to a sudden increase in U. When U decreases, the connection point may keep its position for a day or more, or even move back westwards. In a heliocentric coordinate system (setting a value of $\Phi_0 = 0$), at a typical value of the solar wind velocity of about 400 km s^{-1} one can estimate from (1.9) an optimum connection longitude of 60° to the west from the central meridian.

The heliolatitude of the connection point can be estimated from the heliolatitude projection of the Earth which ranges from −7.25° to +7.25° during a year (Fig. 1.5).

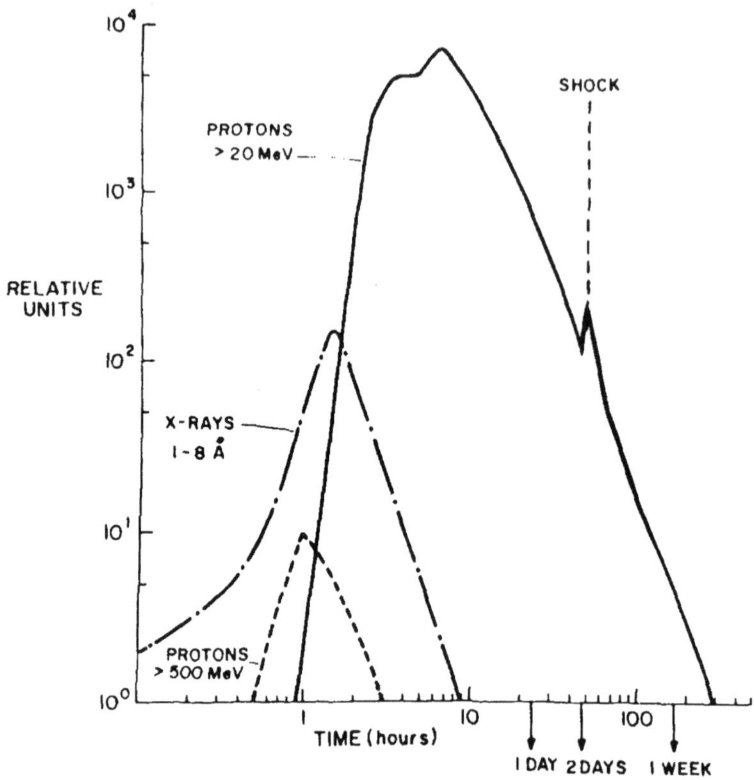

Fig. 1.6 Relative time scales of solar particle emissions at 1 AU (Shea and Smart 1993a, b)

Such estimate, however, should be treated with caution since above regularity may be considerably distorted by the geometry of magnetic field lines convergent at the helioequatorial plane (see Sect. 3.4).

When a major solar flare occurs there is also associated release of enhanced solar plasma into the interplanetary medium. This dense plasma usually propagates to the Earth within 1 or 2 days. When the plasma arrives and interacts with the Earth magnetosphere the resulting energy transfer manifests itself by the occurrence of aurora and geomagnetic disturbances. Their magnitudes depend on the current characteristics of the IMF and solar wind at the time of the arrival of the plasma at the Earth. These traveling interplanetary disturbances (shocks) can severely disrupt the decaying particle flux from the preceding flare. Occasionally, for major solar proton events, the ambient flux can be re-accelerated by interaction with the shock (*e.g.*, Dröge 1994b). Figure 1.6 illustrates the relative time of arrival and duration of solar particle emission at the Earth's orbit (Shea and Smart 1993a). The increase in particle flux at the time of arrival of the interplanetary shock is due to additional acceleration of the ambient particles at the shock front.

There are two important magnetic "barriers" on the way of fast SEPs from the Sun to the Earth – interplanetary magnetic field (IMF) and geomagnetic field

(GMF). The influence of the IMF resulted in that the most of relativistic proton events are quite anisotropic as viewed at a distance of 1.0 AU. It is now generally accepted (*e.g.*, Palmer 1982; Miroshnichenko and Petrov 1985; Bieber et al. 1994) that these particles usually have long mean free path lengths (>0.3 AU) with a variability of a factor of 3 within the range of normal expectations (see Chaps. 9 and 10). Therefore, measuring the SCR fluxes at a distance of 1.0 AU (onboard the "Spaceship Earth"), we may sometimes observe, in fact, the ejection profiles of relativistic solar particles (Shea and Smart 1993a, b).

The analysis of relativistic SPEs by surface detector data is a rather complex procedure dependent of SEP spectrum, their angle distribution (anisotropy) in the IMF, geomagnetic cutoff rigidity for each cosmic-ray station and some other factors (*e.g.*, Shea and Smart 1982; Flückiger and Kobel 1990; Cooke et al. 1991). When high energy solar proton enters the Earth's magnetosphere its path is deflected by the GMF, with the lower energy particles being "bent" more than the higher energy particles. This effect has lead to the concept of "asymptotic cones of acceptance" for cosmic ray detectors on the Earth (McCracken 1962a).

To explain the SCR anisotropy this concept should be described in some details. Asymptotic directions of approach are used to indicate the direction in which cosmic ray (CR) particles were travelling before they come under the influence of the Earth's magnetic field. They relate a specific allowed arrival direction with a unique direction in the interplanetary space. For a CR particle of rigidity R, arriving at the geographic location (λ, Φ) and incident at the angles (θ, φ), the asymptotic direction of approach is given by the unit vector $A = A(R, \lambda, \Phi, \theta, \varphi)$ pointing in the reverse direction to the particle's velocity vector V prior to its entry into the Earth's magnetosphere. The asymptotic direction of approach is represented in geocentric coordinates by the asymptotic latitude, λ, and the asymptotic longitude, Φ; in Cartesian coordinates

$$A = i\cos\lambda \quad \cos\Phi + j\cos\lambda \quad \sin\Phi + k\sin\lambda \tag{1.10}$$

Thus, each cosmic-ray detector has an unique viewing direction in space which is a function of energy. As the Earth rotates, the asymptotic viewing direction will co-rotate.

Asymptotic directions of approach are numerically calculated using a method known as the *trajectory tracing technique* (*e.g.*, Shea and Smart 1975; Gall et al. 1982; Cooke et al. 1991). In general, they are only calculated for vertically incident cosmic rays. For a particular geographic location and incident angle, a given particle trajectory in the magnetosphere is a function of R; therefore, cosmic rays of different rigidities arriving at $(\lambda, \Phi, \theta, \varphi)$ will originate in different parts of the sky. The part of the sky from which a cosmic-ray station observes the majority of the incoming particles is called its asymptotic cone of acceptance.

If a station is "viewing" into the direction toward the Sun, along the IMF lines connecting the Sun with the Earth, when SPE occurs, the station will detect a maximum particle flux increase. If a station is "viewing" in a quite opposite direction it will record a smaller increase depending of the anisotropy of the

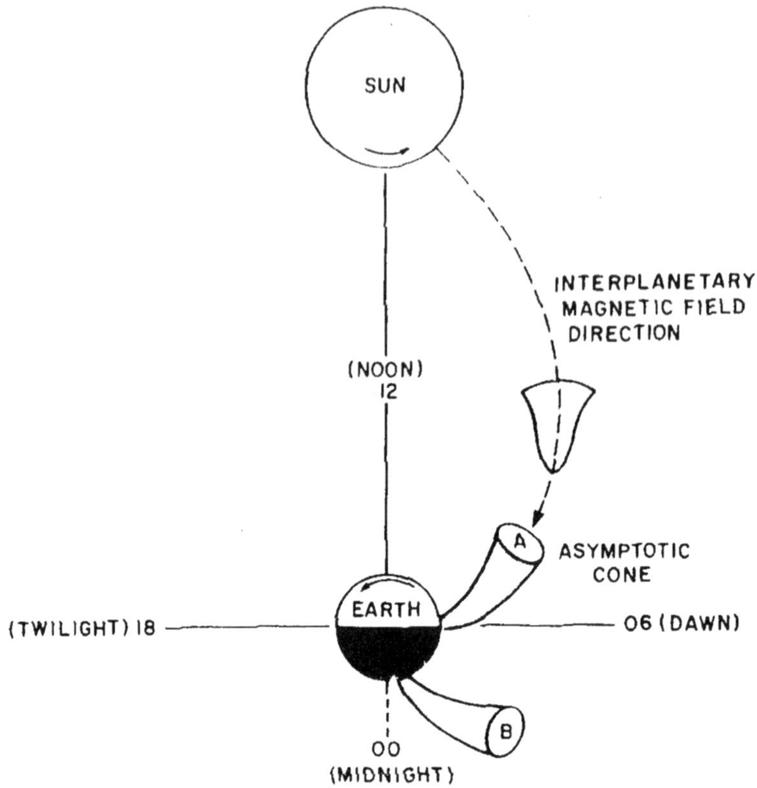

Fig. 1.7 Conceptual illustration of asymptotic cones for two cosmic ray stations and anisotropic particle flux traveling along the IMF line from the Sun to the Earth (Shea and Smart 1993b)

event. A conceptual illustration of asymptotic cones and an anisotropic particle flux traveling along the IMF line from the Sun to the Earth is given in Fig. 1.7 (Shea and Smart 1993b). In this example the station with asymptotic cone "A" viewing into the IMF direction would record a higher increase than the station with asymptotic cone "B" viewing in a completely different direction. If the event occurred a few hours later, when the station with asymptotic cone "B" had rotated such that this cone was viewing away from the Sun, then station "B" would have recorded the larger increase.

The asymptotic directions of approach (or asymptotic cones of acceptance) are, in fact, a mapping of the allowed particle directions in space prior to their interactions with the GMF. These directions can be calculated for each of cosmic-ray station (*e.g.*, Shea and Smart 1975; Gall et al. 1982).

In Fig. 1.8 we present two maps showing selected asymptotic directions of approach for relativistic protons detected during the two largest GLEs of February 23, 1956 and September 29, 1989 (Smart and Shea 1991). In these maps, the sub-solar point at the time of the onset is indicated by a solid dot. By two slightly

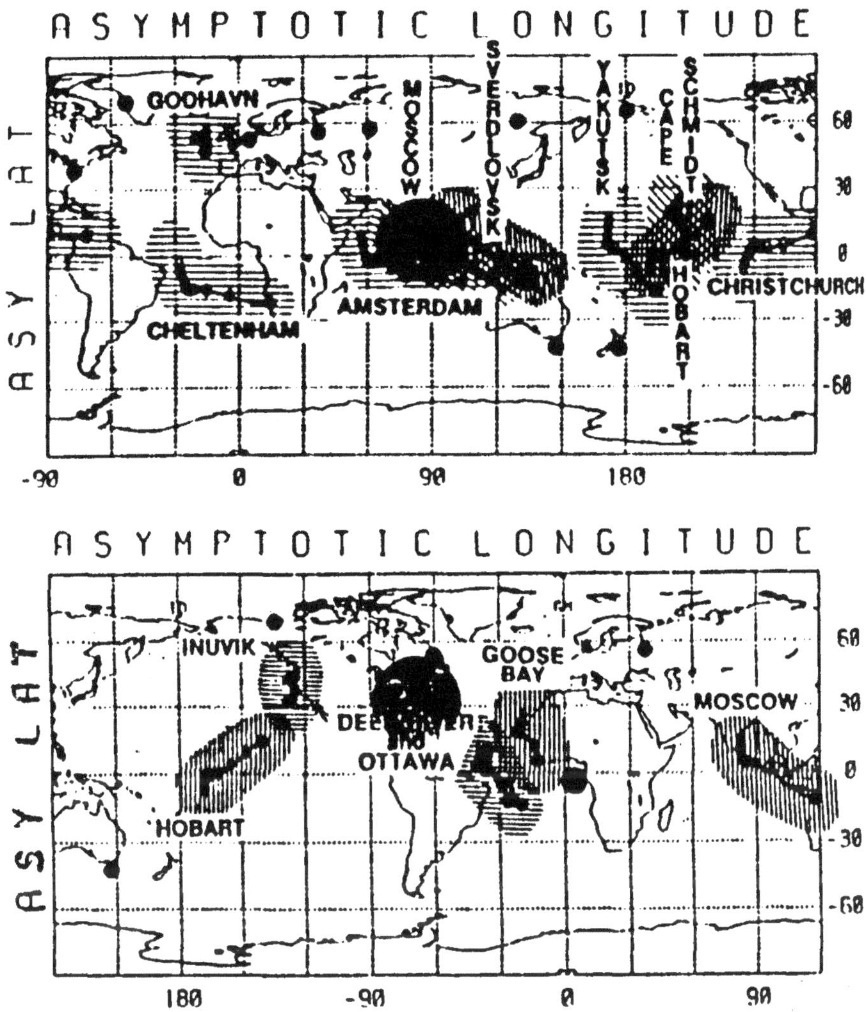

Fig. 1.8 The maps of selected asymptotic directions of approach (or asymptotic cones of acceptance) for relativistic solar protons during the GLEs of February 23, 1956 (*top panel*) and September 29, 1989 (*bottom panel*). The sub-solar point at the flare onset is indicated by a solid dot; the probable IMF direction is indicated by the shaded circle (Smart and Shea 1991)

larger dots, the 5 GV and 10 GV asymptotic directions are accentuated. The probable IMF direction is denoted by the shaded circle. The maps give a certain idea about the pitch angle distribution of relativistic solar particles near the Earth at the event onset.

It is of great importance to note that for each event the maximum increase was observed by the station whose asymptotic cone of acceptance contained or was closest to the probable IMF direction. A comparison of the magnitudes of the observed increase at each station in different times allow, in particular, to watch

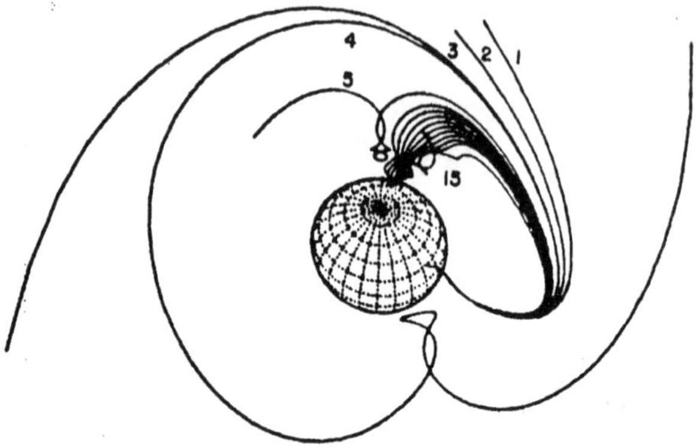

Fig. 1.9 Trajectories of cosmic-ray particles with different rigidities in the geomagnetic field (Smart and Shea 1994). The curves are numerated (*1–15*) from high to low energies (rigidities)

for an evolution of particle flux anisotropy as an indicator of probable changes in the IMF direction or in the direction to apparent particle source (*e.g.*, Smart et al. 1991; Vashenyuk et al. 1995, 1997, 2009, 2011).

The NM network distributed over the globe can be considered as an integral omnidirectional spectrometer for measuring characteristics of relativistic SCR flux at the Earth's orbit. Modeling the NM response to anisotropic SCR flux and solving the inverse problem, we can obtain the characteristics of relativistic solar protons outside the Earth's magnetosphere (Shea and Smart 1982; Humble et al. 1991a, b; Cramp et al. 1997; Vashenyuk et al. 2009, 2011). Data from 25 NM stations and a sufficient ground-level increase ($\geq$10 %) should be used in this basic procedure. Therefore, this procedure is as a rule used to study only rather large GLEs (see also Chap. 9). In this case, the main SCR characteristics – the energy spectrum, anisotropy, and pitch angle distribution – are determined by optimization methods when model NM responses are compared with observed responses. The SCR flux parameters determined at successive instants make it possible to trace the flux dynamics. The analysis methods include the determination of the SCR arrival asymptotic directions by calculating the trajectories of these particles in present-day geomagnetic field models.

A neutron monitor has a specific directional pattern. When the zenith angle increases, the particle flux is weakening due to absorption of secondary neutrons is accompanied by an increase in the device solid angle of reception. This results in the appearance of a pattern maximum, which is reached at zenith angles of $\theta = 20°$ and $\theta = 18°$ for GCRs and SCRs, respectively. The asymptotic directions of arrival for CR particles are calculated by integrating the motion equation for a negative probe particle with the proton mass emitted upward from an altitude of 20 km above a given station (this is the average altitude of production of secondary neutrons contributing to the NM counting). Cosmic ray trajectories in the Earth's magnetic field are illustrated by Fig. 1.9 (Smart and Shea 1994). High-energy trajectories (curves 1, 2, 3...) is seen to be open to interplanetary space, whereas low-energy particles have closed trajectories, i.e., they are trapped inside the magnetosphere.

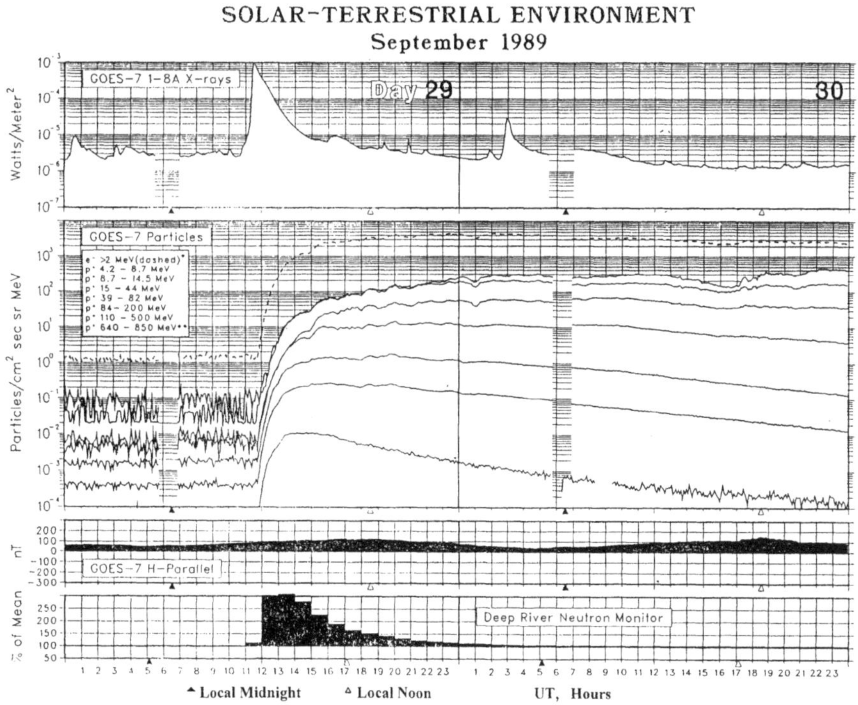

Fig. 1.10 General picture of the September 29, 1989 event (SGD 1989, 1990). *Upper curve*: GOES-7 soft X-rays (1-8Å). *Middle curves*: GOES-7, particles measured: electrons >2 MeV (*dashed*), protons 4.2–8.7 MeV, 8.7–14.5 MeV, 15–44 MeV, 39–82 MeV, 84–200 MeV, 110–500 MeV, 640–850 MeV; GOES-7, geomagnetic field, *H*-parallel. *Lower panel*: Deep River neutron monitor (hourly averages)

Figure 1.10 shows a general picture of a major solar proton event (GLE42) observed on September 29, 1989 as it was observed at the Earth-orbiting satellite and on the terrestrial surface (*e.g.*, Miroshnichenko et al. 2000). In the upper panel one can see a temporal behaviour of the soft X-ray flux at the end of September 1989 by the *GOES-7* measurements. The middle curves are the intensity-time profiles of SEPs (relativistic electrons with the energy >2 MeV and protons in the energy range from 4.2 MeV up to 850 MeV); the lower panel contains the GOES-7 data on the *H*-component of the geomagnetic field and ground-based observations of relativistic solar protons at the Deep River neutron monitor (hourly averages).

Of course, Fig. 1.10 gives only "smoothed" notion about the event. If one goes more deeply into the data it becomes clear that the event of September 29, 1989 is of special interest (see as a review Miroshnichenko et al. 2000). Indeed, since the well-known event of February 23, 1956, it proved to be the most intense in the relativistic range of proton energies. In spite of its occurrence behind the western limb of the Sun, the originating powerful flare could be observed over a wide range of the wavelengths and particle energy spectra from gamma rays to decametric radio

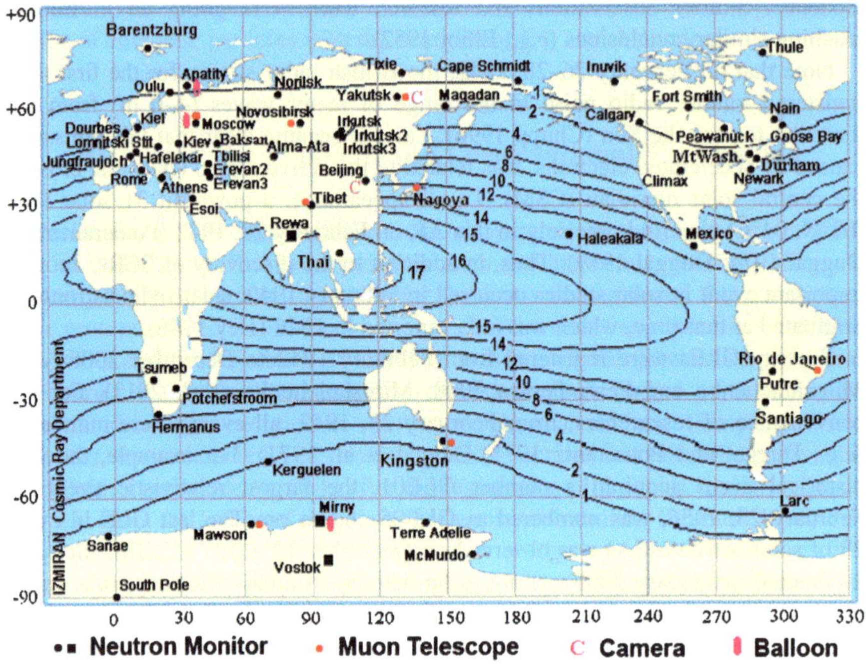

Fig. 2.1 A present-day worldwide network of stations for continuous cosmic ray registration (ftp://cr0.izmiran.rssi.ru/Cosray!/FTP_NM/C/). Figures near the curves correspond to isolines of equal geomagnetic cutoff rigidities for primary GCR or SCR particles (in units of GV)

Krymsky et al. 1990; Swinson and Shea 1990; Karpov et al. 1998; Miroshnichenko et al. 2000).

Ground-level events often give secondary muon intensity bursts registered with substandard instruments, which are designed in order to solve astrophysical problems and study the nuclear effects of GCRs (Karpov et al. 1998). These observations are satisfactorily completed with the network of solar neutron telescopes (SNTs) (Flückiger et al. 1998), which register the arrival of secondary neutrons generated by primary accelerated ions in the solar atmosphere.

While cosmic-ray researchers were developing their instruments, high-frequency communication engineers, particularly those involved in the propagation of electromagnetic signals in the polar regions, noted interference that seemed to be associated with solar activity. It is now known that charged particles interacting with the Earth's ionosphere enhance the ionization and change the electromagnetic propagation characteristics of the medium. In the late 1950s, the development of the riometer (radio ionosphere opacity meter) proved to be very sensitive to particle deposition in the ionosphere directly above the instrument (Little and Leinbach 1959). Even though the riometer could not uniquely distinguish the type of particle, its sensitivity was equivalent to the early satellite instruments. Most of the solar particle flux and fluence data available from the 19th solar cycle (1955–1965) were

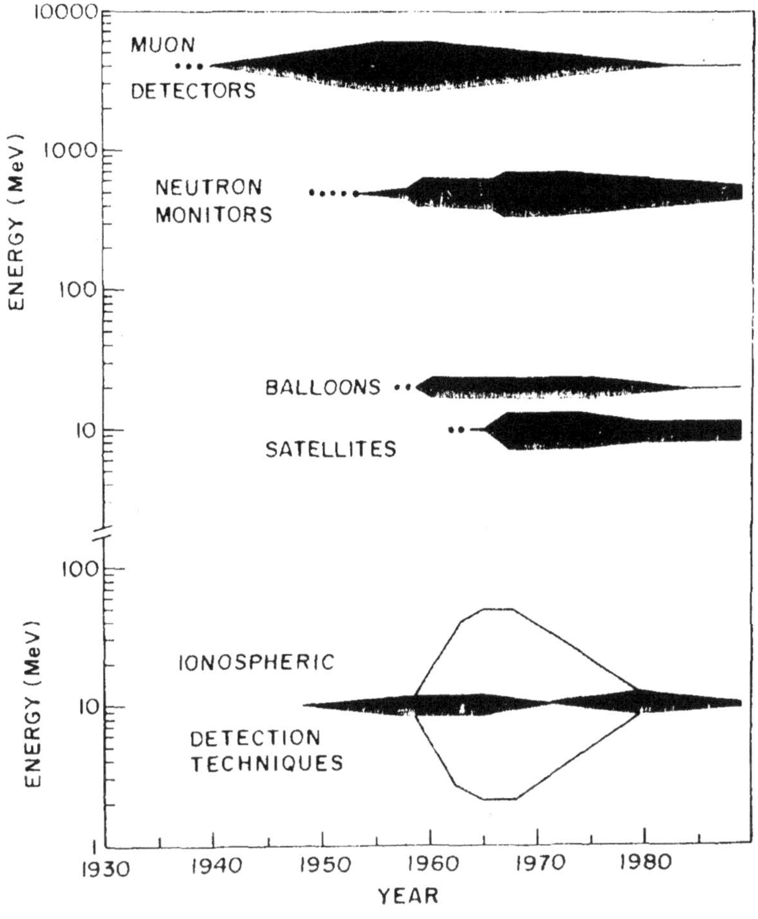

Fig. 2.2 Conceptual history of the detection thresholds of solar proton events. The thickness of the lines indicates the relative number of each type of detector in use. The difference in shading in the ionospheric section indicates changes in detection technique (Smart and Shea 1989b; Shea and Smart 1993a, b, 1994)

derived from riometer measurements in the Earth's polar regions (*e.g.*, Dodson et al. 1975). Even now the ionosphere can be still used as a very sensitive (but nonlinear) particle detection medium, since very low frequency phase and amplitude changes along transpolar propagation paths have the same approximate detection thresholds as particle detectors on spacecraft (Smart and Shea 1989b).

Figure 2.2 gives a summary of observational techniques for SCR study. It illustrates very visually the evolution of detection energy thresholds and detector techniques since 1933 (Smart and Shea 1989a, b; Shea and Smart 1994). The thickness of the lines indicates the relative number of each type of detector in use. The differences in shading in the ionosphere section indicate changes in detection technique. As can be seen from inspection of Fig. 2.2, there are

Earth-based measurements of solar flare generated particles since 1942. However, the indirect detection techniques did not stabilize until approximately 1958, and the spacecraft measurements were not systematic until about 1965.

As it was noted, an historical beginning of SCR observations was a GLE of 28 February 1942. Since then up to now (the middle of 2014) 71 similar events have been recorded by the worldwide network of cosmic ray stations. These events characterize only one, relativistic part of entire energy spectrum of solar cosmic rays (kinetic energy $E \geq 435$ MeV/nucleon, or magnetic rigidity $R \geq 1$ GV). If the energy of primary protons is <100 MeV ($R < 0.44$ GV), neutron monitors are practically do not respond them due to atmospheric absorption of neutrons (so-called "atmospheric cutoff", R_a), a maximum of the NM response being within 1–5 GV. It means that all high-latitude (polar) NM stations start to record secondary neutrons efficiently from the same rigidity of the primary protons about 1 GV (435 MeV), irrespective of the NM nominal (calculated) "geomagnetic cutoff rigidity", R_c.

As it fortunately happened, a rigidity 1.0 GV ($E_k \sim 433$ MeV) is approximately midway between the low rigidity and ultra-relativistic rigidity range, and it turned out to be a convenient reference point as a characteristic rigidity cutoff at the polar NM stations (Smart and Shea 1996). Surface detectors of the secondary muons have their detection threshold Ra about 4 GV determined by air mass absorption for muons. Underground muon telescopes have their own rigidity thresholds, R_u, quite different from the first two and determined by rock mass above the instruments. For a comparison, we cite here the threshold rigidities of 19 and 500 GV, respectively, for two underground telescopes located in Embudo Cave near Albuquerque, New Mexico, USA (Swinson and Shea 1990) at the depth of 35 m of water equivalent (m.w.e.) and in Baksan Valley, Russia, at the depth of 850 m.w.e. under the mountain Andyrchi, Northern Caucasus, Russia (*e.g.*, Karpov et al. 1998).

Relativistic particles are of special interest to understand the maximum capacities of solar accelerators. In particular, the GLEs of solar cycle 22 have renewed interest in the effects of relativistic SPEs. The first 25 GLEs were listed and numbered by Pomerantz and Duggal (1974), later on this list was extended to 54 events (Shea and Smart 1993a). Recently, we present the list of 70 GLEs registered for the period of 1942–2006 (Miroshnichenko and Perez-Peraza 2008). In Table 2.1 we present the list of 71 GLEs registered for the period of 1942–2012 (i.e., for 70 years of ground-based observations) with the relevant information obtained with different types of instrumentation. Table 2.1 has been updated (Miroshnichenko et al. 2013) to include events since the very beginning of ground-based observations up to now (see also http://data.aad.gov.au/aadc/gle/). Notice that the event of November 6, 1997 was the first GLE of the solar cycle 23 (*e.g.*, Smart and Shea 1998; Gomez-Herrero et al. 1998), after the 5-year "pause" coincident with the minimum of the previous solar cycle. Another pause (1998–2000) seems to correspond to the cycle maximum. The event of May 17, 2012 was the first GLE of a new (current) solar cycle 24 (*e.g.*, Li et al. 2013; Papaioannou et al. 2014).

Table 2.1 Ground-level enhancements of 1942–2012

No. GLE	Date of registration	*H*-alpha position	Onset, UT	Importance Hα/X	Reference to data source
1	28 Feb 1942	07N 04E	1228	3+	Duggal (1979)
2	07 Mar 1942	07N 90W	N.O.	–/–	Duggal (1979)
3	25 Jul 1946	22N 15E	1615	3+	Duggal (1979)
4	19 Nov 1949	03S 72W	1029	3+	Duggal (1979)
5	23 Feb 1956	23N 80W	<0334	3	Shea and Smart (1975)
6	31 Aug 1956	15N 15E	1226	3	Shea and Smart (1975)
7	17 Jul 1959	16N 31W	2114	3+	Shea and Smart (1975)
8	04 May 1960	13N 90W	1000	3	Shea and Smart (1975)
9	03 Sep 1960	18N 88E	0037	2+	Shea and Smart (1975)
10	12 Nov 1960	27N 04W	1315	3+	Shea and Smart (1975)
11	15 Nov 1960	25N 35W	0207	3+	Shea and Smart (1975)
12	20 Nov 1960	28N ~ 112W	2017	2	Shea and Smart (1975)
13	18 Jul 1961	07S 59W	0920	3+	Shea and Smart (1975)
14	20 Jul 1961	06S 90W	1553	3	Shea and Smart (1975)
15	07 Jul 1966	35N 48W	0025	2B	Shea and Smart (1975)
16	28 Jan 1967	22N ~ 150W	<0200	–/–	Shea and Smart (1975)
17	28 Jan 1967	22N ~ 150W	<0800	–/–	Shea and Smart (1975)
18	29 Sep 1968	17N 51W	1617	2B	Shea and Smart (1975)
19	18 Nov 1968	21N 87W	<1026	1B	Shea and Smart (1975)
20	25 Feb 1969	13N 37W	0900	2B/X2	Shea and Smart (1975)
21	30 Mar 1969	19N 103W	<0332	1N	Shea and Smart (1975)
22	24 Jan 1970	18N 49W	2215	3B/X5	SGD 323B 19
23	01 Sep 1971	11S 120W	<1934	–/–	SGD 327A 82
24	04 Aug 1972	14N 08E	0617	3B/X4	SGD 342B 06
25	07 Aug 1972	14N 37W	1449	3B/X4	SGD 342B 09
26	29 Apr 1973	14N 73W	2056	2B/X1	SGD 350B 23
27	30 Apr 1976	08S 46W	2047	2B/X2	SGD 386B 11

(continued)

Table 2.1 (continued)

No. GLE	Date of registration	*H*-alpha position	Onset, UT	Importance Hα/X	Reference to data source
28	19 Sep 1977	08N 57W	<0955	3B/X2	SGD 403B 14
29	24 Sep 1977	10N 120W	<0552	–/–	SGD 399A109
30	22 Nov 1977	24N 40W	0945	2B/X1	SGD 405B 13
31	07 May 1978	23N 72W	0327	1N/X2	SGD 411B 11
32	23 Sep 1978	35N 50W	0944	3B/X1	SGD 415B 26
33	21 Aug 1979	17N 40W	0550	2B/C6	SGD 436B 70
34	10 Apr 1981	07N 36W	1632	2B/X2.3	SGD 474B 53
35	10 May 1981	03N 75W	0715	1N/M1	SGD 475B 31
36	12 Oct 1981	18S 31E	0615	2B/X3.1	SGD 481B 54
37	26 Nov 1982	12S 87W	0230	2B/X4	SGD 487B173
38	07 Dec 1982	19S 86W	2341	1B/X2.8	SGD 488B 26
39	16 Feb 1984	–S ~ 130W	<0858	–/–	SGD 476A 99
40	25 Jul 1989	26N 85W	0839	2N/X2	SGD 545B 22
41	16 Aug 1989	15S 85W	0058	2N/12.5	SGD 546B 26
42	29 Sep 1989	24S ~ 105W	1141	1B/X9	SGD 547B 38
43	19 Oct 1989	25S 09E	1229	3B/X13	SGD 548B 19
44	22 Oct 1989	27S 32W	1708	1N/X2.9	SGD 548B 24
45	24 Oct 1989	29S 57W	1738	2N/X5.7	SGD 548B 27
46	15 Nov 1989	11N 28W	0638	2B/X3.2	SGD 449B 20
47	21 May 1990	34N 37W	2212	2B/X5.5	SGD 555B 23
48	24 May 1990	36N 76W	2046	1B/X9.3	SGD 555B 25
49	26 May 1990	~35N,103W	2045	–/–	SGD 555B 61
50	28 May 1990	~35N120W	<0516	–/–	SGD 555B 63
51	11 Jun 1991	32N 15W	0105	2B/X12	SGD 568B 14
52	15 Jun 1991	36N 70W	0633	3B/X12	SGD 568B 20
53	25 Jun 1992	09N 69W	1947	1B/M1.4	SGD 580B 17
54	02 Nov 1992	~25S ~ 100W	0231	–/X9	SGD 580A 28
55	06 Nov 1997	18S 68W	1149	2B/X9.4	SGD 640A 29
56	02 May 1998	15S 15W	1334	3B/X1.1	SGD 646A 28
57	06 May 1998	11S 65W	0758	1N/X2.7	NM Database
58	24 Aug 1998	18N 09E	2148	3B/M7.1	NM Database
59	14 Jul 2000	22N 07W	1003	3B/X5.7	NM Database
60	15 Apr 2001	20S 85W	1319	2B/X14.4	NM Database
61	18 Apr 2001	23S W117	0211	–/–	NM Database
62	04 Nov 2001	06N 18W	1603	3B/1.3	NM Database
63	26 Dec 2001	08N 54W	0432	–/M7.1	NM Database
64	24 Aug 2002	02S 81W	0049	–/X3.1	NM Database
65	28 Oct 2003	16S 08E	1100	4B/X17.2	NM Database
66	29 Oct 2003	19S 09W	2037	–/X10	NM Database
67	02 Nov 2003	18S 59W	1718	2B/X8.3	NM Database
68	17 Jan 2005	15N 25W	0659	3B/X3.8	NM Database
69	20 Jan 2005	14N 61W	0639	2B/X7.1	NM Database

(continued)

Table 2.1 (continued)

No. GLE	Date of registration	*H*-alpha position	Onset, UT	Importance Hα/X	Reference to data source
70	13 Dec 2006	06S 23W	0217	4B/X3.4	NM Database
71	17 May 2012	07N 88 W	0125	?/M5.1	NM Database

SGD XXXB PG refers to Solar-Geophysical Data, v. *XXX*, part *Y*, p. *ZZ*
S. & S. refers to the Catalogue of Solar Proton Events 1955–1969 (Dodson et al. 1975) edited by Z. Svestka and P. Simon
N.O. no optical observations
[a]Flare occurred the day before the onset of the GLE
[b]Flare behind the west limb of the Sun. Position estimated from location of assumed associated active region

The above list is given mainly for reference purposes. Original sources of data contain, of course, a lot of additional information and important comments (*e.g.*, Shea and Smart 1993a) helpful for the more deep studies of certain individual events. For example, it is necessary to explain a "double" GLE of January 28, 1967. Space probe measurements evidenced that two high energy particle events occurred (Dodson et al. 1975). Neutron monitor observations initially indicated that both events were recorded as GLEs. However, recent studies of the first increase make the event somewhat questionable. Position indicated of a flare was taken from location of the McMath region 8687 that assumed to be the source of the event, and this position was approximately 60° beyond the west limb.

A precise magnitude for each GLE has not been included in Table 2.1 for several reasons. The first three events were detected only by ionization chambers (IC); one non-standard neutron monitor was also in operation in Manchester, England, to record the fourth event, GLE04, occurred on November 19, 1949. Using these measurements, Smart and Shea (1991) evaluated relative amplitude of these early events. The event of September 29, 1989 which Smart and Shea (1991) have used as a calibration event would rank third in this "hierarchy". Based on both the muon and neutron monitor data, the GLE of November 19, 1949 is larger than the event of September 29, 1989; however, the well-known event of February 23, 1956 will rank as number 1. The ground-level enhancements where the cosmic ray intensity increased at least 90 % above the background intensity as recorded by NMs located at sea level before 1990 are the following: November 19, 1949; February 23, 1956; May 4, 1960; November 12 and 15, 1960; May 7, 1978; February 16, 1984; September 29, 1989; October 22 and 24, 1989. Note that with the exception of the GLE on November 12, 1960, each of these events has been associated with a flare located from 30°W to beyond the western limb of the Sun. Recently, based on the complete set of data (Table 2.1) we revised and extended the magnitude distribution of GLEs (Table 2.2) for the entire period of SCR observations.

Data of Table 2.2 give some new ideas about maximum increases of SCR intensity for the most powerful GLE events. Evidently, the estimates of GLE magnitude depend on the interval Δt of data averaging, detector sensitivity (response function) and of course, on the geomagnetic cutoff rigidity R_c for a

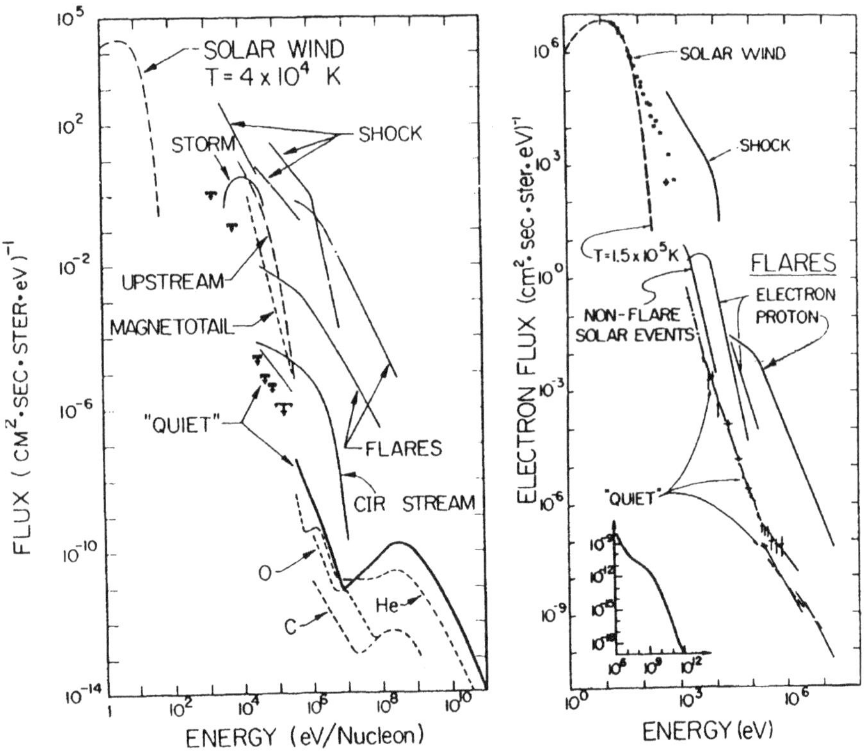

Fig. 2.3 Energy spectra of protons and some other ions (*left plot*) and electrons (*right plot*) observed in the interplanetary space near the Earth's orbit (Lin 1980). For a comparison, at the left plot a typical spectrum of accelerated flare protons is also shown

methodical difficulties in measuring SCR near the Earth (on the background of galactic cosmic rays) and interpretation of the data obtained. For determining the SCR spectrum in a wide interval of energies one has, usually, to align, so to say, the results of a few kinds of measurements (aboard the satellites, at the stratosphere and terrestrial surface, and so on) thus introducing additional errors (uncertainties)into the values of spectral characteristics. Besides, in the low-energy range, there are other sources of the uncertainties due to overlapping of the solar particle flux with ambient particle fluxes of different origin.

Figure 2.3 (*left plot*) gives a summary of the differential energy spectra of protons and some other ions observed near the Earth's orbit in a wide range of energies: from the thermal solar wind protons to relativistic protons of galactic cosmic rays (Lin 1980). From Fig. 2.3 it is seen that below ~10 MeV an observed interplanetary proton flux, in fact, may be a mixture of the flare, shock and storm particle fluxes; above 10 MeV the flare proton fluxes seem to be dominant. Similar observational data for electrons are also represented in Fig. 2.3 (*right plot*). For a comparison, a typical spectrum of accelerated flare protons is also shown. It is

Table 2.3 Extreme SEPs properties

Parameter	Electrons	Ions	Reference
Number of particles	10^{41} (>20 keV)	–	Kane et al. (1995)
	10^{37} (>100 keV)	–	Miller and Ramaty (1992)
	5×10^{44} (>300 keV)	–	Miller and Ramaty (1992)
	–	3×10^{35} (>30 MeV)	Ramaty and Murphy (1987)
	–	$\sim 10^{32}$ (>300 MeV)	Chupp (1996)
	–	6×10^{32} (>500 MeV)	Miroshnichenko (1993)
Rise time, *s*	10^{-2}	>1.0	Chupp (1996)
Duration, *s*	10⇒	60⇒	Chupp (1996)
	–	≤10 (>500 MeV)	Miroshnichenko (1993)
Total energy, *erg*	10^{34} (>20 keV)	–	Kane et al. (1995)
	10^{29} (>100 keV)	–	Miller and Ramaty (1992)
	10^{28} (>300 keV)	–	Miller and Ramaty (1992)
	–	10^{30} (>30 MeV)	Ramaty and Murphy (1987)
	–	3×10^{28} (>300 MeV)	Chupp (1996)
	–	5×10^{29} (>500 MeV)	Miroshnichenko (1993)
Power, $erg\ s^{-1}$	10^{32} (>20 keV)	–	Kane et al. (1995)
	–	2×10^{28} (>30 MeV)	Chupp (1996)
	–	$<5 \times 10^{28}$ (>500 MeV)	Miroshnichenko (1993)

Adapted from Chupp (1996)

noticeable a drastic difference in the intensities of accelerated flare electrons and protons at the same energies starting from $\sim 10^5$ eV. This difference concerns to one of the fundamental problems of flare physics and particle acceleration at the Sun – the role of non-thermal protons as a means of energy transport in the solar atmosphere (as a review see Simnett 1995).

In Table 2.3 we summarize extreme properties of the accelerated ions and electrons inferred from various observations. Due to several reasons, all estimates in Table 2.4 are limited in their validity and accuracy. The main reason is a dependence on underlying model of the event used for the interpreting the data. For example, a number of relativistic protons (>500 MeV) have been estimated with the uncertainty factor of ≥2 (*e.g.*, Miroshnichenko 2001). Nevertheless, in general, they give a certain idea about the upper limit capacities of the solar accelerators. Note that a number of >20 keV electrons and all estimates for relativistic protons (>500 MeV) have been obtained by the data for so-called "giant flares" of June 1, 1991 and February 23, 1956, respectively. Though these events are very rare, they are of great interest from the astrophysical point of view. We discuss this aspect of solar physics in some detail in Chap. 4.

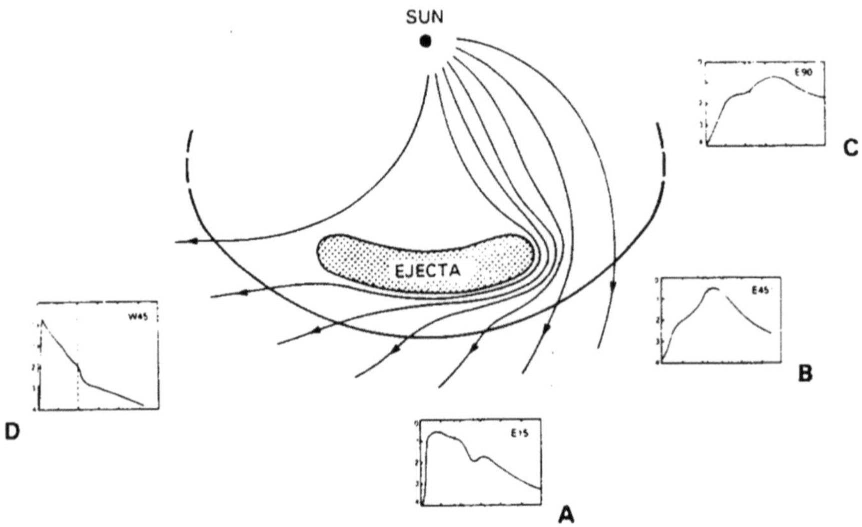

Fig. 2.5 A schematic diagram showing a CME-driven shock and expected time profiles of energetic protons accelerated by the shock (Cane et al. 1988). In each panel the solar longitude of the flare (**A** – E15°; **B** – E45°; **C** – E90°; **D** – W45°) and the time of shock passage near the Earth (*dotted vertical lines*) are indicated

meridian on the Sun. CMEs and coronal shocks accelerate particles over an extended area of the Sun, however, so that solar particles can be observed from flares that occur from about E30° through about W140° (i.e. 50° behind the west limb of the Sun).

Yet another population of particles is accelerated by the interplanetary shock waves that are produced by large CMEs. A schematic diagram of a CME-driven shock is given in Fig. 2.5, together with time profiles expected near the Earth of energetic protons accelerated by the shock (Cane et al. 1988). The high-energy particles from these events usually reach maximum intensity after the passage of the shock, 1–3 days after the event at the Sun. Particle events near central meridian on the Sun usually have a strong component from the interplanetary shock and the particles from flares of E30° are almost entirely shock associated (Cane et al. 1988; Richardson et al. 1991). It is important to distinguish this particle population since it is, in principle, predictable.

The properties of the various solar-particle populations are best observed in small events where a single component dominates. Such studies allow us to understand the relationships of the properties of the particles to those of the acceleration mechanisms. In large events all mechanisms can occur (*e.g.*, Reames 1996, 1999), leading to so-called "Big Flare Syndrome" (Kahler 1982) that, in the opinion by Chupp (1984), causes false correlations among observations.

Kallenrode et al. (1992a, b) have systematized in the form of the table and analyzed the data for 77 SEP events observed by *Helios* 1 and 2 space probes in 1974–1985. These authors classified the SEP parent flares as impulsive (25 cases) or gradual (52 cases) on the basis of their soft X-ray duration. Then, they compared the intensities of the prompt component of ~0.5 MeV electrons, ~10 MeV protons, and ~10 MeV/nucleon helium for the two classes of SEP flares. It was found that SEPs from gradual flares have higher intensities than SEPs from impulsive flares. These differences are most pronounced for protons (about two orders of magnitude) and less for electrons (about one order of magnitude), and helium (about a factor of 5). The SEPs from impulsive flares have a "cone of emission" of $\pm 50°$ versus $\pm 120°$ for gradual flares. In terms of the propagation models, it means that we have two types of the SEP sources – more/less spatially narrow versus rather extended ones.

The smaller cone of emission in impulsive events could also be a reflection of the smaller size of the CME (and piston-driven shocks) in these events. It may be possible in certain circumstances, i.e., in a high-energy density environment, for shocks to produce the non-power-law rigidity spectra of electron events from impulsive flares, but detailed models are not yet available (Dröge et al. 1989). There is another explanation, in addition to different size CMEs/shocks or different acceleration mechanisms, that could account for the different cones of emission for SEP events from gradual and impulsive flares. Recall that all events in this study were selected on the basis of a >20-fold increase in electrons. Thus, at least part of the differences could be explained due to the different event sizes.

2.4 Elemental Abundances and Charge States

The radiation background from SEP events can assume different forms, as the energy and composition of the particles changes from event to event. It has become clear long ago that different particle composition and spectra arise from different classes of events at the Sun (*e.g.*, Bai 1986), and we have learned how to associate the properties of the particles with the radio, X-ray and optical observations of the parent flares or with the related interplanetary shock.

Observations of flare-accelerated ions in the interplanetary space confirm the X-ray characterization of solar flares into two distinct classes: impulsive with duration of minutes and gradual with duration of hours and days (*e.g.*, Pallavicini et al. 1977). Previous measurements were confined to energies of >1 MeV/nucleon, while more recent high sensitivity *WIND* observations have extended the energy range to 20 keV/nucleon (Reames et al. 1997).

The past two decades have brought a significant advancement in our understanding of the relationships between the particles and other solar phenomena such as radio emission, hard and soft X-rays, gamma rays and coronal mass ejections (Reames 1996, 1999). Different modes of particle acceleration at the Sun result in different populations of particles in space. It is important to distinguish these populations since they contribute unequally to different particle species and energy regimes and hence, alter the nature of radiation background.

The event-averaged abundances of elements in gradual events, obtained from low-energy measurements, provide a direct measure of element abundance in the corona and solar wind. These abundances are almost entirely independent of the temperature and ionization state of the source plasma. It has been well known for many years (Meyer 1985) that the ratio of coronal and photospheric abundances of elements is a well-organized function of the first ionization potential (FIP) of the element. A summary of abundances is shown in Table 2.4 (Reames 1994, 1996) in relative units of the oxygen abundances (1,000).

Energetic particles from impulsive flares show element abundances that differ from those in the corona in that elements with $Z > 8$ are strongly enhanced relative to coronal abundances while He, C, N, or O are not. On the other hand, Ne, Mg, and Si are enhanced by almost the same factor, relative to O. This pattern of enhancement is consistent with acceleration of the ions from plasma in the temperature range of 3–5 MK.

Elements with the same charge-to-mass ratio, Q/A, have the same magnetic gyrofrequency, thus they resonate with the same part of the wave spectrum and are enhanced by the same amount. At 3–5 MK, He, C, N and O are fully ionized, with $Q/A = 0.5$, while Ne, Mg and Si all have $Q/A \sim 0.42$. These ions must become fully ionized later in the event; if they were fully ionized before acceleration they would have the same Q/A as He, C and O and could not be relatively enhanced. The pattern of enhancements is discussed in terms of wave absorption in the so-called "He Valley" by Meyer (1996) and is modeled by a theory of cascading waves by Miller and Reames (1996).

A new and very important result reported during recent years concerns the charge state of the energetic ions in SEP events. The charge state of the ions is a critical parameter reflecting not only the particle source properties, but also the particle propagation characteristics in space and through the Earth's magnetosphere. Prior to the middle of 1990s, the available measurements were at energies of ~1 MeV/nucleon. Recently, an extension of these measurements into the 10–100 MeV/nucleon energy range was undertaken (Oetliker et al. 1995). The mean ionic charges found for the large solar-particle events between 30 October and 7 November 1992 (see Fig. 2.6) are consistent with the earlier measurements at ~1 MeV/nucleon. There are still questions about the charge state of iron in the energy range ~0.5–5 MeV/nucleon (Mason et al. 1995a, b): they found a mean charge state $<Q> = 11.04 \pm 0.22$ for Fe, while the average from Luhn et al. (1987) is 14.09 ± 0.09. The investigators, however, preferred to treat this as *a problem* to *be resolved* rather than *a fundamental conflict*.

It is generally agreed that the ion composition of SEP events can be related to the first ionization potential (FIP) of the individual elements (Meyer 1985), and the charge state of the individual ions is related to the temperature of the source plasma (Oetliker et al. 1995; Leske et al. 1995a, b). However, there are still unresolved discrepancies in relating the SEP composition to either the coronal or the solar-wind composition. Boberg et al. (1995) proposed an explanation for the iron (Fe) charge state observed by the HIIS experiment. They suggest that it is possible to model the observed results as a composite of both coronal (90 %) and solar wind composition (10 %).

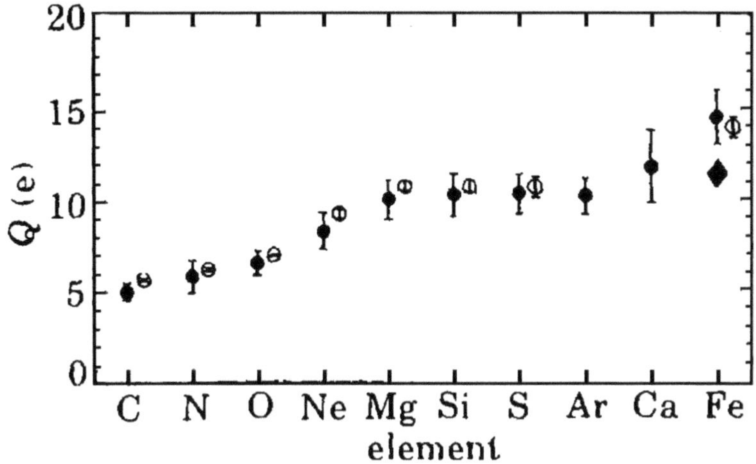

Fig. 2.6 The mean ionic charges of energetic solar particles in the energy range 10–100 MeV/nucleon. These results are a composite of two large solar -particle events which occurred between October 30 and November 7, 1992. The charge state in the 10–100 MeV/nucleon energy range are denoted by the *solid circles*. For comparison purposes the earlier results in the 0.3–3.0 MeV/nucleon are shown by the *open circles*. The *solid diamond* denotes the charge state of iron (Fe) determined by Mason et al. (1995a) (Figure adapted from Oetliker et al. (1995))

The ^{3}He/^{4}He ratio and the energy spectra of SEP events are directly related to understanding the particle acceleration mechanisms and the propagation processes. If the seed populations for large SEP events are indeed interplanetary (solar wind) particles (Gosling 1993), and if these particles are accelerated by a CME-driven shock (Reames 1993), then one would expect a ^{3}He/^{4}He ratio of ~0.0005 (solar wind/coronal value). As it was known earlier, all of extremely large SEP events at high energy (>40 MeV/nucleon) show ^{3}He/^{4}He ratios more than one order of magnitude greater than this solar wind/coronal value. For example, *Pioneer* 10 at ~2.2 AU from the Sun obtained ^{3}He/^{4}He ratio ~0.01 by integrating over four large SEP events from the same active region during 2–11 August 1972 (Webber et al. 1975).

Recently Guzik et al. (1995) examined the temporal variations in the ^{3}He/^{4}He ratios observed at high energy (50–110 MeV/nucleon) during the very large events in March and June 1991. They obtained ^{3}He/^{4}He ratios ranging from 0.003 to 0.22, one to three orders of magnitude different from the solar coronal value. These ratios are independent of the size of event, and are organized by the active region. It is concluded that the large ^{3}He/^{4}He ratios for large SEP events at high energy appear to be inconsistent with the CME driven shock model whose seed particles are the solar wind. Of particular interest is a plot of helium spectral index against time (Fig. 2.7). After the peak flux in most events the spectral index increases (a softer spectrum) until the onset of a new event. This can be understood since particle with higher velocity (or energy/nucleon) arrive first at the Earth.

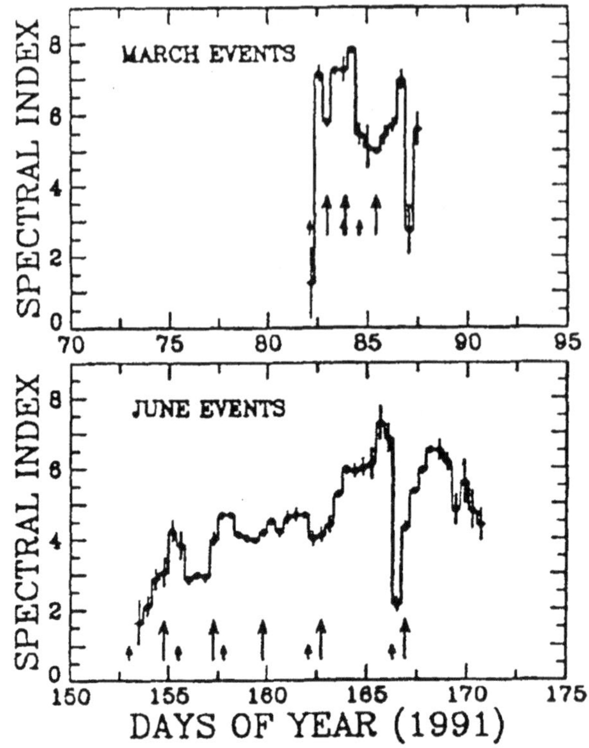

Fig. 2.7 Spectral index *versus* time for the helium nuclei measured at high energy (50–110 MeV/nucleon) in March (*left*) and June (*right*) 1991 (Guzik et al. 1995). The big *arrows* indicate the positions of the peak fluxes of different SEP events, and the *small arrows* indicate the time of onset of different SEP events at 1 AU

Energy spectra of heavy nuclei inside the Earth's magnetosphere have been measured for a series of SPEs in October 1989 onboard the orbital station *Mir* with an orbit inclination of 51.6° (Baranov et al. 1997). The power-law spectral indexes were determined for the nuclei of solar origin as follows: 3.2 for the Fe nuclei in the energy range of 30–130 MeV/nucleon and 2.7–4.3 for the nuclei of the Ne-Si group in the energy range of 25–85 MeV/nucleon (Fig. 2.8). A difference in spectral indexes between heavier and lighter nuclei at low energies may be interpreted as a result of their different charge states.

Of particular importance is the observed enhancement of 3–4 order of magnitude in the $^3He/^4He$ ratio and up to the order of magnitude in heavy ions in impulsive flares, versus coronal abundances in the gradual events. These enhancements constitute one of the largest enrichment in heliospheric physics, and the time scales involved make it one of the most acceleration processes encountered in space physics. In this context, Roth and Temerin (1995) considered resonant plasma effects in the impulsive solar flares, with the changing coronal conditions. It was shown that an interaction with the second harmonic of ion gyrofrequency selectively accelerates specific elements and charge states of the different coronal nuclei. The final abundances depend on the specific coronal parameters.

Two types of SEP events, impulsive and gradual, exhibit different characteristics in the abundances and energy spectra (Reames 1995a, b, c). Impulsive events

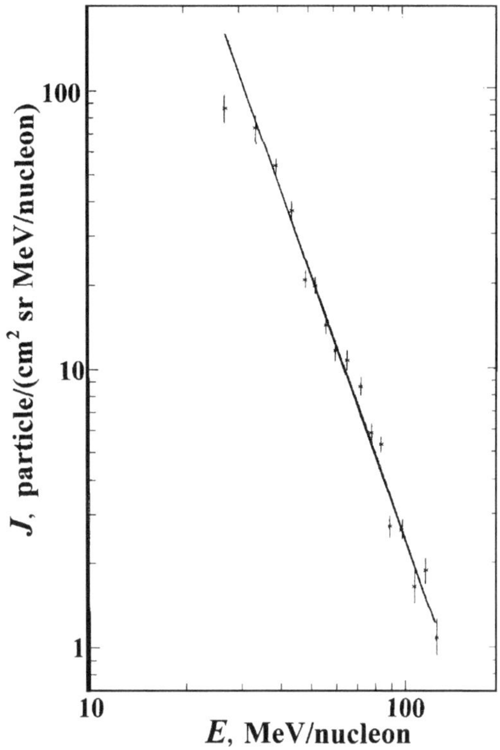

Fig. 2.8 Averaged energy spectrum of the nuclei with the charge of $Z > 20$ in a series of SPEs in October 1989 at the *Mir* station (Baranov et al. 1997)

reveal: (1) electron-rich, (2) strong ^{3}He-rich, (3) O, Ne, and Fe ion-reach and (4) high charge state of Fe ions (Fe + 20). In order to explain the observed characteristics, the scenario was suggested that particles are accelerated by cascading Alvfén wave turbulence (Miller and Reames 1996). Energy spectra of ions from impulsive solar flares have been reported by Reames et al. (1992) and Mazur et al. (1995). These two observations indicated inconsistent results, making it difficult to discuss the spectra in detail. Recently, Reames et al. (1997) observed high-quality heavy ion spectra in five impulsive flares with a high-sensitivity instrument aboard the *WIND* spacecraft. Theses observations showed that the energy spectra of H, ^{3}Hc, ^{4}Hc, C, O, and Fe have more rounded spectra which flatten at low energies, though cnergy spectra in gradual events are represented as power laws over a wide energy range.

Kartavykh et al. (1997) simulated the preliminary energy spectra of Fe ions in an impulsive flare, taking into account only stochastic acceleration by Alfvén waves, direct ionization by ambient electrons and Coulomb losses. Very recently these authors found the important contribution of direct ionization of Fe ions by ambient protons. The simulation results (without possible contribution of ambient protons) do not agree with the observations.

The ionic charge, Q, of SEPs observed in interplanetary space is an important parameter for the diagnosis of the plasma conditions at the source of the SEP in the solar corona. The charge is established through temperature dependent ionization and recombination processes in the corona. When the particles emerge from the Sun, the plasma density decreases rapidly and the charge states "freeze-in". The charge state distribution of the SEP, therefore, reflects the source conditions such as the electron temperature in the corona, T. By observing the charge of SEP, answers to the following questions can be probed: (1) do all elements freeze in at the same coronal temperature, (2) is the measured SEP charge independent of the particle energy and, therefore, not altered by the acceleration processes, and (3) can we observe a variation of the ionization temperature with observed SEP event (*e.g.*, difference between short and long duration events)?

Oetliker et al. (1995) presented measurements on board the polar orbiting SAMPEX satellite for He, C, N, O, Ne, Na, Mg, Al, Si, S, Ar, Ca, Fe, and Ni in the energy range 0.3–70 MeV/nucleon made in two consecutive large SEP events in October-November, 1992. Of all the elements in this data set, Fe has the strongest dependence of $Q(T)$, and thus Fe is the most sensitive indicator of the temperature history of the particles covered in this study. It was found that the particles are highly but not fully ionized in accordance with an equilibrium ionization temperature of ~2 MK. For all elements, the mean charge is constant over the observed energy range except for iron where a strong increase from +11 below 3 MeV/nucleon up to +17 at 60 MeV/nucleon was observed. Contrary to these results, at much higher energies, Tylka et al. (1995) found a charge state of Fe of 14.2 ± 1.4 at 200–600 MeV/nucleon in a series of very large, historic SPEs during September-November, 1989.

Fluxes measured at 1 AU are probably the result of acceleration process that is distributed in space and time. Over the large range of energies studied by Oetliker et al. (1995), these processes yield different behaviour of the flux-time profiles. At high energies, two separate SPEs stand out clearly (on October 31 and November 2, 1992). At low energies (0.2–0.3 MeV/nucleon Fe) the second SPE cannot be clearly identified, and the shock on November 2 produced a very large perturbation on the ambient flux. Such a behaviour would indicate that the high energy particles were accelerated near the Sun, while at the lowest energies much or most of the acceleration took place by the shock in interplanetary space. The lower energy particles would have ionization states more typical of solar wind values, which are in the range ~11–12, similar to the low energy values found here. At the higher energies, the Fe charge states may be indicative of a region closer to the flare site that had higher temperatures, or had been subjected to other processes such as ionization by X-rays (Mullan and Waldron 1986) or by ambient protons (Kartavykh et al. 1997).

A new serious challenge in this field was given rise recently by previously unpublished measurements (Tylka et al. 1997a) of the high-energy time-integrated O and Fe spectra for several large, gradual SPEs observed on board *IMP*-8 in 1977–1989, in particular, for the September 29 and October 24, 1989, events (Fig. 2.9). At high energies (>50–100 MeV/nucleon), the spectra are often surprisingly hard, without the exponential rolls-off suggested by stochastic and diffusive shock fits to lower energy datapoints alone. Moreover, in many cases (for example, in the

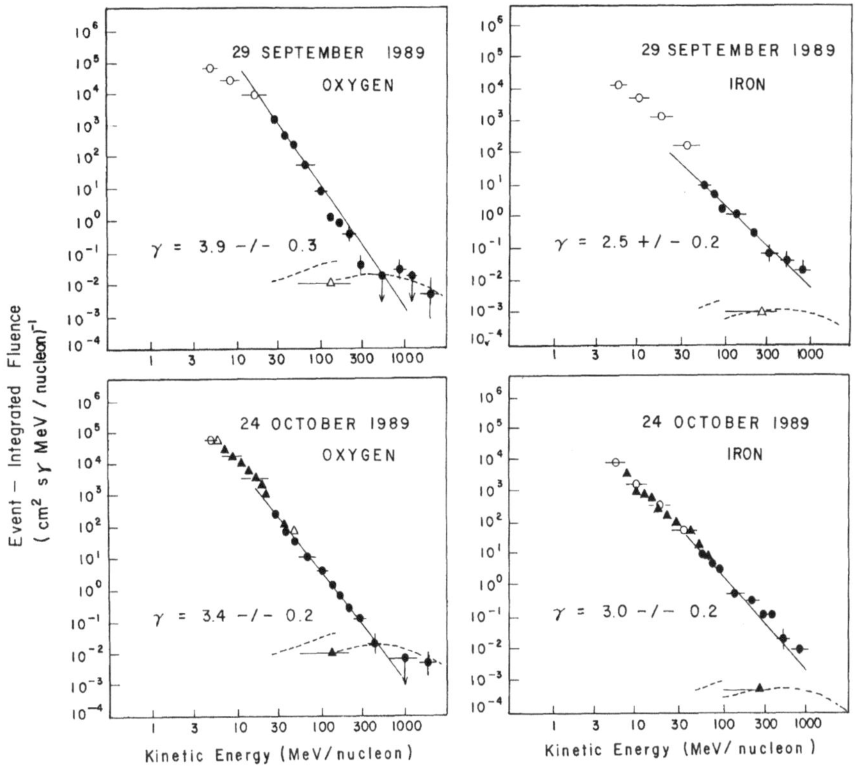

Fig. 2.9 Energy spectra of solar O and Fe ions for the SPEs of 29 September and 24 October 1989 (Tylka et al. 1997a, b) by the data of different detectors (VLET, *open circles*; Galileo, *filled triangles*; and Chicago/CRT, *filled circles*). Galactic background is also shown by *long-dashed curves* (estimates)

September 29 and October 24, 1989, events) the high-energy Fe spectrum is significantly harder than the oxygen spectrum, contrary to the notion that ions with higher mass-to-charge ratio are less abundant at higher energies. Of particular note is the hardness of the Fe spectrum relative to O in the September 29, 1989 event (power-law spectral indices of 2.5 ± 0.2 and 3.9 ± 0.3, respectively). The relatively high statistics in these two very large SPEs allow a clear demonstration of the Fe/O ratio growing with increasing energy and reaching unity at the highest energies of ~800 MeV/nucleon. To our knowledge, present theories of solar particle acceleration are unable to explain observed high-energy Fe/O enhancements.

SEP heavy elemental abundances have provided powerful diagnostics to determine the temperature and first-ionization-potential (FIP) characteristics of the SEP source regions. The ratios of charge to mass, *Q*/*M*, for the heavy elements range from ~0.266 for Fe to 0.475 for C (Reames 1995a, b, c). However, for diagnostics of shock acceleration the optimum elements to use would be a ratio of H and He with their very different *Q*/*M* values of 1 and 0.5, respectively. Several studies have

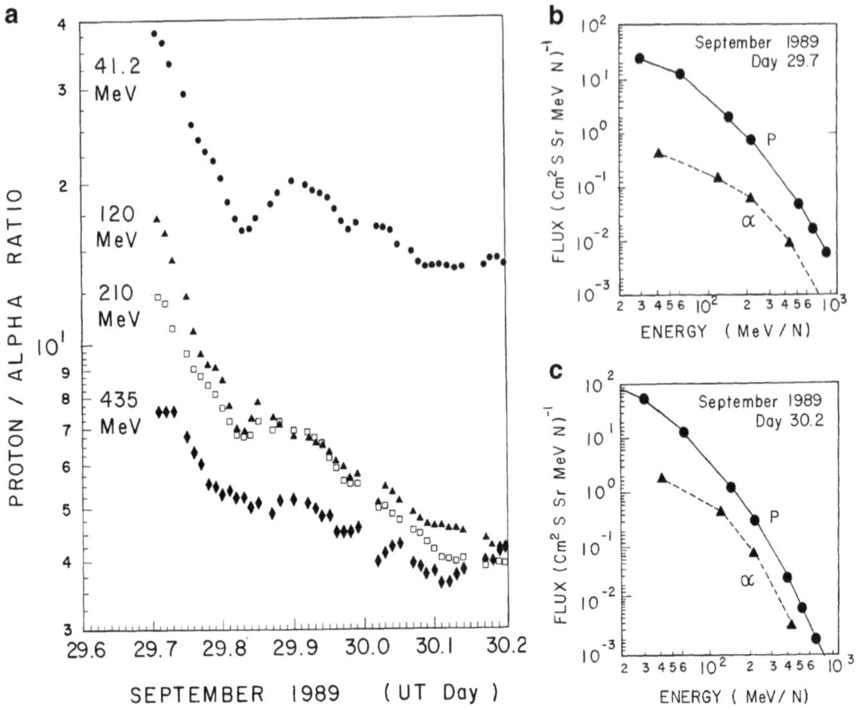

Fig. 2.10 Temporal variations of proton/alpha ratios and their energy spectra during the event of September 29, 1989 (Kahler et al. 1997): (**a**) Proton/alpha ratios for four energy levels during 12 h of the decay of the SEP event (*filled circles* – 41.2 MeV; *filled triangles* – 120 MeV; *open rectangles* – 210 MeV; *filled diamonds* – 435 MeV). While the absolute values are uncertain, all ratios are declining during the event; (**b, c**) Uncalibrated proton and alpha energy spectra at two different times: (**b**) day 27.9, *filled circles* – protons (P), *filled triangles* – alpha-particles (α); (**c**) day 30.2, *filled circles* – protons (P), *filled triangles* – alpha-particles (α).

established that p/α ratio increases with energy and varies substantially from event to event, but generally lies above coronal value. The time variations of p/α through individual events (namely, increases by factors of 2–5) have been established only in the low-energy ($E < 10$ MeV/nucleon) range.

Kahler et al. (1997) calculated the relative p/α ratios in the 100–500 MeV/nucleon energy range during the historical event of September 29, 1989. The p/α ratios show a clear temporal decrease through the event (Fig. 2.10), contrary to the variations observed at $E < 10$ MeV/nucleon in earlier events. At these high energies the effects of interplanetary scattering are minimized, so that the measured p/a values should reflect directly the shock injection characteristics of the SEPs. However, declining values of p/α with time appear inconsistent with the shock acceleration model by Ellison and Ramaty (1985) involved in this study. In particular, both observed proton and alpha spectra are growing steeper with time, in contrast to the expectations of the model.

Ruffolo (1997a) has obtained some constraints on coronal transport and acceleration times imposed by charge states of interplanetary ions from gradual flare events. Recent measurements of the mean charges of various elements after the gradual solar flares of October 30 and November 2, 1992 allow one to place limits on the product of the electron density times the time of acceleration or residence within the corona experienced by the escaping ions. In particular, any residence in coronal loops must be for <0.03 s, which rules out models of coronal transport (*e.g.*, birdcage model) in which escaping ions travel to distant solar longitude within coronal loops. The results do not contradict models of distributed shock acceleration of energetic ions at various solar longitudes followed by prompt injection into interplanetary medium.

2.5 Electrons and Electromagnetic Emissions of Solar Flares

The solar electron events associated with impulsive X-ray events had a characteristic double power-law spectrum which the investigators usually interpreted as indicating a two-stage acceleration mechanism (*e.g.*, Lin et al. 1982). They found that every event shows the same spectral shape: a double power law with a smooth transition around 100–200 keV (Fig. 2.11) and power law exponents of 0.6–2.0 below and 2.4–4.3 above. The solar electron events associated with long-duration X-ray events had a characteristic spectrum typified by a power law in momentum which the authors interpreted as indicating that one stage dominates the acceleration process.

A comprehensive systematic survey of electron data acquired by the *ISSE*-3 (*ICE*) and *Helios* 1 and 2 spacecraft (Dröge 1995) considered the spectral characteristics of solar electron events with respect to the type of associated solar-flare soft-X-ray emission (impulsive or long-duration). An important result was the similarity in the shape and character of the electron spectra observed on widely separated spacecraft (*Helios* and *ICE*). Even though these two spacecraft were widely separated in heliolongitudes, this impulsive/LDE event characterization persisted. The long-duration events had the spectral form of a power law in momentum, and the impulsive events had the double power law even at widely separated heliolongitudinal distances. After adjustment for the flux amplitude, which is a function of relative connection to the solar-activity source, the same characteristics in spectral shape were maintained.

From a survey of 57 electron relativistic (>35 MeV) events observed by the *PHOBOS* 2 space probe during its flight to Mars, Stolpovsky et al. (1995) found a strong correlation (with a correlation coefficient about 0.76) between the maximum electron flux and the peak X-ray flux that was independent of the duration of the X-ray emission. They also found a relation between the times of electron onset to the maximum electron flux that was associated with CME speed.

As it becomes now widely accepted, the proton component in the most of SEP events is causally associated with CMEs. In contrast, the situation with energetic

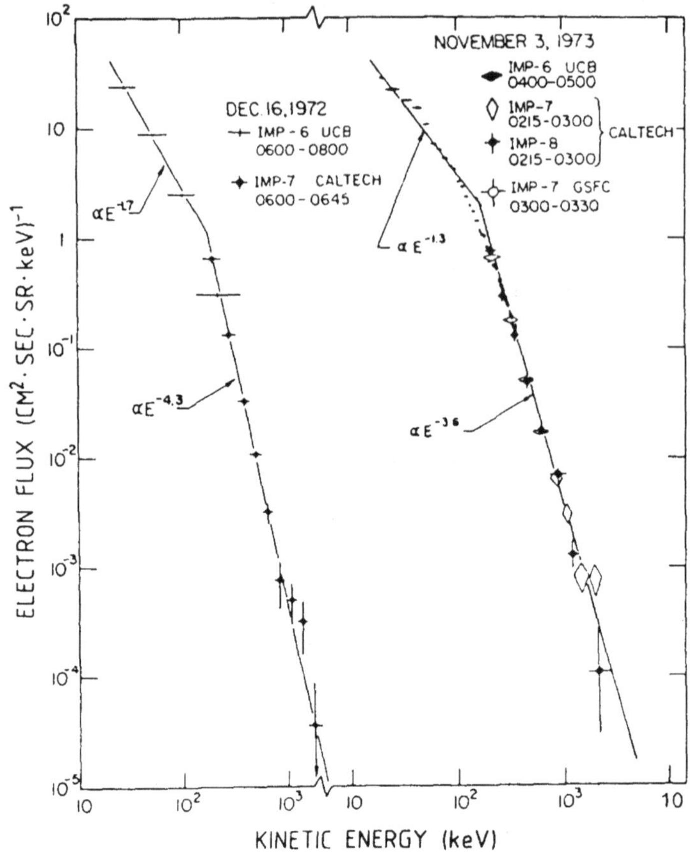

Fig. 2.11 The energy spectra of solar electrons for the events of 16 December 1972 and 3 November 1973 (Lin et al. 1982). Both the events show a smooth transition between ~100 and 200 keV

electrons in SEP events is far from understood. Recently, Stolpovsky et al. (1997) considered the time scales of energetic (>0.3 MeV) electron component in SEP events after CME-associated flares (more than 50 events) by the data of multi-spacecraft observations in 1979–1989. It was found that the onset times of electron events are related to CME speed, the delay between the onset of electron event and the flare onset being close to the time of CME propagation to the observer's magnetic field line. These findings mean that CMEs seem to be an essential factor of electron acceleration in CME associated flares.

The *Yohkoh* satellite recorded a spiky event at 00:18 UT on 22 April, 1994 (Yoshimori et al. 1997). A single-spike hard X-ray emission (>20 keV) with duration of 3 s was observed. It suggests the possibility that strong plasma turbulence, DC electric fields or shock wave were suddenly generated in the flare and electrons were accelerated within a very short time of about 1 s. In addition, both soft and hard X-ray images were observed. They provide information on magnetic loop structures and hard X-ray locations, respectively. From these data, Yoshimori et al. (1997) estimated the electron acceleration times by stochastic and DC electric fields mechanisms. In order to accelerate the electrons to 200 keV, the first of them (stochastic mechanism by Miller et al. 1996; Miller and Reames 1996)) takes 130 ms, and the second one (*e.g.*, Holman 1996) requires 100 ms. Hence, both acceleration processes are thought to explain the observations. The role of shock acceleration of the electrons in a flaring loop has not been clarified so far. In particular, from a study of hard X-ray and wide band radio observations, Klein et al. (1988) concluded that extended coronal shock waves play a minor role in the acceleration of relativistic electrons. Since shocks are not always exist, Huang and Wang (1995) investigated effects of the coronal plasma wake fields (PWFs), which are excited by high energy electron beams in solar flares, on the formation of relativistic electron spectrum. Their calculations show that the PWF model can explain the difference between relativistic electron spectra in impulsive and long-duration flares.

In this context, note also the observations of energetic electrons (44 keV–10 MeV) by the *SOHO* spacecraft instruments on July 9, 1996 (Sierks et al. 1997). An electron event was detected after a solar flare of X2.6/1B class (S10°, W30°). It was found, in particular, that the electron spectrum of the event can be modeled by a single power law between 0.1 and 2 MeV. A relatively sharp break occurs at about 2 MeV, leading to a harder spectrum at higher energies. Such a spectral shape is typical for impulsive flares with associated short duration soft X-ray emission (SDEs) and confirms the findings of Moses et al. (1989). In contrast, flares with associated long duration soft X-ray emission (LDEs) usually have spectra which can be modeled with single power laws in rigidity from 0.2 to 20 MeV. To explain the different classes of electron spectra, and also features of the associated gamma-ray emission, it was suggested by Dröge (1996) that in LDEs electron acceleration above 0.2 MeV occurs in or is dominated by a single stage mechanism which produces a power law in momentum, whereas in SDEs there are two electron components from two different acceleration sites or processes, one with a steep spectrum, which is also consistent with a power law in momentum dominating below ~3 MeV, and one with a flatter spectrum dominating above that energy. Additional analysis of radio observations shows that it is quite possible that in fact two different populations of energetic electrons were detected on July 9, 1996.

Li and Hurley (1995) examined the hard X-ray emission from 2,500 solar flares observed by the *GRO BATSE* instrument and found no significant evidence of photon anisotropy. A further study of the >100 keV photon flux from the most copious hard X-ray emitters (the top 3σ events were selected) found that a single power law does not fit spectra over the 0.03–2 MeV range; a broken power law is a better representation of the data. The average energy of the break point was $E_b = 171.9$ keV. Their conclusion is that no single acceleration mechanism dominates.

As it is known, the soft X-ray emissions generated by solar hot plasma was primarily used as indicators of activity or as a parameter for determining the energy involved in the solar activity. The hard X-ray emission is principally due to bremsstrahlung from the accelerated electrons. During the maximum of the 22nd solar cycle we have been fortunate to have observations from *GAMMA*-1, GRO and *Yohkoh* space platforms during the occurrence of powerful solar flares. These observations have added significantly to our knowledge of solar-flare processes.

2.6 Neutral Flare Emissions

In the following we will focus on hard X-ray, gamma-ray and neutron observations which contribute new insights in understanding of the particle acceleration phenomena associated with solar flares. There is a keen interest in gamma radiation from solar flares which is viewed as a signature of particle acceleration. Accelerated electrons are responsible for the bremsstrahlung continuum; accelerated ions traveling down the magnetic loops, impacting the denser material in the chromosphere and photosphere generate the line emission.

Gamma-rays. Bai (1986) has studied various properties of gamma-ray/proton (GRP) flares, which produced nuclear gamma-rays and/or interplanetary energetic protons. It was found that there exist two classes of GRP flares, each class having many distinct characteristics in common. Gradual GRP flares (so named because of gradual variations of hard X-ray fluxes with duration in spike bursts longer than 90 s) have the following characteristics: long duration (>10 min) hard X-ray and microwave flux, relatively large ratios of microwave to hard X-ray fluxes, large $H\alpha$ areas, long-duration soft X-ray emission (>1 h), hard X-ray emission from extended coronal loops, interplanetary type II emission, coronal mass ejections, and production of large numbers of interplanetary energetic protons. Impulsive GRP flares display directly opposing behavior in the above respects. However, two classes of GRP flares have a few characteristics in common. The author has reached the following conclusions: (1) In both classes of GRP flares protons are accelerated in closed magnetic loops during the first phase by the second-step mechanism, and these protons have a low escape probability and produce gamma-rays interacting with the solar atmosphere; (2) In gradual GRP flares additional protons are accelerated in the high corona by shock waves, and these protons easily escape into interplanetary space. This is the main reason the correlation is poor between gamma-ray fluence and interplanetary proton flux.

Space does not permit to review the relationships of the SEP events to the gamma-ray line producing flares as done primarily by Fomichev and Chertok (1985), Cliver et al. (1989) and Reames (1990a, b). However. it is important to refer, briefly, to the large SEP events observed by *Helios 1* (van Hollebeke et al. 1990), during the large gamma-ray line flares on 21 June 1980 and 3 June 1983, which produced intense high-energy photons (>1 MeV) and neutrons. It is of interest that the SEP intensities and energy spectra of these two events typical of the so-called "large gradual" events (Reames 1990a, b), while the particle composition is like that for the so-called "impulsive, ^{3}He-rich" flares (Reames 1990a, b).

Because of the proximity of *Helios 1* to the Sun (about 0.5 AU) and the excellent magnetic connection to the flare site, small precursor particle increases were observed prior to both events, suggesting particle storage at the Sun from earlier flares (van Hollebeke et al. 1990). This observation may be a vital clue to understand when and how the high-energy particles are accelerated. There is strong evidence that the energy resources of an active region *may not be sufficient* to supply the energetic particles produced in some large flares, *and larger coronal structures* may be involved (Kane et al. 1995).

The great variability between individual flares has been widely discussed (*e.g.*, Yoshimori et al. 1995a, b; Djantemirov et al. 1995; Kotov et al. 1995). In some events, the line emission is clearly evident; in other events it is more difficult to discern the line emission above the bremsstrahlung continuum, and in still other events, the bremsstrahlung continuum totally dominates the emission. The most outstanding gamma-ray events of 1991–1992 under consideration were following: 26 March (Djantemirov et al. 1995; Kurt et al. 1995); 4 June (Debrunner et al. 1995); 6 June (Muraki et al. 1995a, b); 11 June (Chertok 1995); 15 June (Djantemirov et al. 1995); 27 October (Yoshimori et al. 1995b; Djantemirov et al. 1995; Kotov et al. 1995); 15 November 1991 (Yoshimori et al. 1995a, b), and 3 December 1992 (Yoshimori et al. 1995a). A common theme of many studies dealing with these outstanding gamma-ray events was the observation of extended gamma-ray emission with multiple maxima in the emission profile. The authors also note the correspondence between the maxima of the GHz microwave emission and gamma-ray emission. The common feature in these events was the occurrence of multiple episodes of gamma-ray emission which the researchers interpret as episodes of particle acceleration. The longest solar-flare gamma-ray emission to date was observed on 11 June 1991 when the gamma-ray emission persisted for 8 h.

In June 1991 the Sun produced a series of six X-class solar flares, and all of them took place in the same active region 6,659. Rank et al. (1997a, b) presented measurements of gamma-rays that were performed with the Compton Telescope (COMPTEL) during the flares on 9, 11 and 15 June. COMPTEL measured extended emission in the neutron capture line lasting for several hours after the impulsive phase. The time profiles of the flares can be described by a double exponential decay with decay constants of the order of 10 min for the fast and 200 min for the slow component. All three time profiles show the same overall time behaviour. It is interesting to note that there seems to be even a slight enhancement in the emission about 2–3 h after flare onset, producing a late maximum about 4 h later after onset, before decaying more rapidly. This can only be explained by newly accelerated particles, but not by storage alone. The remarkable similarity between the three flares that span a period 6 days raises the question as to whether an expanding post-flare loop system goes through reproducible stages of particle acceleration after each of these flares, or if there are stable loops that remain unchanged during and between the events and can then be populated repeatedly in the acceleration process.

For the flare event on 15 June there exist a rich set of measurements, including *H*α, radio and microwaves, X-rays, gamma-rays, neutrons and interplanetary protons. Using some of these measurements, Rank et al. (1997a, b) were able to obtain a composite photon spectrum spanning the energy range from about 25 keV to 4 GeV. From the gamma-ray and neutron results of COMPTEL and *GAMMA*-1 they could determine the spectrum of accelerated protons. It was found that COMPTEL gamma-ray data are consistent with a proton power law spectrum with index 3.3 ± 0.1 over the energy range 10–200 MeV and a number of interacting protons of $N_p(> 30\ \mathrm{MeV}) = (2.9 \pm 0.4) \times 10^{32}$.

Such a proton spectrum predicts a higher >100 MeV gamma-ray fluence from pion decay than was actually measured by GAMMA-1. To simultaneously fit the higher energy GAMMA-1 measurements, the spectrum must be steepened at a few hundred MeV, having the index of 5.0 above $E_p > 300$ MeV. From spectral studies of the *GAMMA*-1 data alone, a hard spectrum ($\gamma = 3.0$) seems to be required for gamma-ray energies above 1 GeV (Djantemirov et al. 1995). One possible explanation is a two-component spectrum: one is the broken power law, producing gamma rays with energies below 1 GeV and neutrons; the other one, which dominates above 1 GeV, could be either an unbroken $\gamma = 3.0$ power law, or have the turnover at an energy above the sensitivity of the COMPTEL measurements.

As noted above, the long-term time profile of the 15 June flare shows a double-exponential decay of the MeV gamma-radiation, indicating the presence of two different particle populations The two-component spectrum might refer to these populations, with one spectrum turning over at lower energies (as observed), the other one stretching out to higher energies.

Galper et al. (1997) presented a comparative analysis of particle acceleration phase during two powerful solar gamma flares (26 March and 15 June 1991) detected on board the *GAMMA-1* satellite. Although the flares differed significantly in their duration and power, common patterns of the particle acceleration phases have been found. It was shown that separate acceleration acts with duration from 40 ms to 3 s occurred during both impulsive and extended phases of the flares. The extended phase of 15 June gamma flare continued for 2 h, but no separate bursts of radiation are found in this prolonged phase. The observed phases of particle acceleration are suggested to correspond to magnetic structures of different spatial scales.

The properties discovered in impulsive and prolonged flares demonstrate a similarity, both during particle acceleration and in the course of radio wave and gamma-ray generation. It is especially surprising because radio and gamma radiation are produced by different particle populations: synchrotron radio-band is emitted by electrons with MeV energies, for gamma radiation protons of 10 MeV–10 GeV energy interval and electrons up to 100 MeV energy are responsible. According to Galper et al. (1997), different time scales of the flares seem to reflect different sizes of active region where particles are accelerated and regions of generation of the radiation.

The time variation of the energy spectra could be caused by a change of the particle abundances as well as by change of the geometry of magnetic field in the upper solar atmosphere. In particular, some drift of mirror points in an upper solar atmosphere magnetic loop could it provoke. If an active region is located low in the chromosphere the gamma radiation is produced by particle moving downwards. More often it could be realized in flares with small active regions, when the magnetic field change at the transition from the corona into the chromosphere occurs slower, than the increase of density. Flares with large active regions seem to have such magnetic field geometry that the mirror points are relatively high in the bottom corona and the field change occurs faster than that of density. In this case gamma rays are preliminary generated by particles moving upwards, so that the particles have enough time to be reflected before they reached dense layers of the solar atmosphere. Probably, such change of the mirror point position could explain the March 26, 1991 flare properties (Kurt et al. 1995). In this context, Galper et al. (1997) formulated some additional requirements to models of solar gamma flares:

1. Both impulsive and prolonged flares undergo similar development phases, which are reflected in radio and gamma radiations. Time profiles of the flares depend firstly on the size of the generation region rather than on the features of acceleration process.
2. Separate acceleration acts occur at all phases except of the last final recession phase. However, it does not exclude possible acceleration in a slow process, for example, in shock waves.
3. The time variation of the gamma-ray energy spectrum is a result of changing geometry of magnetic field and accelerated particles' beam.

Solar neutrons. The fascination with high-energy solar phenomena is typified by the keen interest in solar neutrons. As mentioned earlier, solar neutrons are diagnostic of interactions of energetic ions above about 20 MeV/nucleon. Neutrons that fail to escape the solar atmosphere can be detected by means of the 2.223 MeV gamma-ray line emission resulting from the $p(n, \gamma)d$ reaction. Neutron that escape into the interplanetary medium can be detected in four ways (Ruffolo 1997b; for detail see also Chap. 6): (1) direct spacecraft detection (Chupp et al. 1982); (2) detection of decay protons (Evenson et al. 1983); (3) detection of decay electrons (Dröge et al. 1996); (4) direct ground-based detection by neutron monitors (*e.g.*, Debrunner et al. 1983; Efimov et al. 1983) or air shower arrays (Chiba et al. 1992).

The measurements of neutron events observed at the Earth, particularly measurements of solar neutrons on the Earth's surface (primarily mountain altitudes), indicates that ions were accelerated to ~GeV energies. The timing sequence between the initial gamma-ray burst and the detection of high-energy neutrons at the Earth strongly suggests that energetic ions can be accelerated to high energies in the order of ~1 s. The major solar-flare neutron events identified to date as being detected on the Earth's surface is discussed in Chap. 6. A common finding of several studies is that the 2.223 MeV gamma-ray emission is an indicator of low-energy neutron generation, but is a poor indicator of the production of high-energy neutrons. The π^0 -generated gamma-ray emission is the best indicator of the generation of high-energy neutrons (Debrunner et al. 1995).

The major findings concerning to solar neutron events are following (for more details see, *e.g.*, Mandzhavidze and Ramaty 1993; Smart 1996; Chupp 1996; Miroshnichenko and Perez-Peraza 2008, Chap. 6 and references therein):

21 June 1980: It was a first direct detection of neutrons from the solar flare (Chupp et al. 1982).

3 June 1982: Two injections of neutrons deduced. This event is probably the best analyzed and has become the base line against which other neutron events are compared (Debrunner et al. 1995). The calculated >100 MeV neutron flux is 8×10^{28} sr^{-1} (Chupp et al. 1987). First detection of the neutron decay protons in space (Evenson et al. 1983).

24 April 1984: Observations of the neutron decay protons in space (Evenson et al. 1983) and of the neutrons above 50 MeV by the SMM/GRS.

16 December 1988: Observations of the neutron decay protons in space (Evenson et al. 1983).

6 March 1989: The total observed neutron fluence (>50 MeV) was ~50 cm^{-2} with a total emissivity at the Sun of 4×10^{28} sr^{-1} assuming isotropic emission (Dunphy and Chupp 1991).

19 October 1989: An upper limit for solar neutrons (>300 MeV) turned out to be ~2.5×10^{27} sr^{-1} at the Sun (Kananen et al. 1997). This corresponds to an upper limit of protons (>600 MeV) stopped in the flare site being ~1.5×10^{29} in the case of isotropic neutron production.

24 May 1990: This is the largest solar-neutron event observed to date. Kovaltsov et al. (1995) deduce that there were two high-energy neutron injections in this event. The calculated >100 MeV neutron flux is 3.8×10^{30} sr^{-1} (Debrunner et al. 1995). Assuming a solid angle of neutron ejection from the Sun of 2 π/3 sr, Belov and Livshits (1995) estimated a total number of escaping >100 MeV neutrons as 2×10^{30}.

1 June 1991: OSSE observations of the 1 June 1991 over-the-limb flare indicate a strong >16 MeV flux which appears to be due predominantly to neutrons (Murphy et al. 1998). GRANAT/PHEBUS also observed this flare and detected GRL emission at <10 MeV which is thought to be produced in thin-target interactions in the Sun's corona. The strong neutron flux therefore is surprising if the neutrons were produced by the same thin-target interactions responsible for the gamma-rays. Using calculations of neutron and gamma-ray yields, Murphy et al. (1998) found that a very hard spectrum for the accelerated particles (assuming photospheric abundances) is required to account for the number of neutrons observed on 1 June.

4 June 1991: Multiple emission of high-energy neutrons was deduced (Debrunner et al. 1995). These authors were also attempting to unfold the responses in different spacecraft detector systems. The calculated >100 MeV neutron flux is ~1.0×10^{29} sr^{-1} (Struminsky et al. 1994; Muraki et al. 1995a).

6 June 1991: Simultaneous observation of solar neutrons at two different sites, Mt. Norikura in Japan and Mt. Haleakala in Hawaii (Muraki et al. 1995b).

Preliminary results suggest perhaps a delta-function injection; however, the results are statistically limited.

9 June 1991: Neutrons were detected with OSSE and COMPTEL on CGRO (*e.g.*, Ryan and McConnel 1996). These instruments are sensitive to lower energy neutrons (15–80 MeV).

11 June 1991: Neutrons were observed by spacecraft and surface detectors. There is a severe problem in interpreting the CGRO observations because of the extreme solar gamma-ray flux (Dunphy et al. 1995, 1999). Nevertheless, their results indicate a neutron flux of 27 cm^{-2} with energies >50 MeV (Smart 1996).

15 June 1991: Neutrons were observed by Earth-orbiting spacecraft and perhaps at the Alma-Ata neutron monitor at 3,340 m altitude (Usoskin et al. 1995). The Alma-Ata observations (if verified) would be consistent with a post impulsive phase injection, but results are statistically limited by the small signal observed.

Summary of neutron observations since 21 June 1980 up to now is presented in Table 2.5. The most notable neutron event was registered on 7 September 2005 by SNT at Mt. Sierra Negra (Mexico), by NM at México City and by NM at Mt. Chacaltaya (Bolivia). Latest results on solar flare neutrons are described in Chap. 12.

The enhanced interest in solar neutrons extends to detection of their decay products (*e.g.*, Evenson et al. 1983). Sakai et al. (1995b) have calculated the expected solar neutron decay flux in interplanetary space as a result of isotropic and pancake angular distribution functions. Kocharov et al. (1995a, b) reported a barely detectable signal in the *GOES* spacecraft P7 channel which they identify as decay protons from the 24 May 1990 solar neutron event. Note, however, that the *GOES* sensor is several orders of magnitude less sensitive than the *ICE*-3 instrument used by Evenson et al. (1983) for the identification of solar neutron-decay protons in the June 3, 1982 neutron event. In analyzing the initial electron flux associated with the 21 June 1982 solar neutron event, Dröge et al. (1995) have identified an excess in observed electron flux above that expected from electron propagation models as likely being from solar-neutron decay. They further support their neutron decay product hypothesis by observing that the spectrum of this electron excess is consistent with that expected from neutron decay (see also Chap. 6).

Solar flare neutrinos. This problem will be briefly discussed in Chap. 4 in the framework of more general problem of maximum rigidity of particles accelerated at the Sun.

2.7 Classification Systems of SEP Events

A great variety of the SPEs observed near the Earth's orbit, in their energy spectra, intensities, elemental abundances, charge composition, spatial and temporal properties make serious difficulties of the classification and analysis of the events. The

Table 2.7 Properties of impulsive and gradual events

Parameters of particles, observation method	Impulsive	Gradual
Particles	Electron-rich	Proton-rich
$^3He/^4He$	~1	~0.0005
Fe/O	~1	~0.1
H/He	~10	~100
Q(Fe)	~20	~14
Duration	Hours	Days
Longitude cone	<30°	~180°
Radio type	III, V (II)	II, IV
X-rays	Impulsive	Gradual
Coronograph	–	CMEs (96 %)
Solar Wind	–	IP Shock
Events/year	~1,000	~10

After Reames (1996)

During two last decades, due to regular spacecraft observations of X-ray emissions from solar flares, a new classification of SEP events raised, depending on the features of originating flare, namely, on the duration of soft X-ray burst. In the light of a new paradigm of particle acceleration in different sources at/or near the Sun (impulsive or gradual flares, CME-driven shocks, etc.), it becomes very keen a problem of SPE identification with different sources. The properties of gradual and impulsive events as they were summarized by Reames (1996) are given in Table 2.7.

The most useful classification of energetic solar events was suggested by Pallavicini et al. (1977). They categorized events as either class I or class II. Class I events had a small volume of $\leq 10^{27}$ cm^3, were low in the corona, possessed impulsive soft X-ray ($\leq$6 keV) emission, had a thermal particle energy density $\geq$100 erg cm^{-3}, and were not associated with white-light coronal transients. Class II events had a volume $\geq 10^{28}$ cm^3, were much higher in the corona, had a thermal particle energy density $\leq$100 erg cm^{-3}, and were associated with coronal transients. Both events were later referred to according to the characteristics of their soft X-ray emission: the compact class I were given the adjective "*impulsive*", while the larger class II events were called "*gradual*".

Historically, the terms *impulsive* and *gradual referred* to the time duration of the soft X-rays in the event (namely, <1 and >1 h, respectively). However, it became clear later that there are other differences as well, in both the radiations that are emitted and the particles that are observed in space. In particular, the X-ray duration gives only a poor, statistical distinction of the underlying mechanisms, while the particle abundances, for example, distinguish them cleanly. Therefore, Reames (1996) suggested to use the terms impulsive and gradual to refer to the underlying acceleration mechanisms, irrespective of the actual X-ray duration in an event. Of course, there are events in which *both,* impulsive and gradual, phenomena occur (Reames 1990a, b; Cliver 1996, 2009). These authors believe that more recent observations allow one to extend this concept to particles of very high energy. In particular, Kahler (1994) suggests that even in GLEs, particles of ~20 GeV have a clear association with CME-driven shocks.

Cliver (1996) has expanded Table 2.6 to include characteristics of the particles that interact at the Sun to produce gamma-ray emission. This addition underscores the contributions of gamma-ray observations to our current understanding. The broad picture that is emerging is remarkable for its simplicity: while SEP events come in two basic types depending on the duration of the associated flares, the interacting particles in impulsive and gradual flares appear to be indistinguishable and resemble the SEPs observed in space following impulsive flares. The expanded classification system was given by Cliver (1996). It includes so-called "*hybrid*" events, *i.e.*, flares in which the gradual/impulsive distinction is blurred and for which the SEP events contain a *mixture* of flare-accelerated and CME/shock-accelerated particles. It is suggested that SEP events associated with long duration flares can be expected to have a temporally and spatially confined "core" of flare-accelerated particles surrounded by a "halo" of CME/shock particles.

The key new features of the system (Cliver 1996) are: (1) introduction of hybrid events (Kallenrode et al. 1992a) referred to as "mixed-impulsive" and "mixed-gradual"; (2) allowance for the temporal evolution of composition (and charge state) of the mixed-gradual events; and (3) a listing of the characteristics of interacting particles for the various classes of the events. Other additions to the system (Cliver 1996) include considerations of: (4) coronal mass ejection (CME) width; (5) electron spectra; and (6) the ratio of interacting (solar) to interplanetary (IP) protons. Revised SEP event classification (Cliver 2009) is presented below (Table 2.8). Observational and physical grounds for this revision are discussed in Chap. 12.

A simple classification of SCR events by fluence magnitude was proposed by Nymmik (1996). Studies of the dependence of event frequency on proton fluence brings the conclusion (King 1974; Feynman et al. 1990a, b) that this distribution is described by log-normal function (see also Chap. 11)

$$\Psi(f) = 1/(2\pi\sigma)\exp\left[-0.5\left\{\left[(f - f_0)/\sigma\right\}\right]\right. \tag{2.1}$$

where $f = \log[F_s(E)]$ is the logarithm of fluence magnitudes, and f_0 and σ are parameters of the normal distribution which, for protons with energies $\geq$30 MeV, have the values $f_0 = 6.93$ and $\sigma = 1.19$ (Feynman et al. 1990a, b). It appears logical to separate this set into groups (classes), according to fluence magnitudes (Nymmik 1996)

$$\Phi_k = 10^{(f + k\sigma/2)} \tag{2.2}$$

where $k = -1, 1, 3, 5$. The group names, symbols, and mean fluence values in the different groups (with account for the probability density) for the suggested classification are given in Table 2.9.

Table 2.8 Revised SEP event classification (Cliver 2009)

Parameter	Flare	Shock Quasi-perpend.	Shock Quasi-parallel
H Upper Limit[a]	~3 pr	~10^3 pr	~10^4 pr
3He/4He[b]	~10^3–10^4	~10^1–10^2	~1
Fe/O[c]	~8	~3	<1
Z(>50)/O[d]	~10^2–10^3	~10^{-1}–10^1	~10^{-1}–10^1
Ion Spectra[e]	–	Power-law	Exp. Rollover
QFe[f]	~20	~20	~11
SEP Duration	<1–20 h	~1–3 days	~1–3 days
Longitude Cone[g]	<30–70°	~100°	~180°
Seed Particles	N/A	Flare STs	Coronal STs
Radio Type[h]	III	II	II
X-ray Duration	10–60 min	~1 h	>1 h
Coronagraph[i]	*	CME	CME
Solar Wind	–	IP Shock	IP Shock

See Cliver (1996) for an early attempt to accommodate the quasi-perpendicular shock events
*Association with CMEs is not precluded (Kahler, Reames, & Sheeley, 2001).
[a]Cliver and Ling (2007); >10 MeV protons and 0.5 MeV electrons
[b]Relative to solar wind at <1 MeV/nuc. The "problem" large SEP events on 6 November 1997, 2 May 1998, and 6 May 1998 had enhanced $^3He/^4He$ ratios
[c]Relative to corona at 5–12 MeV/nuc. For the Flare class, the energy range is 5–12 MeV/nuc (Reames 1999). For the Shock class, the energy range is 30–40 MeV/nuc (Tylka et al. 2005)
[d]Relative to corona at 5–12 MeV/nuc; Reames and Ng (2004)
[e]For a study of ion spectra in Flare events, see Mason et al. (2002). Spectral shapes for Shock events are for 3–100 MeV/nuc (Tylka et al. 2005)
[f]Flare (<1 MeV/nuc; Klecker et al. 1984); Shock (~40 MeV/nuc; Mazur et al. 1999)
[g]Lin (1970), Reames et al. (1991), Kallenrode et al. (1992a, b)
[h]Defining radio type in low frequency range from 14 to 1 MHz
[i]The larger flare events can have associated CMEs (Kahler et al. 2001)

Table 2.9 SEP events classification by fluence magnitude, Φ (>30 MeV), cm^{-2} (Nymmik 1996)

Name	Symbol	Interval	Mean value
Small	S	$<2.0 \times 10^6$	5.5×10^5
Medium	M	2.0×10^6–3.3×10^7	8.6×10^6
Large	L	3.3×10^7–5.2×10^8	1.05×10^8
Very large	VL	5.2×10^8–8.0×10^9	1.34×10^9
Extremely large	EL	$>8.0 \times 10^9$	(3.3×10^{10})

In this classification, SCR events of the VL and EL class are analogues of "anomalously large" events used in some other models (*e.g.*, Adams and Gelman 1984; Smart and Shea 1989b). The only difference is that Nymmik (1996) suggests precise quantitative criteria to analyze the SCR phenomena development.

2.8 Solar Event Databases

In Table 2.1 we represented a list of relativistic SPEs, or GLEs, compiled by Shea and Smart (1993a) and Miroshnichenko et al. (2013) for the entire period of SCR observations (since 1942 up to now). It is timely to also mention here several Catalogues of Solar Proton Events since 1955 up to 2009 (Dodson et al. 1975; Akinyan et al. 1983; Bazilevskaya et al. 1986, 1990a, b; Sladkova et al. 1998; Logachev et al. 2014). Electronic version of the SPE Catalogue data has been prepared by Sladkova (1996). Very important work was carried out by Gentile (1993a, b) who has compiled a database of the GLEs observed during the 22nd solar activity cycle (events Nos. 40–54 in Table 2.1). A database for SCR measurements carried out on board a series of Soviet space vehicles in 1964–1989 has been elaborated by Getselev et al. (1996b).

A list of the rigidity spectra near the Earth for the solar protons above >1 GV derived from observations of 31 GLEs was published by Miroshnichenko et al. (1995c). Since then, the list has been completed to 35 events (see Table 9.2). An extended Catalogue of the energy (rigidity) spectra of SCR in their sources (*i.e.*, reconstructed spectra at the Sun for 80 proton events) has been prepared by Miroshnichenko et al. (1999). In some recent publications one can find also a number of the Catalogues (or lists) of proton, electron, neutron, gamma-ray and other energetic solar events, CME catalogues, the lists of ^{3}He-enriched events and so on. The most of them were compiled by the data of different spacecraft measurements (*IMP*, *GOES*, *SMM*, *Compton GRO*, *GRANAT*, *SOHO*, *CORONAS*, *ACE* and many others). For example, we mention the energy spectra for 55 electron events of the period 1978–1982 (Moses et al. 1989), the list of solar proton events 1980–1985 by Cliver et al. (1989), the data sets of the *IMP* 8 (Armstrong 1993) and *GOES* (Wilkinson 1992), the table of *GRANAT* solar flare data of 1990–1994 by Terekhov et al. (1996) and the comprehensive Atlas of the *SMM* gamma-ray bursts of 1980–1989 by Vestrand et al. (1999). As a summary of the early solar proton events the *Solar Proton Manual* edited by McDonald (1963) also is very helpful.

These Catalogues cover the period of 1955–2009 and contain the main characteristics of SEPs with a threshold intensity of 1 *pfu* for the protons above >10 MeV, including GLEs data for the above mentioned period. A more limited list of the SPE Catalogues (for the other threshold energies and intensities) may be found in the review paper by Vainio et al. (2009). Unique set of data on galactic and solar cosmic rays for about 50 years of stratospheric observations (1957–2008) were collected and systematized by the research group from the Lebedev Physical Institute (FIAN) (Stozhkov et al. 2007, 2009; Bazilevskaya et al. 2010). Data on solar proton events and X-ray flares obtained in the last solar cycles 21–23 have been summarized and statistically analyzed by Belov et al. (2005a, b) (http://dec1.sinp.msu.ru/~osipenko/).

All these collections of data form a very solid base for different kind of fundamental and applied research in the field of solar and solar-terrestrial physics. Although each of the collections is a result of thorough research work, nevertheless, it should be emphasized a limited nature of existing databases because of inevitable methodical problems and instrumental limitations (statistical errors, model uncertainties, sensitivity of sensors, etc.). Below we call the readers' attention only to three of them.

The first problem concerns to the method of SPE data selection, especially during the "pre-spacecraft era" (1955–1965), when there was only indirect ionospheric (riometer) information about the >10 MeV proton intensity. A possibility of systematic effects (errors and artifacts) in routine identification of the SPEs at low intensity was noticed for the first time by Smart and Shea (1989b). These effects are very important, in particular, in studying the SPE size (frequency) distributions (see, *e.g.*, Miroshnichenko et al. 2001, and references therein). It means that one has to keep some quite reasonable precautions in using the data mentioned. As we will see below, the spacecraft era turned out to be burdened of its own methodical defects and uncertainties.

The second problem arises due to a difference in the definition of solar proton events in the different Catalogues. For example, the NOAA Space Weather Operations (SWO) gives a special caution at the end of the SPE list (NOAA SESC, Boulder, Co.). Proton fluxes in the list are integral 5-min averages for energies >10 MeV, given in particle flux units (*pfu*), measured by GOES spacecraft at geosynchronous orbit ($1\,pfu = 1$ proton cm^{-2} s^{-1} sr^{-1}). The SWO defines the start of a proton event to be the first of 3 consecutive data points with fluxes greater than or equal to $10\,pfu$. The end of an event is the last time the flux was greater than or equal to $10\,pfu$. This definition, motivated by SWO customer needs, allows multiple proton flares and/or interplanetary shock proton increases to occur within one SWO proton event. Additional data may be necessary to more completely resolve any individual proton event.

At the same time, a standard threshold intensity $>1.0\,pfu$ at the peak time of the event and threshold energy of >10 MeV was accepted as a criterion of a typical SPE in several Catalogues published in 1975–1998 (Dodson et al. 1975; Akinyan et al. 1983; Bazilevskaya et al. 1986, 1990a; Sladkova et al. 1998; Logachev et al. 2014). In the light of this apparent discrepancy, it was proposed (Smart and Shea 1989b) to use the NOAA intensity threshold as a conventional criterion of a "significant particle event". Such a criterion allowed to compile a list of 218 significant SPEs for the period of 1955–1987 (Shea and Smart 1990a) as homogeneous as possible. To their opinion, this criterion gives the best way to identify a proton event that has the potential of producing perturbations in the Earth's environment. Many statistical studies of solar proton events over the past three solar cycles (1954–1986) were carried out just at the NOAA criterion (*e.g.*, Shea and Smart 1993a, 1994, 1997c; Bazilevskaya and Sladkova 1986, 1997).

At last, we should mention one observational (instrumental) problem. Different detectors, on board different GOES spacecraft, have taken the data since 1976. These proton data were processed using various algorithms. To date, no attempt has been made to cross-normalize the resulting proton fluxes. Moreover, recently it was displayed a very important peculiarity in the *GOES* response to energetic protons of different origin, namely, a presence of secondary channels in *GOES* proton detectors. This adds complexity to the extraction of energy spectra from observed counting rates. The *GOES-6* and *GOES-7* satellites (Geostationary Operational Environmental Satellites) carry on board the energetic particle sensor (EPS), which measures energetic particles coming from directions more or less close to the ecliptic plane with a field of view of 50–120°. Since the satellite spin period, 0.6 s, is much shorter than the accumulation times, the EPS provides a spin-averaged estimate of local particle fluxes (Wilkinson 1992). For protons, the ESP energies are 0.6–500 MeV. GOES-6 carries another detector, HEPAD, which measures energies 355–685 MeV and more (Sauer 1993a, b).

Because the detector is passively shielded, there exist significant secondary responses in the energetic proton channels of EPS (Wilkinson 1992; Kahler 1993). Hence, the determination of energetic proton spectra from the data is a bit complex. Corrections to the data have been made to remove these secondary protons from the data, assuming a power-law spectrum (Zwickl 1992) and applying a special correction to the onsets of relativistic solar particle events. Relying on the corrected data can, however, lead to errors if the energy spectrum deviates a lot from a simple power spectrum $E^{-\gamma}$ with $\gamma \approx 3$. Vainio et al. (1995a, b) have analyzed the practicality of corrected data and demonstrated the conditions when it is necessary to use uncorrected data, taking into account the reported secondary responses. It was shown that when the spectral slope is between -2 and -4, the correction needed for the data is very simple: all one has to do is to multiply uncorrected data in channels of 39–82, 84–200 and 110–500 MeV with ≈ 1.2, ≈ 1.4 and ≈ 3.0, respectively. This corresponds roughly to using the corrected data. However, when the spectrum is hard, the situation is more complicated, because the response in the low energy channels is much too high. This is the situation during the onset of an SCR event: the high energy protons are transported to the Earth faster than the low energy particles, and so the spectrum could even have a positive slope at early times of an event. In that situation it is more preferable direct fitting of uncorrected data.

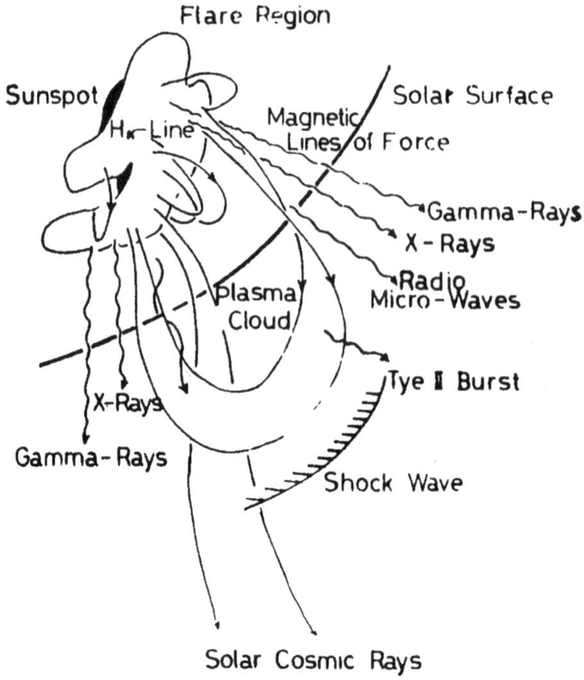

Fig. 3.1 A schematic view of high-energy phenomena associated with a large solar flare (Sakurai 1989)

mentioned above. A schematic view of these phenomena is shown in Fig. 3.1. The characteristics of high-energy photon and particle emissions associated with solar flares are summarized in Table 3.1 (adapted from Sakurai 1989). This table gives a qualitative picture of high-energy phenomena observed after a typical (major) solar flare, together with a short description of principal physical processes involved. In the last line of Table 3.1 we added coronal mass ejections (CMEs) as one of the most powerful phenomena of solar activity.

As is well known, solar flares accompanied by high-energy phenomena mostly occur above or near the sunspot groups defined as the delta-type, in which magnetic polarity distribution is almost reversed from that for the most sunspot groups prevailing in the same solar activity cycle. As schematically shown in Fig. 3.1, the area brightened with the $H\alpha$ emission, for instance, almost covers the whole sunspot group, above which some triggering mechanism seems to initiate the onset of solar flares. Immediately after the onset of a major solar flare, all of the phenomena summarized in Table 3.1 occur. In particular, a plasma cloud heated up as a result of the development of solar flares rapidly expands outward and produces shock wave just ahead of this cloud.

The evolution with time of different phenomena as they are observed near the Earth's orbit is shown in Fig. 3.2. In Fig. 3.2a (adapted from Sakurai 1989) one can see a smoothed, idealized profiles for energetic solar emissions associated with a large solar flare. For a comparison, Fig. 3.2b (taken from Bhatnagar et al. 1996) gives a picture of non-thermal electromagnetic emissions observed during well-

Table 3.1 High-energy phenomena associated with solar flare

Phenomenon	Energy (frequency) range	Physical process
Gamma rays	Line 2.223 MeV	Formation of deuterons by the fusion of neutrons with ambient protons
	0.511 MeV	Electron-positron annihilation
	70 MeV	π°, two gamma-ray photon decay
	4.43 MeV	^{12}C, from excited nuclei
	1.63 MeV	^{14}N, from excited nuclei
	2.31 MeV	^{14}N, from excited nuclei
	7.12 MeV	^{16}O, from excited nuclei
	1.63 MeV	^{20}Ne, from excited nuclei
	Continuum	Bremsstrahlung from relativistic electrons
X rays	Hard X-rays	Bremsstrahlung from relativistic electrons
	Soft X-rays	Mainly thermal radiation from high temperature
		(Line emissions from Fe, Ca, O, and other elements associated with both hard and soft X-ray emissions)
Radio waves	Microwave	(1) A component emitted together with hard X rays. (2) Microwave component of type IV radio bursts (synchrotron radiation from relativistic electrons)
	Decimeter wave	*Type IV radio bursts* (extended to metric waves)
	Meter wave	*Type II radio bursts*
		Plasma radiation excited by the passage of shock waves through the solar corona (relatively slow frequency drift)
		Type III radio bursts. Plasma radiation excited by the keV electrons streams in the solar corona (rapid frequency drift)
		Type IV radio bursts. Synchrotron radiation from relativistic electrons (their sources are usually moving)
	Decameter wave	Extended components of type II, III and IV radio bursts (often kilometric bursts are observed)
High-energy particles		(1) Protons and other heavier nuclei (MeV to GeV). Accelerated in the flare region. Often produce white-light flares. (2) Relativistic electrons (MeV). Gamma-ray continuum. Metric type IV bursts
Plasma clouds		Moving component of metric IV radio bursts
Coronal mass ejections		Expanding through the solar corona with supersonic speed. Responsible for geomagnetic storm

Adapted from Sakurai (1989)

known solar event of September 29, 1989. In general, a real picture is much more interesting and challenging than an idealized one.

It is a question of fundamental interest to estimate a relative share of the energy of flares released in the form of accelerated energetic particles, photon emissions and MHD ejections of solar plasma. The estimates of different forms of energy released by the flare shown in Table 3.2 (Somov 1992) represent only the

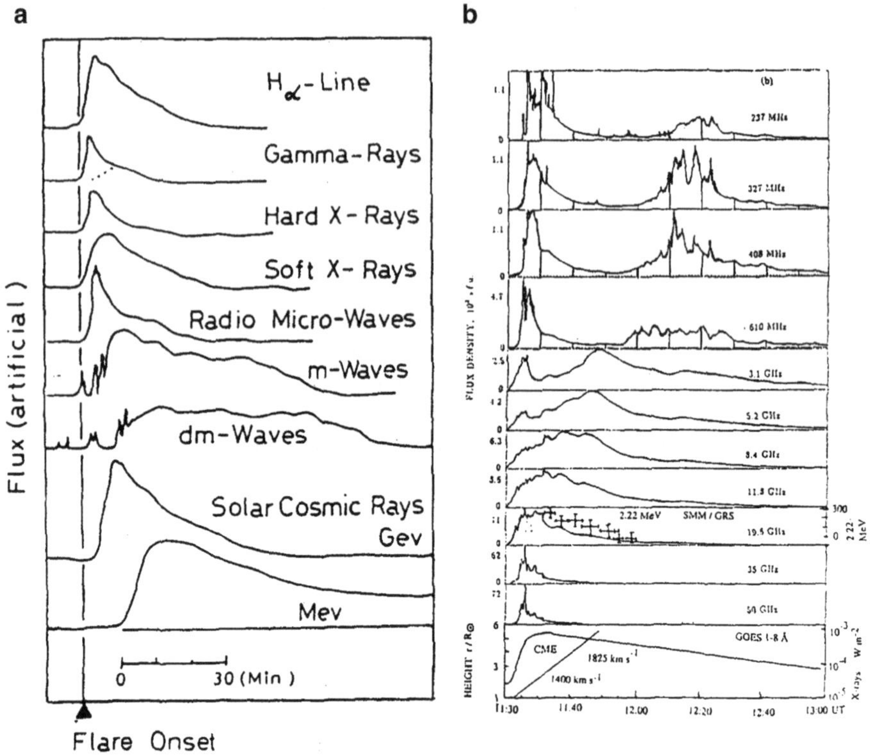

Fig. 3.2 Flux-time profiles for high-energy emissions associated with a large solar flare: (**a**) schematic (idealized) picture with artificial flux scales (Adapted from Sakurai 1989); (**b**) observed temporal evolution of the September 29, 1989 event (compiled by Bhatnagar et al. 1996); the *lower panel* shows the trajectories of the CME (*crosses*) and the shock wave (*solid points*)

characteristic values of the corresponding quantities for the case of the largest and the smallest flares (subflares). They should not be treated as the result of observation of any particular flare. In particular, no gamma-rays and energetic electrons were detected in vast majority of flares.

It is to note that although non-thermal processes at the Sun are very informative from the diagnostic point of view (*e.g.*, Chupp 1996), however, their relative magnitude in the flare energy budget does apparently not exceed 10 % (see Table 3.2). The same seems to be true for the bulk of accelerated electrons and protons (see also estimates of the SCR energetics by Miroshnichenko 1987, 1990). A main part of the energy of flares, undoubtedly, is manifested in the form of hydrodynamic plasma flows.

In conclusion of this short description note that a well-known good association of solar energetic particles with *H*α flares and with flare impulsive radio bursts is most likely not a direct "cause-and-effect" relationship, but rather a manifestation of the "Big Flare Syndrome" (Kahler 1982; Kahler et al. 1985). This phenomenological concept states that, statistically, energetic flare phenomena are more intense

Table 3.2 Energetics of solar flares (Somov 1992)

Form of energy release	Largest flares Energy, erg	Largest flares Power, erg/s	Subflares Energy, erg	Subflares Power, erg/s
1. *Hydrodynamic plasma flows*:				
(a) Interplanetary ejections and shock waves	(1–3) × 10E +32	–	–	–
(b) Surges above photosphere	≈10E+32	≈10E+29	≤10E+29	≤10E+27
2. *Radiation*:				
(a) Soft X-ray and UV	(3–5) × 10E +31	(3–5) × 10E +28	≤10E+29	≤10E+27
(b) Optical continuum	3 × 10E+31	3 × 10E+28	None?	None?
(c) Hα line	3 × 10E+30	3 × 10E+27	≈10E+26	≈10E+24
(d) Hard X-rays	(3–5) × 10E +26	(3–5) × 10E +23	≤10E+24	≤10E+22
(e) Gamma-rays	(1–3) × 10E +25	(1–3) × 10E +22	None?	None?
(f) Radio waves	≈10E+24	≈10E+21	≤10E+21	≤10E+19
3. *Accelerated particles*:				
(a) Electrons (≥20 keV)	(1–3) × 10E +31	(1–3) × 10E +28	10E+27	10E+25
(b) Protons (≥20 MeV)	(1–3) × 10E +31	(1–3) × 10E +28	None?	None?

in larger flares, regardless of the detailed physics of the processes involved. In our opinion, this heuristic concept was a start of our new understanding of the flare-CME associations, of the flare-CME dilemma (see below Sect. 3.2).

3.2 Solar Flare "Myth"?

Many years of research have demonstrated that large, non-recurrent geomagnetic storms, shock wave disturbances in the solar wind, and energetic particle events in interplanetary space often occur in close association with large solar flares. This result has led to a paradigm of "cause and effect" – that large solar flares are the fundamental cause of these events in the near-Earth space. This paradigm, which was called "the solar flare myth" (Gosling 1993), dominates the popular perception of the relationship between solar activity and interplanetary and geomagnetic events.

In 1993, Jack T. Gosling from Los Alamos National Laboratory (Los Alamos, New Mexico, USA) published an extensive paper in which he reviewed the solar-terrestrial relationships and arrived at the conclusion that solar flares are actually

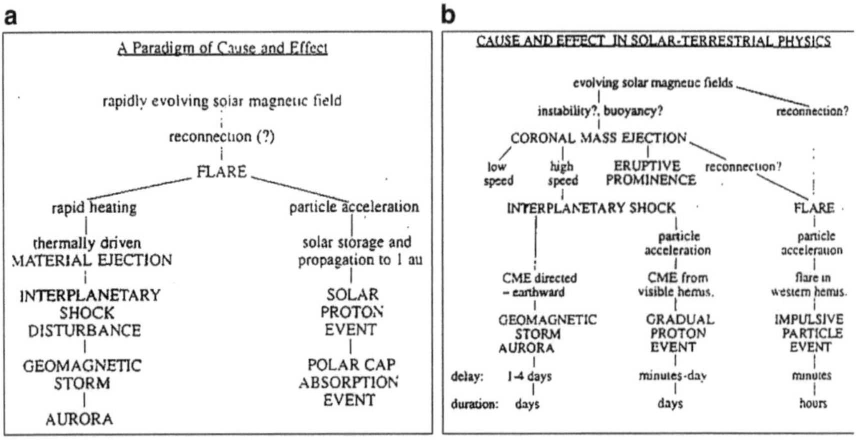

Fig. 3.3 Two paradigms of cause and effect in solar-terrestrial relationships illustrating the supposed central position of solar flares (**a**) or coronal mass ejections (**b**) in producing major disturbances in the near-Earth space environment. Capital letters indicate observational phenomena and lower case letters denote processes or descriptive characteristics (Gosling 1993)

unimportant in this context. In other words, it was stated that the original flare paradigm is wrong and that flares, in general, do not play a central role in producing major transient disturbances in the near-Earth environment. Moreover, the real sources of high-energy particles, shocks in interplanetary space, and large geomagnetic storms at the Earth are (quotation) "coronal mass ejections (CMEs) that have no fundamental association (in terms of cause and effect) with flares". He suggested that "the time has come to lay the solar flare myth in rest". Instead of this, it was outlined (Gosling 1993, 1994a, b) a different paradigm of cause and effect (Fig. 3.3) that removes solar flares from their central position in the chain of events leading from the Sun to near-Earth space. The central role was given to events known as CMEs.

This suggestion has provoked an extensive discussion among the investigators working in the fields of solar-terrestrial physics. For example, a special session "Is "The Solar Flare Myth" Really the Myth?" has been convened to take place during the AGU Spring Meeting in Baltimore (May 29-June 2, 1995). The High Energy Solar Physics Workshop held at the Goddard Space Flight Center (August 16–18, 1995, NASA, Greenbelt, Maryland, USA) also addressed this issue (*e.g.*, Cliver 1996; Dennis 1996; Reames 1996).

First of all, Dennis (1996) has attempted to formulate once more a subject of the dispute, or the «Great Debate» in Eos (Hudson 1995; Miller 1995; Reames 1995c). The myth can be briefly stated as follows: "Solar flares cause the major transient disturbances in the near-Earth space environment". In other words, the myth is that solar flares are responsible for the high energy particles and magnetic disturbances that have such diverse effects as endangering astronauts and satellites in space and causing power outages and communication problems on the Earth. It is now generally accepted that, in fact, coronal mass ejections (CMEs) are responsible for the most of these phenomena. As Reames (1995c) points out, "except for the "sudden ionospheric disturbances" caused directly by photons, flare are not "geo-effective"". Moreover, he believes that so-called "coronal diffusion" is an artifact of the flare myth: "cross-field diffusion is not necessary from an extended CME source" (Reames 1996). He refers, in particular, to a 96 % correlation between CMEs and proton events found by Kahler et al. (1984). Also, contrary to earlier beliefs, it was asserted that flares do not cause CMEs. Some support to the new paradigm Cliver (1996) found in the identification of large SEP events in association with flares without detectable gamma-ray emission (Cliver et al. 1989): to his opinion, this "provided underpinning for the transition from a flare-centered to a CME-centered view of large SEP events".

Significant controversy still surrounds this subject, however, as evidenced by the mentioned "Great Debate" in *Eos*, by discussion at the NASA Workshop, and by some other publications as well (see, *e.g.*, Zirin 1994; Hudson et al. 1995; Gosling 1995; Svestka 1995; Gosling and Hundhausen 1995; Cliver 1995; Reid 1996; Harrison 1996; Sakai and de Jager 1996; Dryer 1996).

In particular, Svestka (1995) considers that the solar flare myth in the Gosling's formulation "is a misunderstanding". He proposed to use the term "eruptive flare" for all solar active phenomena in which an opening of field lines is involved and which lead to magnetic-field and mass ejections resulting in a CME. The process is essentially the same in all events, irrespective of whether only a *disparition brusque* without any chromospheric brightening or major *two-ribbon flare* is involved in it. In their *Reply*, Gosling and Hundhausen (1995) argued against such a generalization of the term "flare" because it goes beyond the observed radiation output (and the thermal energy that produces it) to describe many different dynamical processes on the Sun and in its atmosphere. While it may be incorrect to say that one "causes" the other, Hudson (1995) states that, "Flares, CMEs, and geomagnetic storms all start out in the dynamics of the solar magnetic field, which is a fundamental object of study". Noting the typical sequence of events following large flares, Reid (1996) commented: "to suggest that the flare, the energetic particles, and geomagnetic storm were unrelated flies in the face of common sense".

In its turn, Miller (1995) reminds that the cause and effect issue affects only gradual events, the impulsive events can be called flares without any disagreement. Solar flares occur at a rate that is about of 2 orders of magnitude higher than that of CMEs (see Table 2.6), and release a huge energy for a short time (see Table 3.2); moreover, "Nor are solar flares necessarily small relatives of CMEs". Noting "an unfortunate choice of words" by Gosling (1993), he emphasizes that arising controversy, nevertheless, has initiated a constructive dialog between two related communities that heretofore operated relatively independent of each other. Also, Cliver (1995) believes that the new paradigm focuses attention on a number of open questions involving flares and CMEs. Harrison (1996) considers that the Gosling's phrase "solar flare myth" used in its current sense "is superfluous" because such a thesis decouples flares from CMEs; Sakai and de Jager (1996) support completely this point of view. At the same time, instead of the words like "eruption", "explosion", "transient", "discharge", "transition", or "outburst", Harrison (1996) suggests that the flare/CME/prominence eruption activity should be called a Coronal Magnetic Storm... There are also propositions to call flares "long-decay", "two-ribbon", "dynamic" (Svestka 1995) or "prolonged post eruption" (Chertok 1996).

It would seem that both flares and CMEs are the result of re-arrangements of the coronal magnetic field and that they are often "associated" with one another in some way. As noted by Dryer (1996), "much of the evidence that supports the flare paradigm has been neglected or minimized by some workers". He revised the extended data of apparent central latitudes of 1209 CMEs observed by SMM, of prominence latitudes and bright coronal features, and of latitudes of all classes of optical flares during the 1980–1990 epoch. His conclusion was that the CME associations (flare or non-flare) are at least bimodal, i.e., CMEs can be generated either by flares or by large-scale helmet disruptions. Some MHD studies strongly support this bimodal concept of causes for the origin of CMEs.

In our opinion, the problem reduces not only to "sorting things out" or "reconsidering" the perceived "importance" of flares *versus* CMEs, but rather to trying to understand the underlying physics involved in each case. Apparently, much can be learned from studying them both and exploring their relationship to one another. For example, even if the flare does not produce the CME, nevertheless, it provides a useful diagnostic data, in particular, for determining the longitude on the Sun at which a CME/interplanetary shock originates (Cane 1998).

For the investigators working in the field of solar-terrestrial predictions (*e.g.*, Feynman 1997), there is no doubt that SEP events are clearly associated with processes taking place on the Sun. However, a major controversy still exists as to whether the particle acceleration occurs in the flare itself or the particles are accelerated by associated CME. Concerning the role played by the flare itself or the CME in a large (or gradual) SEP event, the question of principal interest is an origin of the seed particles in acceleration process. As it was postulated (Gosling 1993), the seed populations for large SEP events are interplanetary solar wind particles. These particles are accelerated by the CME-driven shock and would show, in particular, a $^3He/^4He$ ratio of ~0.0005 (Reames 1993, 1996). However, some recent observations contradict this estimate. For example, in Chap. 2 we have already mentioned that for the large SEP events of March 22–28 and June 1–19, 1991 Guzik et al. (1995) obtained the $^3He/^4He$ ratios ranging from 0.003 to 0.22, orders of magnitude greater than the solar wind (coronal) value. One of the key open questions remains also the relative importance of CMEs and flares for the acceleration of relativistic protons (see Chap. 4). In this last case, it is not enough to give a general scenario of the acceleration process involved, – it is necessary to explain some peculiarities of energy spectrum and anisotropy of accelerated particles at the early stage of the GLEs.

3.3 Energetic Solar Particles and Coronal Mass Ejections

In spite of existing controversies in the treating of main SEP sources (see Sect. 3.2), it should be recognized that during the last few years there has been a significant shift in the paradigm of solar particle acceleration (*e.g.*, Reames 1995a, b, c). It happened due to improved measurements of particles and their correlation with solar phenomena. According to a new arising paradigm of SEP acceleration in different sources at/near the Sun (flares, coronal/interplanetary shocks, etc.), it is now widely accepted that the major SPEs seen at the Earth are associated with CMEs. Particles are accelerated from the ambient plasma of the corona and solar wind by the shock wave that expands ahead of a large CME, filling the heliosphere with particles over a wide longitude range. If there is particle acceleration in the hot reconnection region beneath the CME, those particles must remain trapped on the post-flare loops since they do not seem escape the region.

As it evidenced, in particular, by the *Voyager* and *Pioneer* observations, the multiple mass ejections merge to drive shocks that continue to accelerate particles far beyond the Earth's orbit, perhaps all the way to the heliospheric boundary (Reames 1995a, b, c). For the highest energy ions, acceleration does not continue as far from the Sun because these particles are less efficiently contained when the plasma expands as $\sim r^{-2}$. For protons from about 100 MeV to several GeV, peak acceleration occurs when the leading edge of the CME is 6–10 solar radii from the Sun (Reames 1994). In this context, spatial and temporal features of accelerated interplanetary particles are of great interest. As an example, we note here the results obtained by Anglin et al. (1995).

They compared the general characteristics of the low-energy proton flux (0.5–1.0 MeV) between 1 and 5 AU measured at different phases of the solar cycle. It was found that the flux profiles observed at heliocentric distances beyond 1 AU are smoother, the intensity variations smaller and the decay times longer than those observed at 1 AU. In 1978, during the rising phase of solar cycle 21, the trajectories of the *Voyager* spacecraft were approximately on the same IMF field line as the Earth-orbiting *IMP* and *Ulysses* spacecraft, and there were no unexpected time lags in the observed SPE sequence between the Earth and the *Voyager* spacecraft. In 1991, during the maximum phase of solar cycle 22, the *IMP* and *Ulysses* spacecraft were not on the same IMF line, and the time sequence of the proton increases was consistent with the delay expected from the corotating IMF topology.

Figure 3.4a shows the time history of a typical large gradual event observed near the Earth on December 5, 1981 (see observational details in Bazilevskaya et al. 1990a). In accordance with Table 2.6, protons dominate electrons in this event. Low-energy (~1 MeV) protons reach plateau in intensity within 2 days, and 2 days later they reveal additional peak as the shock passes. Profiles of electrons and higher-energy protons decline slowly with time. For a comparison, a typical impulsive (^{3}He-rich) event observed on August 14, 1982 is shown in Fig. 3.4b at the same scale. This event apparently is dominated by electrons, and intensities of all particle species decay rapidly with time. The difference in time behaviour of the

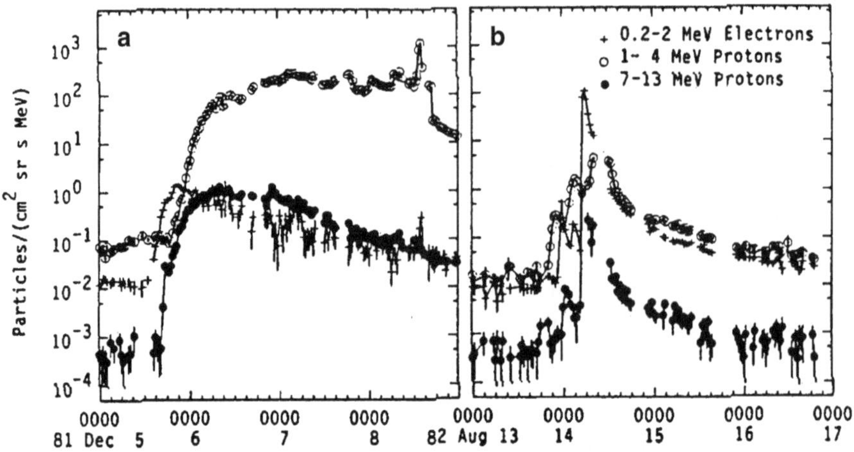

Fig. 3.4 Time profiles of protons and electrons in gradual (**a**) and impulsive (**b**) events (Reames 1995a, b, c). Protons and electrons apparently dominate in the first and second event, respectively

particle intensities implies that the time profiles in both events were controlled by different processes in the interplanetary space, namely, by shock passage and pre-existing scattering, respectively.

In large (gradual) CME-related events, the time profiles depend strongly on the CME longitude relative to the observer. This dependence has been described in detail by Cane et al. (1988) (*e.g.*, Fig. 2.6). In one of the recent reviews Reames (1995a, b, c) illustrates this effect based on the data for protons of different energy. Figure 3.5 shows intensity-time profiles for the events of November 10, 1978, March 7 and December 25, 1982 observed from the three different longitudes.

The highest intensity occurs when the observer is connected to the nose of shock ahead of the CME. Events near central meridian display the intense flat profiles. Behind the shock one can see a second plateau that is characterized, in particular, by bidirectional streaming events (Marsden et al. 1987; Richardson and Reames 1993). For western events, the peak intensity occurs early, when the nose of the shock is best connected to the observer. By the time the shock reaches 1 AU, the observer is connected far around on the eastern flank of the event where the shock is weak (if it is seen at all).

For eastern events, the intensity behaviour seems to be more complicated. The intensity may begin to rise when the coronal shock reaches the base of the observer's IMF line. However, the peak intensity may occur late, after the weak local shock has passed and the observer reaches the IMF lines that connect him to the strong acceleration region near the nose of the shock (at this moment it is far out beyond him). Note that several large events were viewed from three widely separated spacecraft (see Reames 1994).

If the proton component in the most of large SEP events in interplanetary space appears to be causally associated with the CMEs, the situation with energetic electrons in SEP events, in contrast, is far from understood. Stolpovsky

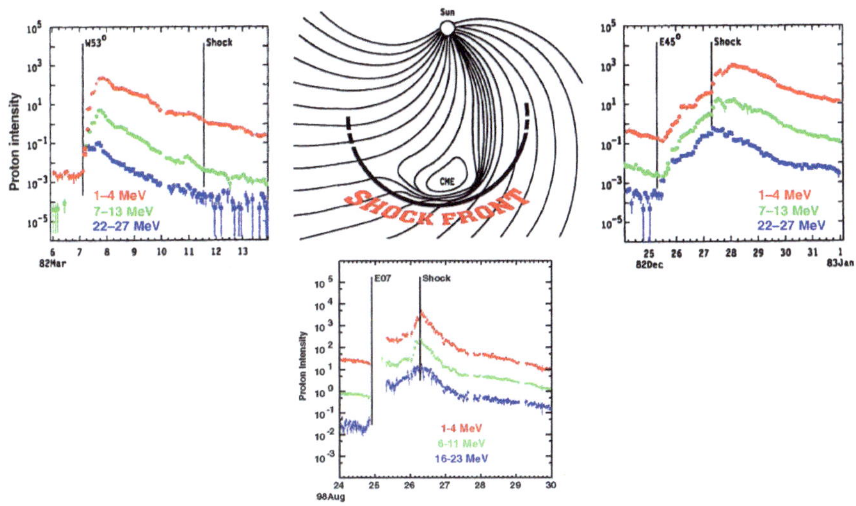

Fig. 3.5 Intensity profiles for protons of different energy for observers viewing a CME from the three different longitudes indicated in the panels (Reames 1995a, b, c). The variation of the profiles with the CME longitude is clearly displayed

et al. (1997) considered the time scales of energetic electrons ($E_e > 0.3$ MeV) in SEP events after CME-associated flares (more than 50 events) using the data obtained in observations on board *Helios* 1, *ISEE* 3, *Venera* 13 and 14, and *Phobos* 2 spacecraft during the period 1979–1989. It was found that the onset time, t, as well as rise time to maximum, of electron events are statistically dependent on the CME speed. Of special interest are the events that occurred outside the fast propagation region, i.e., outside of 25–90 range of the angular distance between the observer's magnetic foot point and the flare-CME site. As shown by Stolpovsky et al. (1997), the delay between t of these events and the flare onset is close to the time of CME propagation to the observer's magnetic field line. They conclude that CME can be considered as essential factor of electron acceleration in CME associated flares.

3.4 Effects of Large-Scale Heliospheric Structures

The thesis that the structure of the interplanetary medium is an important parameter controlling the behaviour of solar particles was explored by many authors. High energy solar protons propagate preferentially along the IMF lines to their detection point in space. For detectors at 1 AU this preferential propagation path is "West" of the Sun-Earth line. For an ideal event with a typical solar wind speed of ~400 km s^{-1} the "footpoint" of this favourable propagation path at the Sun is ~60°W.

Assuming a range of solar wind speeds between 250 and 800 km s^{-1}, the favourable connection longitude on the Sun would be between 30°W and 90°W. However, since 1942 up to now, 20 of the 56 GLEs (see Table 2.1) have been associated with flares occurring outside this range of longitudes. In this context, it was interesting, in particular, to examine the location of solar flares associated with major (primarily GLE) proton events and the location of the Earth at the time of the proton event with respect to the heliospheric magnetic field polarity and distance to the heliospheric current sheet (HCS). The main question arising was as follows: Is proton access influenced by the heliospheric current sheet?

Models of the coronal magnetic field structures through which the large-scale CME-driven shocks must propagate have been developed extensively (*e.g*., Wang and Sheeley 1994; Zhao and Hoeksema 1994a). All these models involve a source surface, usually located at approximately 2.5 solar radii from the Sun's center, above which the magnetic field lines are assumed to be radial and open to the interplanetary medium. The foot points of the open field lines of the source surface lie in coronal holes (CH) which are very asymmetrically located under the unipolar magnetic region of the source surface fields (*e.g.*, Wang and Sheeley 1994).

Coronal helmet streamers, which can be used to map out the "base" of the HCS, are now recognized as one of the sites of CME generation. The question is raised as to the effect of CMEs on the HCS. By numerical calculations, Zhao and Hoeksema (1994b) have examined the change of the HCS after the "streamer CMEs". It was also checked the influence on the IMF sector boundary layers of the counter-streaming suprathermal electron events that are supposed to be the interplanetary counterpart of CMEs. It was found that a streamer CME may locally and temporary expand the HCS width, while passing through the streamer and its interplanetary extension. Besides, this CME may usually displace the location of the HCS locally after the passing.

Some effects of large-scale coronal structures on the SEP transport have been revealed long ago. A detailed statistical study of SPE characteristics at $E_p < 100$ MeV has been carried out by Pereyaslova et al. (1983). A large amount of *Meteor* and *IMP* data for 1965–1982 was used to reveal two types of events differing in the conditions of near-Sun propagation: type I – when the flare site and the longitude of the Sun-Earth conjunction point are situated in one and the same unipolar magnetic region, and type II – when they lie in different unipolar magnetic regions separated by a neutral field line. Type I events are characterized by smaller average values of the peak intensity time, t_m, and spectral index, γ and by larger maximum intensity, I_m, than type II events. Besides, it has been shown that once the separation boundary between the unipolar magnetic regions is crossed, the proton spectrum becomes much softer, *i.e*., this boundary is an effective barrier for the low-energy particles.

The measurements on *Prognoz* 7, *Venera* 11 and 13, and *GMS* 2 spacecraft (Klimenko et al. 1982; Morozova et al. 1985) with significant azimuthal separation have shown time differences in the arrival of electrons (0.3–3.0 MeV) and protons (2–70 MeV) for a number of SPEs when there are unipolar magnetic region boundaries between the flare site and the longitude of the spacecraft-Sun conjunction point. If there is one boundary, protons arrive at the detection point ~40 min later than electrons; when there are 2–3 boundaries, the delay is as large as ~200 min. This fact, in principle, allows to predict the arrival of protons with $E_p = 2$–70 MeV with and accuracy of ±20 min using recent measurements of relativistic electrons.

As it is suggested now, solar energetic protons observed in large, gradual events are accelerated on open magnetic field lines in the solar corona by large-scale shocks driven by fast CMEs. The shock strengths are diminished when they cross current sheets in the corona or in interplanetary space. Recent models of coronal magnetic fields have related the open fields of coronal holes to the source-surface and interplanetary fields. Kahler et al. (1995) have combined these results in a simple model to explain the variations of onset profiles in SEP events.

Kallenrode (1993c) has recently surveyed *Helios* SEP events and found significant effects of the HCS on SEP propagation. Rather than examining all SEP events during some period, Kahler et al. (1995) have sought particular SEP events that serve to support the model. They have examined the period of 1978–1982 for the cases showing either rapid access of SEPs to the Earth from nominally distant source region or delayed access from nominally well connected source regions.

Figure 3.6 (left) shows the time profiles of >4 MeV and >60 MeV proton fluxes and the azimuthal component of the IMF for the delayed event of June 6, 1979 (for details see, *e.g.*, Akinyan et al. 1983; Bazilevskaya et al. 1986). An originating flare occurred at 17°N, 14°E on June 5, 1979 at about 0520 UT, but SEP event onset occurred after 1,000 UT on June 6, and then around the time of the shock arrival at 1,027 UT the flux of >60 MeV protons increased substantially. Earlier, von Rosenvinge and Reames (1983) suggested that protons were unable to propagate to the Earth in this particular event, because they were "intercepted" by a coronal hole between the source region and the observer footpoint.

In fact, several holes in the positive polarity regions are clearly seen at the source surface map in Fig. 3.6 (right). A CME drives a coronal shock across the open field lines of coronal holes. Kahler et al. (1995) concluded, in particular, that the observation of a prompt onset and large SEP flux is provided if the observer is in the magnetic sector occupied by the CME shock. Notice that the effects of coronal structure on SEP flux profiles should be most pronounced during the onsets and for the highest energy particles (see Fig. 3.6) which are accelerated closest to the Sun. Well away from the corona, the effects of the HCS on the shock are diminished, and the SEP fluxes are similar on opposite sides of the current sheet, as found by Kallenrode (1993c).

The concept that the HCS is a factor controlling solar-particle access to different regions of space was advanced by Shea et al. (1995b). They compared the solar magnetic field polarity of the sector, in which solar flares associated with GLEs occur, with the polarity of the plasma domain in which the Earth is located at the time of the event. From the 27 events examined they found that the "parent solar activity" and the Earth are located in the same polarity structure (or within the 10° of the HCS) for 22 events. In addition, the heliocentric angle between the flare location and the nearest position to the HCS was determined. As shown in Fig. 3.7,

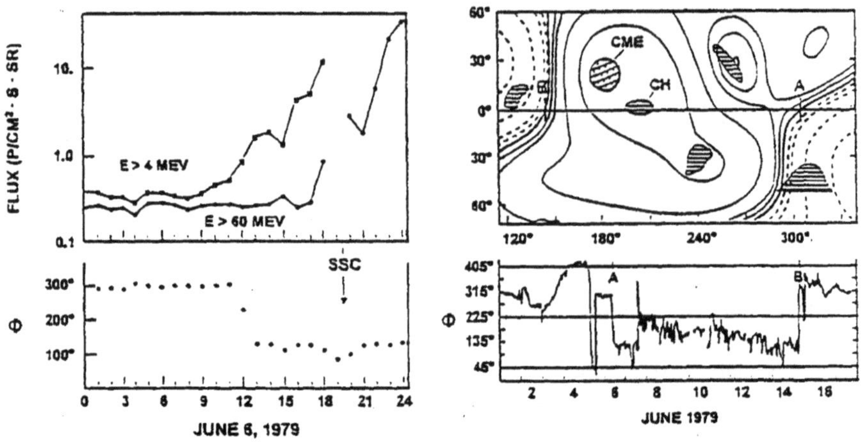

Fig. 3.6 *Left*: Proton fluxes of June 6, 1979 (*top*) and the azimuthal component of the interplanetary magnetic field (*bottom*). The Φ angle show negative polarity until 12:00 UT. The first SEP fluxes appeared at that time. SSC marks the time of the shock at the Earth (Kahler et al. 1995). *Right*: Source surface map (*top*) showing the CME and coronal holes (CH) in the positive polarity regions (*solid lines*) and interplanetary magnetic field (*bottom*) in June, 1979 (Kahler et al. 1995). *A* and *B* mark the sector boundaries on the map and as seen at 1 AU. The Earth lay outside the positive polarity region of the SEP until June 6

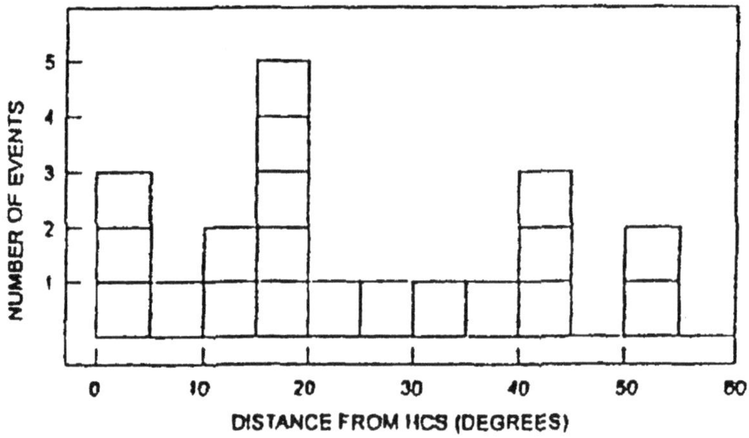

Fig. 3.7 Distribution of solar active region associated with GLEs as a function of the heliocentric angle between the region and the heliospheric current sheet (Shea et al. 1995b)

the values of the heliocentric angle extend to 55°. This was a surprising result from the popular point of view that fast CMEs are associated with major solar flares, and that most CMEs originate close to the HCS (or the streamer belt of the Sun) (Hundhausen 1993).

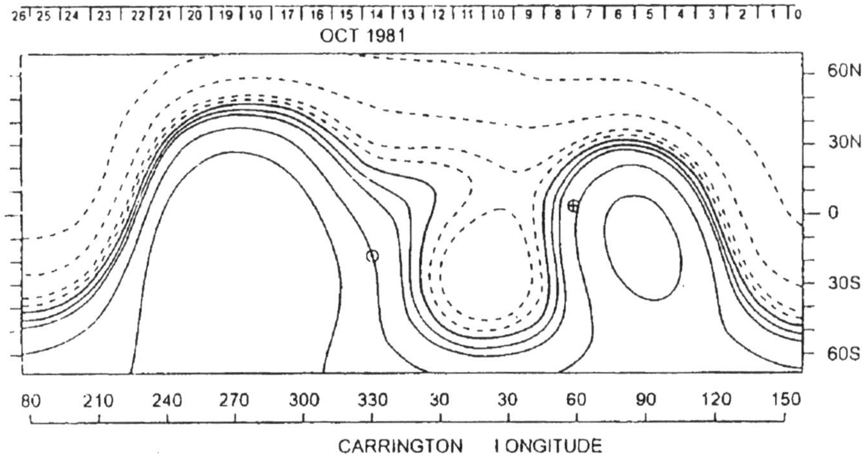

Fig. 3.8 Solar source surface map in October 1981. The location of the active region associated as the source of the October 12, 1981 GLE is shown by the "Sun" symbol at 330° Carrington longitude. The projection of the Earth is shown by the "Earth" symbol at −60° Carrington (Shea et al. 1995b). *Solid lines* = positive polarity; *Dashed lines* = negative polarity

Of particular interest is the event of October 12, 1981 where the location of the active flare region was at 31°E. As shown in Fig. 3.8, both the flare site and the Earth were located in the same polarity domain, in spite of the complex and extremely warped current sheet structure. Although this was a small GLE with an ~10 % increase at high-latitude neutron monitors, the particles were transported relatively quickly between the time of the optical maximum and microwave emission (06:20 UT and 06:33 UT, respectively) and the particle onset at the Earth (06:45 UT) (for details see, *e.g.*, Bazilevskaya et al. 1990a).

A study of this kind, however, has many pitfalls, one of which is the assumption that the solar surface polarity structure as determined at 2.5 solar radii is preserved to 1 AU. Nevertheless, Shea et al. (1995b) believe that there is some preferential particle propagation in the polarity structure in which the parent activity occurs. In other words, if the HCS is between the position of an originating and the observer, there is less efficient access than if there is no interposed current sheet. This result is in overall agreement with the findings by Kallenrode (1993a, b, c) who used lower energy particle data obtained on the two Helios spacecraft.

Additional evidences of this conclusion were obtained by Shea et al. (1995a, b) from an inspection of the *Helios* 1 and 2 data for the very complex event of August 17–19, 1979 associated with the solar activity at 90°E, as well as from the inspection of the >100 MeV proton data for two other events associated with the solar flares at 71°E on March 6, 1989 (X-magnitude X15) and at 75°E on June 4, 1991 (>X12).

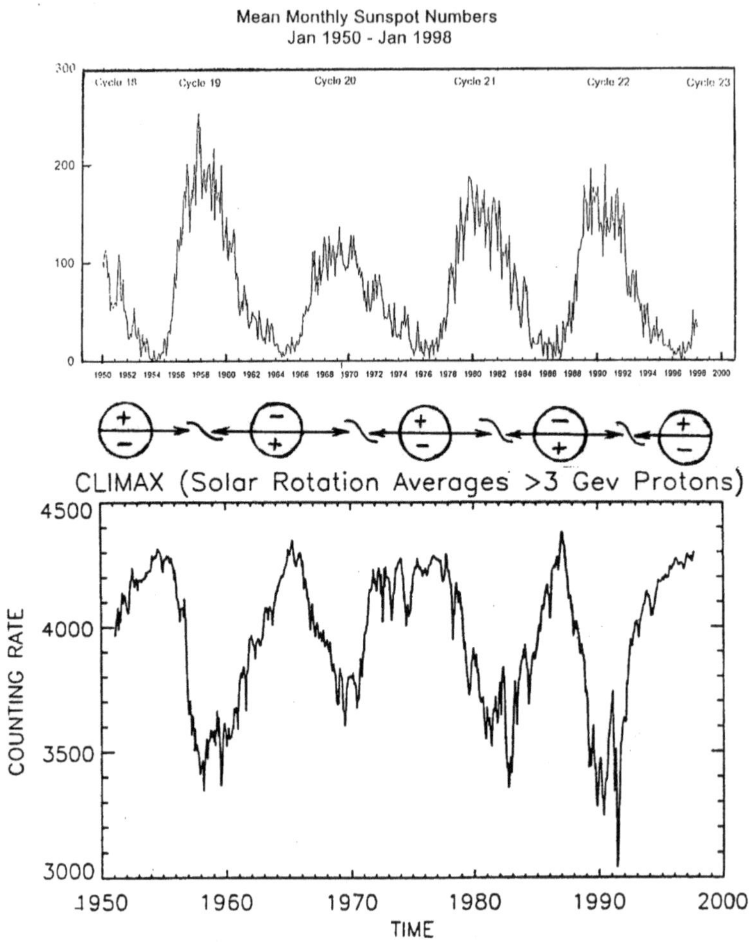

Fig. 3.11 Mean monthly sunspot numbers (*top panel*, SGD, 1998, No. 641), solar magnetic polarity (*center panel*) and solar rotation averages of counting rate at the Climax neutron monitor (*bottom panel*, Jokipii and Kota 1997) since 1951

The periods of the GSMF reversals are also shown: 1957–1958; 1969–1972; 1980–1981. The last reversal seemed to be in 1991–1992. It is clearly seen that a record low in the cosmic ray intensity was observed in the middle of 1991, this period being remarkable for enhanced solar activity with a series of powerful flares and other solar-terrestrial phenomena (see below).

In continuation of this discussion, we describe briefly one irregular phenomenon characterizing the global dynamics of solar activity. It has been recently established that the strongest disturbances in the heliosphere are caused by fast global changes of the solar magnetic fields. A comprehensive analysis of the global magnetic structure dynamics, the total length of quiescent solar filaments and solar wind parameters was carried out by Kovalenko (1994). He succeeded in separating five periods of fast global changes of the background magnetic fields on the Sun (1972, 1978, 1982, 1989, and 1991).

These changes proceed without any marked manifestations in the usually used indices of solar activity (Wolf numbers, solar emission flux, etc.). On the other hand, they are characterized by long-duration (longer than one solar rotation) disturbances in solar wind parameters and the IMF. Besides, they are accompanied by CMEs and typical cosmic-ray intensity variations which correlate with dynamic patterns of large-scale structures of the background magnetic field of the Sun (Kovalenko 1994). Some events are dominated by fast changes in global magnetic structure in the solar corona, while others are typified by changes of the magnitude of low-latitude (<40°) background magnetic fields on the Sun. It is supposedly the fast large-scale changes in electromagnetic structure in the corona and the heliosphere that create favourable conditions for an effective acceleration of a large SCR flux.

All of the five periods mentioned above are well documented. For example, the August 1972 episode of solar activity, with its several significant proton events and strong SSCs, is most often used as a "worst-case model" in solar-terrestrial relationships (*e.g.*, Smart and Shea 1989a, b). Here we describe briefly the period of 1991. An historic cosmic ray intensity minimum, the lowest one since the beginning of cosmic ray observations, occurred on June 13, 1991 (*e.g.*, Shea et al. 1993a).

This month also was the period of numerous powerful flares accompanied by multiple interplanetary shocks propagating through the heliosphere. These effects indicate that propagation conditions in the heliosphere were not quiescent. The giant flare on June 1 (probably the largest flare of the solar cycle 22), released a total energy ~10^{34} erg carried by ~10^{41} electrons with energies above 20 keV (Kane et al. 1995), at a rate of ~10^{32} erg s^{-1}, a rate of energy release ~1,000 time larger than that in the well-studied flares in August 1972. Six sudden storm commencements (SSCs) were recorded at the Earth between 4 and 12 June, and three gamma-flares with the similar time profiles (on June 9, 11, and 15) were produced in the same active region 6659. The remarkable similarity between the three flares, that span a period 6 days, raises the question as to whether an expanding post-flare loop system goes through reproducible stages of particle acceleration after each of these flares, or if there are stable loops that remain unchanged during and between the events and can then be populated repeatedly in the acceleration process (Rank et al. 1997b).

Of particular interest is a dynamics of the solar magnetic field during 1989 which was notable for the series of large SPEs. The main peculiarity of that period is not only due to disturbances in solar wind parameters (although they did also occur), but due to unexpected large GLE of September 29, 1989. This extraordinary event was preceded by an anomalously fast change in the global magnetic structure in the corona. During less than two solar rotations (Nos. 1818–1819) the position of the dipole pole displaced in the heliolongitude as much as $\approx 180°$. Such fast global changes are unique ones for the period from 1976 to 1993 (Zherebtsov et al. 1997).

There are also some evidences (*e.g.*, Vashenyuk et al. 1993) that the Earth was inside the heliospheric current sheet when a well-known flare of September 29, 1989

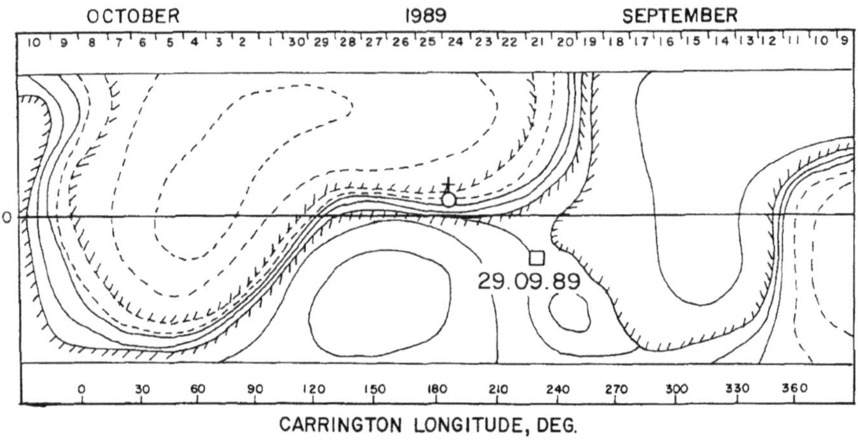

Fig. 3.12 Solar source surface map for September-October 1989 adapted with the Earth-Sun conjunction point (*circle*) and the flare site (*square*) on September 29, 1989 (Miroshnichenko et al. 2000)

occurred behind the west solar limb (~105° W), at angular distance less than 40° (along the arch of a great circle) from the Sun-Earth conjunction point (Miroshnichenko 1997). In spite of the lack of solar wind data, it was possible to determine this point due to the rather quiet interplanetary conditions before the event, when the solar wind speed was ~350 km s^{-1}. In Fig. 3.12 we adapted the source surface map for September-October 1989 with this point and derived flare site (Miroshnichenko et al. 2000). It is seen that the conjunction point of the Earth and the flare site were separated by a neutral line near the solar surface. If so, the angle of the HCS tilt to the ecliptic would be of the same order as the observed IMF inclination (~50°) during the event (Vashenyuk et al. 1995). As a result, some unexpected features have clearly manifested in the behaviour of relativistic protons (see Chap. 5).

A search for similar bursts in the other 17 events from 18 GLEs observed during the BUST operating (since 1981) was undertaken (Alexeyev and Karpov 1994). It was shown that at least three bursts (29, September 1989, 15 June 1991, and 12 October 1981) can be considered as statistically significant ones. These short-term bursts (<15 min) are concentrated in a small solid angle (~0.03 sr) recorded in 1–2 h after the soft X-ray maximum of a proper flare. It was proved that at least three the most significant bursts may be connected with some energetic solar phenomena. The muon bursts associated with the other 15 GLEs had smaller amplitudes. Many of those 15 bursts may be also associated with powerful solar processes, otherwise it is difficult to explain significant distinctions of their spatial and temporal properties from the noise ones (for details see Karpov et al. 1998). In total, 36 GLEs have occurred during the BUST operation since April 1981 up to now (middle of 2014). The data of the muons registration at the BUST are available in 34 cases. The BUST bursts of the end of 21st solar activity cycle, of the 22nd cycle and most events of the 23rd cycle (up to GLE67 on 2 November 2003) have

Table 3.3 BUST muon bursts related to the GLEs of 1981–2005

GLE date	Burst onset, UTC	Excess, standard deviation	Poisson probability	P(3 h)	Delay, min	Long. GSE deg.	Lat. GSE deg.
10.05.81	08:00	4.1	6.7E-5	0.421	29	63W	35
12.10.81	07:45	5.0	2.9E-6	0.024	69	59W	10
26.11.82	04:30	4.2	5.4E-5	0.357	97	141W	47
08.12.82	01:45	4.4	4.1E-5	0.282	111	07W	73
16.02.84	10:00	4.8	2.0E-5	0.153	62	28E	31
25.07.89	10:15	4.1	7.3E-5	0.451	91	15E	71
16.08.89	01:15	4.5	5.3E-5	0.351	−3	94W	32
59.09.89	13:30	5.5	2.2E-7	0.002	117	42W	72
19.10.89	12:45	3.8	1.5E-4	0.696	−13	58W	50
22.10.89	18:45	3.8	2.2E-4	0.832	40	145W	69
24.10.89	20:15	4.2	1.9E-4	0.788	104	140W	15
22.05.90	00:15	3.9	1.4E-4	0.685	116	28E	40
24.05.90	21:30	4.0	1.1E-4	0.581	39	44W	62
26.05.90	21:00	4.0	1.5E-4	0.713	2	93W	89
28.05.90	05:30	3.8	1.6E-4	0.730	57	91W	54
11.06.91	03:45	3.8	1.9E-4	0.789	96	147W	55
15.06.91	10:00	5.0	2.1E-6	0.017	99	20W	51
25.06.92	20:15	4.5	1.0E-5	0.080	4	49E	45
02.11.92	02:30	4.0	2.2E-4	0.836	−38	101W	29
06.11.97	13:15	4.2	4.7E-5	0.321	80	02W	48
02.05.98	14:45	4.2	5.8E-5	0.382	63	130E	73
06.05.98	08:45	4.0	3.1E-4	0.924	36	19W	−02
24.08.98	22:00	4.2	5.6E-5	0.372	−12	165E	53
14.07.00	11:30	4.4	5.9E-5	0.388	66	61E	40
15.04.01	15:30	4.0	1.4E-4	0.681	100	25E	00
18.04.01	03:45	4.9	1.3E-5	0.102	91	19W	49
04.11.01	17:00	4.0	8.1E-5	0.488	40	09W	68
26.12.01	05:30	3.2	1.6E-3	1.000	−10	103E	60
24.08.02	02:15	4.2	6.0E-5	0.395	63	99W	48
28.10.03	13:15	5.1	1.1E-6	0.009	120	127W	66
29.10.03	22:30	4.1	5.8E-5	0.380	101	178E	30
02.11.03	17:30	3.6	2.8E-4	0.901	5	57W	84
17.01.05	11:45	3.7	2.5E-4	0.877	113	151W	72[a]
20.01.05	08:00	4.6	1.6E-5	0.127	59	50W	47[a]

[a]Courtesy by Karpov (2014)

been investigated earlier (*e.g.*, Miroshnichenko and Karpov 2004; Karpov and Miroshnichenko 2008). The two new events, which occurred in 2005 (GLE68 and GLE69), are added now. Results of the statistical analysis of all of 34 BUST bursts are briefly presented below, and the parameters of all those bursts are summarized in Table 3.3.

Table 3.3 contains in the columns: (1) GLE date; (2) time of the burst onset (the beginning of 15-min interval in which the biggest burst was found); (3) burst magnitude, or count rate excess above the background (in standard deviations σ); (4) probability of the chance imitation of the increase due to Poisson fluctuations of the background in the given angular cell ($10° \times 15°$) in the current 15-min interval; (5) upper limit of the probability P(3 h) of chance imitation of such increase in any angular cell (from 680 in total) and in any of 12 15-min intervals for 3 h of observations, real probability is for certain less than this limit (see Karpov et al. 1998); (6) time difference (delay in minutes) between the soft X-ray burst maximum (as an indicator of the flare) and the BUST burst onset; (7) geocentric solar-ecliptic (GSE) longitude of the burst center direction, W – to the west from the Sun, E – to the east from the Sun; (8) geocentric solar-ecliptic (GSE) latitude of the burst center direction ("+" – to the north from the ecliptic plane, "−" – to the south from the ecliptic plane).

To clear up a connection of the bursts with certain solar phenomena, an extended analysis of the data has been performed (Karpov et al. 1995a, b, 1997a, b, c, 1998; Miroshnichenko and Karpov 2004) taking into account angular characteristics of the bursts and sensitivity diagram of the BUST, as well as the position and importance of the proper flares, nominal direction of the interplanetary magnetic field (IMF), anisotropy and spectrum hardness of relativistic solar protons. In particular, an integral energy spectrum of relativistic protons for the event of September 29, 1989 was obtained, as well as an intensity of the primary >500 GeV protons (by the BUST data), have been estimated (Karpov et al. 1998). This spectrum was constructed for a wide range of relativistic energies by all available surface and underground data (Fig. 3.13). The data of the two Baksan arrays – Carpet (black point) and BUST (black square) – are also shown.

These estimations imply that the "BUST particles", most probably, were not accelerated at the main stage of a proper flare, together with a bulk of relativistic protons. At the same time, it was suggested (*e.g.*, Karpov et al. 1998) that the "BUST effect" (or "Baksan effect") is closely linked with the powerful solar processes, implying possible impact of extended coronal structures, CMEs, and heliospheric current sheet. Some later, we re-analyzed the data on the muon burst related to the GLE of 29 September 1989 (Karpov and Miroshnichenko 2008) by improved method of so-called "additional fluctuations" for search of weak signals. Our analysis allowed to reduce an amplitude of the BUST signal from full amplitude of $A = 5.5\sigma$ to the additional fluctuation of $A_0 = 4.9\sigma$. Thus, the true signal will make up only 0.6σ, i.e., about 11 % from full amplitude of the found burst. Therefore, the SCR flux obtained by suggested method is about ten times less, than the authors (Karpov et al. 1998) estimated earlier with use of full amplitude of the burst. This conclusion seems to reconcile the BUST data on the GLE of 29 September 1989 (Fig. 3.13) with integral spectrum for other GLEs in the range of energies above 10 GeV (Fig. 4.4). At any rate, underground data give a certain evidence of that the Sun can produce the particles with the energies above 100 GeV.

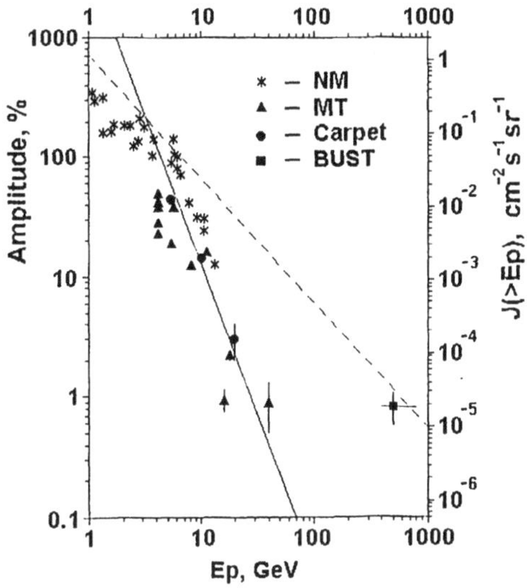

Fig. 3.13 Integral energy spectrum of solar protons near the Earth for September 29, 1989 GLE in a wide range of relativistic energies (Karpov et al. 1998). The solid line is an approximation of the standard neutron monitor (NM) and muon telescope (MT) data above 4 GeV (so-called prompt component of SCRs, see Chap. 5). The *dashed line* is an extrapolation of the satellite data of GOES-7 (SGD, 1989, No. 542, p. 18), Meteor (Nazarova et al. 1992) and GMS-3 (Kohno 1991) up to relativistic energies (delayed component). The data of the two Baksan arrays – Carpet (*black point*) and BUST (*black square*) – are also shown

A new challenging aspect of the problem under consideration raised in connection with the observations of high-energy solar photons ($E_\gamma > 1$ GeV) in several spatially extended or long-lasting gamma-ray events in 1989–1991. We return to this problem in Chap. 5.

In conclusion of this Chapter, we mention briefly the first observation of the seismic wave from a solar flare on board Solar and Heliospheric Observatory (SOHO) on July 9, 1996 (Zharkova and Kosovichev 1998). The flare of July 9 was the only significant X-ray flare observed in 1996. Though this was a fairly moderated flare classified as *X*2.6/1B (10°S, 30°W), it was followed by observations of very spectacular CME (Fig. 3.14) and energetic electrons (44 keV–10 MeV) recorded at the same spacecraft. It was found, in particular, that the electron spectrum of the event can be modeled by a single power law between 0.1 and 2.0 MeV (Sierks et al. 1997). A relatively sharp break occurs at about 2 MeV, leading to a harder spectrum at higher energies. Such a spectral shape is typical for impulsive flares with associated soft X-ray emission. Additional analysis of radio observations shows (Sierks et al. 1997) that it is quite possible that in fact two different population of energetic electrons were detected on July 9, 1996.

As reported by Zharkova and Kosovichev (1998), a circular wave packet propagated after the flare to a distance of 120,000 km from the flare site with an average

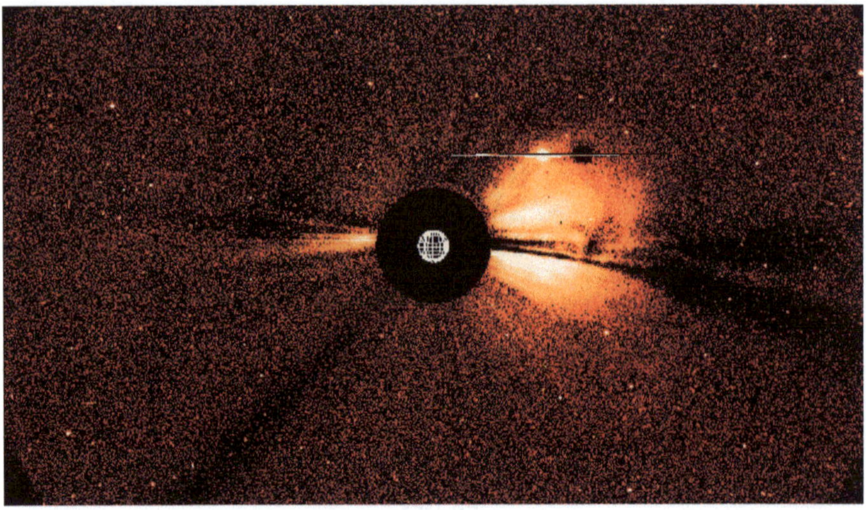

Fig. 3.14 A coronal mass ejection observed by LASCO instrument on board *SOHO* spacecraft at 15:38:41 UT on July 9, 1996 (Cliver and Webb 1997). The *large dark circle* in the center is the occulting disk of the coronagraph (it extends to 3 solar radii). The imbedded *white circle* (with grid) indicates the position of the Sun. The background (*white and black*) dots represent stars and cosmic ray hits

speed of about 50 km s^{-1}. The amplitude of the flare ripples was approximately 50 m s^{-1} which is higher than the amplitude predicted by the thick target model of solar flares. This particular event provides an excellent illustration of very complicated picture of solar energetic phenomena followed, in particular, by production of energetic solar particles.

In the light of some new data on the occurrence rate of GLEs (*e.g.*, Miroshnichenko et al. 2012), the above discussion on fast global changes of heliomagnetic fields acquires some new aspects. In Chap. 4 we return to this intriguing problem in the context of the Sun's proton productivity during different periods of the solar activity cycle, including the periods of heliomagnetic polarity reversal, with its possible effects on SCR variations.

3.6 Energetics of Solar Cosmic Rays

The SCR energetics comprises, at least, three aspects (*e.g.*, Miroshnichenko 1981a, b, 1987; Miroshnichenko and Petrov 1985): (1) the total energy of accelerated particles W_a and their relative contribution to the energetics W_f of the solar flare as a whole, $\delta = W_a/W_f$; (2) variations of the number of accelerated particles $N_a(>E)$ and of the quantity δ with the total energy of the flare; (3) variations of the Sun's proton productivity, or energy release in the form of SCR, due to the changes of the solar activity level. As regards the study of the solar flare physics, particularly interesting are the first two aspects. The third one is important in connection with the problem of the long-term variations in the Sun's proton emissivity (see Sect. 10.7).

The early estimates of W_a and N_a turned out to be rather controversial. Thus, by the data of ground observations of the GLE of February 23, 1956 Meyer et al. (1956) have obtained a value of $W_a \cong 3 \times 10^{31}$ erg for the proton energy range of 1–30 GeV, whereas Dorman's (1957) estimate was $W_a(\geq 1\ \text{GeV}) \cong 10^{32}$ erg. Meanwhile, by the data of stratospheric observations (Webber 1963, 1964) it was calculated that $N_a \cong 10^{35}$ and $W_a \cong 2 \times 10^{31}$ erg at the proton energy $E_p \geq 10$ MeV. For another proton event (July 14–15, 1959) there were deduced $N_a \cong 3 \times 10^{41}$ and $W_a \cong 2 \times 10^{36}$ erg for the protons with the energy of $E_p > 3$ MeV (Dorman 1963a).

Analyzing the energy release of a solar flare, Ellison (1963) has drawn a conclusion that the value of W_a for the total flux of high-energy particles may vary within 10^2–10^{32} erg (at $W_f \sim 10^{32}$ erg); however, these estimates were regarded by this author as extremely unreliable. At the same time, the above estimates appear to have contributed to the origin and spreading of the concept that with respect to the order of magnitude the total energy of SCR is comparable with the energy released in the optical range and with the kinetic energy of the movement of the gaseous masses in the solar flare. Furthermore, on the basis of the results of Dorman (1963a) there were worked out far-reaching assumptions that in the flare there may take place thermonuclear reactions, whilst the density of the SCR energy may be high enough for pushing out so-called "solar corpuscular streams", *i.e.*, for throwing out "coronal transients" or "coronal mass ejections", as they are called today. That is true, it was admitted (Dorman 1963a) that the value of $W_a \sim 10^{36}$ erg appears to be significantly overstated, since in estimating N_a and W_a a possible flattening of the SCR spectrum at low energies was disregarded.

It is to be noted that the estimates of N_a and W_a based on the SCR observations near the Earth are burdened with significant errors and/or uncertainties. The main reason behind the errors is a poor knowledge of the real pattern of SCR ejection and propagation and of the true form of their spectrum in the source (Miroshnichenko 1983b). In particular, all the estimates of W_a published prior to 1981 have been obtained by assuming that the generation spectrum is a power-law one and that being extrapolated into the low-energy range it does not change its form. In the meantime, long ago attention was attracted to the fact (*e.g.*, Dorman 1978) that the power-law index γ displays a tendency towards a decrease in transition to the range of small energies (rigidities). This tendency was afterwards confirmed on more extended data about the SCR spectra near the Sun (Miroshnichenko and Petrov 1985; Miroshnichenko et al. 1999).

Table 3.4 presents estimates of $N_a(>E)$ and W_f obtained for three large SPEs (GLEs) of February 23, 1956; July 14–15, 1959, and January 28, 1967 (Miroshnichenko 1981a, b, 1987) at the energy thresholds of $E_p \geq 10$, ≥ 100, and ≥ 500 MeV. Given therein are also the estimates of N_a and W_a at the threshold of $E_p \geq 23$ MeV and $\gamma = 2.4 \pm 0.3$ for the moderate SPE of September 28, 1961 (Krimigis 1965). The values of W in Table 3.4 are given in *erg*, and the values of

Table 3.4 Energetics of SCR for certain SEP events l (Miroshnichenko 1987)

Flare date	Energy, MeV	≥10	≥23	≥100	≥500
23.02.1956	N_a	3.7×10^{34}	–	6.3×10^{33}	6.1×10^{32}
	W_a	4.3×10^{30}	–	2.7×10^{30}	8.6×10^{29}
14–15.07.1959	N_a	5.0×10^{35}	–	5.0×10^{31}	2.0×10^{30}
	W_a	1.0×10^{31}	–	1.0×10^{28}	8.0×10^{25}
28.09.1961	N_a	2.1×10^{33}	2.5×10^{31}	2.1×10^{31}	8.4×10^{29}
	W_a	6.7×10^{28}	3.3×10^{27}	6.7×10^{27}	1.3×10^{26}
28.01.1967	N_a	7.9×10^{31}	–	2.2×10^{31}	5.4×10^{30}
	W_a	2.3×10^{28}	–	2.0×10^{28}	2.4×10^{27}

N_a correspond to the total numbers of protons with the energy above the given one. Estimates for the event of February 23, 1956 and January 28, 1967 were obtained in the assumption on the steepening form of their spectra at high energy, and for the event of 14–15 July 1959 the spectrum exponent was assumed to be constant (for additional details see Miroshnichenko 1987, 1990). This last event is of special interest. As shown by Filippov and Chirkov (1977, 1978), the event of July 1959 might be caused by relativistic particles accelerated in the interplanetary medium (for details see above Sect. 5.10).

For the sake of comparison, we also give the estimates of $N_a(\geq 10 \text{ MeV}) = 7 \times 10^{40}$ protons and $W_a(\geq 10 \text{ MeV}) = 1.45 \times 10^{36}$ erg obtained for the SPE of February 23, 1956 (Miroshnichenko 1981a, b) on the assumption that the generation spectrum with the exponent of $\gamma = 5.5$ has no flattening in the low-energy range. One can see that the discrepancies in the values of N_a and W_a obtained for different assumption on the form of the source spectrum may attain several orders of magnitude. On the other hand, Friedman and Hamberger (1969) have carried out the numerical integration of the equations of motion of the particles accelerated by the DC electric fields in the Petschek's model of magnetic reconnection. They obtained a spectrum of the form $\sim E^{-\gamma}$ or $\sim \exp(-E/E_0)$ depending on the given configuration of magnetic field B. For the plasma density $n = 2 \times 10^{11}$ cm^{-3}, $B \sim 2 \times 10^2$ G, electric field $\sim 10^2$ V cm^{-1} and characteristic size of the flare $L \sim 10^9$–10^{10} cm there were obtained the values $N_a(\geq 0.1 \text{ MeV}) = 10^{35}$–$10^{36}$ and $N_a(\geq 10 \text{ MeV}) = 10^{31}$–$10^{32}$ protons provided that the total energy contained in the magnetic field was $W_B \sim 10^{31}$ erg.

With $\gamma = 2.4 \pm 0.3$ for the SPE of 28 September 1961, there also was estimated the value of $W_a(\geq 23 \text{ MeV}) \geq 10^{-4} W_B \geq 10^{27}$ erg (Krimigis 1965), which does not contradict our estimate of $W_a(\geq 10 \text{ MeV}) \sim 6.7 \times 10^{28}$ erg $\sim 10^{-2} W_B$. Thus, even overstated values of W_a (*e.g.*, Krimigis 1965; Miroshnichenko 1981a, b) indicate that the contribution of the protons with $E_p \geq 10$ MeV to the flare energetics does not appear to be above 10 %. The proton energy threshold being diminished, this share should gradually increase.

Making use of different combinations of parameters involved in the dynamic model of spectrum formation (Miroshnichenko 1977), we have carried out detailed calculations of $W_a(>E)$ versus E (Miroshnichenko 1983a, b). It has been shown, in particular, that the maximum value of $W_a(\geq 10$ MeV$) \sim 10^{32}$ erg can be attained solely in the case of a very powerful flare with a very hard spectrum ($D_0 = 10^{36}$ proton GV^{-1} and $\gamma = 3$). The value of $W_a(\geq 10$ MeV) calculated by us for the event of February 23, 1956 (the largest one among those observed so far in the range of energies $E \geq 500$ MeV) does not exceed 10 % of the total flare energy $W_f \sim 10^{32}$ erg. Our estimate of the total proton energy $W_a(\geq 1$ GeV) for the same event is about 2.5 orders lower than the one obtained by Dorman (1957). Thus, with taking into account the changing form of the SCR generation spectrum, we get the values of $W_a(\geq E)$ which do not contradict modern estimates of W_f (*e.g.*, Somov 1992), at least, at the proton energy threshold of 10 MeV (see also Table 3.2).

In conclusion of this section, let us compare above estimates with the estimate of $W_a = 2 \times 10^{31}$ erg obtained by Morozova et al. (1977) from the observations of protons in the range of 0.1–240 MeV during the SPE of August 7, 1972. This value of W_a amounts about 7 % of the energy of the flare shock wave. This yields a new confirmation to the inference of Miroshnichenko (1981a, b) that the contribution of SCR at $E_p \geq 10$ MeV to the flare energetics as a whole, obviously, does not exceed 10 %, provided that $W_f \sim (1\text{–}2) \times 10^{32}$ erg, about half of this energy being carried away by a shock wave (Somov 1992).

Chapter 4
Solar Cosmic Rays at High Energies

This Chapter contains available observational data and recent theoretical results concerning astrophysical aspects of particle acceleration at the Sun and extreme capacities of the solar accelerator(s). The list of the problems under consideration includes: relative energy release of solar flare in the form of solar cosmic rays (SCR), their total energetics; maximum production rate of the Sun as to total amount of accelerated particles; their maximum rigidity, R_m, as one of the crucial parameters in any acceleration model; adequate configuration (structure and extension) of the coronal magnetic fields as a decisive condition for effective acceleration of particles to extremely high energy of order 100 GeV and even more (for protons); occurrence probability of gigantic flares; production of flare neutrinos; flare distribution on proton fluences; flare distribution on proton flux at rigidity above 1 GV, *etc*. The most reliable estimates of various parameters are given in some of the research fields mentioned above.

4.1 Largest Proton Events

Proceeding from physical and/or practical reasons some researchers distinguish the most powerful SPEs into a special group. The event data with large fluxes of relativistic protons (for example, February 23, 1956 and September 29, 1989) are used then for evaluation of the extreme possibilities of a solar accelerator (Miroshnichenko 1993, 1994, 1996; Vashenyuk et al. 1993). If the main increase of SCR flux was observed in the non-relativistic region (for example, in July 1959, August 1972, October 1989) then such an event is most suitable for modeling of "a worst case" from the point of view of radiation hazard (Adams and Gelman 1984; Smart and Shea 1989b). The gravity of such a hazard for spacecraft crews and equipment was re-affirmed on October 20, 1989 when the doze on board the orbital *Mir* station increased by 2 rad during 4 h (Tverskaya et al. 1991). Such episodes

L. Miroshnichenko, *Solar Cosmic Rays*, Astrophysics and Space Science Library 405, DOI 10.1007/978-3-319-09429-8_4

emphasize the necessity to develop models of extreme radiation conditions in space. In this context, data on the most powerful proton events were summarized.

We selected the maximum values of integral proton intensity, $I(>E_p)$, obtained for different energy intervals by observations at the Earth's surface, in the stratosphere, ionosphere and near-Earth space environment in 1942–1991 (see Miroshnichenko 1994, 1996 and references therein). As a result of such a selection a set of the 11 most powerful events was formed. Similar data were also collected some later for the events occurred since 1992 and up to now (Miroshnichenko 2001; Miroshnichenko and Perez-Peraza 2008). All those events for the period of 1942–2006 are represented in Table 4.1 by their integral intensities obtained as a rule from the values of proton intensity at the moment of maximum increase, t_m, near the Earth above the energy given. The events of February 1956, August 1972, September and October 1989, 22 March 1991, 30 October 1992, 14 July 2000, 28 October 2003, and 20 January 2005 are of special interest.

Table 4.1 illustrates the range of maximum proton fluxes observed near the Earth. At first sight the data seem to be rather fragmentary, scattered and disordered. However, as it has been shown (Miroshnichenko 1994, 1996, 2001), proton fluxes and their maximum energy are most probably limited. By this means the Sun demonstrates its extreme possibilities (see below Sect. 4.2). For example, the flare of February 23, 1956 is very likely the most intense event as to total number of relativistic protons (see, *e.g.*, Smart and Shea 1991). As to non-relativistic energies the largest events range over a rather narrow intensity interval, mainly between well-known events of February 23, 1956 and August 4, 1972, with the exception of October 20, 1989 event. Moreover, since October 1989 the event of August 4, 1972 would not be considered any longer as "the worst case" from the point of view of radiation hazard.

At the same time, it would be erroneous to ignore some indirect ("archaeological") evidence of the Sun producing a greater amount of non-relativistic protons than was observed, for example, in August 1972 or in October 1989. Thus, for instance, the high-resolution analysis of the content of different nitrogen oxides NO_y in the cores of Antarctic ice have certainly revealed several anomalously large concentration peaks (Dreschhoff and Zeller 1990). The peaks have been dated with confidence and found to correlate with the white-light flare of July 1928 and with two major solar proton events of July 25, 1946 and August 4, 1972 (see Fig. 10.8). The magnitudes of the peaks were about 4, 11 and 7 of standard deviations, respectively. If solar protons with the energies up to 500 MeV are considered to give the main contribution to the concentration jumps, then from the data of Dreschhoff and Zeller (1990) it follows that the fluence F_s (*i.e.*, total event-integrated flux) of low-energy (non-relativistic) protons in July 1946 could be 1.5 times or more the fluence of August 1972. Meanwhile, in the relativistic region, the event magnitude in July 1946 was about 15 times less than in February 1956, according to ionization chamber data (see Table 2.2) compiled by Smart and Shea (1991). It implies that the July 1946 event was similar to (and exceeded) the event of August 4, 1972. On the whole, three major SPEs (July 1946, August 1972, and October 1989) in different solar cycles turned out to produce the comparable amounts of non-relativistic protons.

Table 4.1 Largest solar proton events of 1942–2006

Event number	SPE date	Flare importance	Energy, E_p, MeV	Flux, $I_m(>E_p)$, $cm^{-2}s^{-1}sr^{-1}$	Observation technique
1	19 Nov 1949	3+	>435	4.1×10^1	IC
2	23 Feb 1956	3+	>435	2.5×10^2	IC, NM
3	15 Jul 1959	3+	>88	2.4×10^2	Balloon
4	12 Nov 1960	3+	>10	2.1×10^4	PCA
5	15 Nov 1960	3+	>10	2.1×10^4	PCA
6	12 Jul 1961	3	>10	2.5×10^4	PCA
7	18 Jul 1961	3+	>10	6.3×10^3	PCA
8	04 Aug 1972	3B/X5	>25	1.0×10^4	Meteor
9	09 Jul 1982	3B/X9.8	>10	5.8×10^3	Meteor
10	12 Aug 1989	2B/X2	>10	6.6×10^3	Meteor, GOES
11	29 Sep 1989	–?/X9.8	>10	3.2×10^3	Meteor, GOES
			>600	1.5×10^0	Meteor
12	19 Oct 1989	3B/X13, First max	>10	2.9×10^3	Meteor, GOES
13	20 Oct 1989	Second max	>10	4.0×10^4	Meteor, GOES
			>25	2.2×10^4	Meteor
14	22 Oct 1989	1N/X2.9	>10	5.7×10^3	Meteor, GOES
15	24 Oct 1989	2N/X5.7	>10	3.3×10^3	Meteor, GOES
16	30 Nov 1989	3B/X2	>10	4.4×10^3	Meteor, GOES
17	22 Mar 1991	3B/X9	>10	5.0×10^4	Meteor, GOES
18	11 Jun 1991	3B/X12	>10	8.0×10^3	GOES, Meteor
19	15 Jun 1991	3B/X12	>10	1.2×10^3	Meteor, GOES
20	07 Jul 1991	2B/X1	>!0	2.0×10^3	Meteor, GOES
21	08 May 1992	4B/M7	>10	4.5×10^3	Meteor, GOES
22	30 Oct 1992	2B/X1	>10	1.4×10^4	Meteor, GOES
23	02 Nov 1992	2B/X9	>10	1.8×10^3	Meteor, GOES
24	20 Feb 1994	3B/M4	>10	7.0×10^3	Meteor, GOES
25	06 Nov 1997	2B/X9	>10	5.0×10^2	GOES
26	14 Jul 2000	3B/X5	>10	2.2×10^4	GOES
27	15 Apr 2001	2B/X14	>10	7.8×10^2	GOES
28	28 Oct 2003	4B/X17	>10	3.0×10^4	GOES
29	02 Nov 2003	2B/X8.3	>10	1.6×10^3	GOES
30	20 Jan 2005	2B/X7.1	>10	1.9×10^3	GOES

Notes: *IC* ionization chamber, *NM* neutron monitor, *PCA* polar cap absorption. Flare importance since 1966 is estimated in optical and X-ray ranges

4.2 Upper Limit Spectrum for Protons

One previous attempt to construct an upper limit spectrum (ULS) was undertaken by Adams and Gelman (1984) based on data of the solar cycles 19–21. Using two largest SPEs of that period (February 23, 1956 and August 4, 1972) as models, they found that the integral proton spectra of all the other SPEs (including very intense ones of May 10 and July 14, 1959 and November 12, 1960) fell within an upper bound set by the combined integral spectrum of the February 23, 1956 event and the integral spectrum of the August 4, 1972 event. These two integral energy spectra are shown in Fig. 4.1.

The differential energy spectrum for the SPE of August 4, 1972 is represented in Fig. 4.2. From a combination of the data for the events of August 4, 1972 and February 23, 1956, in the same figure, also a composite differential energy spectrum was constructed for a "worst-case" particle event.

The event of September 29, 1989 was the largest GLEs since February 23, 1956, and was the third largest one since 1942 (Smart and Shea 1991). Figure 4.3 illustrates the differential rigidity spectrum of relativistic protons derived at the moment of the first peak intensity for the event of September 29, 1989 (Smart

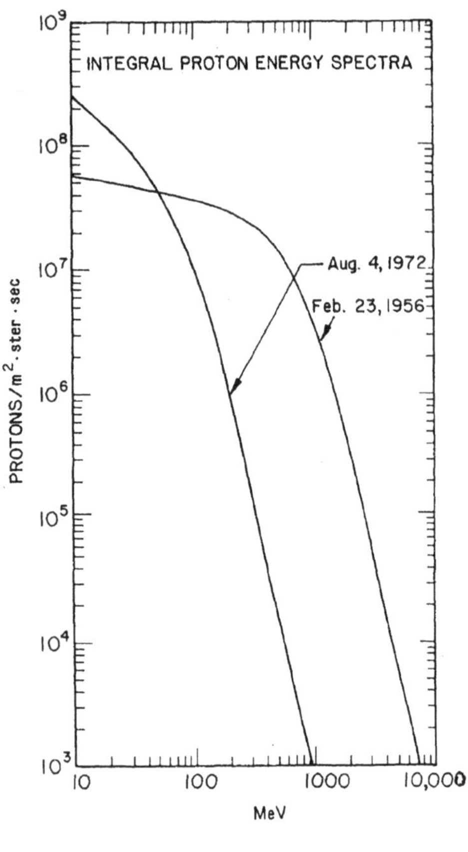

Fig. 4.1 The integral proton energy spectra for the peaks of solar proton events of February 23, 1956 and August 4, 1972 (Adams and Gelman 1984)

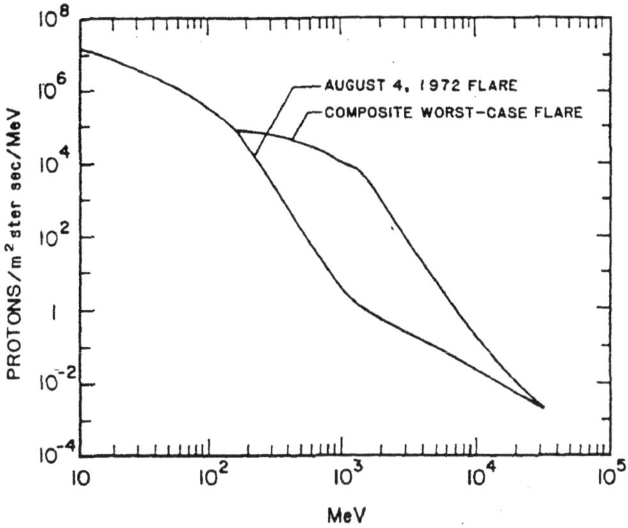

Fig. 4.2 The proton differential energy spectra for the event of 4 August 1972 and for the composite "worst-case" event (Adams and Gelman 1984)

Fig. 4.3 Differential spectra of relativistic solar protons for the events of February 23, 1956 and September 29, 1989 (Shea and Smart 1993b). The galactic cosmic ray spectrum is also shown

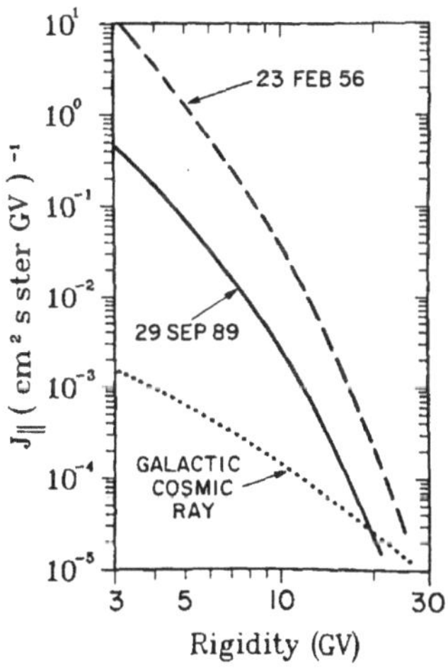

et al. 1991) compared with the similar spectrum for the February 23, 1956 event (Smart and Shea 1990b). The spectrum for galactic cosmic rays (GCR) is also shown for comparison. The area between each event curve and the GCR spectrum represents the excess solar proton flux above the normal cosmic ray background (Shea and Smart 1993a, b).

About 10 years ago, the largest events in the relativistic energy range have been re-examined (Miroshnichenko 1994, 1996), using the above discussed findings by Adams and Gelman (1984) and Shea and Smart (1993a, b), as well as the results of Sarabhai et al. (1956), Webber (1963, 1964), Fichtel et al. (1963), Filippov et al. (1991), Humble et al. (1991a, b), Kolomeets et al. (1993, 1995), Miroshnichenko (1970, 1990), Miroshnichenko et al. (1973), Alexeyev et al. (1992), Nazarova et al. (1992), Perez-Peraza et al. (1992), Shea and Smart (1990a), Smart and Shea (1990b, 1991), Smart et al. (1991), Swinson and Shea (1990), Torsti et al. (1991), Vashenyuk et al. (1993).

Proceeding from these data we were able to model an upper limit spectrum (ULS) for SCR as they may be observed near the Earth's orbit. This spectrum is presented by the upper limit curve indicated in Fig. 4.4 by the shaded zone. All points of the ULS are situated about one order of magnitude above the largest observed (or estimated) values of integral proton intensity at each energy threshold. The ULS may be fitted by a power law function with the exponent depending on proton energy, namely, $\gamma = \gamma_0 E^a$, where $a = 0.1$ and $\gamma_0 = 1.0$ at $E_p > 1$ MeV. The main parameters of the upper limit spectrum are given in Table 4.2.

The uncertainties of exponent values are estimated to be from ± 0.2 to ± 0.5 at the energies below 10^9 eV and above 10^{10} eV, respectively. The factor of 10 was chosen to provide a necessary "reserve" of particle intensity for overlapping the established or assumed range of uncertainties in the measured (or estimated) values of $I_p(t_m)$. Such an empirical approximation of the ULS is far from being a complete model suitable for direct application. However, we believe this simple approach can be improved by some special methods of the statistical treatment of data presented in Table 4.2, with the purpose of constructing a numerical ULS model. When performing such a procedure, those data may be supplemented by refined estimates of the integral fluxes of the >10 MeV protons for the period of 1991–1996 from the Catalogue by Sladkova et al. (1998). Indeed, at least three additional events observed on 22 March 1991, 11 June 1991, and 30 October 1992 should be taken into account, with their peak fluxes of 5.03×10^4, 7.98×10^3, and 1.37×10^4 *pfu*, respectively.

4.3 Search for Extremely High-Energy Particles

If the low-energy threshold of the SCR spectrum turns out to be conditioned by intimate (local) microphysics of the solar plasma (*e.g.*, Miller et al. 1997) then the upper one seems to be determined mainly by the structure, extension and dynamics of the coronal magnetic fields (*e.g.*, Perez-Peraza et al. 1992; Litvinenko and Somov 1995; Miroshnichenko 1997; Karpov et al. 1998). Maximum SCR energy, E_m, GeV (or maximum magnetic rigidity, R_m, GV) would evidently characterize extreme capacities of the solar accelerator, this parameter being a critical one in any acceleration theory. The possibilities of observational discovery of the upper

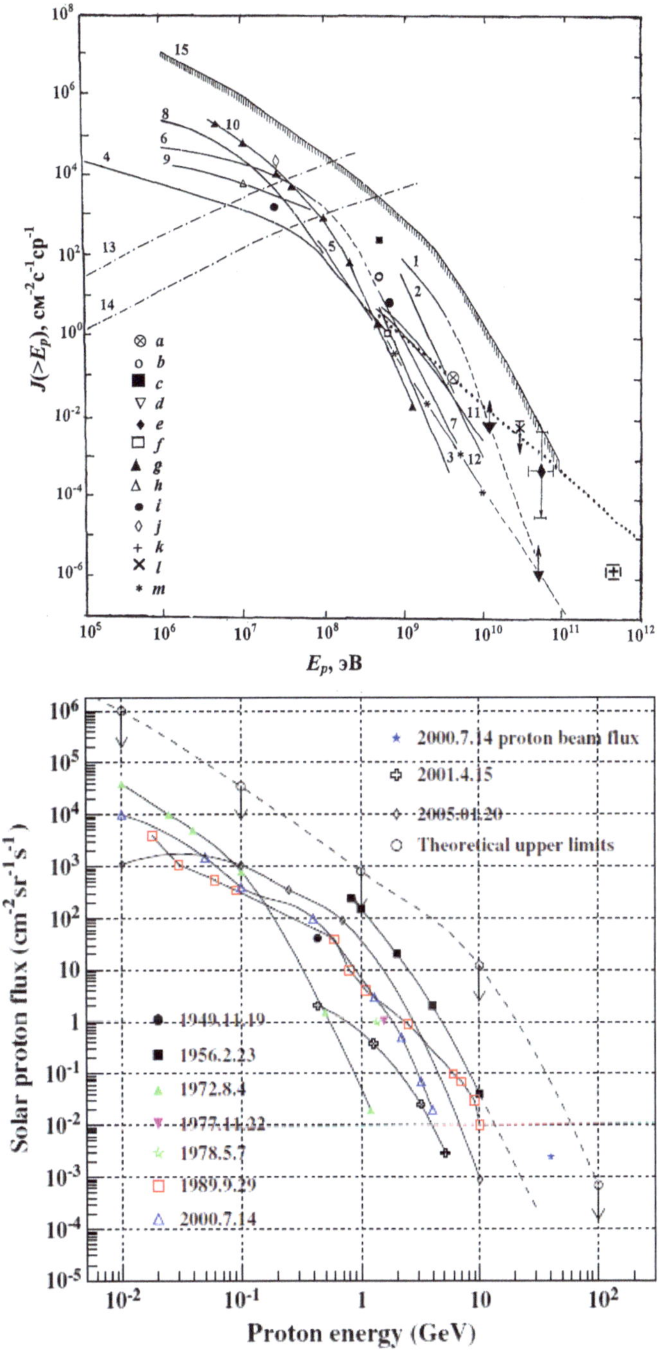

Fig. 4.4 Integral energy spectra of solar protons near the Earth (*top panel*) for the largest proton events of 1942–2002 (Miroshnichenko 2003a): *1–4* – GLE05, spectra at different stages of the event; *5* – event of 15 July 1959; *6–7* – GLE10 and GLE11; *8* – event of 12 July 1961; *9* – GLE13; *10* – GLE24; *11–12* – GLE42; *13–14* – a range of equipartition of the energy densities between

Table 4.2 Parameters of the upper limit spectrum (ULS) for SCR (Miroshnichenko 1996, 2001)

Energy E_p, eV	Exponent, γ	$I(>E_p)$, pfu
$>10^6$	1.00	10^7
$>10^7$	1.45	10^6
$>10^8$	1.65	3.5×10^4
$>10^9$	2.20	8.0×10^2
$>10^{10}$	3.60	1.2×10^0
$>10^{11}$	>4.00	7.0×10^{-4}

rigidity boundary for SCR, however, are limited by the galactic cosmic ray (GCR) background.

Standard observations by the surface detectors allowed to estimate, for example, the magnitude of $R_m = 20$ (+10, −4) GV (Heristchi et al. 1976) by the data on the February 23, 1956 GLE – a largest one since 1942 (historical beginning of regular SCR observations). Meanwhile, by the data of non-standard surface muon telescopes (Sarabhai et al. 1956), solar protons have been recorded in the range of 35–67.5 GeV during initial stage of the same event. Statistical analysis of the worldwide network data of standard neutron monitors (NMs) and muon telescopes (MTs) gave also some evidences (see, *e.g.*, Dorman and Kolomeets 1961, and references therein) that the particles with relativistic energies above 10 GeV are produced even during small solar flares and subflares. These last findings, however, were not supported by similar study (Bazilevskaya et al. 1990b) where no effect of relativistic solar protons after comparatively small flares has been found.

The observations by the underground detectors oriented towards the Sun allow to advance into the energy range of ~100–200 GeV. In particular, very interesting data have been obtained (Schindler and Kearney 1973) by the narrow-angle scintillation muon telescope at a nominal depth of 200 m of water equivalent (m.w.e.) in the Experimental Mine of the Colorado School of Mines, Idaho Springs, Colorado. By the method of epoch superposition (C. Chree technique) there were separated 13 and 6 bursts of muon intensity with the amplitudes alteration from 120 ± 40 to 240 ± 80 %, respectively, within 10 min before the beginning of the proper flare in $H\alpha$ line. These evidences pointed out a possibility of the particle acceleration at the Sun up to the energy of $E_p > 100 \pm 25$ GeV. However, they still needed to be

Fig. 4.4 (continued) solar protons and geomagnetic field. The letters *a–k* are related to different events: *a* – GLE03; *b* – GLE04; *c, d, e, f* – GLE05, estimates for different stages of the event; *g* – GLE24; *h* – event of 9 July 1982; *i* – GLE42, spacecraft *Meteor*; *j* – event of 20 October 1989, spacecraft *Meteor*; *k* – GLE42, Baksan Underground Scintillation Telescope (BUST); *l* – GLE59; *m* – GLE60. Dotted curve is an integral spectrum of GCRs at the energies $E_p \geq 500$ MeV. *Bottom panel* shows similar spectra for a number of large GLEs registered in 1942–2006 (Wang 2009). The upper limit spectrum (ULS) for SCRs (Miroshnichenko 1996) is represented by a rounding curve 15 supported by more reliable observations because the data of Schindler and Kearney (1973) were obtained, in essence, within the limits of 3σ.

This deficiency seemed to be overcome due to the observations by the Baksan Underground Scintillation Telescope (BUST) at the Baksan Neutrino Observatory – BNO (Baksan Valley, Northern Caucasus). It has an effective area of ~200 m^2 and is situated at the effective depth of 850 m.w.e. It should be noted that a research interest to the problem under consideration was extremely enhanced due to the first reliable registration of underground effects of solar flares on September 29, 1989 (*e.g.*, Krymsky et al. 1990; Swinson and Shea 1990). The paper by Alexeyev et al. (1992) described the first (and the largest) burst of muon intensity at the level of 5σ recorded at the BUST during the GLE of September 29, 1989. The energy of the recorded muons is $E_\mu > 200$ GeV (probably, between 200 and 500 GeV), which corresponds to the primary proton energy $E_p > 500$ GeV. A unique increase of 43 % in total counting rate was recorded (Alexeyev et al. 1991) by the "Carpet" detector – a central part of the Air Shower Array at the BNO.

A search for the similar bursts in the other 17 events from 18 GLEs observed during the BUST operating (since 1981) was undertaken (Alexeyev and Karpov 1994). It was shown that at least three bursts (29, September 1989, 15 June 1991, and 12 October 1981) can be considered as statistically significant ones. These short-term bursts (<15 min) are concentrated in a small solid angle (~0.03 sr) recorded in 1–2 h after the soft X-ray maximum of a proper flare. It was proved that at least three the most significant bursts may be connected with some energetic solar phenomena. The muon bursts associated with the other 15 GLEs had smaller amplitudes. Many of those 15 bursts may be also associated with powerful solar processes, otherwise it is difficult to explain significant distinctions of their spatial and temporal properties from the noise ones (for details see Karpov et al. 1998). Toward the end of 1999 the list of the BUST muon bursts has been extended up to 21. The 36 GLE events have occurred during the BUST operation since April 1981 up to now. The data of the muons registration at the BUST are available in 34 cases. The parameters of all those bursts are summarized in Table 3.2. Results of the statistical analysis of all of 34 BUST events are briefly presented in Chap. 3.

4.4 Maximum Rigidity of Accelerated Particles

4.4.1 Determination of R_m *from Observational Data*

The quantity of R_m is one of the key parameters for any model of SCR acceleration. This parameter is apparently determined by a combination of several factors (the limiting properties of the accelerating electromagnetic fields, the rate of energy gain, the time of existence of favourable conditions for acceleration, the increase of energy losses with an increase in the particle energy, *etc.*). Theoretical estimates of R_m, however, are extremely uncertain and differ drastically depending on the

Table 4.3 Maximum rigidity and number of accelerated protons

No.	Date	N(>0.24 GV)	N (>1 GV)	R_m, GV	Reference
1	23.02.1956	1.9×10^{34}	2.3×10^{33}	20(+10, −4)	HTP-1976
2	04.05.1960	1.8×10^{31}	2.0×10^{26}	7.0 ± 1.0	HTP-1976
3	03.09.1960	1.7×10^{31}	6.5×10^{29}	5.0 ± 2.0	HTP-1976
4	15.11.1960	4.6×10^{33}	6.5×10^{31}	4.1 ± 0.8	HTP-1976
5	18.07.1961	1.7×10^{33}	6.5×10^{30}	4.3 ± 0.9	HTP-1976
6	07.07.1966	2.0×10^{31}	2.7×10^{30}	3.2 ± 0.7	HTP-1976
7	28.01.1967	7.8×10^{31}	1.0×10^{31}	5.7 ± 0.7	HTP-1976
8	18.11.1968	2.9×10^{32}	4.0×10^{30}	5.7 ± 1.5	HTP-1976
9	25.02.1969	4.8×10^{31}	1.8×10^{31}	5.7 ± 0.9	HTP-1976
10	30.03.1969	8.9×10^{30}	2.9×10^{25}	4.5 ± 0.7	HTP-1976
11	24.01.1971	5.2×10^{33}	3.0×10^{30}	4.2 ± 0.6	HTP-1976
12	01.09.1971	1.6×10^{32}	4.3×10^{30}	3.4 ± 0.6	HTP-1976
13	07.08.1972	8.1×10^{33}	4.0×10^{29}	6.6 ± 1.0	HTP-1976
14	22.11.1977	8.0×10^{31}	7.8×10^{26}	6.0 ± 1.0	BM-1988
15	10.04.1981	9.7×10^{31}	2.8×10^{27}	1.2 ± 0.2	K-1983
16	12.10.1981	9.7×10^{32}	6.4×10^{29}	9.0 ± 1.0	BM-1988
17	26.11.1982	4.0×10^{32}	2.6×10^{29}	11.9 ± 2.0	ZS-1989
18	07.12.1982	8.5×10^{32}	2.4×10^{31}	10.4 ± 1.5	ZS-1989
19	29.09.1989	8.0×10^{32}	1.0×10^{32}	>20.0	Several authors

Notes: The R_m value for the GLE of September 29, 1989 is still under discussion (for a review see Miroshnichenko et al. 2000, Miroshnichenko and Perez-Peraza 2008). The corresponding references are: HTP-1976 (Heristchi et al. 1976), BM-1988 (Bazilevskaya and Makhmutov 1988), K-1983 (Kocharov 1983), and ZS-1989 (Zusmanovich and Shvartsman 1989)

chosen acceleration model, and values of $E_m > 10^{11}$–10^{12} eV (for protons) are not excluded (see Chap. 5). Here we present available results of a determination of E_m (or R_m) from observational data only.

According to different estimations, during the event of September 29, 1989 solar protons could be accelerated to the energies of $E_m > 20$ (Humble et al. 1991a; Alexeenko et al. 1993), >25 (Swinson and Shea 1990), >150 (Filippov et al. 1991), >900 (Kolomeets et al. 1993), and even >1,000 GeV (Alexeyev et al. 1992). In spite of such a large discrepancy, these results, due to underground effects of the GLE, are of paramount importance from the point of view of estimates the upper capabilities of solar accelerator(s) (*e.g.*, Karpov et al. 1998; Miroshnichenko and Perez-Peraza 2008): the maximum number of relativistic particles, their upper limit rigidity, R_m, maximum rate and total duration of particle acceleration up to relativistic energies, *etc*.

Up to 1990 it has been possible to determine the quantity E_m (R_m) for 18 events only (Heristchi et al. 1976; Bazilevskaya and Makhmutov 1988; Kocharov 1983; Zusmanovich and Shvartsman 1989). It is still under discussion several estimates of E_m for the event of September 29, 1989 (Miroshnichenko et al. 2000). In order to verify a possible relation between R_m and the number of accelerated protons, N_a, we have compiled the Table 4.3 which includes the values of R_m, N_a(>0.24 GV) and

$N_a(>1.0 \text{ GV})$ (E_p >30 and >433 MeV, respectively) for all 19 proton events. The estimates of N_a have been obtained by involving the data on source spectra of solar cosmic rays of 1949–1991 (Miroshnichenko et al. 1999).

The procedure of the R_m estimates was based on the distribution of the increase effect over the Earth's surface. Thus, if a large number of cosmic ray stations are used with different geomagnetic cutoff rigidities R_c, then the value of R_m can be obtained by estimating by the method of least squares for what R_c the SCR flux is comparable to the background of GCR. If the increase effect was detected at two stations with similar cones of acceptance, then R_m can be determined from the ratio of the increase amplitudes A_1/A_2 at these stations by means of an iteration method (or method of sequential approximations). For a pronounced latitude effect, the value of R_m was determined from the latitude curve of the SCR using the integral multiplicities, *m*(*R*), of the generation of the neutron component (for details see, *e.g.*, Bazilevskaya and Makhmutov 1988, and Chap. 9). These methods are suitable during the isotropic stage of the event. They are also usable during the anisotropy period, but in this case they give somewhat poorer results. Thus, for the January 24, 1971 event the following values were obtained (Heristchi et al. 1976): $R_m = (3.5\text{–}4.0) \pm 0.6$ GV at the spectral exponent $\gamma(R) = (3.7\text{–}3.9) \pm 0.4$ from the data for the isotropic stage and $R_m = (4.0\text{–}4.6) \pm 0.6$ GV at $\gamma(R) = (4.0\text{–}4.6) \pm 0.4$ from the data for the period of anisotropy.

Employing the iteration method, Bazilevskaya and Makhmutov (1988) used as a rough approximation the value of the exponent $\gamma(E)$ of the differential energy spectra, obtained from the data of stratospheric observation for $E_p > 100$ MeV, and the value of E_m +500 MeV, where E_m corresponded to the value of R_c of the lowest latitude of all the monitors that detected the event being analyzed. In this manner the E_m values from 1971 to 1982 were estimated. For the events of 1971–1972 the estimates by Bazilevskaya and Makhmutov (1988), the accuracy of which is ±500 MeV, differ from the estimates of Heristchi et al. (1976) by no more than 300 MeV.

In this context, a very important question arises about the restrictions of the accuracy in the E_m determination. The statistical accuracy of modern neutron monitors amounts to ~0.15 % in terms of hourly data and ~0.5 % in terms of 5-min readings. The actual width of the distribution of hourly values under undisturbed geomagnetic conditions exceeds the width of the Poisson distribution by a factor of ~1.5. This means that the accuracy of the E_m determination is limited significantly by the sensitivity of the monitors to the minimum measurable fluxes of SCRs near the energy of $E_p > 500$ MeV. As shown by Bazilevskaya and Makhmutov (1988), this sensitivity, on the one hand, is comparable to the sensitivity of the stratospheric experiment of FIAN (Physical Lebedev Institute), where the fluxes $I(>E_p) = 0.03 \text{ cm}^{-2}\text{s}^{-1}\text{sr}^{-1}$ in the $E_p = 100\text{–}500$ MeV interval are the minimum measurable values. On the other hand, it is inferior to the best measurements in interplanetary space, where for the same spectral indices the measurable particle fluxes with $E_p = 10\text{–}400$ MeV are $\sim 10^{-3} \text{ cm}^{-2}\text{s}^{-1}\text{sr}^{-1}$ (see, *e.g.*, Ellison and Ramaty 1985). Thus, the absence of an increase in the counting of the neutron monitors does not yet indicate the presence of an upper limit to the energy of SCR in the range of $E_m = 500–1,000$ MeV.

Another fundamental restriction is due to the uncertainty of the $m(R)$ values, used in the iteration method for determining E_m. As shown by Dorman and Miroshnichenko (1966, 1968), the accuracy of the $m(R)$ calculation from ground data, due to the presence of different nuclei in the primary flux of cosmic rays, does not exceeds a factor 2 (for details see also Chap. 9). Bazilevskaya and Makhmutov (1988) used the multiplicity values $S(E)$ that had been calculated (Lockwood et al. 1974) with the elemental composition of the SCR taken into account. Nevertheless, when comparing their calculations with the results of other works, Lockwood et al. (1974) also found discrepancies from 50 % to an order of magnitude in the $S(E)$ values.

One of the serious common basic flaws in all above procedures consists in that the response functions of different (standard) ground detectors are known insufficiently. In particular, this is true for NMs in the region of comparatively low ($\leq$2 GeV) SCR energies (Struminsky and Belov 1997). The latter circumstance was mentioned again by the authors of the PAMELA direct space experiment (Adriani et al. 2011), when they tried to coordinate the spectral data of different detectors at energies varying from 80 MeV/nuc to 3 GeV/nuc based on the measurements performed during GLE70 (December 13, 2006). Taking into account the accuracy in estimating the absolute intensities of accelerated solar particle fluxes based on the NM data, Adriani et al. (2011) managed to reach a reasonable agreement between the fluxes measured during international PAMELA experiment onboard the Russian spacecraft *Resurs-DK1* and those estimated using the NM data. However, the PAMELA spectra were always harder than the spectra obtained from the NM data at low energies. This can indicate that the response functions for NMs are understated at energies of $\leq$700 MeV. During the second satellite pass over the polar cap, the indicated difference between the PAMELA and NM fluxes became larger, whereas the PAMELA data remained in very good agreement with the data of the IceTop ground-based experiment (Antarctica). Direct measurements of the SCR fluxes in the stratosphere also confirmed that the PAMELA data are correct.

Finally, let us remember a distinct tendency of the SCR spectra to become steeper in the range of energies $E_p > 100$ MeV (see, *e.g.*, Fig. 4.4). Such a tendency was confirmed, in particular, by Bazilevskaya and Makhmutov (1988) based on the data from the SPE Catalogue 1970–1979 (Akinyan et al. 1983) for 59 events, which were reliably identified with solar flares. In their study, however, the effect of increase in the counting rate at neutron monitors after weak flares at the Sun, pointed earlier (Dorman and Kolomeets 1961), was not confirmed. Moreover, it is impossible to exclude the possibility for the formation of an extremely rigid spectrum of protons with an upper limit of $E_m >> 10$ GeV at the Sun. Some evidences of such a possibility have been obtained in the event of September 29, 1989 (see above). If such particles arrived at the Earth in the form of a narrow (anisotropic) beam and experienced a deflection in the geomagnetic field, then they could give a ground increase effect event at night. Such an increase with an amplitude greater than 3σ was found (Martinic et al. 1985) from the data of the Chacaltaya neutron monitor ($R_c = 13.1$ GV, $h = 5{,}220$ m above sea level) by the Chree method for 16 X-ray and gamma flares, only one of them having been accompanied by a weak enhancement of the flux of protons with $E_p > 100$ MeV (Bondarenko et al. 1986). The arrival of the prompt component of SCR at the Earth in some events (Perez-Peraza et al. 1992) seems to corroborate the existence of narrow beams of relativistic protons.

4.4.2 Temporal Variations of Maximum Rigidity

Although the value of R_m in Table 4.4 changes from one event to another, no distinct relationships has not yet been found between this parameter and the amplitude of the proton event near the Earth and the amplitude and time profile of X-ray and microwave bursts (Heristchi et al. 1976). On the other hand, from the data of Table 4.4 one can see a slight tendency for R_m to increase in the case of the largest SPEs. In our opinion, the accuracy of the determination of the values of E_m and limited statistics of Table 4.4 are not yet adequate for investigating the correlation or physical relationship between this parameter and other parameters of the flares and the solar activity indices. Nevertheless, it appears to deserve attention an attempt (Makhmutov 1983; Bazilevskaya and Makhmutov 1988) to compare the occurrence rate of GLEs with the largest value of E_m in a year and the values of the exponent of the integral spectrum of protons with $E_p = 100$–500 MeV with the smoothed values of the number of sunspots W during the period of 1956–1985 (Fig. 4.5). It is evident that the slope of the spectrum is practically independent of the phase of the solar cycle, whereas the largest values of E_m for each year have a double-hump behaviour in the solar-activity cycle, reaching values >5 GeV during the years before and after maximum activity. Such a dependence of E_m on the phase of the solar cycle can be explained by a coronal magnetic field structure that varies during the cycle.

Similar results were obtained by Nagashima et al. (1991). Using the data of NMs and MTs during the 1942–1990 period these authors analyzed the well-known tendency of GLEs to be grouped preferentially during the ascending and descending phases of the 11-year solar cycle (*e.g.*, Miroshnichenko 1992b). It was shown that flares causing such increases are essentially forbidden during the transitional phase when a change occurs in the sign of the global magnetic field of the Sun (see Fig. 3.12) near the periods of solar activity maxima. Nagashima et al. (1991) suggest that the absence of GLEs near the maximum is explained not by the suppression of proton production by the Sun because of strong magnetic fields but by a deterioration of the efficiency of proton acceleration during the structural re-arrangement of the fields in the transitional period. On the whole, however, the question of the magnitude and nature of the parameter R_m remains unanswered. In order to separate the effects of SCR acceleration and their escape

Table 4.4 Integral fluxes of solar protons at rigidity above 1 GV

No.	Date of GLE	Time, UT	Rigidity ΔR, GV	D_0, $sm^{-2}s^{-1}GV^{-1}$	γ	I_m, $sm^{-1}s^{-1}sr^{-1}$
1	28.02.1942	1300	>1.0	8.33×10^2	4–5	1.21×10^1
2	07.03.1942	0600	>1.0	1.04×10^3	4–5	1.53×10^1
3	25.07.1946	1853	>1.0	Int. IC data	–	2.26×10^1
4	19.11.1949	1200	>1.0	2.78×10^3	4–5	4.14×10^1
5	23.02.1956	0500	1.5–5.0	1.25×10^4	6.8	2.55×10^2
6	04.05.1960	1050	2.0–5.0	6.30×10^1	3.4	8.27×10^0
7	12.11.1960	2000	1.0–3.5	1.70×10^2	5.2	1.02×10^1
8	15.11.1960	0400	1.5–4.0	1.55×10^2	5.0	1.18×10^1
9	28.01.1967	1200	0.5–10	1.25×10^1	4.5	4.45×10^{-1}
10	18.11.1968	1100	1.6–5.0	1.57×10^1	5.0	1.27×10^0
11	25.02.1969	1000	1.0–4.4	9.50×10^0	4.1	9.86×10^{-1}
12	30.03.1969	1400	1.0–3.0	2.45×10^0	4.0	2.60×10^{-1}
13	24.01.1971	2400	1.0–5.0	1.66×10^1	5.0	1.34×10^0
14	01.09.1971	2200	1.0–5.0	1.57×10^1	5.0	1.14×10^0
15	04.08.1972	1600	1.0–1.6	2.04×10^1	8.0	9.23×10^{-1}
16	07.08.1972	1700	1.0–3.0	7.00×10^0	4.0	1.02×10^0
17	29.04.1973	2215	>1.0	Int. NM data	–	1.52×10^{-1}
18	30.04.1976	2140	1.0–1.7	1.40×10^0	3.7	5.09×10^{-1}
19	19.09.1977	1400	>1.0	2.40×10^{-1}	4.0	1.90×10^{-1}
20	24.09.1977	1012	1.0–6.3	4.00×10^0	3.4	5.41×10^{-1}
21	22.11.1977	1200	2.3–4.0	5.00×10^2	5.5	1.05×10^0
24	21.08.1979	0700	>1.0	5.73×10^0	4.6	5.09×10^{-1}
25	10.04.1981	1730	>1.0	1.72×10^0	4.5	1.55×10^{-1}
26	10.05.1981	1000	>1.0	2.00×10^0	4.3	1.90×10^{-1}
27	12.10.1981	1000	>1.0	1.37×10^1	4.4	1.30×10^0
28	26.11.1982	0455	>1.0	5.67×10^0	4.1	5.72×10^{-1}
29	08.12.1982	0045	>1.0	8.62×10^1	5.5	6.05×10^0
30	16.02.1984	0915	>1.0	7.25×10^0	4.3	1.02×10^1
31	29.09.1989	1217	1.0–4.0	9.33×10^0	2.9	3.02×10^1
32	22.03.1991	0439 (24.03)	>1.0	Int. Meteor Data	–	1.10×10^{-1}
33	11.06.1991	0156	1.0–4.0	1.55×10^1	5.5	1.11×10^0
34	15.06.1991	0810	1.0–4.0	6.19×10^1	6.0	4.14×10^0
35	25.06.1992	0032 (26.06)	>1.0	Int. Meteor Data	–	1.20×10^{-1}

Notes: Flux estimates for the events Nos.3, 17, 32 and 35 were obtained by integral data due to measurements by ionization chambers (IC), neutron monitors (NM) and satellite *Meteor*

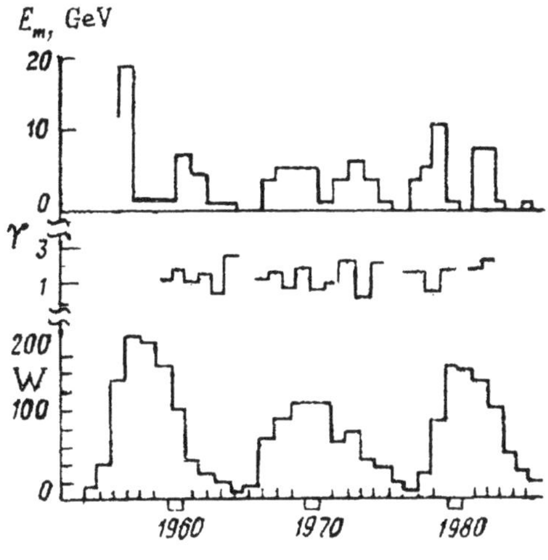

Fig. 4.5 Temporal behaviour of the largest value of E_m for each year (*top*), the indices of the integral energy spectrum of solar protons with $E_p = 100–500$ MeV (*middle*), and the smoothed values of sunspot numbers (*bottom*) (Bazilevskaya and Makhmutov 1988)

from the solar atmosphere it is necessary to investigate the structure of the coronal magnetic fields in individual events.

4.4.3 Recent Estimates and Measurements

Many researchers illustrated the state of this problem (Karpov et al. 1998; Miroshnichenko 2001, 2003a; Miroshnichenko and Perez-Peraza 2008). In spite of the experimental limitations, scarce observational data, and theoretical difficulties, researchers are still interested in the problem because of its fundamental character. The BUST results gave a new impetus to the search for the SCR energy upper limit based on the data of substandard CR detectors (Falcone and Ryan 1999; Ryan et al. 2000; Ding et al. 2001; Tonwar et al. 2001; Poirier and D'Andrea 2002; Wang 2009). Below, we present some results that have been achieved by different researchers during the last years. We mainly consider the most outstanding GLEs during cycle 23, including the events of November 6, 1997 (GLE55); July 14, 2000 (GLE59 or BDE); April 15, 2001 (GLE60); October 28, 2003 (GLE65); and January 20, 2006 (GLE69).

For example, the EAS experiment (AGASA, Japan) indicated that neutrons with energies no lower than $\geq$10 GeV, which corresponds to the accelerated proton energy (at least, $E_p \geq 10$ GeV), could be produced on the Sun during the flare of June 4, 1991 (Chiba et al. 1992). At the same time, measurements with GRAPES-III giant muon detectors (Ooty, India) in March 1988–January 1999 did not give statistically significant results (Kawakami et al. 1999). On the contrary, the Milagrito (water Cherenkov detector) measurements during GLE55 made it possible to detect a certain effect in a channel with a high energy threshold (Falcone and

Ryan 1999). Although the registration thresholds for this detector were not known very precisely, we can state that the energy of coming solar protons was a priori higher than 10 GeV.

4.4.4 Giant Detector Experiments

A group of researchers at CERN (Tonwar et al. 2001) tried to register the solar flare effects with an array of 50 EAS scintillation counter detectors located above an L3 muon detector (the international collaboration of the L3 + C experiment). Specifically, it was mentioned that the count rate of scintillation detectors pronouncedly increased on July 14, 2000, close to the instant when the ground network of NMs registered GLE59. However, this increase, as well as other 42 episodes during 353 days of EAS registration, cannot be unambiguously interpreted (purely atmospheric effects (in particular, air humidity) can contribute to this increase). Collaboration was also reported for the muon fluxes measured during the same event (Ding et al. 2001; Achard et al. 2006). The measurements were performed with a high-precision spectrometer of high-energy muons. The spectrometer made it possible to directionally register muons with energies higher than 15 GeV, which corresponds to the energy of primary protons higher than 40 GeV. The authors reported that a certain excess of muons (4.2σ) was registered simultaneously with an SCR flux enhancement peak at lower energies. The probability that the excess of muons was a random fluctuation in the background is 1 %. Similar fluctuations were not observed during 1.5 h after the solar flare.

To all appearance, the flare of April 15, 2001, had to cause a much more distinct effect, which was actually observed (Poirier and D'Andrea 2002; Karpov et al. 2005). However, the EAS detectors at CERN did not register any increase in the count rate in this case (Tonwar et al. 2001), most probably, because the solar zenith angle was large (>60°). Based on the NM data, we estimated the maximal values of the relativistic proton integral flux for the events of July 14, 2000 (BDE), and April 15, 2001. On July 14, 2000, the SCR spectrum was very soft; therefore, it is not surprising that the BDE event did not cause statistically significant effects at substandard detectors.

The event of 15 April 2001 (GLE60) had a harder spectrum (γ ~ 3.0). Solar proton effects were particularly registered with the Project GRAND Array (an increase in the muon intensity with an amplitude larger than 6.0σ) (Poirier and D'Andrea 2002) and Andyrchi (~10σ) (Karpov et al. 2005) instruments. According to (Poirier and D'Andrea 2002), during this event, the most probable energy of SCR primary protons was close to 100 GeV at a differential spectral index of ~2.0. We assume that such an index value is unrealistic and the *Ep* value is also too large. It is difficult to interpret these data mainly because reliable response functions are absent for the GRAND facility. The same difficulty is typical of the Andyrchi facility (Karpov et al. 2005) and other non-standard detectors.

Using the method and optimization parameters (Achard et al. 2006) for selecting events, Wang (2009) found an excess of muons (5.7σ) in the same sky area as the authors of the experiment based on high-energy muons measured with an L3 + C experiment detector. In this case, the effect duration coincided with the time when the peak flux of lower-energy protons and X and gamma rays were observed. The numerical simulation by the Monte Carlo method indicated that the burst of muon intensity was caused by primary protons with energies $E_p > 40$ GeV and the most probable energy about 82 GeV. Based on the simulation results, Wang (2009) estimated that the upper limit for the flux of such protons is ~2.5×10^{-3} *pfu*. The author assumed that protons with such high energies were accelerated during the impulsive stage of the flare that occurred on July 14, 2000, 2 min after the bursts of hard X and gamma rays.

The last giant SCR GLE was observed on January 20, 2005 (GLE69). This extreme event, which is a second rank event after GLE71 (Table 2.2), made it possible to estimate once again the maximal possibilities of the solar accelerator. In particular, the Aragats neutron monitor and muon detector (3,200 m above sea level, geomagnetic cutoff rigidity $R_c = 7.6$ GV) registered small, but pronounced enhancement (Bostanjyan et al. 2007). Small enhancements were also registered with the Tibet NM and SNT ($R_c = 14.1$ GV, 4,310 m above sea level) (Miyasaka et al. 2005; Zhu et al. 2007) and with GRAND MT (D'Andrea and Poirier 2005). These detectors confirmed that very small fluxes of protons with energies of >15 GeV are present. Bombardieri et al. (2008) simulated the response of sea-level NMs to this event based on output functions (Debrunner et al. 1984). As a result, they concluded that high-rigidity SCR fluxes in the GLE69 event were small and could not cause a substantial increase in the count rate of other NMs with high geomagnetic cutoff rigidities. This is in agreement with the data on the spectra, pitch angle distribution, and SCR arrival direction obtained by the authors themselves (Bombardieri et al. 2008) for the same event.

Recently, Karpov and Miroshnichenko (2008) have managed to progress in understanding the nature of muon bursts at BUST (the Baksan effect). We also estimated again the maximal intensity of primary protons ($I_p(\geq 500$ GeV) ~ $(1.5 \pm 0.2) \times 10^{-6}$ *pfu*) that generated the muon burst of September 29, 1989. This value can apparently be satisfactorily coordinated with the PC spectrum for GLE42 (Miroshnichenko et al. 2000). This estimate at least agrees with the value $I_p(>82$ GeV) ~ 2.5×10^{-3} *pfu* for the BDE event (Wang 2009), if the integral spectral index is >4.0 (Table 4.2). Thus, we for the first time generalized the data of substandard detectors on the upper limits of relativistic solar proton fluxes and maximal SCR energy. These data are fragmentary and cannot be unambiguously interpreted; nevertheless, they put forward fundamental problems: can particles be actually accelerated to energies $E_p \geq 500$ GeV on the Sun or we deal with any specific effect of GCR solar modulation? These problems were also raised previously but only with respect to individual GLEs. The acceleration theory still cannot adequately describe the entire SCR spectrum, especially at $E_p \geq 100$ GeV, although very simple maximal energy estimates (Perez-Peraza et al. 1992) are based on the current sheet model. Thus, it was found that $E_m \approx 250$ GeV for GLE05. Meanwhile, such events that were observed on September 29, 1989; November 6, 1997; and April 15, 2001, with non-standard detectors clearly demonstrate that solar protons with energies $E_p \geq 10$ GeV (and even $\geq$100 GeV) are available. However, the number of detectors that can register secondary muons from such protons is still insufficient. We note that information on the anisotropy of coming particles can only be obtained during single point-by-point measurements. It is difficult (although possible) to perform such measurements. However, no muon detector can measure the SCR anisotropy during GLEs. Therefore, Ryan et al. (2000) consider that several muon detectors with sufficient sensitivities in different directions could ideally be added to the worldwide network of NMs.

4.5 Production of Flare Neutrinos

The data of Table 4.4 may be used to estimate possible contribution of SCRs to the production of solar flare neutrinos. When estimating, it should be taken into account the different sensitivity to the flare neutrinos of radiochemical detectors (of the type of well-known chlorine detector by R. Davis, in Homestake Gold Mine, South Dakota, USA) and direct count detectors (of the Kamiokande type, Japan). During its almost 30-year operation (since October 1970 to the present time) the Homestake detector has recorded several events (see, *e.g.*, Bahcall 1990) when some excess of solar neutrino flux (in comparison with the average values) could be related with the certain powerful proton events (August 4–7, 1972; October 12, 1981 and others). However, direct calculations (Kovaltsov 1981) carried out by the data of source spectra (Ramaty et al. 1975; Miroshnichenko 1979) led to the negative result (see also Bahcall 1990, and references therein). For example, calculated flux of the flare neutrinos turned out to be >2 orders of magnitude low as observed one in August 1972. Meanwhile, a number of the counts in the detector of the Kamiokande type could be 2 orders of magnitude high as in the Homestake detector.

In the light of the given estimates, it is worth to discuss briefly the probability of recording flare neutrinos by means of existing and projected detectors. Decisive parameters for such recording are, on the one hand, the intensity and orientation of the relativistic proton beam ($R > 1$ GV) in the Sun's atmosphere and, on the other hand, the sensitivity of the specific detector to high-energy neutrinos. From the generation conditions, flare neutrinos of electron type ν_e have maximum intensity at the energy $E_\nu \sim 10$ MeV with an isotropic distribution and in the energy range $E_\nu \sim 10$–100 MeV – at various angles θ relative to the orientation of the original proton beam (Kocharov et al. 1990). As a result, the isotropic neutrino flux turns out to be a factor of 5–10 smaller than the anisotropic one. Generation rates and spectra of muon neutrinos ν_μ and antineutrinos $<\nu_\mu>$ slightly differ from those of ν_e, and the flux of electron antineutrinos $<\nu_e>$ proves to be much less than that of ν_e. The probability of recording will evidently depend on the kind and energy of neutrino and on the value of θ as well.

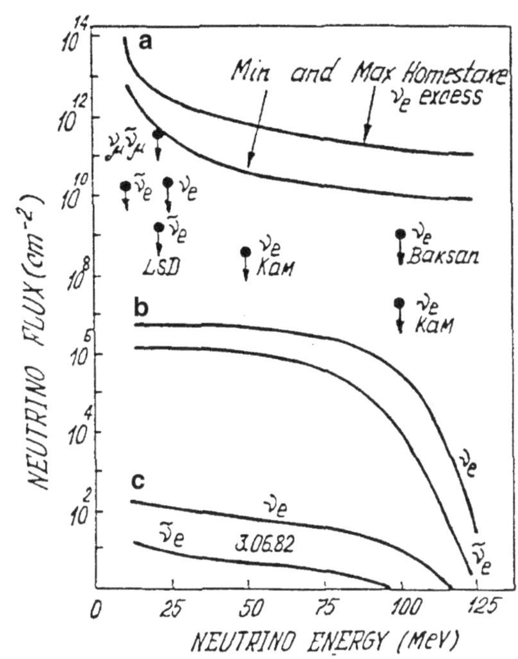

Fig. 4.6 Estimated fluxes of solar flare neutrinos: (**a**) upper limits obtained for different detectors (Aglietta et al. 1991); (**b**) the case of the most restricting suggestions: neutrinos are generated by a beam of relativistic particles moving downwards the Sun; a flare is on the invisible side of the Sun; the energy spectrum of particles is a power law with the spectral index $\gamma = 1.0$; $E_m = 100$ GeV; N_p (>500 MeV) $= 3 \times 10^{32}$ (Kocharov et al. 1991); (**c**) results of theoretical considerations for the flare of June 3, 1982 (Kocharov et al. 1991)

It was not surprisingly that a powerful solar flare of September 29, 1989 has called a steady attention of many researchers of solar neutrinos (see, *e.g.*, Kocharov 1991, and references therein). In fact, it was a good possibility to testify some theoretical aspects of the production of flare neutrinos and a rare occasion to detect them. For example, background of Kamiokande detector for high energy "events" in the solar direction is extremely small and thus even one "event" within a narrow time gate – between 1120 and 1135 UT of September 29, 1989 – could be a brilliant signature of the solar flare neutrino. However, as far as we know, no positive results were reported since then. Meanwhile, Aglietta et al. (1991) presented the results of a search for flare neutrinos and antineutrinos during the period August 1988 – April 1991, performed by the Mont Blanc Liquid Scintillation Detector (LSD). In all, 27 large flares have been analyzed, including the two powerful ones which occurred on September 29 and October 19, 1989. No significant signal was found in time coincidence with any solar flares.

The obtained upper limits on neutrino fluxes are presented in Fig. 4.7. As analysis includes two large solar flares (the first of them was located on the hidden solar side), Aglietta et al. (1991) concluded that obtained results do not support the hypothesis of the Homestake excess being due to solar flare neutrinos with $E_\nu > 25$ MeV. This statement completely confirms the conclusion by Kovaltsov (1981) based on the theoretical consideration.

With the aim to understand existing experimental possibilities, Kocharov (1991b) combined the observational restrictions of Aglietta et al. (1991) (Fig. 4.6a) with theoretical estimates of expected fluxes of flare neutrinos (Kocharov et al. 1991) (Fig. 4.6b, c). One can see that the sensitivities of existing radiochemical detectors in

South Dakota (^{37}Cl) and Baksan Valley (^{71}Ga) and direct count detectors (Kamiokande II and LSD) are several orders of magnitude below the threshold necessary for recording flare neutrinos, even in the most "optimistic" conditions of their generation (narrow beam of relativistic protons with a rather hard spectrum from the flare on the invisible side of the Sun). Therefore, recording of flare neutrino depends on the creation of neutrino detectors of a new generation.

A possible type of detector of direct registration was examined theoretically by Erofeeva et al. (1983). A water detector with a mass of 10 t can record muon neutrinos by Cherenkov radiation of muons generated in the interaction between ν and the target nucleons (H_2O). Estimates by Erofeeva et al. (1983) show that the necessary number of relativistic protons for recording a significant neutrino flux (for a sufficient ν flux generation at a flare) is N_p (>1 GV) > 10^{32} (assuming isotropic generation of neutrinos). In the case of an anisotropic generation (narrow proton beam from a flare on the invisible side of the Sun), the estimated required number of protons can be decreased by a factor of 5–10 (Kocharov et al. 1990).

From the data on the ejection spectrum for the SPE of February 23, 1956, without separating the prompt and the delayed SCR components, Miroshnichenko (1990) obtained N_p (>1 GV) = 6.1×10^{32} (the accuracy of this value is within a factor of $\geq$2). This estimate is evidently compatible, within the uncertainty limits, with the value of N_p(>1 GV) < 2.3×10^{33} obtained by Perez-Peraza et al. (1992) from the calculations for the prompt component only (see also Table 4.4). As shown by Vashenyuk et al. (1993), for the event of September 29, 1989 the value N_p (>1 GV) should be less by 1–2 orders of magnitude. It means that for this detector, a flare of September 29, 1989 still could not be observed, whereas a flare of the February 23, 1956 type would be observed, especially at the "optimum" orientation of the proton beam (Miroshnichenko 1993).

In our opinion, the most "efficient" orientation occurs for a strictly antipodal flare (on the Sun's invisible side), provided for the geometry of the coronal magnetic fields near the source of the SCR also satisfies optimal criteria (Perez-Peraza et al. 1992; Miroshnichenko 1997). In other words, besides enhanced detector sensitivity for recording flare neutrino, it is also necessary to have a rare auspicious geometry of magnetic fields in the source region. In spite of this pessimistic conclusion, we stress the importance of the search for flare neutrinos. Their detection may answer a number of crucial questions in flare physics, such as the acceleration mechanism, the maximum rigidity, R_m, of accelerated particles, the source location (altitude) in the solar atmosphere, and the time needed for particle acceleration up to relativistic energies.

4.6 Occurrence Probability of Giant Flares

How large an event can the Sun produce? How frequently the largest events occur? As shown below, both of these questions are not trivial. Meanwhile, it would be very interesting, in particular, to estimate the effect of solar flares on the evolution of life (*e.g.*, Reid et al. 1976). Besides, the extreme features of the Sun's proton productivity are very important not only for fundamental research, but also for the planning and protection of future space missions (see Chap. 11).

Notice that the upper limit spectrum (ULS) model (see Sect. 4.2) deals with the largest proton fluxes observed (or expected) near the Earth's orbit at the moment t_m, but not with the fluences (event-integrated fluxes). Therefore, the ULS seems to be not very representative as to determining largest particle fluences. For example, based on the limit intensity I_p (>10 MeV) $= 10^6$ cm^{-2}s^{-1}sr^{-1} (see Table 4.3) one can obtain a limit fluence, Φ(>10 MeV) $= 1.25 \times 10^7 \Delta t$ cm^{-2}, where Δt is the integration time interval. Hence, to obtain the fluence values of $>10^{10}$ cm^{-2}, it is necessary to integrate the peak proton intensity over $\Delta t > 10^3$ s. On the other hand, proceeding from the largest fluence Φ(>10 MeV) $= 3.2 \times 10^{10}$ cm^{-2} estimated for the single event of November 12, 1960 (Feynman et al. 1990a), our model gives $\Delta t = 2.5 \times 10^3$ s. Although both estimations of Δt are very similar it should be emphasized that the ULS model is hardly able to characterize thoroughly a single proton event because of rather complicated correlation between its time profile, peak intensity and duration.

In this context the estimates of Sakurai (1979) for occurrence probability of extremely large flares are of great interest. The occurrence rate of the flares during solar cycle 19 at the Wolf number $W > 100$ turned out to be approximately proportional to the value of W, independent of flare importance. A number of flares for this cycle diminished exponentially with increasing of flare importance from 2 to 4. The extrapolation of such a dependence indicates that during the cycle 19 one gigantic flare of hypothetical importance 5 could occur. The most real candidate for such a case is the flare of February 23, 1956, though this event turned out to be not extreme one as regards, for example, the fluence of >30 MeV protons (Webber 1963, 1964). According to estimates of Sakurai (1979), the flares of importance four or more release about 50 % of their total energy in the form of SCR with energy $E_p > 10$ MeV which in turn is expected to result in very large enhancement of proton energy density near the Earth. However, the data of SCR observations already carried out for more than 50 years still give no grounds for such expectations (see, *e.g.*, curve 4 in Fig. 4.5 for the late phase of the February 23, 1956 event). Moreover, according to our estimates (Miroshnichenko, 1981, 1983a, 1990), the contribution of protons with $E_p > 10$ MeV to the flare energetics seems to be <10 % for the most powerful SPEs, this portion being slowly increased at $E_p < 10$ MeV (for details see Sect. 5.10).

The occurrence rate of giant flares can be estimated also from some circumstantial data. For example, it is suggested (Beland and Russel 1976) that the recently discovered four cases of extinction of *Radiolaria* for the last 2.5 million years were due to the occurrence of such giant flares with a frequency ~10^{-4} year^{-1}, coinciding with the geomagnetic inversion period. As to the SPE distribution in terms of proton fluence, Φ, per single event, the observation data are controversial. On the one hand, Lingenfelter and Hudson (1980) have revealed an abrupt cut-off in the distribution of proton events, ~$\Phi^{-1.5}$, at $\Phi > 10^{10}$ cm^{-2}. This result was also confirmed by McGuire et al. (1983). On the other hand, more recently Feynman

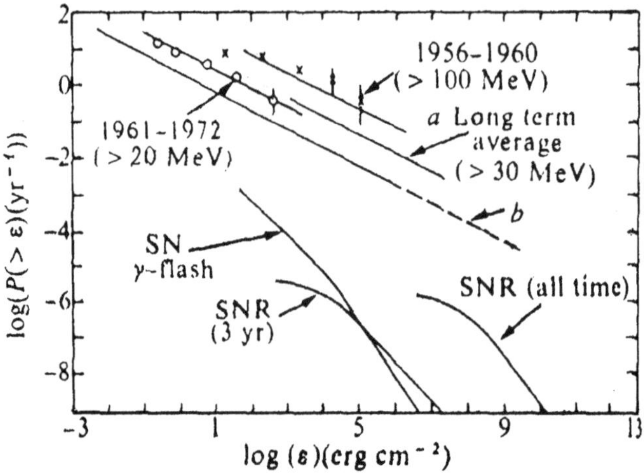

Fig. 4.7 Integral frequencies of SCR events at the Earth (Wdowczyk and Wolfendale 1977): The abscissa is the energy density in the event and relates to the top of the atmosphere. Line *a* is a rough estimate of the long-period average event frequency (for energies above ~30 MeV) and derived from measurements made on protons during the very active period 1956–1960 and during the most recent solar cycle 1961–1972, the latter being of apparently rather average solar activity as judged by the mean sunspot numbers. Most of the particles under consideration normally arrive in the polar regions. Line *b* represents the frequency distribution when the event energy density is averaged over the Earth's surface. *SN γ-flash* denotes the frequency distribution of energy deposition from the gamma-ray flash from Supernovae at 10 pc. *SNR (3 years)* and *SNR (all time)* represent energy deposition over a 3-year period, and integrated over the whole time, respectively, from protons when the Earth is immersed in a Supernova remnant. *P* is a probability and ε the energy density

et al. (1990b, 1993) showed that the fluences for events in solar cycles 19–22 all fitted in one continuous log-normal distribution. Anyway, at the present level of solar activity the largest fluence is apparently confined to the value of 10^{10}–10^{11} cm^{-2} (see also Shea and Smart 1990a). Some recent findings for a number of "ancient" large SPEs seem to change the situation radically (see Chap. 11).

A detailed study of expected catastrophic effects from cosmic rays (primarily the depletion of atmospheric ozone layer) was undertaken by Wdowczyk and Wolfendale (1977). In terms of the energy density ε of cosmic ray particles (in erg/cm^2) received at the top of the atmosphere, they endeavoured to estimate, in particular, the likely frequency of solar flares of sufficient strength to have significant effect. The frequency distributions were constructed for two periods, 1956–1960 and 1961–1972, with different average levels of solar activity. The results obtained for solar energetic particles are summarized in Fig. 4.7, together with the corresponding estimates for the frequency of gamma flashes from the Supernovae.

As it has been shown by Crutzen et al. (1975), a prominent effect is a destruction of the ozone layer by nitrogen oxide NO produced after ionization of the stratosphere, and the incident proton energy necessary to reach the appropriate levels in the stratosphere is about 30 MeV. Thus, an energy threshold of 20 MeV in Fig. 4.7 is only a little low for the necessary limit. In spite of an evident disparity in the absolute frequencies, $P(>\varepsilon)$, the slopes of the variations are very similar in the two periods. There seems to be evidence for an empirical power law for $P(>\varepsilon)$ over seven orders of magnitude. The authors suggest that such a distribution can probably be extended by at least several orders of magnitude more. Anyhow, one can clearly see that among the bursts at a given energy density, solar energetic particles in the range of 20–100 MeV drastically exceed in occurrence rate such an exotic source of radiation as a Supernova remnant.

These conclusions, however, have been seriously questioned by Mullan and Kent (1979). They argued against the proposed extrapolation (Wdowczyk and Wolfendale 1977) of the frequency distribution function of solar flares to time intervals of the order of 10 years. Mullan and Kent (1979) propose that the power law spectra, which have been fitted by Wdowczyk and Wolfendale (1977) to the SCR data, in fact, could not be extended to arbitrarily high energies. Instead of this, the spectra fall-off rapidly beyond the last data point. In their discussion, Mullan and Kent (1979) refer to certain similarities in the energy distribution functions obtained by Rosner and Vaiana (1978) for three different classes of flaring objects: solar X-ray bursts, optical flares in dwarf M stars of spectral class dMe, and X-ray bursts from a cosmic X-ray source (burster) MXB 1730–335. In all three cases, there is observed to be a range of flare energies, E, in which the flare frequency, $f(>E)$, can be fitted by a power law in E.

Rosner and Vaiana (1978) developed a general model for flaring in which stored energy is built up in a short time scale, and the rate of energy storage, dE/dt, is assumed to be proportional to the energy already stored, $dE/dt - aE$. The release of the stored energy is thought to constitute the flare event. In this context, Mullan and Kent (1979) proposed the following physical argument for a rapid cutoff of the flare frequency distribution for the Sun at about 11 years.

The point is that solar flares energies are derived, ultimately, from the toroidal magnetic field which is created inside the Sun by the action of solar differential rotation on the poloidal field. It is known, however, that after 11 years elapsed, the poloidal and toroidal fields reverse sign, the toroidal field having been decreased to zero. From this point of view, each 11-year cycle begins with an emptying out of the energy reservoir.

This suggests that the equation derived by Rosner and Vaiana (1978) for the amount of stored energy $E(t) = E_0[\exp(at) - 1]$ is applicable only up to a maximum time of approximately 11 years. Hence, the frequency distribution $f(E) \sim E^{-\gamma}$ applies as long as f^{-1} does not exceed 11 years. Therefore, in the opinion by Mullan and Kent (1979), extrapolation of the power-law behaviour beyond 11 years is not valid, and ancient catastrophes should not on this account be related to extremely high level of solar activity.

4.7 Flares on the Sun and Other Stars

Flares on the Sun and other stars are important to astrophysics because they originate in out-of-equilibrium magnetic field-plasma interactions rather than in gravitational, thermonuclear, or radiative processes in near equilibrium. According to Haisch et al. (1991), flare stars constitute about 10 % of the stars in the Galaxy. The Sun is an invaluable proving ground to test predictions of flare theories and to develop analytical techniques for future stellar application. In turn, extreme flare star conditions impose the limits of models. In this context, a flare may be defined as a catastrophic release of magnetic energy leading to particle acceleration and electromagnetic radiation, bearing in mind that the magnetic energy release has never been directly observed. Since flare-like physical processes occur in diverse astrophysical regimes, the field of solar and stellar flares can serve as an astrophysical "touchstone" (*e.g.*, Haisch 1989; Shakhovskaya 1989; Haisch et al. 1991). On the other hand, solar flares release a considerable portion of their energy (as a rule, ~1–10 %) in the form of SCRs, mainly protons with the energy range 1 MeV–10 GeV (Miroshnichenko, 1981, 1983a, 1990, 2001). These particles are observed near the Earth's orbit as a solar particle event (SPE).

Recently, some indirect evidences of particle acceleration at some other stars have been reported. Using data from the COMPTEL experiment on the Compton Gamma-Ray Observatory (CGRO), McConnel et al. (1997) have obtained all-sky map in the neutron capture line 2.223 MeV. They found a significant signal (at the level of 4σ) from a point-like source which is located in the southern part of the sky. It seems to be a first observational indication of the neutron production processes in flares at the other stars, through energetic particle interactions with a deuterium production (see Chap. 5).

There is also another indirect method to search for energetic protons in the atmosphere of some other stars. As shown by Simnett (1995), the most sensitive diagnostic of protons in sub-MeV energy range is red-shifted La emission of the relevant excited state of hydrogen. Notice, however, that this method, unfortunately, has never been applied successfully to solar observations (see Simnett 1995, and references therein). Although the *SMM* Ultraviolet Spectrometer and Polarimeter was designed with a suitable capability, its response degraded before definitive measurements were undertaken. On the other hand, observations by the Goddard High Resolution Spectrograph on the Hubble Space Telescope turned out to be more successful. Woodgate et al. (1992) have used its data to search for a *Lα* red-wing enhancement during a flare from red dwarf star AU Microscopii on September 3, 1991. They found an event lasting 3 s, supposedly attributed to a low energy proton beam; this occurred a few seconds after the start of observations. From the strength of the *Lα* red-wing they derived an integrated beam power of $>10^{30}$ erg s^{-1}. Using simultaneous observations of the Si III line, Woodgate et al. (1992) estimated the flare energy. If AU Microscopii has an elemental abundance similar to the Sun, the total energy radiated by the plasma from which the Si III line originated was 6×10^{28} erg s^{-1}. In spite of considerable systematic uncertainties involved in these estimates, Simnett (1995) believes that, if taking the measurements at face value, this flare was consistent with a dominant energy input from a low-energy proton beam (proton beam hypothesis, see Sect. 5.9). As he notes, it remains to be seen if these signatures are found in other stellar, or solar, flares.

This discussion reverts us to existing or assumed restrictions in the maximum energy and intensity of SCR, those parameters being of great significance for the formulation of self-consistent model of particle acceleration at/near the Sun. The main problems of fundamental interest in the theory of particle acceleration at the Sun lie now at two boundary domains of SCR spectra, namely, in low-energy (non-relativistic) and high-energy (relativistic) ranges. The most important of them are: initial acceleration from the thermal background (*e.g.*, Vlahos 1989; Vlahos et al. 1989; Miroshnichenko 1995; Miller et al. 1997), and final stage of acceleration to extremely high energies of $E_p \geq 100$ GeV (*e.g.*, Podgorny and Podgorny 1990; Miroshnichenko 1994, 1996, 2001; Karpov et al. 1998 and references therein).

Initial stage of acceleration is discussed at length in Chap. 5; some aspects of final stage are treated in Chap. 7. Here, we consider some peculiarities of events with relativistic solar protons (RSP) that are concerned with their energy release and size (frequency) distribution. In particular, observational data on Ground Level Enhancements (GLE) of SCRs in the 22nd solar cycle (since September 1986) are of special interest due to unusually high occurrence rate in 1989–1991 and large energy content of the events (Smart et al. 1991). Another exciting finding of recent flare studies turned out to be a registration of long-lasting flux of high-energy (pion) gamma-rays (>1 GeV) on March 26 and June 15, 1991 (Akimov et al. 1991; Leikov et al. 1993) and on June 11, 1991 (Kanbach et al. 1993). In this last event, the observations revealed for the first time the existence of pion radiation as late as 8 h after the impulsive phase of the flare. The problem of SCR generation in relativistic range ($R > 1$ GV) was unusually actualized due to the first confident observations of underground effects correlated with solar flares. As mentioned above, significant increases of counting rate at several muon telescopes (for example, in Yakutsk and Embudo) were registered during GLE of September 29, 1989 (Krymsky et al. 1990; Swinson and Shea 1990), including one very peculiar muon burst (Alexeyev et al. 1992) at the Baksan Underground Scintillation Telescope (BUST). All these new findings give a challenge to our present understanding of utmost capacities of particle accelerators at the Sun. In this context, SCR spectral data (in absolute units of proton flux) at rigidity $R > 1$ GV are of paramount importance.

Since February 28, 1942 (an historical beginning of the SCR observations) the generous data have been obtained on the SCR fluxes, and their spectra have been intensively studied in the energy range from ~1 MeV to 10 GeV and even more. Hitherto, there are ground-based data for 71 GLEs (see Table 2.1), however, spectral data at the rigidities above 1 GV (>435 MeV) are fairly scarce, rather uncertain and/or controversial. Based on GLE observations of SCRs since 1942 we summarize available data on absolute spectra of relativistic protons at the Earth's

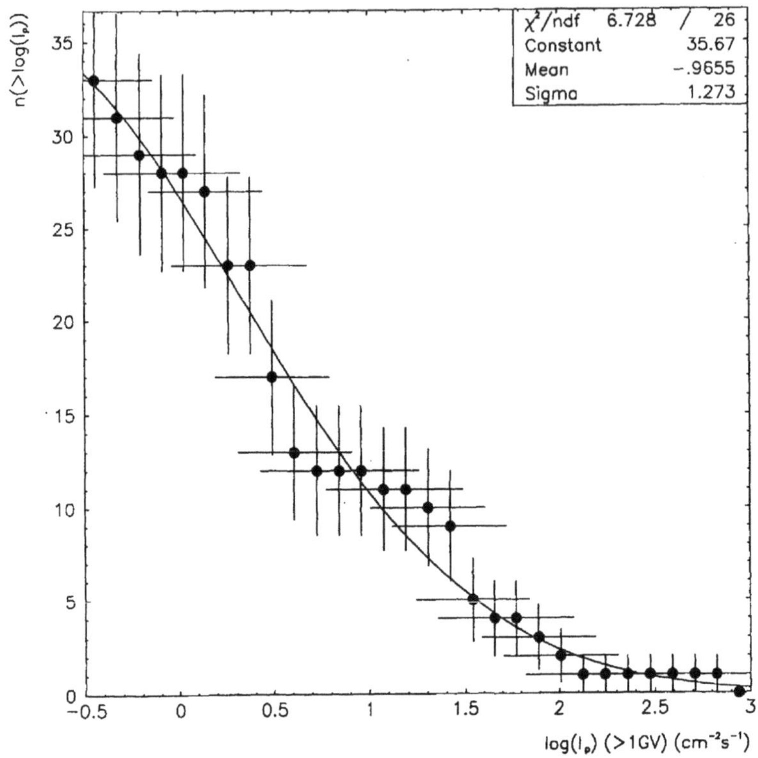

Fig. 4.8 Distribution of the GLEs observed in 1942–1992 on the integral flux of solar protons with the rigidity above 1 GV (Miroshnichenko et al. 1995c)

orbit (for details see Sect. 9.4). By the present time absolute SCR spectra above 1 GV have been estimated for 35 events of 1942–1992 (Table 4.4).

The data compiled in Table 4.4 are rather limited and not very impressive. Nevertheless, a question arises about the GLE distribution on maximum absolute fluxes of solar protons above 1 GV (Miroshnichenko et al. 1995c). This problem is of great interest being extend our knowledge of upper limit capacity of solar accelerators (maximum values of R_m and a number of accelerated relativistic particles). Because of the poor statistics of the relativistic events, we were able to construct a distribution function only for an integral number of GLEs with the integral flux of solar protons in the rigidity range above 1 GV (Fig. 4.8). The results show that in spite of significant methodical uncertainties the distribution may be fitted by Gaussian curve with the proper parameters: constant = 35.67; means = –0.9655, and sigma = 1.273.

Obviously, the total statistics of GLEs with estimated maximum flux of RSP is rather poor for more comprehensive study. Nevertheless, it would be interesting to compare a power of their energy release with the suitable distributions of stellar flares on their characteristic parameters (see Haisch et al. 1991 and references

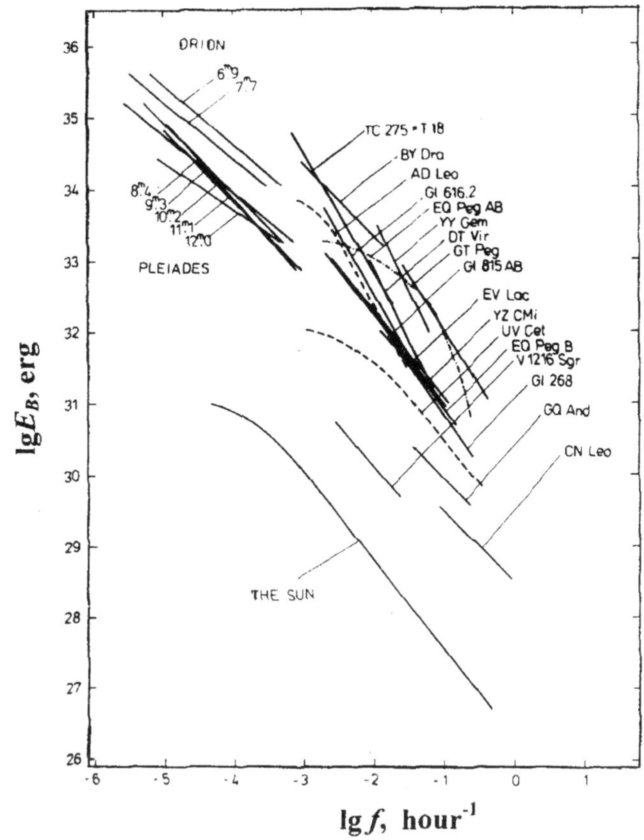

Fig. 4.9 Energy spectra of flares on red dwarf stars and the Sun (Shakhovskaya 1989). Total energy in the B-band (Balmer emission lines) flare radiation, E_B, is plotted versus frequency, f, of flares with energy exceeding E_B

therein). Here we only note that the energy distributions of stellar flares in the B-band (Balmer emission radiation) are power laws and similar to that for the solar flares, suggesting a similar scenario on other stars (Shakhovskaya 1989). The spectral indices in the energy spectra of flares have a rather narrow range of values: from 0.4 to 1.4. To illustrate present situation in this field, we show in Fig. 4.9 energy spectra of flares of 23 red dwarf stars in the solar vicinity, several groups of flare stars in clusters, and, for comparison, on the Sun (see for details and references Shakhovskaya 1989).

In general features, the curves of stellar spectra and solar ones are similar, though a difference in their amplitudes may be of several orders of magnitudes in energy. Quite recently, Schrijver et al. (2012) have estimated the occurrence rate of extremely energetic solar events, based on solar, stellar, lunar and terrestrial records. Concentrations of selected radionuclides measured in natural archives may prove useful in extending the time interval of direct observations up to ten millennia, but as their calibration to solar flare fluences depends on multiple poorly known properties and processes, these proxies cannot presently be used to help determine the flare energy frequency distribution. Being thus limited to the use of direct flare observations, the authors evaluated the probabilities of large-energy solar events by combining solar flare observations with an ensemble of stellar flare observations.

They conclude that solar flare energies form a relatively smooth distribution from small events to large flares, while flares on magnetically active, young Sun-like stars have energies and frequencies markedly in excess of strong solar flares, even after an empirical scaling with the mean coronal activity level of these stars.

In order to empirically quantify the frequency of uncommonly large solar flares extensive surveys of stars of near-solar age need to be obtained, such as is feasible with the *Kepler* satellite. Because the likelihood of flares larger than approximately X30 remains empirically unconstrained, Schrijver et al. (2012) present indirect arguments, based on records of sunspots and on statistical results, that solar flares, at least in the past four centuries, have likely not substantially exceeded the level of the largest flares observed in the space era, and that there is at most about a 10 % chance of a flare larger than about X30 in the next 30 years.

Chapter 5
Particle Acceleration at the Sun

As an introduction to this chapter we would like to give several relevant comments of Dennis (1996) (quotation):

> With the dispelling of "The Solar Flare Myth", there is a danger that the following new myth will take its place: "We don't understand impulsive flares but who cares". Well, Reames cares. He has gone on record, stating that "The physics of particle acceleration in impulsive flares is no less interesting because the events are not geo-effective" (Eos, 1995). Indeed, understanding particle acceleration is one of the most important problems in plasma physics and astrophysics. It is particularly challenging in impulsive flares because of the rapidity with which such a large number of particles are accelerated so efficiently. One of the great successes of flare observations over the last two cycles has been enormous strides that have been taken in the observations of the impulsive phase in many different energy or wavelength regions. The continued study of impulsive flares, with our new abilities to resolve the high energy processes on relevant physical scales in space, time, and energy, will hopefully dispel this insidious myth before it gets off the ground.

In this context, we focus below on some global and local aspects of particle acceleration at the Sun: the main acceleration processes; coherent, stochastic, and shock wave acceleration; acceleration in a fibrous corona; threshold effects and hierarchy of acceleration mechanisms; size (frequency) distributions of solar flare phenomena; a role of low-energy protons in solar flares; energetics of solar cosmic rays. We will estimate existing acceleration models from the point of view of their ability to give an explicit form of energy spectrum and to explain observed chemical composition of energetic solar particles. Such an approach corresponds to our basic concept (see Chap. 1) that energy spectra of all accelerated particles (electrons, protons and heavier ions) and their chemical composition (chemical, isotopic and ionic species, or elemental abundances, isotopes and charge states) are two key points, two parts of the acceleration problem "core".

L. Miroshnichenko, *Solar Cosmic Rays*, Astrophysics and Space Science Library 405, DOI 10.1007/978-3-319-09429-8_5

5.1 Global and Local Aspects of Particle Acceleration

Generation of non-thermal particles is known to be a widespread phenomenon in space and laboratory plasmas. This fundamental process, termed particle acceleration, is the subject of a great deal of study. The mechanisms of acceleration, however, are still rather far from being understood completely. This is especially true for the particle acceleration in solar flares, because of severe restrictions imposed by observations on the models of acceleration. A successful flare model should quantitatively explain a number of characteristics of energetic flare particles, both non-relativistic and relativistic.

As mentioned in Chap. 1, the energy and charge distributions of SEPs are two main "cores" of the problem under consideration. In turn, acceleration of electrons, protons and heavier ions at the Sun is an essential part of the "flare problem". The other parts of the "flare problem" are (*e.g.*, Vlahos 1989, 1994; Somov 1992, 2012): (1) global structure of the active region and energy storage; (2) energy release; (3) interaction of energized plasma with ambient plasma; (4) reaction of the atmosphere, and (5) radiation signatures. In fact, all these parts are one complex ("global") problem that is very hard to "split" into independent ("local") pieces. On the other hand, it should be recognized that solving the global problem apparently goes beyond our instrumental and theoretical capabilities today.

Those circumstances being given, current numerical simulations rely upon rather controversial base. Indeed, some aspects of the flare problem demand "global" MHD modeling at typical lengths $L \geq 10^{10}$ cm and time scales $\tau \geq L/V_A >> 1$ s, where V_A is Alfvén speed; others require "kinetic" modeling at $L \geq c/\omega_{pe} << 10^5$cm, $\tau \cong 1/\omega_{pe}$, and $1/\Omega_e << 1$ s (ω_{pe} and Ω_e are electron plasma frequency and electron gyrofrequency, respectively). Besides, the "global modeling" imposes dynamic boundary conditions on the "kinetic" part, and the "kinetic" modeling, in turn, determines characteristic parameters (*e.g.*, resistivity) for the global modeling. In practice, however, many researchers ignore these important peculiarities and continue their work on the global or local level imposing artificial boundary or local conditions. A similar controversy concerns also the treatment of particle acceleration at the Sun.

The physical processes that control acceleration lie on the interface of the global and local phenomena. Just the first phase of acceleration from thermal distribution is well-known to be the most interesting since it provides a preliminary energization (injection) and determines essentially a charge composition of accelerated particles. This phase proceeds on the local level when only kinetic "micro-processes" in the solar plasma are important (*e.g.*, Miroshnichenko 1995; Miller et al. 1997). On the other hand, in order to be accelerated to relativistic energies, suprathermal particles need quite different spatial scales, at least comparable to their gyroradii in the coronal magnetic fields. At the same time, numerous observations indicate that acceleration to relativistic energies occurs on time scale of order 1 s for protons and a few second for electrons (*e.g.*, Miller et al. 1997). In other words, at the early and late stages of particle spectrum formation we deal with different time and spatial scales of acceleration (*e.g.*, Simnett 1995).

Several theoretical attempts made so far emphasize the global or the local aspect of the acceleration process. Vlahos (1989) mentioned a few characteristic examples. A number of articles have proposed a DC electric field as a possible mechanism for electron acceleration in solar flares (*e.g.*, de Jager 1986, and references therein). The strength of the electric field is estimated from the linear evolution of the tearing mode (no temporal or spatial dependence is considered). However, it is well known that the appearance of the electric field inside the plasma, as well as its temporal and spatial structure depends critically on the boundary conditions.

Another example of this approach is the acceleration of ions by MHD turbulence (Fermi acceleration) or shock waves. In this context, Vlahos (1989) made a serious effort to match the spectrum of the turbulence with the spectrum of the accelerated, but little attention is placed on the mechanism that excites the turbulence or drives the shocks (*e.g.*, Forman et al. 1986; Ramaty and Murphy 1987, and references therein). As a final example, Vlahos (1989) considers the concepts of so-called "localized hot spots" or "conduction fronts". Although these concepts are used for a long time, however, it is known rather little on the physics of their origin, stability and evolution.

A large number of review articles have appeared on particle acceleration in solar flares (de Jager 1986; Forman et al. 1986; Ramaty and Murphy 1987; Sakai and Ohsawa 1987; Miroshnichenko 1987; Scholer 1988; Vlahos 1989; Vlahos et al. 1989; Mandzhavidze and Ramaty 1993; Chupp 1996; Miller et al. 1997). These articles review the observed data and the mechanisms that can accelerate charged particles, but the most of them omit the process of fitting to the existing data or the connection to the global energy release processes in solar flares.

A comprehensive description of particle acceleration in space plasmas is out of the scope of our present consideration (see Chap. 10 in Berezinsky et al. 1990). In what follows we will give briefly only main (conceptual) ideas and discuss a few key observational points and theoretical aspects of particle acceleration at the Sun.

5.2 Main Acceleration Processes

When considering the problem of acceleration, the difference between the acceleration mechanism (process) and acceleration model must be kept in mind (Korchak 1978). The problem of the formation of the energy and charge distributions in the source reduces just to the construction of a model. Following mainly Vlahos (1989) and Miller et al. (1997), we "split" the acceleration processes up into three broad classes: (1) stochastic acceleration; (2) shock acceleration, and (3) coherent acceleration. This last class can be the result of a direct electric field E (DC) (when acceleration time is shorter than the time of change of the E-field), or narrow-band electromagnetic wave (Vlahos 1989).

5.3 Stochastic Acceleration

In a broad sense, stochastic, or Fermi, acceleration is usually defined as the process in turbulent plasma in which a particle can gain or lose energy in a short interval of time. These changes of particle energy are considered to occur in a random manner, due to particle collisions with moving scattering centers (magnetic clouds, turbulent waves, shock fronts, etc.). Such a process, with many increases and decreases, lead finally, over longer times, to acceleration. The most important example of this is acceleration by waves.

If the scattering centers are moving toward each other in the rest frame of the plasma, a particle gains additional energy at each elastic encounter, and there is a rather fast first-order Fermi process of acceleration. In the original Fermi mechanism (Fermi 1949) the acceleration occurred more slowly since a particle gained its energy only on average, due to many encountering and overtaking reflections from randomly moving magnetized clouds, and this true stochastic process was called a second-order Fermi acceleration. Stochastic acceleration can also result from resonant pitch-angle scattering from Alfvén waves with wavelengths of the order of the particle gyroradius. To accelerate particles these waves must propagate both parallel and anti-parallel to the average magnetic field. Other modes of stochastic acceleration, called magnetic pumping and transit-time damping, occur through interaction with magnetosonic waves whose wavelengths are much longer than the particle gyroradius. Langmuir (plasma) waves with phase velocities of the order of the particle speed will also accelerate particles stochastically (see, *e.g.*, Forman et al. 1986; Scholer 1988; Miller et al. 1997, and references therein). Therefore, a central key to understanding stochastic acceleration are the normal modes which may exist in magnetized plasma.

A second key issue for understanding stochastic acceleration by waves is resonant wave-particle interaction. When the wave amplitude is small, stochastic acceleration is a resonant process that occurs when the condition (*e.g.*, Miller et al. 1997)

$$x \equiv \omega - k_{\parallel} v_{\parallel} - l\Omega/\gamma = 0 \tag{5.1}$$

is satisfied. Here ω is the wave frequency, $v_{\parallel}$ and γ are the parallel particle velocity and Lorentz factor, $k_{\parallel}$ is the field-aligned component of the wave vector k, Ω is the cyclotron frequency of the particle, and x is referred to as the frequency mismatch parameter. For harmonic numbers $l \neq 0$ (gyroresonance), Eq. (5.1) is a matching condition between the particle's cyclotron frequency and the Doppler-shifted wave frequency in the particle's guiding center frame. It means that the frequency of rotation of the wave electric field is an integer multiple of the frequency of gyration of the particle in that frame and that the sense of rotation of the particle and electric field is the same.

The convention that is usually employed (see, *e.g.*, Miller et al. 1997) is that Ω is always positive and the sign of l depends upon the sense of rotation of the electric field and the particle in the plasma frame: if both rotate in the same sense (right or left handed) relative to B_0, then $l > 0$ (normal Doppler resonance); if the sense of rotation is different, then $l < 0$ (anomalous Doppler resonance). Hence, when the resonance condition is satisfied, the particle sees an electric field for a sustained length of time and will either be strongly accelerated or decelerated, depending upon the relative phase of the field and the gyromotion. The most effective resonance is $|l| = 0$, and $l = +1$ is usually referred to as cyclotron resonance. For $l = 0$ the resonance condition specifies matching between the parallel components of the wave phase velocity and particle velocity. This resonance is sometimes referred as the Landau or Cherenkov resonance.

Formally, stochastic acceleration is described as the solution of the diffusion equation in phase (or momentum p) space:

$$df/dt = \left(1/p^2\right)dp^2/dpD_{pp}df/dp \tag{5.2}$$

where D_{pp} is the diffusion coefficient in the phase space. When the random energy gains are small compared to the particle energy, stochastic acceleration results in a diffusive current in momentum space, $S = -D_{pp}(\partial f/\partial p)$, where p is the magnitude of the momentum, $f(p)$ is the number of particles per unit volume in phase space, and S is measured in cm^{-3} momentum^{-2} s^{-1}. Particle injected at some momentum p will diffuse in momentum to larger and smaller p. In terms of f, the differential particle intensity per unit energy per nucleon is given by $dJ/dE = Ap^2 f$, where A is the nuclear mass number. Additional non-diffusive energy changes can be added to S (Forman et al. 1986)

$$S_p = -D_{pp}(df/dp) + dp/dt \tag{5.3}$$

where dp/dt represents convection in momentum space due to processes which change the energy of all particles (*i.e.*, ionization or Coulomb losses). It should be emphasized that the physical modes of stochastic acceleration are involved in the acceleration models only through the momentum diffusion coefficient D_{pp}. Neglecting spatial convection, averaging over some volume of space and introducing an escape time T from this volume one can reduce the particle conservation equation to the transport one (*e.g.*, Forman et al. 1986):

$$(\partial f/\partial t) + \left(1/p^2\right)\partial\left(p^2 S_p\right)/\partial p + f/T = Q(p,t) \tag{5.4}$$

where $Q(p, t)$ is the particle source in momentum space. Equation (5.4) has been applied by many workers, in fact, as a principal equation of stochastic acceleration theory (see, *e.g.*, Tverskoi 1967a, b, 1968; Korchak 1978; Forman et al. 1986; Miroshnichenko 1987; Vlahos 1989, 1994; Perez-Peraza and Gallegos-Cruz 1993, 1994; Gallegos-Cruz and Perez-Peraza 1995; Miller et al. 1997, and references therein).

The diffusion coefficient can be estimated from the wave spectrum. The solution of the transport Eq. (5.4) for the particle distribution in non-relativistic energy range

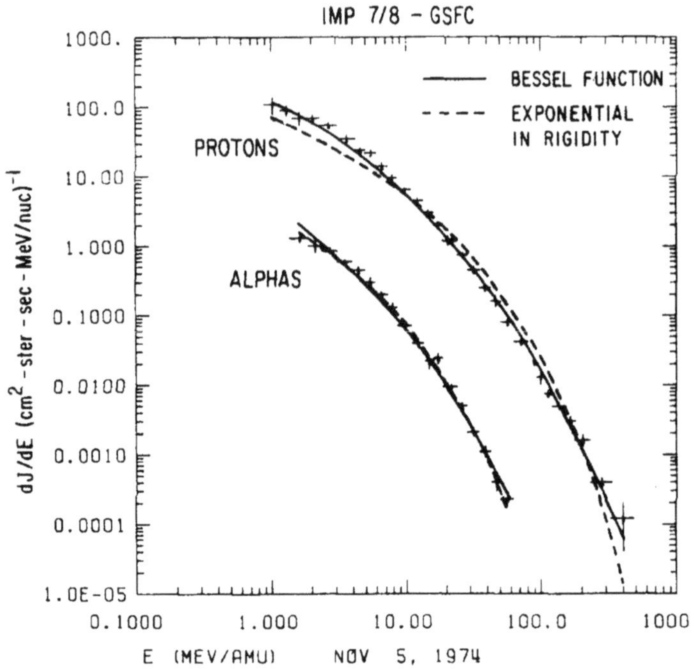

Fig. 5.1 Typical proton and alpha particle spectra during a solar flare event of November 5, 1974 (McGuire et al. 1981). The best fits to the observed spectra are Bessel functions in velocity (*solid lines*) and exponential in rigidity (*dashed lines*)

can be expressed in terms of modified Bessel functions for the protons or an exponential for electrons (*e.g.*, Forman et al. 1986). Assuming that the particles stay inside the acceleration volume for a finite time T, a series of characteristic spectra of accelerated particles were calculated and then used for fitting the observed data, mainly on gamma-ray and neutron radiations of the flares (Ramaty and Murphy 1987).

The form of the proton and alpha particle spectra has been extensively investigated by many workers (*e.g.*, McGuire et al. 1981). When constructing source spectra at the Sun it has to keep in mind that the spectra may be considerably distorted by coronal and interplanetary propagation effects. These effects can be minimized by selecting only particle events from flares that are magnetically well connected (see Chap. 1) and by constructing the spectra at times of maximum intensity (time-of- maximum, or TOM method) at each energy (see, *e.g.*, Fig. 4.4). Here we show an example of the spectrum fitting for the non-relativistic protons and alpha-particles observed during the SEP event of November 5, 1974 (Fig. 5.1). It is seen that both spectra have been fitted with a Bessel function and exponential in rigidity. Note that in general the ion spectra are not given by power laws over the observed energy range (McGuire et al. 1981).

The maim problems with stochastic acceleration in solar flares are the turbulent spectra and injection of the "seed" particles. For example, gyroresonant acceleration of ions by Alfvén waves takes place only when the particles move with velocities $v >> V_A$ (the velocity of the moving scatterers). In other words, injection energy is needed for the process to be effective. Also, very little is known on the processes that generate the turbulent spectra used in the diffusion Eq. (5.1). Nevertheless, due to some recent theoretical findings (*e.g.*, Miller and Reames 1996, 1997), the hopes raise to explain a few key observations by stochastic mechanism.

For example, Miller and Reames (1997) propose that heavy ion enhancements in impulsive solar flares result from stochastic acceleration by cascading Alfvén waves produced at some large-scale cascade to higher wave number k. In this process the waves resonate with ions of progressively higher gyrofrequency ω. The consideration concerns to a plasma consisting of Fe, Ne, and He group ions, with taking into account wave cascading, wave damping by accelerated ions, ion acceleration, and ion escape from the acceleration region. Their preliminary quantitative estimates show that cascading along with cyclotron resonant acceleration is a viable mechanism for yielding the observed heavy ion abundance enhancements in impulsive solar flares. There is no need for separate pre-acceleration and main acceleration mechanisms: the waves accelerate ions directly out of the background plasma up to tens of MeV/nucleon on very short time scales. The ions that are left out of this picture are protons, since the Alfvén waves are damped before reaching the proton gyrofrequency. However, Alfvén wave energy may be supplied to the protons by another way – through specific conversion of fast mode waves.

5.4 Shock Wave Acceleration

Particle acceleration at fast mode shocks is an attractive mechanism since, in distinction from the case of stochastic acceleration, shock accelerated ions have been observed directly in association with corotating interplanetary shocks, solar transient shocks, and planetary bow shocks (see, *e.g.*, Scholer 1988, and references therein). There are, in fact, two mechanisms that can accelerate at fast mode shocks: the first order Fermi (or diffusive) shock acceleration (DSA) and the shock drift mechanism.

If one assumes that there is no plasma disturbances upstream and downstream of the shock, then the main acceleration mechanism is the drift of ions and electrons along the convective electric field $E = -V_{sh}B$, where B is the value of the magnetic field and V_{sh} is the *upstream flow velocity* as it is measured in the shock frame (the shock frame is the frame moving with the shock discontinuity). Examples of ion motion in the shock frame are shown in Fig. 5.2 (Decker 1988). It is obvious that when $V_{sh} \parallel B$ (parallel shock) the electric field approaches zero, and drift acceleration is not important.

This mechanism is most effective for quasi-perpendicular shocks, i.e. shocks with an angle of the upstream magnetic field to the shock normal larger than 45°.

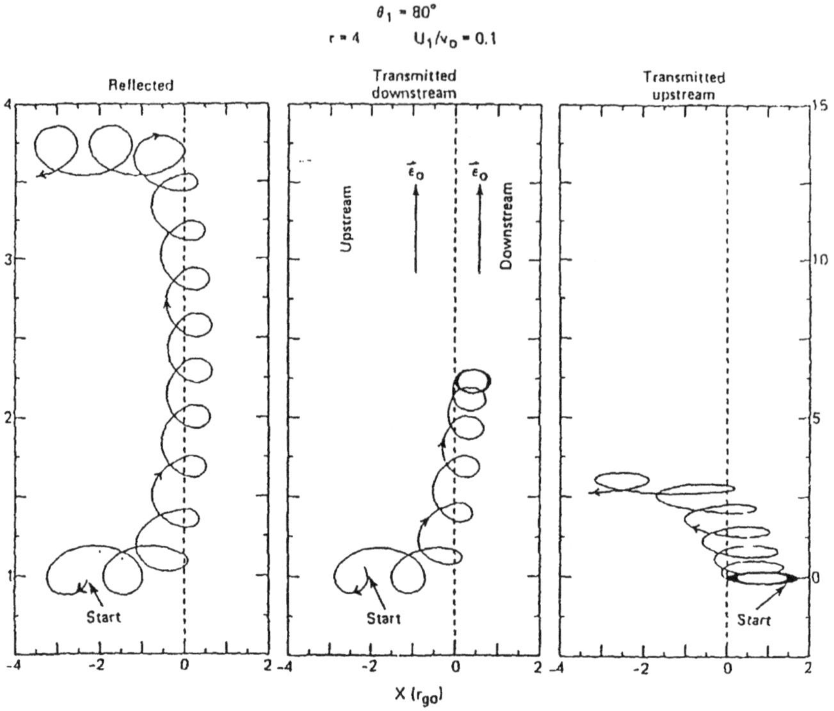

Fig. 5.2 Calculated particle trajectory in shock frame at quasi-perpendicular shock with $\theta = 80°$ (Decker 1988)

Energy gains due to drifts are fast, but in the absence of a return mechanism particles will escape the shock and never return. This will limit the energy gain to about five times the initial energy. Pitch angle scattering in the upstream medium will eventually bring the particle back to the shock, so that it can interact with the shock several times. In the case that scattering is strong, so that the distribution function remains quasi-isotropic, the shock drift acceleration can be included in the diffusive acceleration model (see below). Different aspects of shock acceleration have been considered or reviewed by many workers (for details see, *e.g.*, Ellison and Ramaty 1985; Decker and Vlahos 1986; Scholer 1988; Vlahos 1989; Simnett 1995; Miller et al. 1997). In Fig. 5.3 the results of Decker and Vlahos (1986) are shown for acceleration in a turbulent, oblique shock. Commenting these results, Simnett (1995) noted that as the acceleration proceeds, upstream scattering is produced through Alfvén wave turbulence where the Alfvén waves are generated by the accelerated ions themselves.

If the upstream and downstream plasma is turbulent, then ions and electrons are intensively scattered. In the first order Fermi model (or diffusive shock acceleration, DSA) it is assumed that particles are scattered approximately elastically in the plasma frame. This occurs due to particle pitch-angle scattering on small angles by MHD waves that are propagating approximately with the local flow speed. The

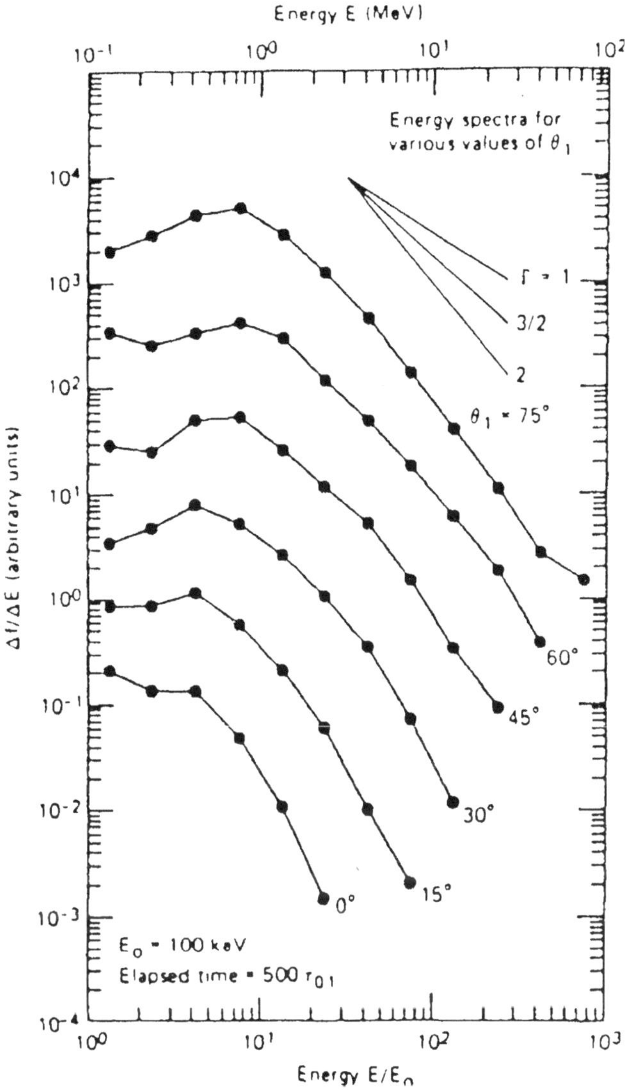

Fig. 5.3 Predicted proton energy spectra from acceleration at turbulent shocks (after Decker and Vlahos 1986). The proton injection energy is 100 keV; θ is the angle that the upstream magnetic field makes to the shock normal; τ_{01} is the upstream gyroperiod

particles which are scattered toward the shock in the upstream medium can gain considerable energy in the shock frame. The particles are possibly reflected back from the shock front or are scattered back by downstream waves, so that the particles can re-encounter the shock many times (Fig. 5.4).

Diffusive shock acceleration (*e.g.*, Krymsky 1977; Axford et al. 1977; Bell 1978; Blandford and Ostriker 1978; Völk 1981; Berezhko and Taneev 2003; see

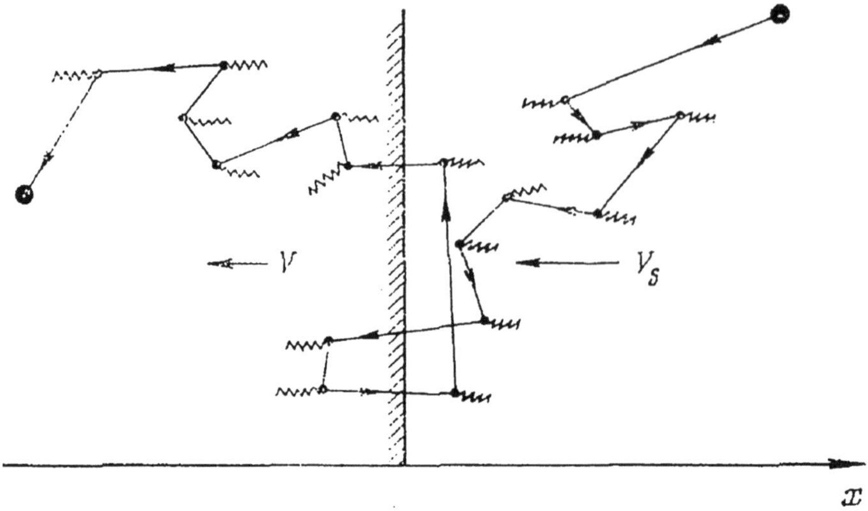

Fig. 5.4 Particle scattering by wave turbulence near the shock front (*vertical shadow line*) in DSA model (Völk 1981), V_s and V are the plasma velocities in the upstream and downstream regions, respectively. The particles can re-encounter the shock many times

also Zank et al. 2000, 2007 and the reviews by Berezinsky et al. 1990, Priest 2000) is a promising mechanism, first of all, for resolving the problem of cosmic rays origin in the Galaxy. It becomes also applicable for interpretation of particle acceleration in the interplanetary medium (*e.g.*, Zank et al. 2000, 2007) and in the nearest solar environment (Berezhko and Taneev 2003). In general, diffusive shock acceleration, unfortunately, is rather slow process and requires a number of special conditions. For example, it is necessary to satisfy a condition of $v \gg V$ (injection problem) where v and V are the velocities of particle and plasma, respectively. Otherwise the particles will simply flow through the front from the upstream region to the downstream region. At the same time, DSA provides an effective mechanism of particle confinement near the shock front. In the case of a CME-driven shock it allows seed solar particles (protons) to gain rather high energy, up to $\geq$1 GeV (*e.g.*, Zank et al. 2000).

The transport equation which describes diffusive shock acceleration is similar to Eq. (5.4) without stochastic term proportional to D_{pp}, but with additional terms due to convective transport, spatial diffusion and adiabatic compression of particles in the plasma flow. This equation is given by

$$\partial f/\partial t + V\nabla f - \nabla(\kappa\nabla f) - (\nabla V/3)p(\partial f/\partial p) + f/T + (1/p^2)\partial[p^2(\mathrm{d}p/\mathrm{d}t)\,\mathrm{f}]/\partial p = Q(p, r, t) \tag{5.5}$$

where f, T, and dp/dt have been defined in connection with Eq. (5.4), V is the plasma velocity, and κ the spatial diffusion tensor which couples the energetic particles to the plasma converging at the shock (see, *e.g.*, Toptygin 1985; Forman et al. 1986). The terms containing V and κ are essential for the description of acceleration across the shock front. The injected particles are explicitly introduced in Eq. (5.5) by the source term Q; the injection may also be treated as a boundary condition such that f approaches a given value f_0 far upstream. The losses due to particle escape can be treated via the escape time T, or as diffusive escape. In the latter case the scattering becomes negligible at a finite distance from the shock.

For deriving a steady-state ($\partial f/\partial t = 0$) particle spectrum it is usually solved first Eq. (5.5) separately on each side of the shock and then the two solutions are matched at the shock by imposing boundary conditions. These conditions are that both the energetic particle density and the normal component of the spatial streaming of these particles ($S = 4\pi p^2(Vp(\partial f/\partial p) + \kappa\nabla f)$) be continuous at the shock. As it was shown (Toptygin 1985), this is an appropriate approach even though Eq. (5.5) is not valid very close to the shock. The time scale in order to reach the steady state depends on the diffusion coefficients in the upstream and downstream medium.

If one view the picture in the shock frame and assumes that $V_{sh} \| B$, then an energetic particle crossing the shock from upstream will be scattered back upstream since the randomly moving turbulence upstream has now a systematic velocity (the downstream fluid velocity) away from the shock. This scattering will change the energy E of the particle by $(V_2/c)E$, where V_2 is the downstream fluid velocity. The particle will then move backward cross the shock again and propagate upstream gaining $(V_1/c)E$, where V_1 is the upstream fluid velocity. The total energy gain is $(3/4)V_{sh}E$, if the upstream velocity is V_{sh}, and the downstream for a strong shock $(1/4)V_{sh}E$. In other words, the parallel shock organizes the upstream and downstream turbulence such that the rate of energy is the first-order power of the turbulent velocity. This is in contrast with the stochastic acceleration which is proportional to the second-order power in V_{sh} (Vlahos 1989). Here $V_1/V_2 = \mathrm{r}$ is the shock compression ratio and $V_1(V_2)$ is the upstream (downstream) bulk plasma flow velocity. The corresponding differential intensity $j = dJ/dE$ is then given by the power-law function $\sim E^{-\gamma}$, where

$$\gamma = (1/2)(r + 2)/(r - 1) \tag{5.6}$$

$$\gamma = (r + 2)/(r - 1) \tag{5.7}$$

in the non-relativistic and relativistic regime, respectively.

Shock wave acceleration in solar flares have been used extensively (*e.g.*, Ramaty and Murphy 1987; Miller et al. 1997; Berezhko and Taneev 2003) since this process enables to accelerate electrons and ions rather quickly and form the observed spectrum provided for we know how to generate shock waves and turbulence with the "correct" characteristics (wave energy, wave spectrum, etc.). In solar flares shock waves have been associated with acceleration of particle in the upper corona since they were related to metric and decametric type II bursts. The problem with acceleration at quasi-perpendicular shocks is the high injection energy: unless the

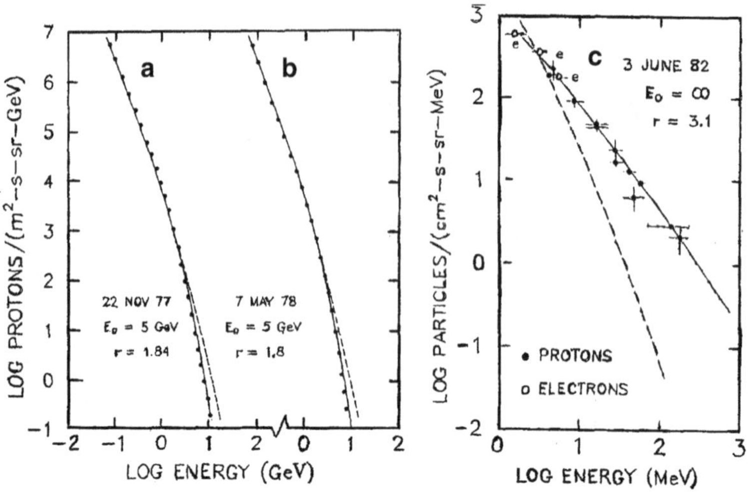

Fig. 5.5 Differential energy spectra for the solar particle events observed on November 22, 1977 (**a**), May 7, 1978 (**b**) and June 3, 1982 (**c**). The *solid lines* are the shock model fitting of those spectra using the shock compression ratio r and characteristic energy E_0 shown. The *dashed lines* assume $E_0 = \infty$ (Ellison and Ramaty 1985)

initial energy is not high enough the particles are simply transmitted through the shock into the downstream medium.

The difference in the spectral exponent of the differential intensity for relativistic and non-relativistic particles (see Eqs. (5.6) and (5.7)) impelled Ellison and Ramaty (1985) to suggest simultaneous first order Fermi acceleration for all particle species in solar flares, *i.e.*, for ions and electrons. They assumed that the differential intensity is given by

$$j(E) = j_0 \exp(-E/E_0) \tag{5.8}$$

where E_0 determines a high-energy cutoff, j_0 is the spectrum for an infinite one-dimensional shock with compression ratio r, and E is energy for electrons and protons and energy per charge for ions. The value of E_0 is supposedly determined by the finite spatial extent of the shock. Ellison and Ramaty (1985) were able to fit observed electron, proton and alpha particle spectra reasonable well with the two parameters r and E_0. As an illustration we show in Fig. 5.5 their fittings of several spectra for the events of November 22, 1977; May 7, 1998; and June 3, 1982. They could not produce the flatter slope of the electron spectrum below ~100 keV (*cf.* Fig. 2.11), but assumed instead that the injection is at 100 keV. They also obtained acceleration times of a few seconds as observed during the impulsive phase.

5.5 Coherent Acceleration

The third important class of acceleration models deals with a DC electric field in the solar atmosphere. The origin and strength of the electric field in solar flares are not well known. There are, at least, two possible ways that an electric field will appear in solar flares (Vlahos 1989): (1) magnetic reconnection or (2) double layers. The suggestion that magnetic field line reconnection could be the cause of particle acceleration in solar flares was first made by Giovanelli (1947). Reconnection has later been applied to geomagnetic phenomena and in situ spacecraft measurements at the Earth's magnetopause have indeed provided the most definitive evidence for the occurrence of reconnection in nature (see, *e.g.*, Scholer and references therein). Most models for solar flares incorporate in some way or the other (for the reviews concerning magnetic reconnection at the Sun see, *e.g.*, Priest 1982; Vlahos 1989, 1994; Démoulin et al. 1994; Miller et al. 1997; Priest and Forbes 2000; Somov 1992, 2012).

There have been several attempts to estimate the electric field produced by the resistive tearing mode instability but the results disagree. The principal reason for the lack of agreement is that the induced E-field depends critically on the small-scale structure of the magnetic field and the transport properties of the instability as it nears the point of saturation, and such nonlinear behaviour is poorly known. There are two distinct mechanisms available in a reconnecting field, for accelerating particles (*e.g.*, Vlahos 1989):

(a) the electric field in the tearing layer itself

$$\mathrm{E}_0 = \eta J_0 e_z \tag{5.9}$$

where η is the resistivity, J_0 is the current, e_z is the unit vector, and

(b) the electric field due to the flow velocity v outside the resistive layer which converts magnetic flux into the tearing layer:

$$\mathrm{E} = -v \times B \tag{5.10}$$

The strength of these fields is still open question. The answer depends sensitively on the time development of the field structure at the reconnection point, which, in turn, depends upon the local resistivity and on the external boundary conditions.

A number of articles have calculated the detailed changes of local reconnection due to radiation losses and thermal conduction (see references in Vlahos 1989, 1994). Bulanov and Sasorov (1975) have estimated the E-field from the rapid changes in a magnetic field structure in the course of the breaking of a current sheet, which gives rise to an induced electric field (5.10), $E \sim (V_A/c)B$, where V_A is the Alfvén velocity (see also Chap. 10 in Berezinsky et al. 1990). They have given also an *approximate estimate* of the dimensions of the current sheet, and then the maximum energy gain by particles and their spectrum was estimated assuming conservation of particle flux in phase space. Depending on the structure of magnetic field near the reversal, the energy spectrum can be power-*law* or exponential form. Similar estimates were performed also by Perez-Peraza et al. (1977). The authors have simulated the motion of particles in a typical picture of the slow reconnection (see Priest 1982). It was suggested that the inflow into the reconnection region occurs at a small fraction of the Alfvén velocity and the fluid is ejected out of the reconnection region with Alfvén velocity. In the frame of these limitations, Perez-Peraza et al. (1977) have derived an analytical form for the differential energy spectrum of accelerated particles

$$N(E) = N_0(E/E_0)^{-0.25} \exp\left\{-1.12\,(E/E_0)^{-0.75}\right\} \tag{5.11}$$

$$N_0 = 1.47 \times 10^7 \left(nL^2/BE_0\right) \text{proton/MeV} \tag{5.12}$$

$$E_0 = 8.236 \times 10^{-3} \left(B^3L/n\right)^{2/3} \text{MeV} \tag{5.13}$$

where E_0 is characteristic energy of the spectrum; B, n and L are the magnetic field magnitude, plasma density at the flare site and linear dimension of the current sheet, respectively. The constant N_0 and characteristic energy E_0 are related to the parameters of the source by the Eqs. (5.12) and (5.13). Based on these simple expressions it turned out possible to estimate the source parameters for several relativistic proton events (*e.g.*, Perez-Peraza et al. 1992). There were also a number of articles (see Scholer 1988; Vlahos 1989, 1994 and references therein) which treated the stochastization of orbits near the magnetic field reversal and the consequences of this stochastization for the reconnection efficiency and particle acceleration.

All these attempts are important steps towards our understanding of particle acceleration near the reconnection sheet, but as we mentioned above, depends so critically on the magnetic structure, the boundary conditions and the time evolution of the resistive instability. Thus, we feel that it is not yet easy to construct detailed models based on this acceleration mechanism for solar flares.

Double layers were initially proposed more than 30 years ago by Alfvén and Carlqvist (1967). They assumed that in current flowing through the plasma, a density depression may rise while the induction of the total circuit is large enough for the current to be maintained. A DC electric field must appear to adjust the velocity of the electron flow, $v(x)$, in such a way that the current density, J, remains constant, $en(x)v(x) = J$. One can easily estimate the potential drop and the energy gained by the particles. It is possible to show rigorously (Vlahos 1989) the existence of self-consistent solutions of the Vlasov equation able to sustain large potential drops, and able to accelerate electrons and ions to high energies. The role of double layers appears very important in understanding of the large-scale structure circuits in solar flares, as well as the local conditions at the point that the circuit breaks down. Although it is an open question whether double layers are good candidates for particle acceleration, the whole subject, as noted by Vlahos (1989), should be re-examined, in particular, for a fibrous corona since the conditions for double-layer formation are easier to achieve inside the fiber due to the stronger current localization.

The presence of an electric field inside the plasma (independent of its origin) is a subject that needs careful study. As is well known, if an electric field, E, is applied to a plasma to drive a current, the electrons are accelerated by this field while simultaneously being dragged by dynamic friction due to collisions. A critical electron's velocity $V_e = V_c$ is reached when the electric force is balanced by the dynamic friction. By equating the first force with second one and setting $V_c = V_{Te}$ (where V_{Te} is the electron thermal velocity) the critical velocity may be easily obtained (see, *e.g.*, Spicer 1979)

$$V_c = V_{Te}(\mathrm{E}_D/\mathrm{E})^{1/2} \tag{5.14}$$

The value of electric field where the drag force at the thermal speed equals the electric field force is called the Dreicer field E_D (Dreicer 1959, 1960) and is given by

$$\mathrm{E}_D = e \ln\Lambda/\lambda^2{}_{De} = (e/4\pi\sigma_0)(\omega_{\mathrm{pe}}/V_{Te})^2 \ln\Lambda, \quad \mathrm{V\,m^{-1}} \tag{5.15}$$

where $\ln\Lambda = \ln(12\pi n_e \lambda^3{}_{De})$ is the Coulomb logarithm with $\lambda_{De} = (kT_e/4\pi n_e e^2)^{1/2}$ being the Debye length, and n_e and e are background number density and electron charge, respectively, and all quantities are in SI units. Hence, even in the presence of a weak electric field E some electrons in the higher energy portion of the electron distribution (at $V_e > V_c$) will run away. In this context, the parameter E_i is the threshold energy corresponding to the critical velocity V_c. In other words, all the electrons retaining to high-energy tail of thermal distribution (at $E_e > E_i$) are accelerated in practice without collision losses.

It is worthwhile to estimate the magnitudes of E_D expected from (5.15) with different values of the plasma parameters in the solar corona. The results are shown in Table 5.1. As a typical value of the Dreicer field in the solar corona Miller et al. (1997) give $E_D \cong 10^{-4}$V cm^{-1}. It is to compare two extreme values of E_D in Table 5.1 with the tentative value $E_D = 1.2 \times 10^{-1}$ V cm^{-1} calculated by Syrovatsky (1976, 1981) for typical parameters in a neutral current sheet. This value was derived from data on magnetic field variations in solar active regions before major flares. A few evaluations of E at coronal heights differ by several orders of magnitude – from ~3×10^{-2} to $>10^2$ V cm^{-1} (Miroshnichenko 1987, 1990, 1995; Berezinsky et al. 1990; Litvinenko and Somov 1995; Litvinenko 1996a; Somov 1992, 2012). These values are large enough to accelerate the runaway electrons from thermal to sub-relativistic energies (~100 keV) in fractions of a second (Vlahos et al. 1989). We will not touch here upon complex problems associated with the initial acceleration of electrons, such as reverse currents and propagation of particle beams in the solar atmosphere (see, *e.g.*, Melrose 1990; Simnett 1995).

If neglecting Coulomb pitch angle scattering and electron-ions collisions, for $E > E_D$ (super-Dreicer field) the electric force exceeds the drag force on all electrons, which will then be freely accelerated to higher energies. For $E < E_D$

Table 5.1 Electric fields in the solar corona

T_e, K	2×10^6	3×10^7	2×10^6	3×10^7	Derived by indirect data:
n_e, cm^{-3}	10^9	10^9	10^{11}	10^{11}	
λ_{De}, cm	3.1×10^{-1}	1.2×10^0	3.1×10^{-2}	1.2×10^{-1}	
E_D, V cm^{-1}	3.1×10^{-5}	2.5×10^{-7}	2.8×10^3	2.1×10^{-4}	$\geq 10^{-1}$–10^2

(sub-Dreicer field), there is a critical velocity V_c, below which the drag force overcome the electric force. Above V_c the situation is reversed. Electrons with speeds $<V_c$ will then be heated, while those with speeds $>V_c$ will be freely accelerated.

For solar flare acceleration, models which employ both super-Dreicer and sub-Dreicer fields have been proposed. The most advanced model in the former category is that of Litvinenko (1996a) (see also Martens 1988). The geometry of this model is that of a large reconnecting current sheet (RCS) above a bipolar magnetic loop or arcade. The sheet has the height x and the length z of ~10^9 cm, the width y of a 100 m, and contains an electric field E_z along the length of the sheet whose strength is ~10 V cm^{-1}. This is several orders of magnitude higher than the Dreicer field (~10^{-4} V cm^{-1}), but is a reasonable $V \times B$ field for quasi-steady magnetic reconnection in the corona.

Magnetic field in this sheet model has a constant longitudinal B_z component along the electric field, a reconnecting B_x component normal to electric field and parallel to the height of the sheet, and a transverse B_y component normal to the electric field and parallel to the width of the sheet (Fig. 5.6). The geometry is similar to that found by in the Earth's magnetotail when the interplanetary magnetic field has an east/west component, except that B_y in the magnetotail corresponds to B_z in Litvinenko's model and vice versa. The longitudinal component (~100 G) is much larger than the transverse component (~1 G). Particles will therefore be magnetized mostly along the direction of the electric field and be able to gain large energies.

However, the energy corresponding to the potential drop along the length of the sheet (~10 GeV) will not be realized as a result of the finite B_y). This component slightly magnetizes particles in the transverse direction and causes them to escape from the sheet over distances of order 10^4 cm. Typical maximum energies are then about 100 keV. While limiting the particle energy, this rapid transverse escape prevents the current in the sheet from reaching values where the self-induced magnetic field would exceed observational limits. The model provides electron energies and fluxes consistent with hard X-ray observations, with few problems, associated with replenishment of the acceleration region (for details see Miller et al. 1997). At the same time, this model gives interesting consequences concerning the composition of accelerated particles and the dynamics of acceleration process itself (see Sect. 5.9).

The question remains, however, whether the RCS with $B \neq 0$ can describe the generation of relativistic protons (up to a few GeV) in some flares. The affirmative answer was given by Litvinenko and Somov (1995), who proposed that the protons

Fig. 5.6 Electric and magnetic fields inside the reconnecting current sheet with a nonzero longitudinal magnetic field component (Litvinenko 1996a)

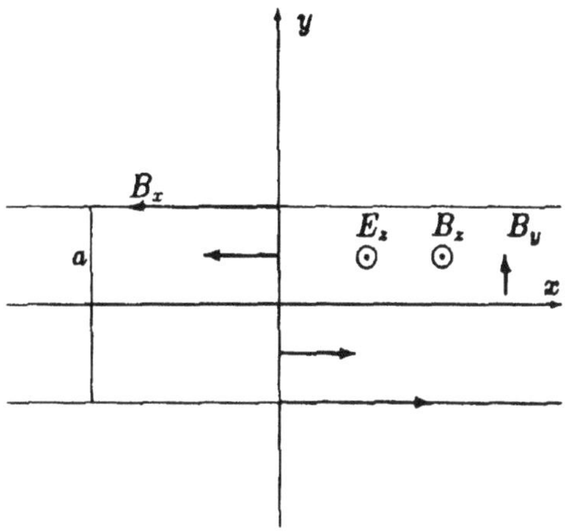

interact with the RCS more than once, each time gaining a finite, relatively small amount of (see below). The cumulative effect was shown to be the required fast acceleration to relativistic energies.

It is well known that if the electric field is less than the Dreicer field, a small fraction of electrons $(n_r/n_0) \cong 0.5 \exp(-E_D/2E)$, where n_r is the number density of runaway particles, n_0 is the ambient density, and E_D is the Dreicer field, will run away. In the absence of magnetic field (or if $\omega_{\rm pe} >> \Omega$) the energy gained by the runaway particles will be limited only by the scale length of the potential drop. In the presence of a magnetic field the scenario of the runaway particles changes since the electrons can excite an instability (the anomalous Doppler resonance instability) which scatters the electrons perpendicular to the magnetic field direction. The final result is that the distribution tail will be isotropized and eventually thermalized. As it was shown earlier (Moghaddam-Taaheri et al. 1985), for $E_{\parallel} < 0.2E_D$ the anomalous Doppler resonance scattering is weak and the tail is possible to be accelerated to very high energies.

If the electric field exceeds the E_D inside the plasma the whole distribution will runaway and drive currents. Depending on the details of the ambient plasma parameters a number of current driven instabilities can be excited. However, some difficulties arise when we attempt to accelerate all the necessary electrons for a hard X-ray burst from a single potential drop (see Vlahos 1989; Miller et al. 1997).

In addition to the plasma ejected out of the reconnection region there exists the possibility of direct acceleration of particles in the electric field along the neutral line (*e.g.*, Scholer 1988). The problem with accelerating particles along a single neutral line is that only particles injected very close to the neutral line will get accelerated to high enough energies. Away from the diffusion region there is a magnetic field component B_z normal to the current sheet, and the particles will perform so-called Speiser's orbits (Speiser 1965), *i.e.*, they will essentially drift half

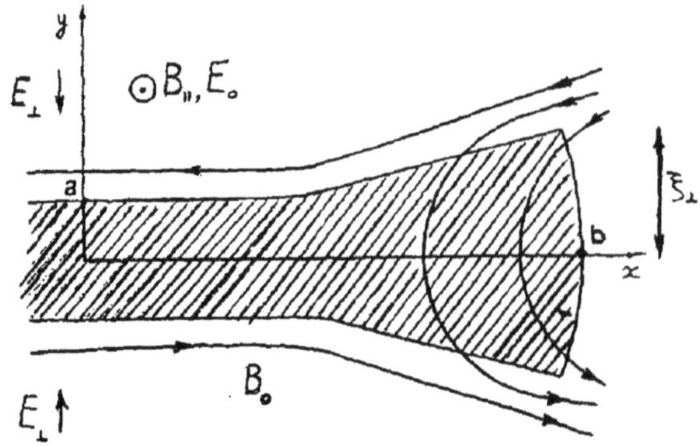

Fig. 5.7 Sketch of the non-neutral reconnecting current sheet (not to scale) from Litvinenko and Somov (1995). B_0 is the main (reconnecting) magnetic field component, $B_\perp$- is the transverse field, E_0 is the main electric field related to the reconnection process inside the sheet, $E_\perp$ is transverse electric field outside the sheet due to electric charge separation; (*a*) is the half-thickness and (*b*) the half-width of the current sheet

a gyroorbit in the small normal magnetic field parallel to the electric field and will then get ejected along the magnetic field out of the current sheet region.

Speiser (1965) was the first to treat the charged particle acceleration in current sheets, taking into account not only the reconnecting field B_0, but also a small transverse (perpendicular to the plane of the RCS – reconnecting current sheet) magnetic field component $B_\perp = \xi_\perp B_0$. A typical relative value of the transverse field, penetrating into such RCS, termed *non-neutral* (Somov 1992; 2012), is $\xi_\perp = 10^{-3} - 10^{-2}$. In Fig. 5.7 we show a sketch of the non-neutral reconnecting current sheet, as it was drawn by Litvinenko and Somov (1995).

The basic Speiser's result is that both the energy gain ΔE and the time that particle spends in the non-neutral RCS Δt are finite. The transverse magnetic field makes the particle turn in the plane of the sheet, and then a component of the Lorentz force expels it from the RCS plane almost along the magnetic lines of force. The distance that the particle can travel along the sheet equals the Larmor diameter determined by the transverse field and typical speed of the particle.

Litvinenko and Somov (1993) generalized the results of Speiser (1965) by including into consideration the longitudinal magnetic field $B_{\|}$ in the sheet. This component, however, while efficiency magnetizing electrons in the RCS, cannot influence the motion of the relativistic protons and heavier ions that are of primary interest to us there. This is because the "critical" longitudinal field, necessary to magnetize a particle in the RCS, is proportional to the square root of the particle mass (see Litvinenko and Somov 1993). Thus, on the one hand, electrons acquire the relativistic energy in RCSs with a nonzero longitudinal field $B_{\|}$. On the other hand, a nonzero $B_\perp$ radically restricts the energy of heavier particles: ΔE for protons cannot exceed 20 MeV if a typical value of $\xi_\perp = 3 \times 10^{-3}$ ($B_\perp = 0.3$ G) is assumed (Litvinenko and Somov 1995). Therefore, the relativistic energies cannot be reached after a single "interaction" of the particle with the sheet (*cf.* Martens 1988).

To overcome this difficulty, Martens (1988) conjectured that the relativistic acceleration could take place in RCS region where $B_{\perp} \Rightarrow 0$ (the neutral current sheet approximation), and the protons are freely accelerated by the electric field. This conjecture, however, does not seem to be adequate for actual RCSs, where reconnection always occurs in the presence of a nonzero transverse magnetic field. Though we do expect the latter to vary somewhat along the RCS (Somov 1992), the region with a vanishing $B_{\perp}$ is so small that a particle will quickly leave the region (and hence the RCS) before being accelerated. Thus Litvinenko and Somov (1995) were led to modify the classic Speiser's model significantly.

They proposed that the protons interact with the RCS more than once, each time gaining a finite, relatively small amount of energy. The cumulative effect would result in the required acceleration to relativistic energies. The factor that makes positively charged particles to return to the RCS is the transverse electric field directed toward the sheet (see Fig. 5.6). Physically, the transverse electric field outside the RCS, $E_{\perp}$, is a consequence of electric charge separation. Both electrons and protons are deflected by the magnetic field when they move out of the sheet. The trajectories of electrons, however, are bent to a greater degree owing to their small mass. As for protons and much heavier ions, they stream out of the RCS almost freely. Hence the charge separation arises, leading to the electric field that detains the protons in the RCS region. The following point is worth to emphasizing here. The charge separation that gives rise to the potential

$$\Phi = e \int \mathrm{E}_{\perp} \mathrm{d}y \tag{5.16}$$

mainly stems from the motion of protons perpendicular to the RCS plane. At the same time, some protons are known to leave the RCS almost along its plane. This property is characteristic feature of the above mentioned Speiser's mechanism of particle acceleration. It seems that even a small transverse electric field will considerably influence the motion of these particles because they always move almost perpendicular to this field. Hence, the transverse electric field efficiently "locks" non-thermal ions in the RCS, thus allowing their acceleration by the electric fields in the RCS. Taking into account both the main components of electromagnetic field (B_0 and E_0) and the transverse ones ($\mathrm{B}_{\perp}$ and $\mathrm{E}_{\perp}$), Litvinenko and Somov (1995) have estimated the energy rate (and maximum energy for the protons being accelerated in the RCS. It was clearly demonstrated, in particular, that protons can actually be accelerated to about 2.4 GeV in the high-temperature RCS (Somov 1992). Even larger energies can be reached in RCS regions with a smaller transverse magnetic field (*cf.* Martens 1988). An interesting feature of the mechanism considered is that neither the maximum energy nor the acceleration rate depends on the particle mass. Hence, the mechanism may play a role in the preferential acceleration of heavy ions during solar flare. It was successfully applied in constructing a model for generation of relativistic ions in large gamma-ray event of June 15, 1991 (Akimov et al. 1996).

5.7 Brief Summary of Acceleration Models

The results briefly reviewed above apparently suggest that particle acceleration processes in solar corona may be more diverse than previously supposed. They raise as many questions, at least, as they answered (see, *e.g.*, Vlahos et al. 1989, 1994; Litvinenko and Somov 1995; Miller et al. 1997, and references therein). Nevertheless, it is timely to formulate some critical issue for understanding particle acceleration, at least, in impulsive solar flares. The best summary of this kind was proposed recently by Miller et al. (1997). Below we set forth their conclusions.

Flare observations and our ability to model physical processes in magnetized plasmas have developed enough that now may makes sense to strive toward a comprehensive model for particle acceleration in impulsive flare. Although the observational data and our knowledge of plasma processes are still not extensive enough to settle upon one (or more) acceleration mechanism(s), we have been able to identify a number of issues that must be addressed by a successful model of flare particle acceleration. Such a model must be able to satisfy following requirements.

1. Acceleration of electrons and ions to energies in excess of 100 keV and 100 MeV, respectively, in order to account for hard X ray and gamma ray emission. It should also allow the possibility of energizing electrons to about 10 MeV and protons to about 1 GeV, in order to account for the less common ultra-relativistic electron bremsstrahlung and pion radiation.
2. Acceleration of electrons and ions to the lower energies in less than 1 s and to the higher energies over several seconds (*cf.*, *e.g.*, Somov and Oreshina 2011).
3. For a large flare, the model must provide a production, at least, of 2×10^{35} electrons s^{-1} (hybrid model), and possibly as many as 10^{37} electrons s^{-1} (nonthermal model), above 20 keV and over a period of 10–100 s. It must also produce $\sim 3 \times 10^{30}$ protons s^{-1} above 30 MeV and $\sim 10^{35}$ protons s^{-1} above 1 MeV over the same time.
4. Resulting electron and ion distributions should be consistent with (*i.e.*, can be used to successfully fit) high-resolution X-ray and nuclear gamma-ray line spectra, respectively.
5. The model must reproduce the observed enhancements of ^{3}He, Ne, Mg, Si, and Fe relative to ^{4}He, C, N, and O.
6. The model must describe how the accelerated electrons and ions are pulled out of the thermal plasma.
7. The model must describe the relationship between electron and ion acceleration and heating and, in particular, provide observed relationships between the evolution of hot plasma an accelerated particles.
8. The model should describe the relationship between electron and ion acceleration, and, in particular, it should account for the simultaneity to within ~1 s of the acceleration of these two particle species.
9. It should be clear how the local acceleration model can be incorporated into the large scale coronal structures, as they were observed, for example, by *Yohkoh* spacecraft.

Table 5.2 Summary of acceleration models

Observations	Sub-Dreicer E[a]	Super-Dreicer E[b]	MHD turbulence[c]	Shocks
~100 keV electrons	Yes	Yes	Yes	Yes
~10 MeV electrons	No	?	Yes	?
~100 MeV protons	No	?[d]	Yes	Yes
~1 GeV protons	No	?[d]	Yes	Yes
~1 s acceleration time[e]	Yes	Yes	Yes	Yes
<100 keV electron flux	Yes	Yes	Yes	?
<100 MeV proton flux	No	?	Yes	?
Electron distribution	Yes	?	?[f]	?
Proton distribution	No	?	?	?
^{3}He enhancement[g]	No	No	No	No
Heavy ion enhancement	No	No	Yes[h]	No
Electron acceleration from thermal plasma	Yes	Yes	Yes	?
Ion acceleration from thermal plasma	Yes	Yes	Yes	Yes
Hyperfilamentation essential	Yes	No	No	No
Current complexity	High	Low	Low	Media/low
Strength of link to large-scale reconnection	Weak	Good	Good	?

Adapted from Miller et al. (1997)

Notes:

[a]Based on the work of Holman et al. (1989)

[b]Based on the work of Litvinenko (1996a)

[c]Based on the work of Miller et al. (1996)

[d]But does not appear promising

[e]To those energies given in the above 4 rows accessible by the mechanism

[f]However, Hamilton and Petrosian (1992) have shown that whistlers can produce correct distributions

[g]None of these models will directly give the ^{3}He enhancement. However, if any generate a bump-on-tail electron distribution, then the models of Temerin and Roth (1992) and Miller and Viñas (1993) may be applicable

[h]But the parameters are restrictive

This last point becomes of highest importance in the context of global and kinetic modeling of particle acceleration at the Sun (see Sect. 5.1), as well as in connection with a number of challenging peculiarities of the GLEs and gamma-ray bursts observed in the solar cycle 22 (see Chaps. 2, 3 and 7).

Table 5.2 summarizes the results of the review paper of Miller et al. (1997) for the three main acceleration processes: stochastic acceleration by MHD waves, sub- and super-Dreicer DC electric fields, and shocks. The top 13 rows deal with the properties discussed above. Each of the mechanisms has successes and failures. For example, none can account for the enhancement of ^{3}He in flares; this requires a separate process. All can account for the observed acceleration times. With the

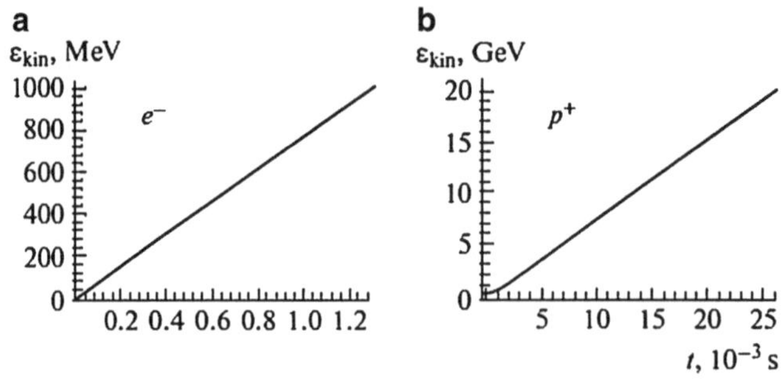

Fig. 5.9 Kinetic energies of (**a**) electrons and (**b**) protons in a Reconnecting Current Layer model versus acceleration time (in seconds) (Somov and Oreshina 2011)

possible exception of shock electron acceleration, particles can be extracted from the thermal plasma in each case.

Miller et al. (1997) have focused mainly on the "typical" flares and non-relativistic ions (up to <100 MeV/nucleon). At the same time, they also pay a considerable attention to relativistic ions, but specifically in the connection with the variety of high-energy neutral emissions and energization rates of accelerated particles in solar flares (see Sect. 5.9 below and Chap. 6). In the context of numerous restrictions imposed on the acceleration models, it is timely to mention here one important requirement concerning to relativistic protons: a spatial scale of acceleration region should be much more than particle gyroradius. This constraint is valid for the shock wave acceleration and for the acceleration by DC electric field produced by a Reconnecting Current Layer (RCL), or Reconnecting Current Sheet (RCS).

According to the most recent investigations (*e.g.*, Somov and Oreshina 2011; Somov 2012), a RCL model with 3D-component magnetic field and DC electric field really provides fast acceleration of both species (electrons and ions) to relativistic energies for very short times: ~$2 \times 10^{-7} \div 10^{-3}$ s for electrons and ~$10^{-4} \div 2 \times 10^{-2}$ s for protons. The authors consider the mechanism of acceleration by electric field in the RCL as a main one. The reconnection-induced electric field in the RCL attains rather large values (up to ~30 Vcm^{-1}) during the impulse stage of a flare (*e.g.*, Somov 2012). If we apply the values of $B_0 \approx 100$ G, $B_\perp \approx 5 \times 10^{-4} B_0$, and $B_\parallel \approx 0.1 B_0$, we get the results depicted in Fig. 5.9. In this case, the size (length) of the acceleration region must be on the order of 7×10^2–3×10^7 cm for electrons and appropriate region length is 3×10^5–7×10^8 cm – for protons.

The above results have been obtained by Somov and Oreshina (2011) based on numerical integration of relativistic equation of particle motion in the RCL. The authors conclude that proposed mechanism of particle acceleration by the electric field in a RCL can explain the observational data obtained, in particular, during the event of 13 December 2006 (GLE70). This event was observed in the PAMELA experiment onboard the Russian satellite *Resurs-DK1*. Protons in the GLE70 were accelerated to energies of the order of several GeV/nucleon (*e.g.*, Adriani et al. 2011).

According to recent observations, solar flare electrons and protons are accelerated to high energies almost simultaneously in each elementary flare burst. Some evidences of this were found, in particular, by Kuznetsov et al. (2011) by the data of measurements of gamma-rays and high-energy neutrons onboard *CORONAS-F* during the solar flare of 28 October 2003 (GLE65). In this context, a concept of Multiple Acceleration Processes (MAP) at the Sun is of special interest (see, *e.g.*, Miroshnichenko 2003a; Miroshnichenko and Perez-Peraza 2008).

5.8 Recent Developments of Shock Acceleration

Shock acceleration in solar flares has been considered previously in many papers. In particular, Ellison and Ramaty (1985) have modeled the simultaneous acceleration of protons, alpha particles, and relativistic electrons by first-order Fermi (or diffusive) shock acceleration (for details and references see, *e.g.*, Priest and Forbes 2000). In all cases examined, Ellison and Ramaty (1985) found for any given event that a single shock compression ratio in the range ~1.6–3.0 simultaneously produces reasonably good fits to the observed electron, proton, and alpha-particle spectra. The differential intensity of accelerated particles is given by

$$dJ/dE \propto (dJ/dE)_0 \exp(-E/E_0) \tag{5.17}$$

where E and E_0 are energy for electrons and protons and energy per charge for ions. As energy and gyroradius increase, it becomes less probable that a particle can be contained within the shock region. Ellison and Ramaty (1985) suggested that this escape would cause the energy spectra of shock-accelerated particles to roll over more or less exponentially, with *e*-folding energy E_0 directly proportional to the ion's charge-to-mass (Q/A) ratio.

As noted by Zank et al. (2000), there is increasing evidence to suggest that energetic particles observed in "gradual" SEP events are accelerated at shock waves driven out of the corona by coronal mass ejections. Energetic particle abundances suggest, too, that SEPs be accelerated in situ solar wind or coronal plasma rather than from high-temperature flare material. In this context, the authors presented a dynamical time-dependent model of particle acceleration at a propagating, evolving interplanetary shock (IP shock). The theoretical model includes the determination of the particle injection energy, the maximum energy of particles accelerated at the shock, energetic particle spectra at all spatial and temporal locations, and the dynamical distribution of particles that escape upstream and downstream from the evolving shock complex. Note that injection here refers to the injection of particles into the diffusive shock acceleration mechanism.

As the shock evolves, energetic particles are trapped downstream of the shock and diffuse slowly away. In the immediate vicinity of the shock, broken power-law spectra are predicted for the energetic particle distribution function. The escaping distribution consists primarily of very energetic particles initially with a very hard power-law spectrum (harder than that at the shock itself) with a rollover at lower energies. As the shock propagates further into the solar wind, the escaping ion distribution fills in at lower energies, and the overall spectrum remains hard. Downstream of the shock, the shape of the accelerated particle spectrum evolves from a convex, broken power-law shape near the shock to a concave spectrum far downstream of the shock.

Maximum particle energies accelerated at IP shocks result from a competition between the decelerating shock speed, the weakening IMF, and the shock age. Unless a shock is accelerating, the maximum energy to which particles are accelerated at an IP shock decreases monotonically with increasing radial distance. Nonetheless, according to the model by Zank et al. (2000), substantial maximum particle energies are possible in the early stages of shock evolution. In particular, energies of order of 1 GeV are possible for young shock waves, and this decreases to ~100 MeV at 2 AU. Higher-energy particles tend to escape more easily from the shock complex, but a small number can remain trapped for an extended period.

As noted by Berezhko and Taneev (2003b), in both above papers (Ellison and Ramaty 1985; Zank et al. 2000) the authors have considered a case of plane wave approximation that does not allow to take into account a finite size of the shock wave and its temporal dependence. Such an approximation is applicable to a bulk of accelerated particle in the vicinity of the shock, but it is broken in the range of ultimate energies where the spectrum undergoes to exponential cut-off. In fact, this approach results in significant softening of particle spectrum and decreasing of their maximum energy. To substantiate their model of acceleration of SCR up to relativistic energies by the shock waves produced by CMEs, Berezhko and Taneev (2003b) proposed to use some new observational data.

They used the Alfvén turbulence data at the distances from the Sun above 3 $R_{\oplus}$ (Andreev et al. 1997) and semi-empirical model of proton density distribution in the low-latitude corona (Sittler and Guhathakurta 1999). Berezhko and Taneev (2003b) have performed detailed numerical calculations of the spectra for the SCR produced during the propagation of shocks in the solar corona in terms of a model based on the diffusive transfer equation using a realistic set of physical parameters for the corona. The resulting SCR energy spectrum

$$N(E) \propto E^{-\gamma} \exp[-(E/E_{\max})^{\alpha}] \tag{5.18}$$

is shown to include a power-law portion with an index $\gamma \cong 2$ that ends with an exponential tail with $\alpha \cong 2.3 - \beta$, where β is the spectral index of the background Alfvén turbulence. The maximum SCR energy lies within the range $E_{max} =$ 1–300 MeV, depending on the shock velocity. Because of the soft spectrum of the SCR, their reverse effect on the shock structure is negligible.

The decrease in the Alfvén Mach number of the shock due to the increase in the Alfvén velocity with heliocentric distance r causes the effective SCR acceleration to terminate when the shock reaches a distance of $r = (2-3)R_{\oplus}$. In this case, the velocity of diffusive SCR propagation exceeds the shock velocity. As a result, SCR particles intensively escape from the vicinities of the shock. Berezhko and Taneev

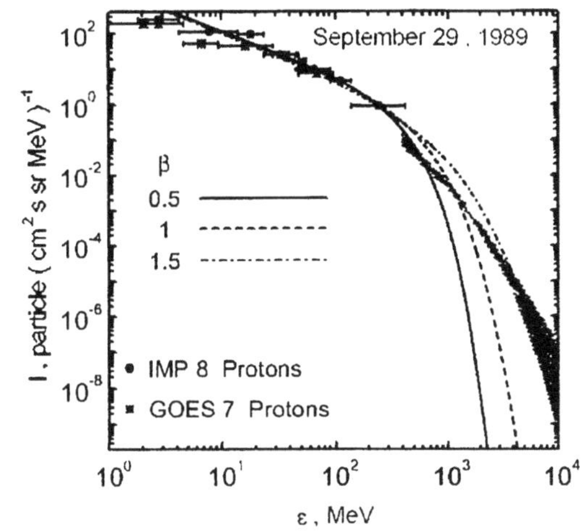

Fig. 5.10 Proton energy spectrum in the event of 29 September 1989 including data points from IMP 8, GOES 7, the neutron monitor spectrum by Lovell et al. (1998) (*shaded area*) and calculated spectra by Berezhko and Taneev (2003b) for different Alfvén wave spectral index β

(2003b) performed a comparison of the calculated SCR fluxes expected near the Earth's orbit with available observational data (*e.g.*, GLEs of 7 May 1978 and 29 September 1989). Their results indicate that the theory may explain well enough some of the main observed features (absolute intensities, spectrum slopes *etc.*) of non-relativistic solar protons. As an example we demonstrate in Fig. 5.10 the results by Berezhko and Taneev (2003b) for the GLE of 29 September 1989.

More deep analysis of their calculations, however, shows that their model still fails in the description of relativistic proton spectrum. In particular, Berezhko and Taneev (2003b) applied their model to the observed spectrum by Lovell et al. (1998) related only to rather late period of the GLE of 29 September 1989. Meanwhile, as it was certainly shown (Miroshnichenko et al. 2000), this event distinctly revealed two-component (two-peak) structure with quite different spectra in two peaks. There are also some other methodical disadvantages of this model that requires to examine it more thoroughly. In this respect, a new good opportunity arises from the observations of proton events in October-November 2003 (including three GLEs) and on 20 January 2005. In particular, the fastest shock wave in October has overcome the Sun-Earth distance for 19 h, with a shock speed about 2,754 km s^{-1}; estimated CME speeds on 20 January 2005 were from 2,500 km s^{-1} (Simnett and Roelof 2005) to 3,675 km s^{-1} (Gopalswamy et al., 2005). Taking into account, additionally, a temporal evolution of the accompanying CMEs, this model may provide a new insight on the problem of separation of the SCR sources (flares or CME-driven shocks).

Alternative numerical model has been recently suggested by Roussev et al. (2004) for CME-driven shock acceleration. These authors were based on a fully three-dimensional, global MHD code for the initiation and evolution of the coronal mass ejection which occurred on 2 May 1998. This event was followed by rather small GLE56 (see Table 2.1). In their model, the solar eruption reaches a critical point where a magnetic rope is ejected with a maximum speed in excess of 1,000 km s^{-1} (Fig. 5.11). The shock that forms in front of the rope reaches a

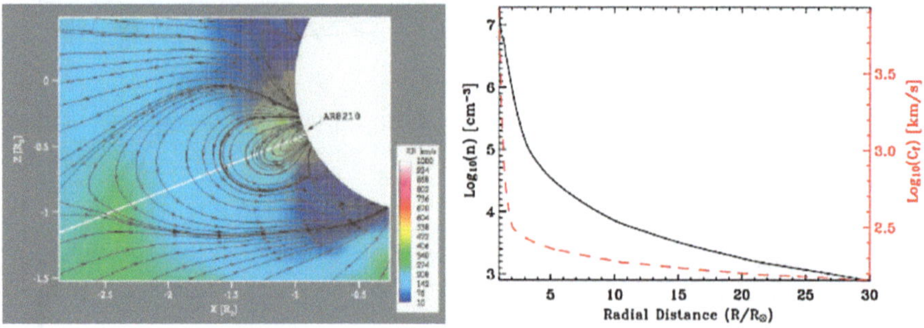

Fig. 5.11 Shock acceleration model for the SEP event of 2 May 1998 (GLE56). *Left:* Magnetic field configuration and flow pattern prior to the eruption (at $t = 230\tau_a$) in the $y = 0$ plane (τ_a is a characteristic time of acceleration). *Right:* Curves of the number density and the fast-wave speed (*red curve*) as derived along the white line in the left panel at $t = 0$ (Roussev et al. 2004)

compression ratio greater than 3 by the time it has traveled a distance of 5 solar radii from the Sun's surface. For such values, diffusive shock acceleration theory predicts a distribution of SEPs with a cutoff energy of about 10 GeV. Whether similar results will be obtained for other events or other assumptions about the initiation mechanism remains to be questionable. However, Roussev et al. (2004) believe that for this event there is no need to introduce an additional mechanism to account for SEPs with energies below 10 GeV.

In conclusion of this subsection it should be noted some recent reviews on acceleration mechanisms and models: Priest and Forbes 2000 (see their Chap. 13); Miroshnichenko and Perez-Peraza 2008; Zharkova et al. 2011; Somov 2012. In particular, Zharkova et al. (2011) reviewed basic theoretical concepts in particle acceleration, with particular emphasis on processes likely to occur in regions of magnetic reconnection. Several new developments are discussed, including detailed studies of reconnection in three-dimensional magnetic field configurations (e.g., current sheets, collapsing traps, separatrix regions) and stochastic acceleration in a turbulent environment. Fluid, test-particle, and particle-in-cell approaches are used and results compared. While these studies show considerable promise in accounting for the various observational manifestations of solar flares, they are limited by a number of factors, mostly relating to available computational power. Not the least of these issues is the need to explicitly incorporate the electrodynamic feedback of the accelerated particles themselves on the environment in which they are accelerated. A brief prognosis for future advancement is offered. The exact acceleration scenario will depend on the magnetic field topology and on the absolute magnitudes of physical parameters. Further, it is possible, or even likely, that all the acceleration processes discussed above play some role in solar flares and elsewhere in astrophysics. The strict demands imposed by the *RHESSI* observations have necessitated a thorough and critical review of all these models, leading to substantial new understanding along the way.

5.9 Rogue Events and Acceleration in the Interplanetary Space

In distinction of the near-Sun environments, in the inner heliosphere valuable information about the physics and dynamics of the heliospheric plasma can be derived from the studies of so-called Forbush-decreases in the intensity of galactic cosmic rays, coronal mass ejections and propagation of SCR in the interplanetary magnetic field (IMF). Possible particle acceleration at interplanetary plasma turbulence and shock waves is also of great interest, especially in the case of so-called "super-events" or "rogue events" (*e.g.*, Kallenrode and Cliver 2001a, b), in analogy to rogue ocean waves having unusually large amplitudes.

Rogue events are associated with multiple shocks and CMEs. In particular, from observations of two converging shock waves accompanied by an energetic particle event with unusual high and long-lasting intensities in August 1972, it was proposed (see references in Kallenrode and Cliver 2001a, b) the first order Fermi acceleration between converging interplanetary shocks as a fast and highly efficient acceleration mechanism. Subsequently, time periods with unusually high intensities related to multiple CME-driven shocks as well as other particle events with high intensities between pairs of shocks have been identified. Well-known examples of rogue SEP events at the Earth occurred on 14 July 1959, 4 August 1972, 19 October 1989, and 14 July 2000. Rogue events also have been observed in the inner heliosphere – with *Helios* 1 on 4 November 1980 at 0.5 AU and with *Ulysses* in March 1991 at 2.5 AU.

Rogue events cannot result in a simple compression of the medium between two converging shocks: although the distance between converging shocks would decrease with r as the shocks propagate outwards, the cross-section of the flux tube increases $\sim r^2$, leading to a net increase in volume between shocks. Figure 5.12 shows different geometries for a pair of shocks and the IMF. The left hand side shows the simplest case: two shocks and a background Archimedean spiral magnetic field. Particles traveling from the upstream region towards the shock partly are reflected at the shock because the magnetic compression across the shock front creates a magnetic mirror. Thus particles can be swept by the shock. However, reflection of particles approaching the shock from its downstream medium cannot be understood because the particle then experiences a diverging field. Therefore, although the shocks would sweep-up the particles, there would be no particle storage in the volume between them and consequently no first order Fermi acceleration.

To avoid this problem, it was proposed the mechanism of a Gold bottle (middle section of Fig. 5.12): a large loop extends ahead of the shock and ones in the upstream medium particles have a chance (depending on their pitch angle) of being reflected back and forth along the field line. As the shock expands, the length of the field line in the upstream medium is reduced and the particles are accelerated by a Fermi I process. Although large loops extending beyond 1 AU have been observed, the authors think that this configuration might not be the only one to explain rogue

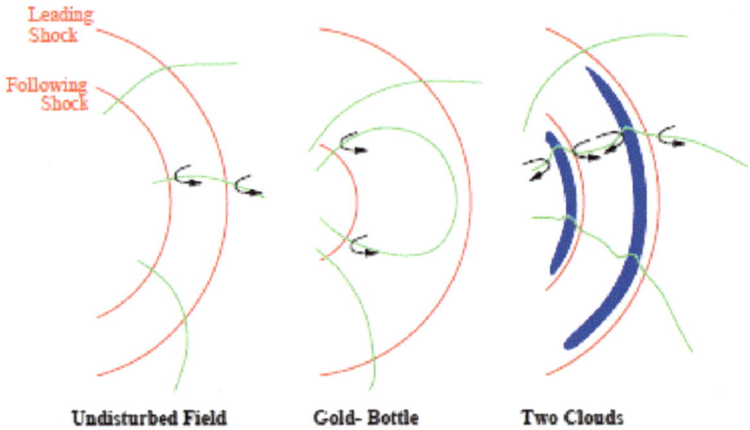

Fig. 5.12 Acceleration of energetic particles in the interplanetary space between 2 converging shocks produced by the Sun (Kallenrode and Cliver 2001)

events, in particular, since a typical rogue event has been observed by *Ulysses* at a distance of 2.5 AU (see references in Kallenrode and Cliver 2001a, b).

Considering the calculations regarding the capability of a magnetic cloud to separate the particle populations upstream and downstream of a magnetic cloud, the authors suggest the scenario depicted on the right hand side of Fig. 5.12: particles then are reflected repeatedly between the following shock (particles with small pitch angles passing the shock at the cloud) and the magnetic cloud behind the leading shock. The authors proposed a numerical model that allows simulating the effect of pairs of CME/shock on particle populations. They concluded that (a) the magnetic cloud following the leading shock is of outmost importance for the creation of high particle intensities, (b) the shocks need not to converge to create an intensity enhancement, and (c) the trailing cloud is required to reduce intensities after the passage of the shock pair.

A possibility of particle acceleration in the interplanetary space up to relativistic energies has been noted, for the first time, in 1973–1974 by Yakutsk research group (see, *e.g.*, Kozlov et al. 1974; Chirkov and Filippov 1977; Filippov and Chirkov 1978 and references therein) and by Pomerantz and Duggal (1974) after analysis of the event of 4 August 1972 (GLE24). In particular, Kozlov et al. (1974) found that at the end of the event a compression of interplanetary plasma increased of several times in comparison with undisturbed period. This may affect the modulation of relativistic particles up to the energies ~40 GeV. Unlike usual prompt arrival of relativistic solar protons in typical GLEs, the onset of GLE24 has been registered only about 6 h after parent solar flare. Later on (see Fig. 9.4), energy (rigidity) spectrum of solar protons was found to be very soft. Retrospective analysis of some past SPEs (GLEs) showed that similar phenomena might take place also on 17 July 1959 and 12 November 1960. The corresponding data are summarized in Table 5.3 (Filippov and Chirkov 1978).

Table 5.3 Specific large SPEs with production of relativistic particles in interplanetary space

Flare date	Flare time, UT	Flare coordinates	Flare importance	Magnetic storm date	Magnetic storm time, UT	Shock wave velocity, km/s
2 Aug 1972	03:16	13N, 35E	2N	4 Aug 1972	01:10	900
2 Aug 1972	18:38	14N, 26E	2B	4 Aug 1972	02:20	1,400
4 Aug 1972	06:20	14N, 08E	3B	4 Aug 1972	20:54	2,800
10 Jul 1959	02:10	20N, 60E	3+	11 Jul 1959	16:25	1,100
14 Jul 1959	03:42	17N, 04E	3+	15 Jul 1959	08:02	1,400
16 Jul 1959	21:00	11N, 30W	3+	17 Jul 1959	16:35	2,200
10 Nov 1960	10:00	28N, 28E	3+	12 Nov 1960	13:48	800
11 Nov 1960	03:10	28N, 12E	3+	12 Nov 1960	18:44	1,050
12 Nov 1960	13:22	28N, 02W	3+	13 Nov 1960	10:21	1,980

Analysis of the data from Table 5.3 allowed to conclude that observed peculiarities of dependence of the event size on geomagnetic cutoff rigidities may be explained under assumption on particle acceleration between two converging shock waves (Filippov and Chirkov 1978). The interplanetary conditions on 17 July 1959, 12 November 1960 and 4 August 1972 seem to be similar to that depicted in Fig. 5.12 (right side, two converging clouds).

In conclusion of this Section we return again to one of the most disputable questions in this field. It was widely believed that the most-energetic and longest-lasting SEP events observed in interplanetary space result from acceleration by the bow shocks of coronal mass ejections (CMEs). However, using gamma-ray, X-ray and radio diagnostics of interacting (with the solar plasmas and magnetic fields) particles and spaceborne and ground-based detection of $\geq$20 MeV protons at 1 AU during two large events (1989 September 29 and October 19), Klein et al. (1999) demonstrated that time-extended acceleration processes in the low and middle corona, far behind the CME, leave their imprints in the proton intensity time profiles in interplanetary space for one or several hours after the onset of the solar flare. So the bow shock is not the main accelerator of the high-energy protons (*e.g.*, Somov 2012, Part 2, Chapter 11.4, p.315).

5.10 Threshold Effects and Event Distributions

As it is known (see, *e.g.*, Miroshnichenko 1987, 1992a; Vlahos et al. 1989; Miller et al. 1997) under flare conditions two extreme stages of particle acceleration are of most interest. The first is an initial acceleration starting from thermal energies, and second one is a final acceleration to relativistic energies within the framework of the same (or different) acceleration scenario(s). The latter is supposed to be determined by coronal magnetic structures (and/or shock waves), and the former by the fundamental properties of solar plasma and the most basic problems of flare physics. In particular, Lu and Hamilton (1991) and Lu et al. (1993) develop the idea that the energy release process in flares can be understood as avalanches of small reconnection events. They predict that the power law flare frequency distributions will be found to continue downward with the same logarithmic slopes to energy of ~3×10^{25} erg and duration of ~0.3 s.

These lower limits are the characteristic energy and time scale of an elementary instability, which is estimated to have a length scale $l = 400$ km. In terms of particle acceleration by plasma turbulence this length may be considered as a "scale barrier" preventing the acceleration process (Sturrock 1974). To move across this obstacle a particle with the energy E must satisfy a condition r_g $(E > E_i) > l$, where r_g and E_i are the gyroradius and so-called "injection energy", respectively.

On the other hand, in terms of particle collisions in the plasma the value of E_i is usually defined from the equality of energy gain rate to Coulomb loss rate ("Coulomb barrier"). In the course of his study of electron acceleration to energies of order several hundred keV, Korchak (1978, 1979) concluded that the use of the term "injection energy" for the electrons has no justification. Since plasma heating to temperatures $T > 10^7$ K (soft X-ray emission) as a rule precedes a non-thermal flare, it is natural to think that the thermal distribution is initial one, the injection being as a matter of fact the process of plasma heating or energization. As to the acceleration of heavy particles, only Coulomb drag on the electron component of the plasma is usually taken into consideration.

Based on the reconnection theory of the flare we examine below the problem of initial acceleration of solar cosmic rays (SCR) under the following assumptions: (1) the electric fields in the solar corona are enough large to accelerate electrons and ions to high energies; (2) all the particle species are accelerated equally, without any selection; (3) threshold effects may exist in a certain sense at SCR generation; (4) differences between proton and electron spectra are inevitable from the very beginning of the acceleration process. The problem is discussed in terms of the size (frequency) flare distributions on various parameters as well as in terms of accelerated particle production and flare energetics. Additionally, we propose to use the data on SCR rigidity spectra at the Sun as an independent source of information.

As it stated above, for considering the acceleration of heavy particles with the charge Ze and mass Am_p, only Coulomb drag is usually taken into account. The natural and dominant characteristics of the Coulomb barrier then are not the injection energy but the position of the Coulomb loss maximum, $\varepsilon_{\max} = 5 \times 10^3$

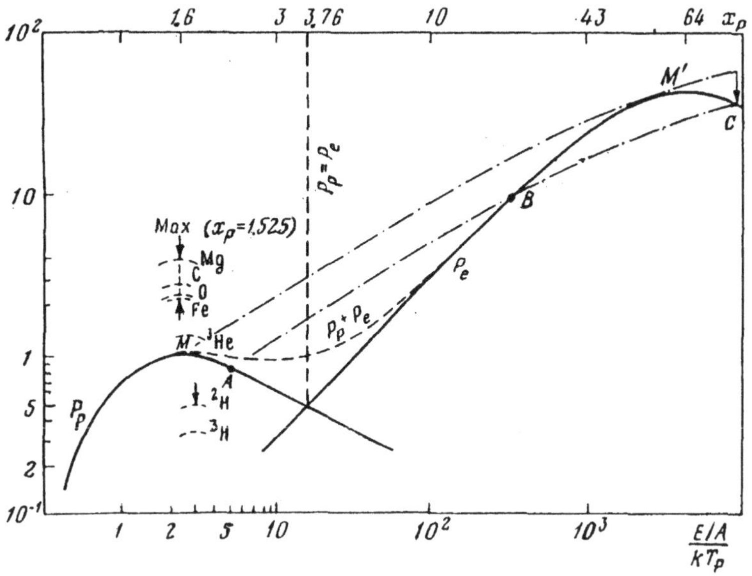

Fig. 5.13 Dependence of the Coulomb losses on the energy per nucleon for the ions ^{4}He in the two-component coronal plasma (After Korchak 1979)

$kT_e A$ erg, as a function of particle energy, and the loss rate for ions, $P_{max} = -(dE/dt)_{\max} = 1.6 \times 10^{-22} Z^2 n/A(kT_e)^{1/2}$ erg/(nucleon s) on the electron component of plasma (*e.g.*, Korchak 1979). At $T_e = 2 \times 10^6$–3×10^7 K we get $\varepsilon_{\max} = 0.8$–11 MeV/nucleon.

In Fig. 5.13 we show a complete curve of the Coulomb losses (Korchak 1979) for the ions ^{4}He in two-component plasma at high temperature (solar corona). The curves P_p and P_e correspond to the drag of helium ions on the proton and electron components of plasma, respectively. Loss curves for the other elements distinguish from the helium curves mainly by vertical shift proportionally to the ratio Z/A. The curves of energy gain MM' and ABC are given by dashed lines for the case of Fermi acceleration. The curve M' separates two acceleration regimes, one of them being more intensive (without injection) and the second one corresponding to slow ("preliminary") acceleration. Obviously, a proton with the energy E_p can penetrate the energy region $E_p > \varepsilon_{\max}$ only when the value of E_p exceeds the thermal energy (~10^3 eV) by 3–4 orders of magnitude. If after this moment the acceleration mechanism continues to operate, the proton will continue to be accelerated. Its final energy will depend only on the rate and duration of the acceleration process.

These results imply that certain threshold effects in SCR generation and/or escape may exist in the energy range of ~1–10 MeV (for the protons). In particular, we anticipate a constraint on the ratio of the total proton energy, W_p, to the global energy of the flare, W_f, this ratio being dependent on flare importance. To quantify this, one can assume a power-law dependence of peak proton flux, J_p, with W_f:

$$J_p(W_f) = B_p W^b{}_f, \tag{5.17}$$

where the power-law exponent $b > 0$ varies within certain limits. This refined problem is of fundamental interest because of its close relation to reconnection theories of solar flares.

In search for the effect (5.17) we looked first at the available data on size (frequency) distributions extensively reported for various solar flare phenomena (parameters). Studies have been done on radio microwave bursts, type III bursts, soft and hard X-rays, interplanetary electron and proton events, *etc.* (*e.g.*, van Hollebeke et al. 1975; Hudson 1978; Kurt 1989, 1990; Kuznetsov and Kurt 1991; Cliver et al. 1991; Crosby et al. 1993; Miroshnichenko et al. 2001; Cliver et al. 2012). All these distributions can be represented above the sensitivity threshold by differential power laws as

$$f(J) = (dN/dJ) = A_j J^{-a}, \tag{5.18}$$

where dN is the number of events recorded with the parameter J of interest between J and $J + dJ$, and A_j and a are constants determined from a least-squares fit to the data. Integral size (frequency) distributions of the form

$$N(>J) = [A_j/(a-1)]J^{-a+1}, \quad \text{for } a > 1 \tag{5.19}$$

are often used for studies with poor statistics. As an excellent example of the distribution functions obtained with a good statistics, in Fig. 5.14 we show the frequency distributions of the peak rate for 7,045 hard X-ray bursts (*a*) and for all 12,776 flares (*b*) detected by the Hard X-ray Burst Spectrometer (HXRBS) on board the Solar Maximum Mission (SMM) satellite in 1980–1982 and 1980–1989, respectively (Crosby et al. 1993). One can see that both of these distributions are consistent with the same power-law index of 1.8. The turnover of the plots below 30 count s^{-1} is treated in Crosby et al. (1993) as a manifestation of the HXRBS sensitivity limit.

Unlike flare electromagnetic emissions, the data on interplanetary particle events are still rather poor and discrepant, their distribution functions being discernibly different from those for flare electromagnetic emissions. For instance, the frequency distributions of electron events (the peak electron flux) reveal the following slopes: 1.50 ± 0.20 at $E_e > 17$ keV and >45 keV; 1.35 ± 0.15 ($E_e > 70$ keV); 1.46 ± 0.15 ($E_e = 0.5$–1.1 MeV); and 1.30 ± 0.07 in the interval $E_e = 3.6$–18.5 MeV (see details in Kurt 1989, 1990; Cliver et al. 1991; Crosby et al. 1993). These values, however, may be reconciled with those for energy fluences of flare electromagnetic emissions. This is true, at least, for electrons with energy >70 keV, which in the non-thermal interpretation are considered to be responsible for hard X-ray generation. In her summarizing the results on size distributions of

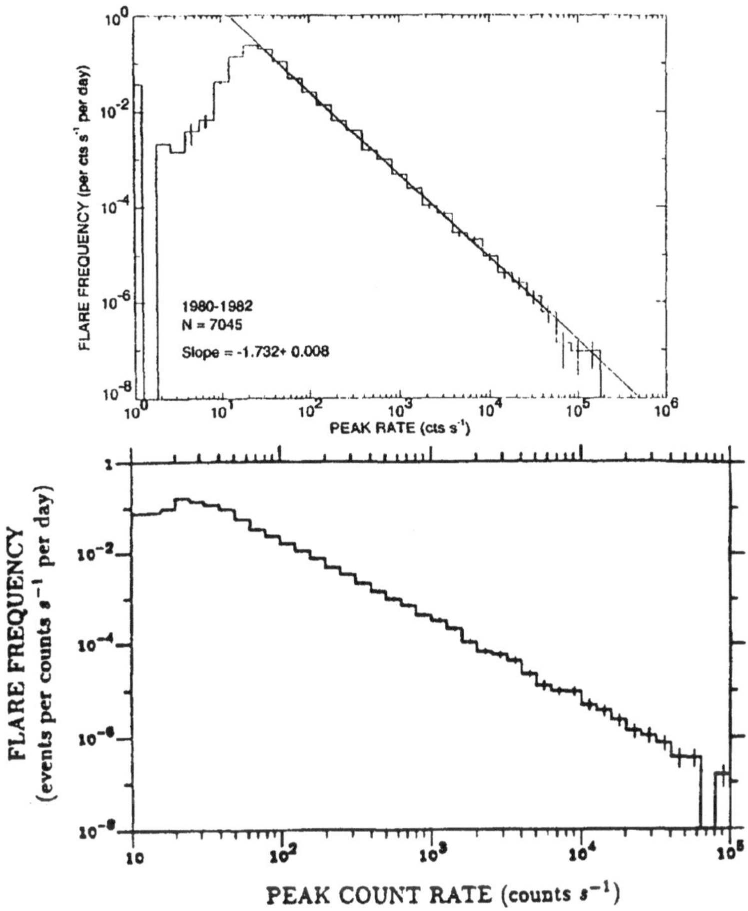

Fig. 5.14 Frequency distributions of the peak rate for 7,045 X-ray bursts (*top*) and for all 12,776 flares (*bottom*) recorded with HXRBS/SMM in 1980–1982 and 1980–1989, respectively (Crosby et al. 1993). The error bars represent $\pm 1\sigma$ uncertainties based on Poisson statistics on the number of flares in each bin. The straight line through the points above 30 count s^{-1} (*top figure*) represents the least-squares fitted power-law function with a slope of $-a = 1.732 \pm 0.008$; the turnover below 30 count s^{-1} corresponds to the HXRBS sensitivity limit. The distribution in *bottom figure* has logarithmic slope 1.8 (Figure provided by B. Dennis, see Lu and Hamilton 1991)

electromagnetic fluences and particle fluxes, Kurt (1990) has concluded that both types of emissions can be described in general by a differential power-law function with a slope of -1.45 ± 0.15. More accurate and extended analysis of all available data (Crosby et al. 1993) shows that solar flares, indeed, exhibit very similar distributions at different wavelengths, such as in radio, soft X-rays or hard X-rays. The slope of the distribution functions, however, is dependent on the flare parameter under study. Typically, the slopes are 1.7–1.8 for the peak count rate (or peak flux), 1.4–1.6 for flare energies, and about 2.0 for flare duration.

As to the proton peak flux distributions at the Earth's orbit, they turn out to be significantly flatter than those obtained for other parameters of solar flare more representative of the total flare energy. Setting the differential distribution in a power-law form, the following slopes have been reported: 1.15 ± 0.10 in the energy range of 20–80 MeV (van Hollebeke et al. 1975); 1.46 ± 0.15 at >10 MeV (Belovsky and Ochelkov 1979) and at >25 MeV (Kurt 1989); 1.13 ± 0.04 in the range of 24–43 MeV (Cliver et al. 1991). In addition, using the integral numbers of the >10 MeV proton events per solar cycle, Smart and Shea (1997) have obtained the slopes of 1.47 and 2.42 in the intensity ranges below and above 10^3 pfu, respectively. One can see apparent distinction between the slopes of distributions for differential energy intensities and those for integral energies.

The clear differences between the slopes of size distributions for proton, electron and electromagnetic flare emissions were shown to be very important (*e.g.*, Miroshnichenko 1995) when interpreting an initial stage of acceleration of solar cosmic rays. More recently, in the light of a new arising paradigm of particle acceleration in different sources at/near the Sun (*e.g.*, Reames 1995a, b, c, 1996, 1999), we started an extended statistical study of solar proton events (Melendez-Venancio et al. 1998) based on more abundant SPE statistics than in previous works. Relying upon the data of several SPE Catalogues 1955–1996 (see Chap. 2) and using peak fluxes for >10 MeV protons, we separated, first of all, a group of 320 events associated with flares (flare-related events). Then, within this sample, a second group (subgroup) was formed of 159 events which have, additionally, a certain or probable sudden storm commencement (SSC) association (SSC-related, or shock associated events).

In Fig. 5.15 the size distributions for the 320 flare-related events and 159 shock-associated events (*diamonds* and *triangles*, respectively) are plotted at the threshold intensity of >1 pfu (Miroshnichenko et al. 2001). The straight lines through the data points correspond to the least-square fitted power-law function (5.19). A differential plot for all 320 events is consistent with a single slope of 1.37 ± 0.05 over entire range of the proton intensities. Such a conclusion evidently contradicts the result by Smart and Shea (1997) obtained at the threshold intensity >10 pfu. Unlike the plot for 320 basic event, the 159 shock-associated events display two power-law behaviour, with the slopes of 1.00 ± 0.04 and 1.53 ± 0.05 below and above 10^3 pfu, respectively, the difference between the slopes being evidently out of the limits of approximation errors. For comparison, in Fig. 5.15 also is represented one additional plot constructed by Miroshnichenko et al. (2001) using the list of Smart and Shea (1997) for the 45 large events (>300 pfu) observed in 1965–1996. Though with different slope (2.12 ± 0.03), this plot is qualitatively consistent with a sharp break in the slope of distribution for shock-associated events at about 10^3 pfu. At present, in terms of acceleration mechanisms, there is still no possibility to separate distinctly the proton events according to their sources (flares, shocks, *etc.*), though the difference in the distributions between the flare- and SSC-associated events in Fig. 5.15 may evidence their different origins. Meanwhile, there is an obvious interplanetary effect to explain, at least, the change in

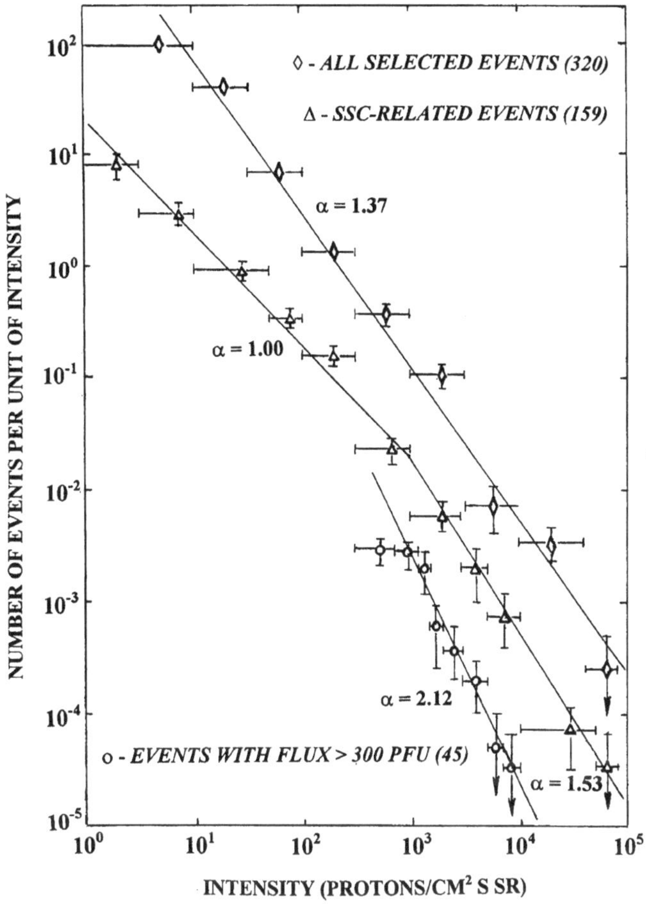

Fig. 5.15 Differential size distributions of 320 flare-related SPEs (*diamonds*) and of 159 SSC-related proton events (*triangles*) from 1955 to 1996. For a comparison, a size distribution of 45 large events of 1965–1996 (*circles*) by the list of Smart and Shea (1997) is also shown (Miroshnichenko et al. 2001)

slope at the 10^3 pfu value for shock-associated events. That is so-called "streaming-limited saturation" of SEP events (Reames and Ng 1998, see Sect. 8.5).

The results by Smart and Shea (1997) and, partly, by Miroshnichenko et al. (2001) are qualitatively consistent with those obtained by Reedy (1996) for the fluence distribution, $N(>F_s)$, of solar proton events from 1954 to 1991. The integral distribution of the number of events, N, *per year* was shown to have a form of $F_s^{-0.4}$ in the range of low fluences (up to ~10^{10} cm^{-2}) and of $F_s^{-0.9}$ at high fluences ($\geq 10^{11}$ cm^{-2}) of the >10 MeV protons. A similar tendency was found by Nymmik (1999a, b, c) for the >30 MeV protons: their fluence distribution in the solar cycles 20–22 can be described by a power-law function with exponential steepening for large fluences.

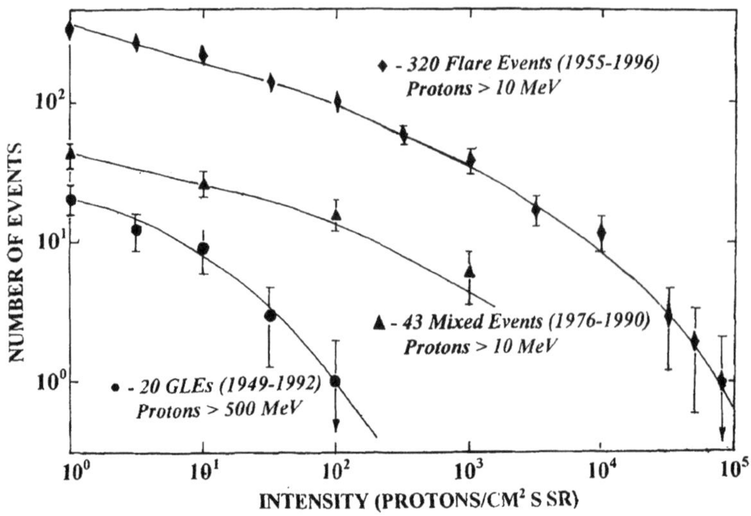

Fig. 5.16 Integral size distributions constructed by the large database (*diamonds*, 320 proton events), by the data of Kahler et al. (1991) for the >10 MeV protons (*triangles*, 43 events), and by the data from Table 4.5 for the >500 MeV protons (*circles*, 20 events) (Miroshnichenko et al. 2001)

In view of an evident distinction between the slopes for the size distributions of proton events detected in differential and integral energy ranges, it is of great interest to compare the distribution slopes at different proton energies. In Fig. 5.16 we present three integral distributions using the large database of 320 events (*diamonds*), 43 events from the paper by Kahler et al. (1991) for the ≥10 MeV proton events (*triangles*), and 20 GLEs for the ≥500 MeV protons (*circles*) from Table 4.5. Manifestly, the middle plot (43 events) is similar to the upper one (320 events), and both of them display rather smooth fall over entire range of comparable intensities between 1 pfu and 10^3 pfu. At the same time, the lower curve (20 GLEs) steeply slopes down between 1 pfu and 10^2 pfu. This may point out to a certain dependence of slope on the proton energy range under consideration.

A serious problem arises to convert from the size (frequency) distributions of solar flare events to expected threshold parameters. In particular, the distribution of partial energies amongst different emissions in a certain flare is of special interest. In the same vein, it would be important to derive the intensity dependencies of various emissions on total flare energy. In this context, it should be mentioned the following assertion (Crosby et al. 1993) (quotation):

> The frequency distributions of two flare parameters X and Y do not reveal, by themselves, whether the parameters are correlated or not. However, if a correlation exists, a one-parameter functional dependence can be derived from the slopes of the two frequency distributions. In particular, if the correlation function follows a simple power law, i.e., Y (X) ~ X^{-c}, the exponent c obeys the relation

$$c = (a - 1)/(b - 1) \tag{5.20}$$

where a and b are the power-law slopes of the X and Y frequency distributions, respectively, i.e., $N(X) \sim X^{-a}$ and $N(Y) \sim Y^{-b}$. This can be shown by substituting the function Y (X)dX = N[Y(X)](dY/dX)dX.

Proceeding from assumption (5.17) and considering the first (incoherent) data samples, Hudson (1978) pointed out that an exponent b_p, in accordance with (5.20), should be at least >4 for large flares. Meanwhile, large uncertainties in the size distribution of proton events permit an interpretation *in terms of an absolute threshold; this would lead to* $b_p \Rightarrow 1.0$ *for the smallest events.*

Such a conclusion derived from different (independent) data sets was called in question by some researchers (see Daibog et al. 1985; Kurt 1989, and references therein). These authors have constructed the size distributions on flare parameters (peak fluxes of electrons, J_e(>70 keV), soft X-rays, J_t, and hard X-rays, J_h) relying on the same (coherent) set of solar event data obtained by extensive measurements on board the spaceprobes *Venera* 13 and 14 in 1981–1983. It was found that a good correlation ($r_c > 0.8$) does exist between the energy fluences of electrons and X-ray emissions (Daibog et al. 1985). The same is true for the correlations among the five flare parameters studied by Crosby et al. (1993) on the SMM satellite data: peak rate, peak photon flux at 25 keV, peak photon flux above 25 keV, peak electron energy flux ($E_e > 25$ keV), total energy in non-thermal electrons (>25 keV), and total flare duration.

These results imply that the energy lost for electron acceleration is approximately proportional to W_f, *i.e.*, $b_e = 1.0$, and hence, $b_t = b_h = b_e = 1.0$, in accordance with the "Big Flare Syndrome" concept (Kahler 1982). Moreover, these values do not contradict the original suggestion (Hudson 1978) of an absolute threshold ($b \Rightarrow 1.0$) for accelerated protons. However, in general, based only on the frequency distributions it is difficult to derive a real threshold effect in flare data, probably excepting the possible existence of minimum solar events in the interpretation of Lu and Hamilton (1991). Indeed, they predicted the following slopes for the power-law parts of the frequency distributions: ~1.8 for the peak dissipation rate, ~1.4 for the dissipated energy, and ~2.0 for the duration of energy dissipation process. These values are in close agreement with those obtained from the flare wave data (Kurt 1990; Crosby et al. 1993) but carry almost no news in the proton threshold problem (5.17).

On the other hand, Crosby et al. (1993) have presented the results of their analysis of peak-frequency distributions for various flare-associated phenomena in the frame of a stochastic flare model of Rosner and Vaiana (1978). Following this model the ratio of average dissipated energy W_f per flare to the quiescent level W_Q may be expressed by the exponential relation

$$W_f/W_Q = \exp[1/(a - 1)] - 1 \tag{5.21}$$

i.e., by a one-parameter function of the power-law slope a. Hence, for the case of interplanetary proton events ($a_p = 1.15 \pm 0.05$) (van Hollebeke et al. 1975) it was found a ratio of $W_f / W_Q > 200$. This may suggest that these events are preferentially associated with very energetic flares. For electron events this ratio is one order less, for instance, a value of $a_e = 1.3$ for 3.6–18.5 MeV electrons (Cliver et al. 1991) corresponds to $W_f / W_Q = 20$.

Such a separation of solar energetic events seems to be more adequate from the point of view of flare physics. In any case, it would be reasonable to expect that the energy released in the form of accelerated particles is not bound to obey to linear dependence of total flare energy, as stated by Kuznetsov and Kurt (1991). Their conclusion that protons with the energy $E_p < 20$ MeV account for a proportionate fraction of the total flare energy budget was not confirmed by recent findings of Cliver et al. (1991). In any case, the flatter size distribution found by Cliver et al. (1991) for >20 MeV protons negates the argument that similar size distributions for flare electromagnetic and proton emissions imply a single class of flares. Besides, if taking the values of $a_e = a_p = 1.4$ (Kuznetsov and Kurt 1991), then from (5.20) it follows that the exponent b becomes about 1.0, this value being in accordance with the threshold effect in Hudson's formulation.

In the light of these discrepancies, it is quite appropriate to apply to the possible independent sets of data. One of them could be the source proton spectra reconstructed by different techniques to the moment of particle acceleration or their ejection from the Sun. Recently, existing data for 80 SPEs were compiled by Miroshnichenko et al. (1999). In search for the possible threshold effects, the source spectra should be treated within a frame of a certain acceleration model. In such a case there is a possibility to relate a total number of accelerated particle, $N_s(R)$, to a set of source parameters, in particular, to the source power W_f. Using the source spectrum data, within a simple acceleration model it was estimated (Miroshnichenko 1995) that

$$N_s(R), \tilde{\ } Wf^{0.5-3.0} \tag{5.22}$$

the Eq. (5.22) being valid, at least, for the proton rigidities $R \geq 1$ GV ($E_p \geq 500$ MeV). The interval of $b_p = 0.5$–3.0 is determined by the admitted range of power-law exponents in the source rigidity spectra ~$R^{-\gamma}$. This might be treated as an evidence for a specific threshold effect $b_p > 0$ for the protons overcoming the Coulomb loss maximum. Although the dependence (5.22) differs from that derived by Hudson (1978) the general tendency of b_p increasing with flare energy W_f seems to remain. In other words, our estimates of b_p corroborate the concept of "Big Flare Syndrome" (Kahler 1982) as well.

Thus, based on existing statistical and semi-empirical findings we have tried to scrutinize the f flare threshold problem in different formulations (minimum flare energy or time scale, peak rate of energy release, peak flux dependence on total flare energy for accelerated particles, *etc*). From our point of view, if considering the problem in terms of flare distribution functions on various parameters, the previous approaches may provide some tentative estimations of minimum flare energy, duration and spatial dimension. However, this way seems to be still deficient in explanation of possible threshold effects in the behaviour of energetic flare particles. Therefore, it was suggested (Miroshnichenko 1995) to treat the problem in terms of particle acceleration and flare energetics, provided some physical conditions can be taken into account: (1) absence of injection threshold for acceleration of electrons from the tail of thermal distribution; (2) existence of Coulomb barrier for acceleration of protons; (3) inevitable differentiation between proton and electron spectra nearly from the very beginning of the acceleration process. The last topic is of fundamental interest for flare physics and acceleration theory (see Sect. 5.9).

Judging from incessant and hard discussions throughout recent decades, all those findings were recognized to be very helpful for the resolution of some problems related to flare modeling (*e.g.*, Rosner and Vaiana 1978; Lu and Hamilton 1991; Litvinenko 1996b, 1998; Wheatland and Sturrock 1996; Wheatland and Glukhov 1998) and particle acceleration (*e.g.*, Hudson 1978; Miroshnichenko 1995, 2001; Litvinenko 1996a, b; Aschwanden et al. 1998a, b). In particular, it has been found (Crosby et al. 1993) that the frequency distributions of various solar flare phenomena show a power-law shape consistent with the stochastic model of Rosner and Vaiana (1978), suggesting that the flare energy build-up is governed by exponential growth. The measured distributions of flares are also consistent with those predicted by computer simulations of avalanche models (Lu and Hamilton 1991) that are governed by the principle of self-organized criticality (SOC).

On the other hand, in the development of the avalanche model of solar flares, Wheatland and Sturrock (1996) suggested to take into account the finite size of the active regions and then compared their model to the distribution of hard X-ray bursts observed by the *ICE* spaceprobe. Later on, this work has been modified by Wheatland and Glukhov (1998) to include a growth rate of free energy in active regions. The energy release through magnetic reconnection in multiple current sheets is used by Litvinenko (1996b) as an alternative suggestion to the avalanche model for flares (Lu and Hamilton 1991). Notably that a power-law flare distribution with the slope of 1.5 can be deduced only from scaling law arguments as it follows from dimensional analysis by Litvinenko (1998).

A new interesting application of the data on size (frequency) flare distributions seems to arise in connection with a giant flare on June 1, 1991 (Kane et al. 1995). In particular, Dennis (1996) did not exclude a cutoff for the largest flares (see also Sect. 4.6). Occurrence of the flare of June 1, 1991 suggests that the size distribution might extend to even more powerful flares than had previously been suspected, perhaps so large, in fact, that a single active region could not have provided all of the energy (Kane et al. 1995). At the same time, as noted by Dennis (1996), the large energy estimate for this flare should be taken with some precaution because of significant saturation effect during the observations by Kane et al. (1995). There is other evidence relating to a possible end or high-energy cutoff in the flare size distribution (Kucera et al. 1997). These authors have plotted the peak counting rates of the X-ray flares recorded with the HXRBS/SMM as a function of the size of the sunspots in the active regions from which they originated. It was found evidence for a cutoff in the size distribution of flares from active regions that have the sunspots with areas of <500 millionth of the visible hemisphere. Taking this result at face value, according to Dennis (1996), one can assume that an active region does have a maximum energy that it can release during a flare as would be predicted by the avalanche model (Lu et al. 1993).

On the other hand, Aschwanden et al. (1998a, b) explored elementary time scales in solar flares by the wavelet analysis and logistic models. They applied a multi-resolution analysis (using a triangle-shaped wavelet transforms) to 647 solar flares observed with the Compton Gamma Ray Observatory (CGRO) at hard X-ray energies >25 keV with a time resolution of 64 ms. They inferred a distribution of time scales $N(T)$ for each flare and found a cutoff for the shortest detected time scales T_{min} that is independent of the Poisson noise for strong flares. These shortest times are found to correlate with the flare loop radius r_l (measured with *Yohkoh*) and the electron density n_e (determined by the trapping time measured from energy-dependent time delays with CGRO). From this analysis, *spatial sizes* of 60–600 km have been estimated for elementary acceleration cells.

It was also determined the frequency distribution of elementary time structures in over 600 flares, based on some 10^4 hard X-ray pulses at 25 and 50 keV, 4,000 radio type III bursts, 4,000 decimetric quasi-periodic broadband pulsation events, and 10^4 decimetric millisecond spike events. All elementary time structures are shown to have a quasi-Gaussian shape and can be modeled with the *logistic equation*, which describes the exponential growth phase and nonlinear saturation (caused by the limited amount of available free energy) of a general instability. Aschwanden et al. (1998a, b) have derived a theoretical description of frequency distributions in terms of this *logistic avalanche model*. They conclude that the power-law slope of observed distributions provides a powerful diagnostic on coherent versus incoherent instabilities.

An important diagnostic parameter is a ratio of the number of electrons to the number of protons (*e*/*p* ratio) at the same energy (*e.g.*, Ramaty and Murphy 1987). In interplanetary space, the ratio *e*/*p* for escaping particles can be directly measured; in the interaction region at the Sun one can deduce this ratio for interacting particles by comparing the bremsstrahlung continuum with emissions resulting from ion interactions, *e.g.*, 4–7 MeV nuclear line emission, neutron fluxes, and pion-decay emission. As noted in Ramaty end Murphy (1987), the *e*/*p* ratio of the escaping particles at energies greater than several MeV tends to be higher for flares which are observed to emit gamma rays than for flares which only produce interplanetary particles (Evenson et al. 1984; Cane et al. 1986). All events with $e/p > 3.5 \times 10^{-3}$ ("electron rich") were from flares between W12° and W90°, so good magnetic field connection from the spacecraft to the source is clearly important for the electrons (remind that for a solar wind speed of ~400 km s^{-1} the Earth is magnetically connected to about W50° solar longitude).

For example, for the flare of June 21, 1980, the ratio of the numbers of electrons of energies greater than 30 MeV to the numbers of protons above the same energy was $\sim 1.5 \times 10^{-2}$, while the same ratio for flares from which no gamma rays were seen could be lower by many order of magnitude (Evenson et al. 1984). At the same time, for interacting particles, this ratio deduced for the same flare was ~7×10^{-4} (see Ramaty and Murphy 1987).

During the first peak of the flare of June 3, 1982, the ratio *e*/*p*(>30 MeV) was estimated to be ~1.7×10^{-3}, and during the second phase it was $<10^{-2}$. The *e*/*p* ratio in interplanetary space has only been measured at a few MeV, but the extrapolation of the electron spectrum using shock-acceleration theory (Ellison and Ramaty 1985) yields *e*/*p*(>30 MeV) ~1.3×10^{-2}. Since the *e*/*p* ratios for these two particle populations are consistent, it means that both populations were accelerated on open magnetic field lines (Ramaty and Murphy 1987). Possible anisotropy of the interacting particles, of course, can complicate the *e*/*p* calculations. The main questionable point, however, seems to be rather artificial (eventual) choice of the >30 MeV threshold for determining this ratio, this choice being only common to identify strong gamma-ray line flares with protons >30 MeV (for details see Chap. 6).

Forman et al. (1986) have discussed the correlation between 0.5 and 1.1 MeV electrons and 10 MeV protons observed in interplanetary space (Fig. 5.12). As can be seen, for large events the two populations are well correlated, but for smaller events there is an overabundance of electrons. Evenson et al. (1984) have examined the relationship between protons and relativistic electrons at nearly the same energy (~10 MeV). They found that these two particle populations are very poorly correlated. In particular, the majority of the proton events had very low ($\leq 10^{-3}$) electron-to-proton ratios at ~10 MeV. This is in contrast to the correlation seen in Fig. 5.17 where all proton events are accompanied by 0.5–1.1 MeV electrons. A few of the events, however, show larger e/p ratios, and some of them are as high as 0.2 at 10 MeV. At present, these electron enhancements may be explained in the framework of the new paradigm for SEP events (see, *e.g.*, Table 2.4, and Reames 1996).

Let us conclude this section by discussing briefly the acceleration of protons in the model of the RCS proposed by Litvinenko (1996a). According to his finding, the magnetizing longitudinal field (see Fig. 5.6) is proportional to the square root of the particle mass, being about 40 times larger for protons than for electrons. This gives interesting consequence that concerns the composition of accelerated particles. For a small longitudinal field, the Speiser's mechanism is applicable for both electrons and protons, these particle gaining the same speed in the RCS. Thus, the energy release mainly occurs in the form of protons with the energy about 0.1–1.0 MeV. Protons and electrons leave the RCS with the same speed almost parallel to the magnetic field. Thus, a neutral beam is created, traveling down the flare loops. Because the energy resides mainly in protons, they might be responsible for the chromospheric heating and flare X-ray emission. Experimental evidence and theoretical arguments in favour of the neutral beams were reviewed, *e.g.*, by Martens and Young (1990).

As the longitudinal magnetic field increases, one returns to the standard picture of acceleration, in which the electrons carry the bulk of particle energy. The model of Litvinenko (1996a), therefore, relates the properties of accelerated particles to the structure of the reconnection region. This approach may be a step toward a

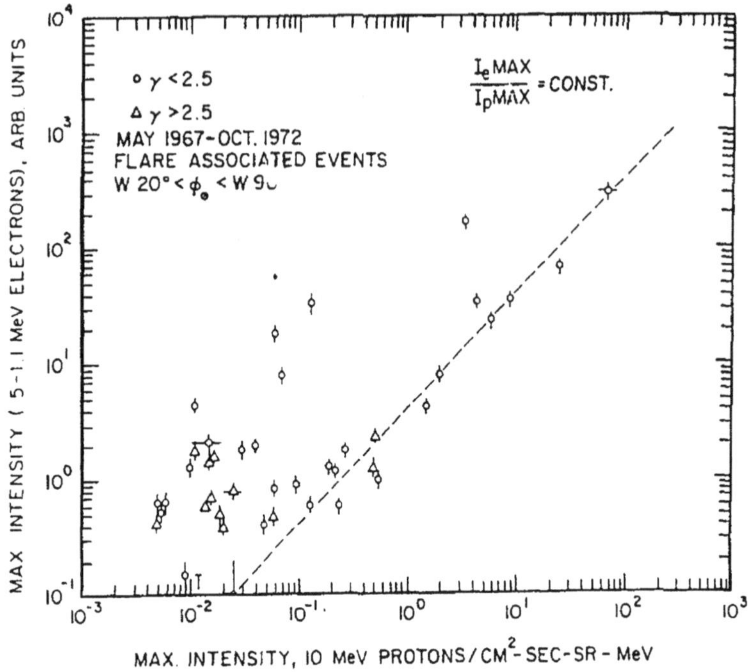

Fig. 5.17 Correlation between solar flare electron and proton intensities observed in interplanetary space (After Forman et al. 1986)

unified description of particle acceleration in flares and may resolve the existing controversy between the proton and electron beam models (for more details see Litvinenko 1996a). Though this study focused on the electron acceleration, the RCS model also allows for the generation of neutral beams with energy primarily residing in protons. The relative efficiency of the two processes was found to be determined by the magnetic field structure inside the RCS. This approach might resolve the existing controversy between the electron and proton beam models (*cf.* Simnett 1995).

Observations from the spacecraft *Yohkoh*, *Compton Gamma Ray Observatory*, *GRANAT*, *SOHO*, and re-analysis of older observations from the *Solar Maximum Mission*, have led to important new results concerning the location, timing, and efficiency of particle acceleration in flares. In particular, the review of pertinent observations and their implications (Miller et al. 1997) allowed to deduce the average rates $\partial N/\partial t$ at which particles are energized above a given energy. These rates are summarized in Table 5.4, along with the total energy content of the particles. The electron energization rates are for large flares, such as those which have detectable gamma-ray emission.

As noted by Miller et al. (1997), there is evidence that electron acceleration in impulsive flares occurs in small bursts, which have been termed “energy release fragments (ERFs)”, the accelerated electron energy content in an ERF being of

Table 5.4 Summary of typical energization rates and total energy contents (Miller et al. 1997)

	Electrons >20 keV	Electrons >20 keV	Electrons >20 keV	Proton >1 MeV	Proton >1 MeV
	ERF	Entire flare	Entire flare	Entire flare	Entire flare
Quantity[a]	Nonthermal model	Nonthermal model	Hybrid model[b]	Pre-1995[c]	Present
$\partial N/\partial t$, s^{-1}	5×10^{34}	10^{37}	2×10^{35}	3×10^{33}–2×10^{34}	10^{35}
U_p, erg	5×10^{26}	3×10^{31}	6×10^{29}	10^{29}–10^{30}	10^{31}

Notes:
ERF energy release fragments
[a]The quantities $\partial N/\partial t$ and U_p denote, respectively, the energization rate and the total energy content above either 20 keV (for electrons) or 1 MeV (for protons)
[b]$\partial N/\partial$, and U_p are taken to be a factor of ≅50 lower than those resulting from the nonthermal model. This factor is based on an application of both thermal and nonthermal models to one flare
[c]The lower limit results from stochastic acceleration proton spectra (specifically K_2 Bessel function), while upper limit results from power-law proton spectrum

10^{26}–10^{27} erg. In ERFs, the average rate of energization must be sustained for about 400 ms, while in the entire flare it must occur over several tens of seconds. In light of recent observations (see references in Miller et al. 1997), about 5×10^{34} electron s^{-1} need to be energized above 20 keV over 400 ms in order to account for an ERF. For protons, Miller et al. (1997) suggest rates and energy contents obtained by both pre-1995 and present calculations.

In conclusion of this discussion, we again return to the proton hypothesis of Simnett (1985, 1986). This hypothesis has become the focus of serious debate in literature. Numerous researchers are continuing to put forward arguments in favour and/or against the arguments of Simnett (1986) (for greater details see the reviews of de Jager 1986; Miroshnichenko 1987; Simnett 1995). For example, de Jager (1986) considers certain assumptions and proofs of Simnett (1985, 1986) to be unconvincing, but, on the other hand, he admits that observations have not yet specified the upper limit of the energy contained in fast ions; this has still to be done. Analyzing the pre-flare accumulation of energy, Hudson (1985) has drawn at the conclusion that the proton hypothesis is acceptable as far as flare energetics is concerned. As a test for its checkup could be used, in particular, high-sensitivity observations of gamma-rays generated by captured protons in a thin target. A number a key flare observations and energy arguments were debated by Simnett (1995) from the viewpoint of protons versus electrons (see above), and the conclusion was that primary non-thermal protons are much more important, in terms of total energy, than non-thermal electrons in flares, the bulk of the energetic electrons being secondary.

As one can see from Table 5.4, Miller et al. (1997) do not consider the energetics of solar particle below 1 MeV (for protons). On the other hand, these authors give several important estimates concerning the protons above 1 MeV. For the stochastic acceleration spectrum, the energy content of these protons is ~10^{29} erg, while for the power law this content is nearly 10^{30} erg. It is interesting that the energy contained in the heavier ions is roughly equal to the energy contained in the protons. The ion energy content is then more than order of magnitude lower than the energy contained in the electrons. This result has lead to the notion that energetic ions are not the main players in the overall energy budget of flares. However, note that for a flare volume of 10^{27} cm^3, the flare must still produce of order 10^2–10^3 erg cm^{-3} of accelerated protons, which is much larger than the thermal plasma energy density and still a sizable fraction of the estimated magnetic field energy density.

However, as noted by Miller et al. (1997), the conclusion that ions are energetically unimportant has changed recently. Using data by Share and Murphy (1995) from 19 gamma ray flares observed during a 9-year period with the Gamma Ray Spectrometer on *SMM*, Ramaty et al. (1995) have used the fluence ratio of the 1.63 MeV ^{20}Ne de-excitation line to the 6.13 MeV ^{16}O de-excitation line to determine energetic ion spectra. It was shown that this technique is a good diagnostic for energetic ions above about 1 MeV/nucleon. The new ratio turned out to be lower than previous estimate, and this leads to an increased number of ions at low energies (for further details see Sect. 6.4).

At the beginning of this Chapter we have already concerned the question: What is the relationship between the flare micro-processes associated with the acceleration of particles and the ambient medium where macro-processes of the MHD nature take place? For instance, what part of magnetic field energy is transmitted to fast particles? Our simple estimate (5.22) indicates an effective (though not clear completely) relationship between the number of accelerated particles and the total flare energy, or, more generally, between the acceleration model (through the spectrum parameters) and the physical conditions in flares. It calls for the new approaches to the estimates of the SCR energetics (*e.g*., Miroshnichenko 1981a, b, 1983a, b, c, 1987; Simnett 1985, 1986, 1995; Miller et al. 1997). We have realized an empirical method of estimates in Sect. 3.6.

Chapter 6
Interactions of Accelerated Particles with the Solar Atmosphere

As one can see from above considerations, in no other situation except for that during solar flares the acceleration of charged particles can be explored in such details, because (*a*) events can be studied in their temporal history and (*b*) the Sun is near enough to investigate the phenomenon in a very wide energy range from X-rays to gamma rays – two main kinds of flare neutral radiation, where the accelerated particles leave their "fingerprints" more clearly.

6.1 Accelerated Particles and Solar Neutral Radiation

The most unambiguous signature of energetic protons in the solar atmosphere comes from the variety of neutron capture line and de-excitation gamma-ray lines (GRL) produced through nuclear reactions. A schematic of nuclear reactions in the solar atmosphere is shown in Fig. 6.1. These processes have been presented in detail by Ramaty et al. (1979), and a comprehensive review was given by Ramaty and Murphy (1987). The interaction cross-sections start to become significant at proton (or ion) energies above ~10 MeV/nucleon, but it is common to identify strong gamma-ray line flares with protons >30 MeV. This in part stems from the observation of neutron capture line at 2.223 MeV; an important source of neutrons is from the break up of ^{4}He nuclei, which have a binding energy ~28 MeV. While gamma ray lines are produced whenever energetic protons are present, the energy content of the part of the spectrum >10–30 MeV is relatively small compared to the total flare energy. Some researchers (*e.g.*, Simnett 1995) believe that, in fact, gamma rays can tell us little about the presence of protons below ~10 MeV (see, however, Sect. 6.2).

The direct evidence that energetic protons are present in solar flares comes from observations of nuclear de-excitation gamma-ray lines (Ramaty and Murphy 1987). In addition, in some flares, gamma rays resulting from pion decays are observed. The pions produced predominantly by protons and alpha particles in the energy

L. Miroshnichenko, *Solar Cosmic Rays*, Astrophysics and Space Science Library 405, DOI 10.1007/978-3-319-09429-8_6

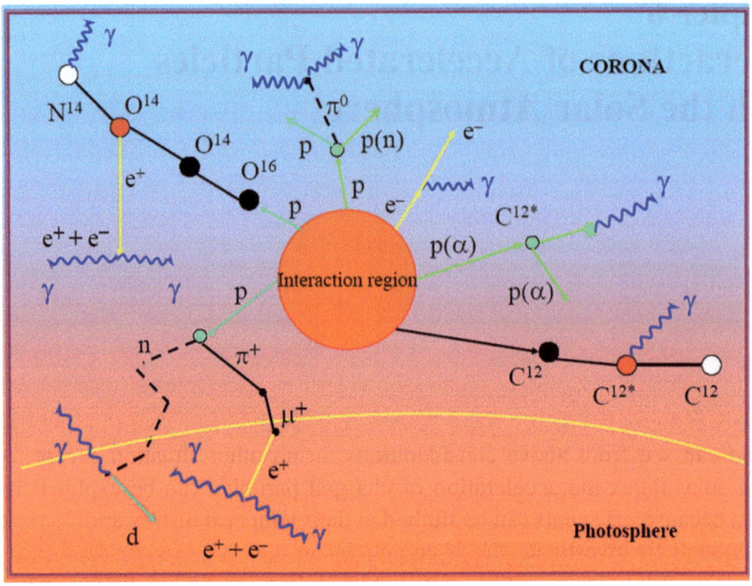

Fig. 6.1 A schematic of nuclear reactions in the solar atmosphere (Courtesy by Yu. D. Kotov, MEPhI, Moscow, 2009)

range around 1 GeV/nucleon. Figure 6.2 illustrates the time dependence of the 100 MeV and 4.1–6.4 MeV gamma-ray intensities following an intense flare of 3 June 1982. The 4.1–6.4 MeV band covers the strong gamma-ray emission lines from excited states of ^{12}C and ^{16}O. The intensity-time history of the two energy bands is quite different. Ramaty and Murphy (1987) interpreted this behaviour as indicating two different populations with different energy spectrum. It was also suggested that relativistic protons were interacting in the atmosphere for over 10 min following the flare onset.

Data on solar gamma-ray (GR) flares, included those with gamma ray lines (GRLs), recorded up to June 1982 were analyzed, in particular, by Fomichev and Chertok (1985). They considered data on 24 flares with GRLs recorded on the *SMM* and *Hinotori* satellites up to June 1982, as well as on 64 flares in the period from 14 February 1980 to 22 November 1981. The latter were accompanied by a GR continuum of >0.3 MeV, but did not have detectable emissions in GRL, according to *SMM* data. It was shown that from the point of view of radio emission, the differences between flares with and without GRLs has a purely quantitative character: The former are accompanied by the most intense microwave bursts. Meter type II bursts are not a distinctive feature of flares with GRLs. Impulsive flares, regardless of the presence or absence of GRL, are not accompanied by significant proton fluxes at the Earth's orbit. On the whole, contrary to the popular opinion in the literature, flares with GRLs do not display a deficit of proton flux in interplanetary space in comparison with similar flares without GR lines.

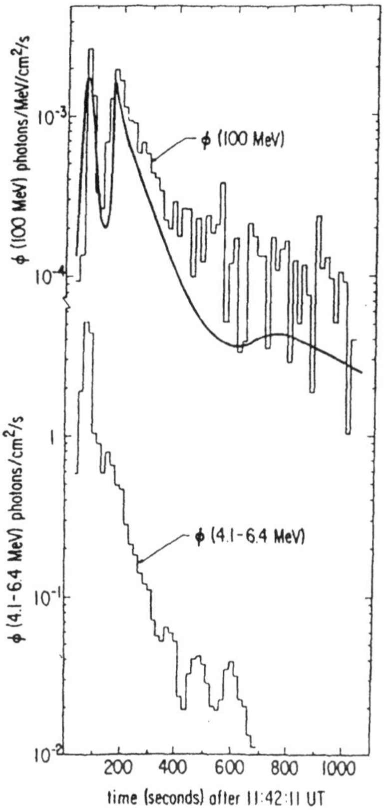

Fig. 6.2 The time dependence of the 100 MeV and 4.1–6.4 MeV photon intensities from the 3 June 1982 flare (Ramaty and Murphy 1987)

Frequently, in flares the intensity-time histories of the various energy emissions are complex (for example, in the flare of 11 June 1991), and reliable correlations in such events are difficult (Mandzhavidze et al. 1996). However, occasionally they are unambiguous. For example, in an event of 8 February 1982 bursts of photons from ~40 keV to ~40 MeV (the highest energy channel) were coincident to ~1 s (Kane et al. 1986). If the highest energy photons are from pion decay this would prove that the very fast acceleration mechanism is operating up to GeV energies, such as that discussed by Sakai and Ohsawa (1987) and Sakai et al. (1995a).

Flares with evidence of pion production, however, are rather rare. Nevertheless, on 11 and 15 June 1991 two major events were seen from GOES class X12 flares in AR 6659 (Akimov et al. 1991; Kanbach et al. 1993; Leikov et al. 1993). In particular, from the 11 June flare pion decay gamma rays were detected, for the first time, during 8 h. According to working hypothesis of Mandzhavidze and Ramaty (1992), those gamma rays were most likely caused by trapping of relativistic protons in the corona, followed by gradual loss due to pitch-angle scattering to the chromosphere. Trapping for up to 8 h is required as gamma-ray flux took this time to return to background levels (Kanbach et al. 1993). This point is important in connection with the availability of seed particles for input to proton acceleration mechanisms. However, some later, Ramaty and Mandzhavidze (1994b) and Mandzhavidze et al. (1996) re-examined their preliminary conclusions (see below, Sect. 6.5).

Results from the *SMM* showed that gamma-ray flares were not particularly rare. From a total of 258 flares of different importance (C, M, and X classes) from 1980 to 1989 listed by Vestrand et al. (1999), 185 events were *SMM*/GRS bremsstrahlung bursts detected above 300 keV with sufficient statistics for spectral analysis and with good background subtraction; 72 event presented gamma-ray emission, including 22 distinctive bursts with >10 MeV photons. From them, 67 gamma-ray (GR) bursts registered by *SMM*/GRS exhibited a "narrow" gamma-ray line (GRL) component: 10 narrow lines, corresponding to nuclear de-excitation of heavy elements in the energy range 1–10 MeV, positron-electron annihilation line at 0.511 MeV, and neutron capture line at 2.223 MeV.

Of course, the total *SMM* statistics of the GR events looks rather poor and not very impressive, especially in comparison, for example, with the statistics of the soft X-ray events (Crosby et al. 1993) and/or the SPEs (*e.g.,* Sladkova et al. 1998; NOAA SESC 2014). Nevertheless, complex analysis of available GR and SPE data for the same period of observations allowed to obtain a set of interesting results (Perez-Enriquez and Miroshnichenko 1999). They analyzed 67 GRL events, whose fluences are given in the Atlas by Vestrand et al. (1999), together with the data on the bremsstrahlung bursts and energetic solar particles in interplanetary space. To select SPEs, it was used a standard threshold intensity >1 *pfu* at the peak time of the event and a threshold energy >10 MeV (*e.g.*, Bazilevskaya et al. 1990a; Sladkova et al. 1998). It was found, in particular, that 74 flares from 258 were accompanied by SPEs near the Earth's orbit; 39 events from 67 GRL bursts (*i.e.*, about 58 %) turned out to be related to SPEs. It is notable that, with this standard criterion, 17 GR events from 19 analyzed by Share and Murphy (1995), *i.e.*, about 90 %, were accompanied by SPEs.

Perez-Enriquez and Miroshnichenko (1999) also used data on 110 high-energy solar flares observed in X-ray and gamma-ray ranges (100 keV–100 MeV) by the PHEBUS device on board the orbital space *GRANAT* Observatory during 1990–1995 (Terekhov et al. 1996). In 10 % of the observed events this device detected photons with energy above 10 MeV, and 4 flares emitted photons above 100 MeV. In particular, a giant behind-the-limb flare of June 1, 1991 (112°E, *GOES* class >X12), probably the largest flare of the 22nd cycle of solar activity (Kane et al. 1995), was observed with the large flux of >10 MeV photons. At least 19 flares out of 110 can be identified with subsequent SPEs (NOAA SESC 2014.; Sladkova et al. 1998). In some flares, the installation of the *GRANAT* Observatory also recorded photons of neutron capture line at 2.223 MeV (in particular, after the flare of 22 March 1991).

According to Rieger (1996), the flare of 16 December 1988 was one of the biggest gamma-ray line events recorded by the GRS on SMM. The X4.7/1B flare which occurred in NOAA AR 5278 at a heliographic position of 26°N, 37°E was very much extended in time and proceeded in several well separated bursts (Fig. 6.3). Due to this peculiarity, the spectrum evolution of accelerated particles

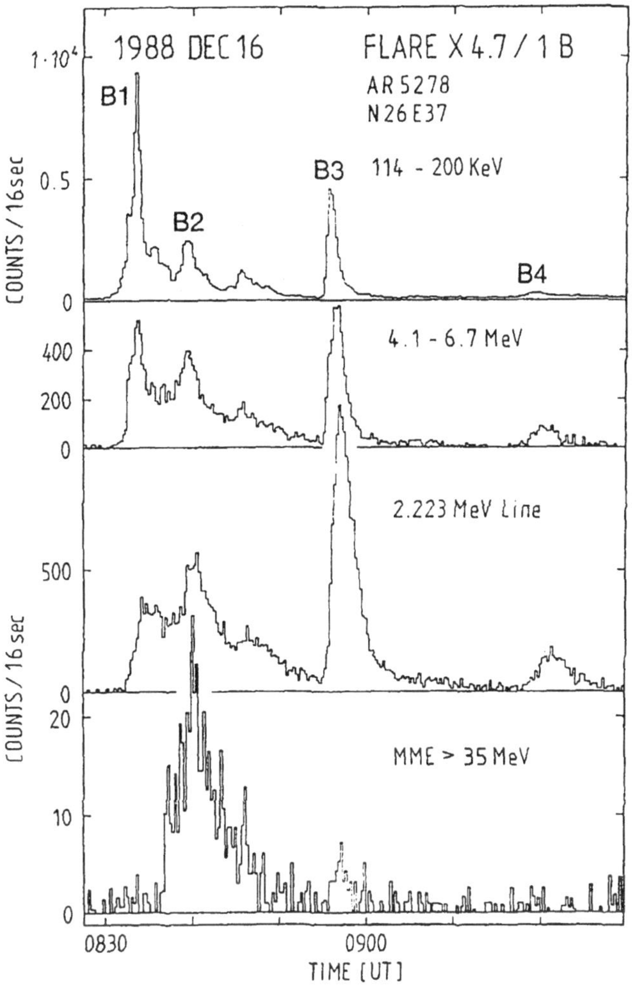

Fig. 6.3 Temporal history of the December 16, 1988 flare in different energy bands (Rieger 1996)

from burst to burst could be studied by taking the gamma-ray fluences of selected energy bands. From an inspection of panel 2 and 3 of Fig. 6.3 one can see that the spectrum of energetic particles must have hardened progressively from burst 1–3. It was shown (Rieger 1996) that the flare as a whole was very hard, but the 2.223 MeV to 4–7 MeV ratio of burst 3 (about 3.1 ± 0.2) was the highest value observed so far for a flare or a burst within a flare. It means that the parent particle spectrum must have been extremely hard.

It is not a big surprise that a solar flare extended in time exhibits spectral variations. But it is of interest to note how dramatic the changes are from burst to burst for this flare, even suggesting (Rieger 1996) the action of different acceleration mechanisms. This result once more demonstrates the importance to observe solar flares with detectors sensitive in a wide energy range. Then insight into the phenomenon of particle acceleration can be gained which is out of the reach of particle detectors in space.

6.2 Generation of Neutral Radiation

High energy solar flare emissions (gamma rays and neutrons) results from the interaction of flare accelerated particles with the ambient solar atmosphere (see Fig. 6.1). The photon and neutron productions mechanisms are by now quite well understood (*e.g.*, Ramaty and Murphy 1987). A considerable amount of research has also been carried out on the relevant particle transport processes in the solar atmosphere. New interest in these processes has been stimulated by observations of a series of six X-class flares in June 1991 with instruments on the *Compton Gamma Ray Observatory* (*CGRO*) and *GAMMA*-1.

Of special interest are the observations of GeV gamma ray emission that lasted for hours (*e.g.*, Akimov et al. 1991; Kanbach et al. 1993; Leikov et al. 1993). These observations are raising questions on the nature of the fundamental transport processes (adiabatic motion, pitch angle scattering by plasma turbulence, drifts) as well as on the structure of the coronal magnetic field. In addition, the possibility of particle acceleration to GeV energies over long periods of time has also been brought up. Such acceleration should take place under markedly different physical conditions than the acceleration of the ions responsible for the gamma ray emission observed during the impulsive phase of flares.

The photon and neutron production processes have been recently reviewed in detail by Ramaty and Mandzhavidze (1994a). The principal mechanisms that produce high energy photons and neutrons in solar flares are summarized in Table 6.1.

6.2.1 Bremsstrahlung

Interactions of the accelerated electrons with ambient gas in the flare region produce a continuum of X- and gamma rays via non-thermal bremsstrahlung. This continuum extends from about 20 keV to over 100 MeV. At the low energy it merges into the thermal bremsstrahlung produced by hot flare plasma. There is no known high energy cutoff; the highest energy observed bremsstrahlung is around several hundreds of MeV.

The bremsstrahlung yield in ionized gas is generally lower than that produced in a neutral gas because of the higher rate of energy loss in the ionized gas. Ramaty et al. (1993) have used an isotropic bremsstrahlung model to fit the observed 0.3–1 MeV continuum spectra of 10 flares and 6 individual emission episodes during the 6 March 1989 flare. Although the angular distribution of the electrons could be

Table 6.1 High energy photon and neutron production mechanisms (Ramaty and Mandzhavidze 1994a)

Emissions	Processes	Observed photons or neutrons	Primary ion or electron energy
Continuum	Primary electron bremsstrahlung	20 keV–1 MeV	20 keV–1 GeV
		>10 MeV	
Nuclear de-excitation lines	Acceleration ion interactions, e.g.,	Lines e.g.,	1–100 MeV/nucl.
	$^{4}He(\alpha, n)^{7}Be^{*}$	0.429 MeV	
	$^{4}He(\alpha, p)^{7}Li^{*}$	0.478 MeV	
	$^{20}Ne(p, p)\ ^{20}Ne^{*}$	1.634 MeV	
	$^{12}C(p, p)\ ^{12}C^{*}$	4.438 MeV	
	$^{16}O(p, p)\ ^{16}O^{*}$	6.129 MeV	
Neutron capture line	Neutron production by accelerated ions followed by $^{1}H(n, \gamma)^{2}H$	Line at 2.223 MeV	1–100 MeV/nucl.
Positron annihilation radiation	B^{+} emitter or π^{+}	Line at 0.511 MeV	1–100 MeV/nucl.
	Production by accelerated ions, e.g.		
	$^{12}C(p, pn)^{11}C \rightarrow {}^{11}Be + e^{+} + n$		
	$p + p \rightarrow \pi^{+} \ldots, \pi^{+} \rightarrow \mu^{+} \rightarrow e^{+}$		
	followed by $e^{+} + e^{-} \rightarrow 2\gamma$	Ps – positronium	
	$e^{+} + e^{-} \rightarrow Ps + h\nu$		
	or $e^{+} + {}^{1}H^{-} \rightarrow Ps + p$		
	$Ps \rightarrow 2\gamma, 3\gamma$		
Pion decay radiation	π^{0} and π^{+} Production by	10 MeV–3 GeV	0.2–5.0 GeV
	Accelerated particles, e.g.,		
	$p + p \rightarrow \pi^{0}, \pi^{\pm} \ldots$		
	followed by $\pi^{0} \rightarrow 2\gamma$,		
	$\pi^{\pm} \rightarrow \mu^{\pm} \rightarrow e^{\pm}$		
	$e^{+} \rightarrow \gamma$(brem), γ(ann. in flight), $e^{-} \rightarrow \gamma$(brem)		
Neutrons	Accelerated particle interactions, e.g.,		
	$^{4}He(p, pn)^{3}He$	Neutrons in space (10–500 MeV)	10 MeV–1 GeV
	$p + p \rightarrow \pi + n + \ldots$	Neutron induced atmospheric cascades (0.1–10 GeV)	0.1–10 GeV
	$^{22}Ne(\alpha, n)^{25}Mg$	Neutron decay protons in space (20–200 MeV)	20–400 MeV

anisotropic, the use of isotropic model is justified since in this energy range the bremsstrahlung angular pattern is not strongly beamed and Coulomb collisions will nearly isotropize the electrons.

The results were combined with data on nuclear line emission, and the ratio of the electron flux at 0.5 MeV to the proton flux at 10 MeV was derived. This ratio was extensively studied earlier for solar flare particles observed in interplanetary space (Kallenrode et al. 1992). For these interplanetary particles, on the average, the 0.5 MeV electron to 10 MeV proton flux ratio is much larger for impulsive flares (in which particles are thought to be accelerated from hot flare plasma near the site of flare energy release) than for gradual flares (in which particles are accelerated from cooler coronal gas). The gamma ray results, pertaining to the particles which interact at the Sun, reveal an even higher electron to proton ratio, regardless of whether the flare is impulsive or gradual. This result suggests that the particles responsible for gamma ray production and the particles observed in interplanetary space from impulsive flares are probably accelerated by the same mechanism. It is argued that this mechanism is stochastic acceleration due to gyroresonant interactions with plasma waves.

For many flares, the gamma ray spectrum between about 1–8 MeV is dominated by nuclear line emission. Above 10 MeV bremsstrahlung can become important again. There are, however, only two flares (21 June 1980 and 3 June 1982) for which there are published data on the continuum below 1 MeV and continuum above 10 MeV extending to around 100 MeV (Ramaty and Mandzhavidze 1994a).

6.2.2 Line Emission

Nuclear de-excitation lines result from the bombardment of ambient C and heavier nuclei by accelerated protons and α particles, and from the inverse reactions in which ambient hydrogen and helium are bombarded by accelerated carbon and heavier nuclei (Ramaty et al. 1979). Because of their low relative abundances, interactions between accelerated and ambient heavy nuclei are not particularly important. Furthermore, since H and He have no bound excited states, p-p and p-He interactions can also be ignored. However, interactions of α particles with ambient He (hereafter $\alpha\alpha$ interactions) produce two strong lines, at 0.478 MeV from ^{7}Li and at 0.429 MeV from ^{7}Be. As the shape of the spectral feature resulting from the superposition of these $\alpha\alpha$ lines is strongly dependent on the angular distribution of the interacting a particles, measurements with good spectral resolution in the energy range 0.4–0.5 MeV could turn out to be particularly useful in the study of the anisotropy of the interacting particles.

The gamma-ray lines from ^{7}Be and ^{7}Li produced when flare-accelerated alpha particles interact with ambient He (in particular, ~0.45 MeV line, see Ramaty and Mandzhavidze 1998) have been found to be surprisingly intense from measurements made by gamma-ray spectrometers on the *SMM* and *COMPTON* satellites (Share and Murphy 1997; Murphy et al. 1997; Share and Murphy 1998; Ramaty and Mandzhavidze 1998). These high intensities suggest either accelerated α/p ratios >0.5 and/or a He/H abundance >0.1 in the sub-coronal regions where the particles interact (Share and Murphy 1998). In this context, it should be noted that

Mandzhavidze et al. (1997) outlined how to distinguish between the two possibilities by resolving and comparing intensities of additional line at 0.339, 1.00, 1.05, and 1.19 MeV produced by interactions of α-particles on Fe with the intensity of the 0.847 MeV line produced by proton interactions on Fe. They also note that the ^{16}O (^{3}He, p)^{18}F reaction produces lines at 0.937, 1.040, and 1.080 MeV. These lines complicate the analysis but offer the possibility of making an in-situ measurement of accelerated ^{3}He in flares. Preliminary analysis of Share and Murphy (1998) suggest, in particular, that the accelerated $^{3}He/^{4}He$ ratio is significantly less than unity in most flares observed by SMM (see Vestrand et al. 1999).

6.2.3 The 2.223 MeV Gamma-Ray Line

This very important line, as shown in Table 6.1, is produced in solar flares by neutron capture on hydrogen. High energy neutrons in the solar atmosphere are slowed down by elastic scattering. When they reach the thermal energies they are captured. Therefore, the line of 2.223 MeV is delayed by the thermalization and capture of the neutrons. The study of neutron transport and neutron capture in the solar atmosphere generally requires the use of Monte Carlo methods due to the complex nuclear interactions of high energy neutrons with the ambient solar material. However, recent observations of solar flares (see, *e.g.,* Young and Ryan 1997, and references therein) suggest that the 2.2 line can be produced by lower energy neutrons (below 10 MeV). The distribution of these lower energy neutrons can be calculated using analytical methods. Young and Ryan (1997, 1998) first present a rigorous solution of the Boltzmann equation describing neutron transport in the solar atmosphere. They compute the resulting form of the 2.223 MeV photon flux as a function of heliocentric angle. Because some of the flares registered by the COMPTEL instrument on board *CGRO* have a paucity of 4–7 MeV emission, a spectral index from the 2.223/4–7 flux ratio would have large errors. Young and Ryan (1997) believe that the study of the 2.223 MeV line from *low energy* neutrons may enable the development of a separate measure of the spectral index.

6.2.4 Pion Decay Radiation

In the energy range above 10 MeV, along with the bremsstrahlung from primary electrons, there can also be a significant contribution from pion decay radiation. The theory of pion decay was treated in detail, and Mandzhavidze and Ramaty (1993) have reviewed the observations.

6.2.5 Positrons

Positrons in solar flares result from the decay of radioactive nuclei and charged pions. The contribution from radioactive nuclei is closely related to 4–7 MeV nuclear de-excitation emission. The ratio of this positron production to the 4–7 MeV photon production is not strongly dependent on the ion spectrum and composition. In addition to the positron production, the 0.511 MeV line flux also depends on the fraction of the positrons which annihilate via positronium and possible attenuation of the 0.511 MeV line in the solar atmosphere.

The 0.511 MeV line is delayed by the finite lifetime of the parent radioactive nuclei and by the subsequent slowing down and annihilation of the positrons. Data on 0.511 MeV line emission are available for a few flares (4 and 7 August 1972, 21 June 1980, 1 July 1980, 27 April 1981, 3 June 1982). For the 21 June 1980 flare it was shown that the observed 511 keV line flux is consistent with that expected to accompany the observed 4–7 MeV nuclear de-excitation emission. The bulk of the positrons responsible for the 511 keV line emission in this flare resulted from the decay of radioactive positron emitters. On the other hand, in the 3 June 1982 flare, the 0.511 MeV line emission resulted from positrons from both charged pions and radioactive positron emitters. The 3 June 1982 flare is the only one for which simultaneous pion decay emission and 0.511 MeV line observations were reported.

An excellent measure of the atmospheric density is provided by the $3\gamma/2\gamma$ ratio from positron-electron annihilation. The 3γ continuum comes from annihilation of the triplet state of positronium, while the 0.511 MeV line comes from either free annihilation or from the singlet state. The triplet state is depleted at hydrogen densities $>10^{14}$ cm^{-3} (Share et al. 1996).

6.2.6 Abundances of Ambient Gas and Accelerated Particles

As it was discussed in detail earlier (Ramaty and Murphy 1987), solar gamma rays and neutrons result from the interaction of accelerated ions and relativistic electrons with the ambient solar atmosphere. Since August 1972, satellite observations by hard X-ray and gamma-ray spectrometers provided an avalanche of new information on solar flares. In particular, abundant data have been obtained with the SMM/GRS on the long-duration gamma-ray flare of 27 April 1981 (Murphy et al. 1990, 1991). Figure 6.4 shows the observed gamma-ray spectrum of this flare fitted with the calculated spectrum for the case with the accelerated α/p ratio $[^4\mathrm{He}/^1\mathrm{H}]_{acc} = 0.5$ (Murphy et al. 1991). It is seen that most of the lines are due to the de-excitation of the abundant heavy elements – C, N, O, Ne, Mg, Si, and Fe. Remind that the effective energies of the particles producing this emission are between 10 and 30 MeV/nucleon, where the nuclear excitation cross-sections have pronounced peaks. Also, in this spectrum there are two strong delayed gamma-ray lines – at 2.223 MeV from neutron capture by photospheric hydrogen, and at

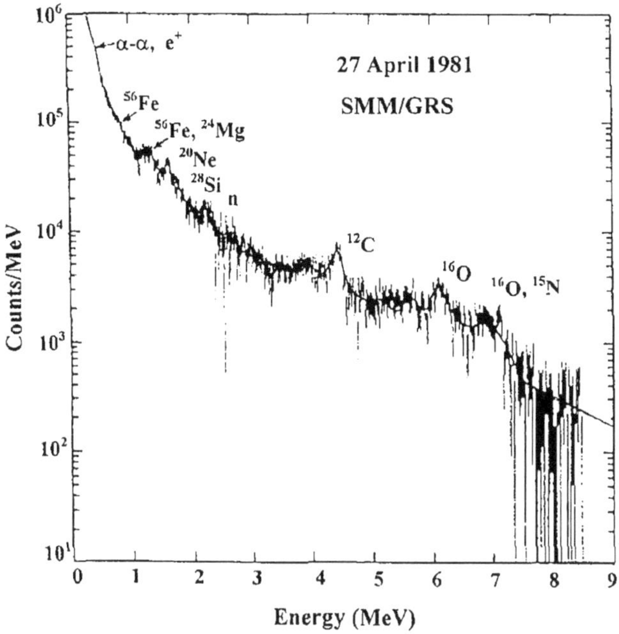

Fig. 6.4 Observed gamma-ray spectrum of the 27 April 1981 flare fitted with the calculated one for the case with the accelerated α/p ratio $[^4He/^1H]_{acc} = 0.5$ (Murphy et al. 1991)

0.511 MeV from positron annihilation. Line emission is superposed on a bremsstrahlung continuum produced by primary electrons. This continuum dominates below 1 MeV and in most cases above 10 MeV (*e.g.*, Mandzhavidze and Ramaty 1993; Chupp 1996).

The observed gamma ray spectrum of the 27 April 1981 flare has been used to derive abundances of both ambient gas and the accelerated particles (Murphy et al. 1991). The derived accelerated particle abundances indicate a very significant enhancement of heavy element abundances, similar to the heavy element enhancement observed in interplanetary particles from impulsive flares (Reames 1990a, b). This supports the conclusion mentioned above that the particle responsible for gamma ray production and the particles observed in interplanetary space from impulsive flares have a common origin.

The derived ambient gas composition points to enhanced Ne, Mg, Si and Fe abundances relative to C or O. The enhanced Mg, Si and Fe abundances (elements with low first ionization potential, FIP) could be understood in terms of a charge dependent ambient gas transport process from the photosphere to the chromosphere and corona which favours the collisionally ionized, low FIP elements in the photosphere. The enrichment of Ne (a high FIP element) could be due to photoinization by soft X-rays. This interpretation of the Ne enhancement predicts that S should also be enhanced. Both the Ne and S enhancements have been confirmed by observations with the Flat Crystal Spectrometer on *SMM*.

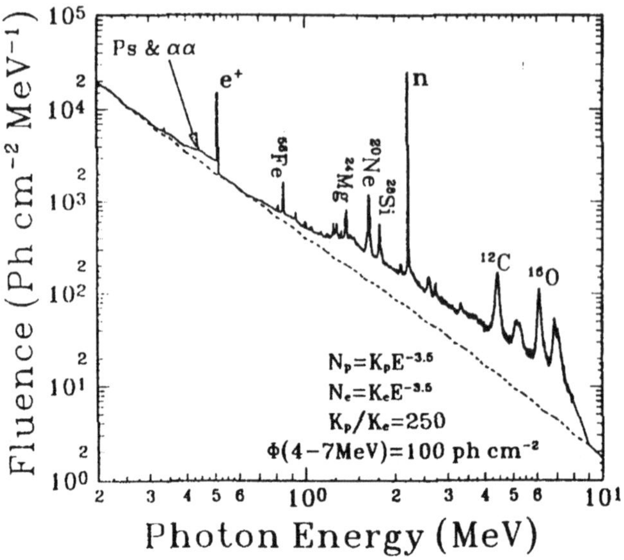

Fig. 6.5 Theoretical solar gamma ray spectrum calculated by Ramaty and Lingenfelter 1995) from the ion and electron spectra with the same spectral index. The *dotted line* is bremsstrahlung from the electrons, and the *solid line* is the total gamma ray emission. The principal nuclear de-excitation lines, the neutron capture line, the positron annihilation line, the positronium continuum (Ps), and the broad de-excitation lines from α-α reactions are marked. Positrons result from β^+ decay of radionuclides generated in reactions between incident and ambient ions

Furthermore, it is possible that the feature at about 2.26 MeV observed from the 27 April 1981 flare contained a significant contribution from the ^{32}S line at 2.230 MeV.

6.2.7 *Theoretical Spectrum*

A theoretical spectrum similar to what is needed to model a neutral emission from large flare events is shown in Fig. 6.5 (Ramaty and Lingenfelter 1995). This spectrum has been calculated using the nuclear code of Ramaty et al. (1979) with some recent updates (see also Ramaty et al. 1995), under the assumption that the ion and electron spectra incident on a thick-target emission region are power laws of the same spectral index. The narrow de-excitation lines result from the interaction of protons and alpha particles having energy between ~1 and ~100 MeV/nucleon with ambient heavier nuclei (Ramaty et al. 1979). Positrons result from β^+ decay of radionuclides generated in reactions between incident and ambient ions. Note that inverse reactions between energetic heavy nuclei and ambient H and He yield de-excitation lines of width ~1 MeV. This emission, together with many closely spaced and weak narrow lines, constitute broad unresolved features in the gamma ray spectrum. The neutrons which yield the capture line also result from reactions of ions having energies between ~1 and ~100 MeV/nucleon with the ambient nuclei. The maximum energy determined from GRL emission is thus consistent with that determined by direct inspection of the ions in space. Relying upon these findings, Miller et al. (1997) conclude that the absence of detectable GRL emission from the majority of smaller flares may be a consequence of detector sensitivity (Miller et al. 1997).

6.2.8 Energy Content in Accelerated Particles

Using data of Share and Murphy (1995) from 19 gamma ray flares observed during a 9-year period with the Gamma Ray Spectrometer on *SMM*, Ramaty et al. (1995) have applied the ratio of the 1.63 MeV ^{20}Ne de-excitation line fluence to the 6.13 MeV ^{16}O de-excitation line fluence (see Fig. 6.5) to determine energetic ion spectra. This technique relies on the fact that the lines are produced by ions of different energies: the cross section for the ^{20}Ne line becomes nonzero above ~2 MeV/nucleon and peaks around 7 MeV/nucleon, while that for the ^{16}O line becomes nonzero above ~7 MeV/nucleon and peaks around 12 MeV/nucleon. These energies are for incident protons; for incident alpha particles they are somewhat lower. The Ne line is therefore a good diagnostic tool for energetic ions above ~1 MeV/nucleon.

The observed ^{20}Ne and ^{16}O line fluences (Share and Murphy 1995) imply that the energetic ion spectra $N(E)$ are relatively steep power laws (spectral index $\gamma \sim 4$) down to ~1 MeV/nucleon, with the number of protons above 30 MeV still remaining at about 10^{32} (Ramaty and Mandzhavidze 1996). However, as a result of the steep spectra, the number of protons above 1 MeV now rises to typically 3×10^{36}. For 30 s flare duration, the rate at which protons are energized above 1 MeV is then nearly 10^{35} s^{-1}, and can rival the electron energization rate above a few tens of keV (see Table 5.4). The total ion energy content for these flares is shown in Fig. 6.6.

While there is significant scatter, one can see that a typical energy content is about 3×10^{31} erg, more than an order of magnitude above previously derived values (see Miller et al. 1997, and references therein). The protons and heavier ions each have approximately the same energy contents. In addition, the ion energy for these flares is now comparable to the typical non-thermal electron energy (~3×10^{36} erg) given in Table 5.4 and is also comparable to the energy contained in a ~1,000 G coronal field in a volume of ~10^{27} cm^3. A case-by-case comparison can also be made for 12 flares (Ramaty and Mandzhavidze 1996) for which hard X ray data are also available from the *SMM*/HXRBS. The energy contained in >20 keV electrons for these 12 flares is shown in Fig. 6.6, too.

Again, while there is significant scatter, a typical electron energy content is about 3×10^{31} erg, consistent with the findings summarized by Miller et al. (1997). Moreover, a few flares even have more ion energy than electron energy. Hence, at

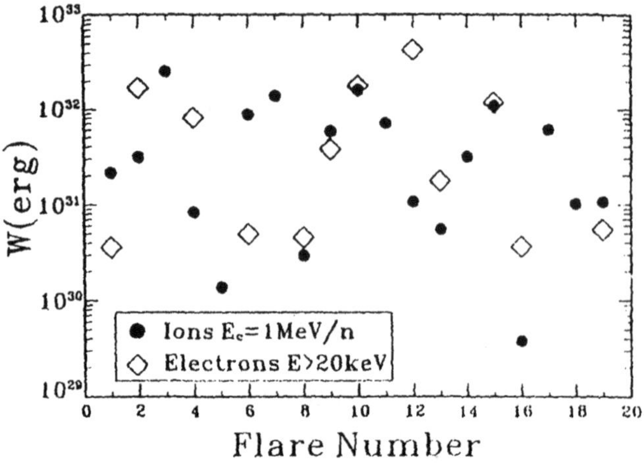

Fig. 6.6 Energy contained in >1 MeV/nucleon ions (*solid dots*) for 19 gamma ray flares observed from 1980 to 1989 (from Ramaty et al. 1995). The *diamonds* denote the energy contained in >20 keV electrons for 12 out 19 flares for which hard X ray data was also available (From Mandzhavidze and Ramaty 1996)

least for flares with detectable gamma ray emission, there is evidently a near equipartition in energy between ions and electrons, within uncertainties in the low-energy cutoffs of the ion and electron energy distributions. Such an equipartition, however, seems to be rather conventional, because at the energies of electrons and ions under consideration, their velocity ratio is $V_e/V_i \sim 6.0$, and in the light of our discussion in Sect. 5.10 (see Fig. 5.15), this prevents of any far-reaching conclusions. In other words, the results of Fig. 6.6 do not remove the problem of energy content of accelerated ions and electrons and their relative contributions to the flare energetics.

6.3 Neutron Production in Solar Flares

As well-known from observations of different kinds, higher-energy ions (above 100 MeV/nucleon)̓ are present in many flares. For example, according to Mandzhavidze and Ramaty (1993), six gamma-ray flares have exhibited a hardening or "bump" in the photon spectrum near ~70 MeV, which is due to mainly neutral pion decay radiation (see Table 6.2). This pion excess immediately indicates that protons (which dominate the pion production) were accelerated above the pion production threshold ~300 MeV. Moreover, some of these flares had pion-decay emission up to a few GeV (*e.g.*, Akimov et al. 1993; Kanbach et al. 1993), which then pushes the proton energy upper limit to a few GeV as well. Modeling of pion-decay emission for one flare (11 June 1991) indicates that a high-energy cutoff of

Table 6.2 Flares with pion decay emissions and neutrons detected in space

Date	Position	Pions	Neutrons	GLE
21 Jun 1980	W90 N20	–	*SMM*/GRS	No
03 Jun 1982	E72 S09	*SMM*/GRS	*SMM*/GRS, *ISEE*-3 (NDP)	No
24 Apr 1984	E43 S12	*SMM*/GRS	*SMM*/GRS, *ISEE*-3 (NDP)	No
16 Dec 1988	E37 N26	*SMM*/GRS	*SMM*/GRS, *ISEE*-3 (NDP)	No
06 Mar 1989	E69 N35	*SMM*/GRS	*SMM*/GRS	No
24 May 1990	W76 N36	*GRANAT*/PHEBUS	*GRANAT*/PHEBUS	Yes
04 Jun 1991	E70 N30	–	*CGRO*/OSSE	No
09 Jun 1991	E04 N34	–	*CGRO*/COMPTEL	No
11 Jun 1991	W17 N31	*CGRO*/EGRET	–	Yes
15 Jun 1991	W69 N33	*GAMMA1*/GAMMA	*CGRO*/COMPTEL	Yes
28 Oct 2003	E08 S16	*CORONAS-F*/SONG	*CORONAS-F*/SONG	Yes
20 Jan 2005	W61 N14	*CORONAS-F*/SONG	*CORONAS-F*/SONG	Yes

10 GeV in the proton spectrum is most consistent with the data (Mandzhavidze et al. 1996).

Neutrons are also a signature of very high-energy protons and are generated mostly by protons and alpha particles interacting with ambient H and He. They usually accompany pion decay radiation in the largest flares. Neutrons between ~50 and 500 MeV can be directly observed in space (*e.g.*, Chupp et al. 1982) and are in turn produced by protons with energies up to ~1 GeV (*e.g.*, Ramaty and Mandzhavidze 1994a). The very high-energy (~1 GeV) neutrons can be detected by ground-based neutron monitors (*e.g.*, Debrunner et al. 1983; Kudela 1990), and indicate the presence of protons of roughly the same energy. Hence, while the most gamma ray flares exhibit evidence for ions up to 100 MeV/nucleon, some of the largest appear capable of accelerating protons up to at least ~1 to 10 GeV.

Neutron production in solar flares has been studied in detail (see Ramaty and Mandzhavidze 1994a, and references therein). Solar flare neutrons have been observed directly with detectors on spacecraft, and indirectly with detectors on the ground. Solar flare neutrons have also been studied indirectly by observing neutron decay protons in interplanetary space (*e.g.,* Mandzhavidze and Ramaty 1993; Ramaty and Mandzhavidze 1994a; Chupp 1996, and references therein).

The bulk of the neutrons which move to downward to the photosphere are captured on H and ^{3}He in the photosphere. Capture of H produces the 2.223 MeV line. The ratio of the fluence in this line to the 4–7 MeV nuclear de-excitation fluence is used to determine the spectral index of the accelerated ions. This technique was discussed in detail by Ramaty et al. (1993). Studies of the 2.223 MeV line have also been used to determine the photospheric ^{3}He abundance.

The analysis of the measured energy spectra of >10 MeV gamma rays has so far revealed the presence of pion decay emission in only 12 flares (Kurt et al. 2010a, b). The flares for which pion decay emission and neutrons were directly detected in space are listed in Table 6.2 (Mandzhavidze and Ramaty 1993; Kurt et al. 2010b). Note that only 5 flares listed in Table 6.2, were followed by registered GLEs.

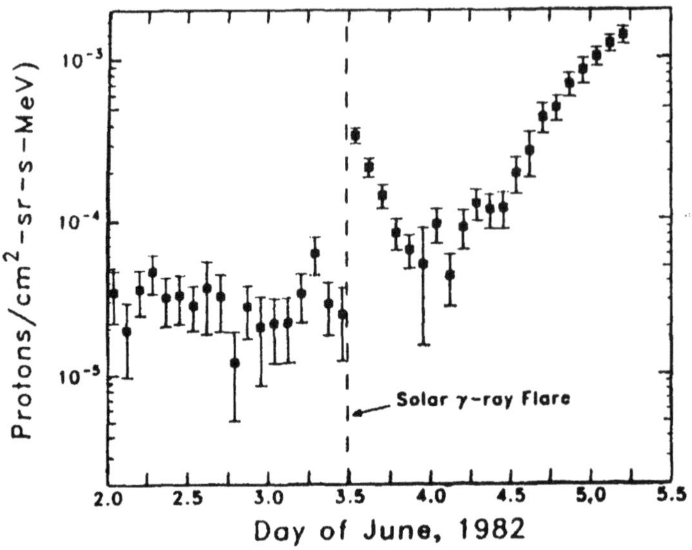

Fig. 6.7 Flux of 25–45 MeV neutron decay protons observed at *ISEE* 3 in the event of June 3, 1982 (Evenson et al. 1983). Two hour averages are plotted. Arrival of gamma rays from an intense solar flare is indicated by a *dashed line*

In five flares the observations were made with *SMM*/GRS, which is sensitive to neutrons with energies above 50 MeV. In three of this flares neutron decay protons (NDP) with energies 20–200 MeV were also detected on *ISEE* 3 (Evenson et al. 1983, 1990). This is illustrated by Fig. 6.7 that shows the flux of 25–45 MeV protons observed at *ISEE* 3 in the event of 3 June 1982. Theoretical (Monte Carlo) simulations of solar neutron decay protons were performed by Sakai et al. (1995b) using a simple slab model of the interplanetary magnetic field (IMF). It was shown that the observed time profiles of decay protons (Evenson et al. 1990) during the flares of 03 June 1982 and 24 April 1984 can equally be fitted with isotropic and pancake models of angular distribution of solar neutrons near at flare site. This means that both models give almost equal neutron emissivity at the Sun.

It is interesting to note that Dröge et al. (1995) have found evidence for additional ("forestalling") fluxes of energetic electrons in interplanetary space on board the *ISEE-3*/*ICE* spacecraft which they interpret as the decay products of neutrons generated in a solar flare of 21 June 1980. The decay electron arrived at the spacecraft shortly before the electrons from the flare and can be distinguished from the latter by their distinctive energy spectrum (Fig. 6.8). The time profile of the decay electrons is in a good agreement with the results from the a simulation based on a scattering mean free path derived from a fit to the flare electron data. The comparison with simultaneously observed decay protons (Chupp et al. 1982) allowed Dröge et al. (1995) to determine the parent neutron spectrum; the total

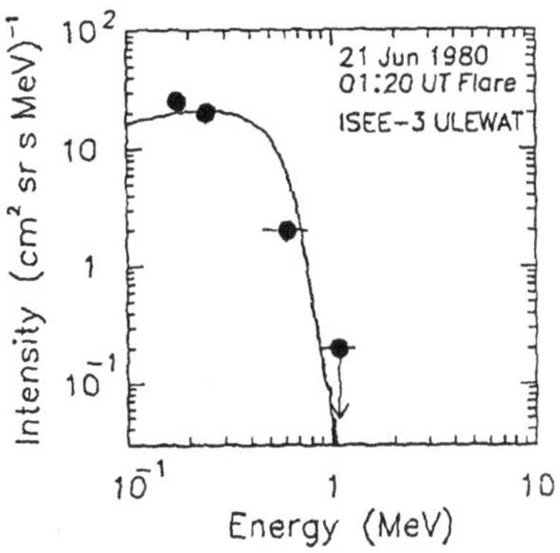

Fig. 6.8 Energy spectrum of the excess electrons observed on board *ISEE/ICE* spaceprobe on June 21, 1980 (*filled circles*) and theoretical prediction for neutron decay electrons (*solid line*) (Dröge et al. 1995)

flux and $N(E_n > 50$ MeV) are consistent with a power law index of 2.7–3.4, depending on the horizon-to zenith emission ratio.

Neutrons were distinctly detected from the 4, 9, and 15 June 1991 flares with OSSE and COMPTEL on *CGRO*. These instruments are sensitive to lower energy neutrons (15–80 MeV). It is important to note that COMPTEL is so far the only instrument in space capable of measuring neutron energy; determination of neutron energy spectra with *SMM*/GRS was based on the Sun-Earth transit time method (*e.g.*, Chupp et al. 1982; Chupp 1984, 1996). According to Mandzhavidze and Ramaty (1993), the heliocentric angle distribution of these 8 flares gives some indication of a limb brightening. However, the number of flares is still not sufficient to allow a statistically significant conclusion. Observations of high-energy emission from the June 1991 flares have been summarized by Ramaty and Mandzhavidze (1994a).

There are a number of studies dedicated to the search for solar neutrons in ground-based neutron monitor (NM) data (Usoskin et al. 1997). For example, during the 3 June 1982, simultaneous increases occurred in the count rates of three European NMs coincident in time with the gamma ray and neutron emission observed with *SMM*/GRS (*e.g.*, Kudela 1990). Positive signals were also found in Japanese and Siberian NM data from the 21 June 1980 and 24 April 1984 flares (*cf.*, however, Smart et al. 1990). Some other neutron events were identified in the analysis of the Tokyo and Siberian NM data, however, no neutrons and no >10 MeV gamma ray emissions have been detected on *SMM* during the corresponding flares.

The largest ground level neutron event was observed by nine North American NMs during the 24 May 1990 flare (*e.g.*, Shea et al. 1991a, b; Pyle et al. 1991; Belov and Livshits 1995). Another very large event, possibly associated with solar neutrons was detected at the Mt. Haleakala NM (Hawaii) on 22 March 1991 (Pyle and Simpson 1991). Both the 24 May 1990 and 22 March 1991 neutron events were coincident with the >10 MeV gamma ray emission measured with SIGMA on *GRANAT* Observatory (Pelaez et al. 1991, 1992).

Recently, a new type of solar neutron detector, a neutron telescope designed to measure neutrons directly, has been installed at the Mt. Norikura Cosmic Ray

zenith angle. The asymmetry of the angular distribution enhances the solar neutron refraction effect suggested in Smart et al. (1995). This effect makes the effective zenith angle to become smaller as the solar neutrons propagate downwards. In other words, the transport of solar neutrons through the Earth's atmosphere for arrival at inclined zenith angles is essentially different from that for vertical arrival. These calculations show (Dorman and Valdes-Galicia 1999) that the result of Shibata (1994) with respect to the expected solar neutron flux for inclined arrival should be corrected.

This learning discussion started from doubtful interpreting the 24 May 1990 event due to incorrect taking into account an angular diagram of the NM response. A source of this error lies in the studies of the 80s (see, *e.g.*, Takahashi 1989) when it was suggested to use an empirical angular dependence of the NM counting rate, I, in the form:

$$I(\alpha)\tilde{}\exp(-h/(\lambda\cos\alpha)) \tag{6.1}$$

where h is the NM atmospheric depth, α is initial zenith angle of the Sun, and λ is the attenuation length for solar neutrons in the atmosphere. Such an approach resulted in a great underestimation of the NM response at large zenith angles. According to calculations of Dorman and Pakhomov (1979) for monodirectional beams of nucleons incident at the atmospheric boundary (the case of solar neutrons), the NM response is well fitted by the relation:

$$I(\alpha)\tilde{}\cos^n(\alpha)\exp(-h/\lambda) \tag{6.2}$$

Note that in the case of galactic cosmic rays (when an incident of the nucleons at the atmospheric boundary is near isotropic) such a dependence for the monodirectional beams gives the NM response diagram obtained in Yanke (1980), Dorman and Yanke (1981).

The first increase in counting rate of the North American neutron monitors in the 24 May 1990 event was interpreted by Shea et al. (1991a, b) as a response to arrival of solar neutrons. However, they did not carry out an analysis of the NMs data in detail. Further, a great discrepancy between observed NM responses and calculated ones by the Eq. (6.1) discovered by Debrunner et al. (1992) was an argument against the neutron nature of the event. Some later, based on the calculations of Dorman and Pakhomov (1979), it was shown (Kovaltsov et al. 1993; Efimov et al. 1993a,b) that the NM data are in a good accordance with the Eq. (6.2) at $n\sim 4$ and $\lambda\sim 100$ g cm^{-2}, and the flare of 24 May 1990 acted undoubtedly as a source of high-energy neutrons. Parameter $n\sim 4$ was derived from the data analysis on the events of 3 June 1982 and 24 May 1990 (Kovaltsov et al. 1993; Efimov et al. 1993a,b).

Simultaneously, Debrunner et al. (1993) revised their position and also recognized the neutron nature of the event. This was the end of the discussion concerning the nature of the first increase in the NM counting rates during the event under consideration (for a review see, *e.g.*, Stoker 1994). Some later, after the end of the discussion, Smart et al. (1995) have published some comments on the apparent attenuation length and refraction effect. Unfortunately, their work does not contain all necessary references. Nevertheless, they emphasize that the study of the apparent attenuation length of the 24 May 1990 neutron event has led us to change fundamental concepts regarding solar neutron transport through the atmosphere.

The observations at the Earth of solar neutrons generated during powerful solar flares (in combination with X-ray, gamma-ray and other data) allows us to obtain unique information on the Sun's flare processes and particle acceleration mechanisms. The first solar neutrons were observed near the Earth by the GRS/SMM on 21 June 1980 (*e.g.*, Chupp 1996). The first simultaneous measurements of solar neutrons by space and ground based detectors were made during prominent event on 3 June 1982.

Table 6.3 gives a summary of existing data on registration of high-energy neutrons produced in solar flares in 1980–2005. Solar neutrons were registered by different detectors on board the spacecraft, by surface neutron monitors (NM) and solar neutron telescopes (SNT) at the mountains (Alma-Ata, Aragats, Chacaltaya, Gornegrat, Haleakala, Jungfraujoch, Mauna Kea, Norikura, Sierra Negra, and Tibet). In some events also pion decay emission was detected. To compile this Table, we used several papers where those data are discussed in more details (Mandzhavidze and Ramaty 1993; Flückiger et al. 2001; Matsubara et al. 2005; Watanabe et al. 2005; Sako et al. 2006, 2008).

There are also 19 bursts, the most probable candidates for registration of solar neutrons, as they were identified by the data of mountain NM Alma-Ata (Aushev et al., 1999): 24 July 1979; 7 April 1980; 9 August 1982; 4 September (two bursts), 14 September, and 2 October 1989; 17 September 1990; 1 June, 6 June, 12 June, 15 June, 17 June, 11 July, 14 July, 22 July, 5 August, and 27 October 1991; 8 May 1998.

Five solar neutron events (SNEs) were detected by the ground-based neutron monitors in association with five solar flares with deviations greater than 5σ from the background fluctuations in solar cycle 23 (Watanabe et al. 2005): 24 November 2000, 25 August 2001, 28 October, 2 and 4 November 2003 (the authors did not include yet in their analysis the event of 7 September 2005). Also, five SNEs were detected by NM s in previous solar cycles: 3 June 1982, 24 may 1990, 22 March 1991, 4 and 6 June 1991.

Using these data, the authors report on some properties of the SNEs as neutron and proton spectra, flare positions where solar neutron events were produced, and the relation between neutron flux and flare class. An extensive statistical discussion on the properties of SNEs was made. The results of this work can be summarized as follows: (1) the spectral indices of solar neutrons range between -3.0 and -4.0, the corresponding proton index is softer by about -1.0; (2) the numbers of accelerated protons are $10^{30} \sim 10^{33}$, that is 100–1,000 times more than the neutron flux; (3) there is no correlation between the longitude of solar flares and SNEs. Hence, a solar flare model must account for the acceleration of ions away from the solar surface, or produce neutrons moving away from the solar surface; (4) the class of solar flare is not the main indicator of the magnitude of the ion acceleration.

Table 6.3 Summary of solar neutron observations in 1980–2005

No.	Date	Position	Pions	Neutrons
1	21 Jun 1980	W90 N20	–	*SMM*/GRS
2	03 Jun 1982	E72 S09	*SMM*/GRS	*SMM*/GRS, Decay Protons (*ISEE*-3)
				NM, Jungfraujoch
3	24 Apr 1984	E43 S12	*SMM*/GRS	*SMM*/GRS, Decay
				Protons (*ISEE*-3)
4	16 Dec 1988	E37 N26	*SMM*/GRS	*SMM*/GRS, Decay
				Protons (*ISEE*-3)
5	06 Mar 1989	E69 N35	*SMM*/GRS	*SMM*/GRS
6	19 Oct 1989	E09 S25	–	Several NMs
7	24 May 1990	W75 N33	*GRANAT*/ SIGMA	Several NMs
8	22 Mar 1991	E28 S26	*GRANAT*/ SIGMA	NM, Haleakala
9	04 Jun 1991	E70 N30	–	*CGRO*/OSSE; SNT, MN, Norikura; EAS, Akeno
10	06 Jun 1991	E54 N32	–	NM, Norikura, Haleakala
11	09 Jun 1991	E04 N34	–	*CGR*O/COMPTEL, SNT, Norikura
12	11 Jun 1991	W17 N31	*CGRO*/EGRET	*CGRO*/COMPTEL, SNT, Norikura,
13	15 Jun 1991	W69 N33	*GAMMA*-1/ PHEBUS	*CGRO*/COMPTEL, NM, Alma Ata
14	06 Nov 1997	W43 S16	–	SNT, Chacaltaya
15	17 Aug 1998	Unknown	–	SNT, Mauna Kea
15	18 Aug 1998	E87 N33	–	SNT, Mauna Kea
16	28 Nov 1998	E32 N17	–	SNT, Tibet
17	12 Jul 2000	E27 N17	–	SNT, Gornegrat, Aragats
18	24 Nov 2000	N22 W07		NM, Chacaltaya
19	29 Mar 2001	W19 N20	–	SNT, Gornegrat
20	02 Apr 2001	W61 N17	–	SNT, Gornegrat
21	09 Apr 2001	W04 S21	–	SNT, Gornegrat
22	10 Apr 2001	W09 S23	–	SNT, Aragats

(continued)

Table 6.3 (continued)

No.	Date	Position	Pions	Neutrons
23	12 Apr 2001	W04 S21	–	SNT, Gornegrat, Aragats
24	25 Aug 2001	S17 E34	–	NM, Chacaltaya
25	28 Oct 2003	S16 E08	*CORONAS-F/ SONG*	NM, Tsumeb
26	02 Nov 2003	S14 W56		NM, Chacaltaya
27	04 Nov 2003	S19 W83		NM, Haleakala
28	07 Sep 2005	S06 E89		NM, SNT, Chacaltaya, Mexico, Sierra Negra

Notes: The neutron bursts Nos.17–24 have been registered at the level of statistical significance from 2.7σ to 4.9σ (see for details Flückiger et al. 2001)

As a brilliant illustration of the solar neutron registration we present recent data on the SNE from the giant flare of 3B/X17 (S06, E89) on 7 September 2005. The event was registered with high statistical significance's by the Solar Neutron Telescopes located at Mount Chacaltaya (Bolivia) and Mount Sierra Negra (Mexico) as well as by the neutron monitors at Chacaltaya and Mexico City (Sako et al. 2006, 2008). These observations are illustrated by Fig. 6.9. The maximum of soft X-ray burst was observed by *GOES* 11 at 17:40 UT and Type II onset at 17:42 UT. By the *GEOTAIL* satellite data, the hard X-rays (>50 keV) peaked at 17:36:40. At the same time an orbiting space laboratory *INTEGRAL* detected gamma rays in the MeV range. Because there was no clear evidence of nuclear lines, high energy radiation is considered to trace the high-energy electrons. Preliminary analysis of the data shows that a model of the impulsive neutron emission at the time of the X-ray/gamma-ray peak can explain the main peaks of all the detected neutron signals, but failed to explain the long tailed decaying phase.

Alternative model that the neutron emission follows the X-ray/gamma-ray profile also failed to explain the long tail. The results by Sako et al. (2006) indicate that the ions were accelerated at the same time with electrons but they were continuously accelerated or trapped longer than the electrons in the emission site. The authors also believe that the neutron data observed by multi-energy channels of SNTs can constrain the emission models in more detail.

6.4 Particle Acceleration and Solar Elemental Abundances

Knowledge of the cosmic abundances of elements is critical for testing theories of the early Universe, stellar and galactic formation and dynamics, and nucleosynthesis. The Sun has been one of the primary sources of information on these cosmic

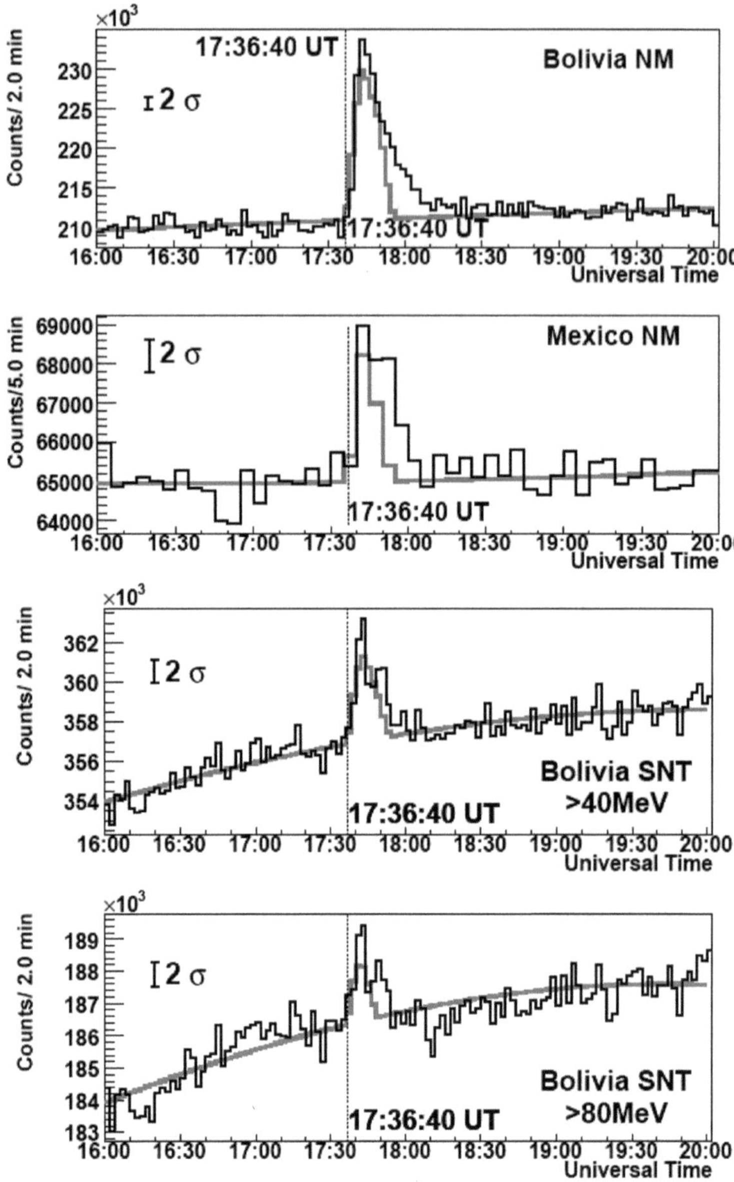

Fig. 6.9 Solar neutron burst on 7 September 2005 (Sako et al. 2006): 2-min counting rate of the Bolivia NM (first, *top panel*); 5-min data of the Mexico City NM (*second panel*); 2-min data of different channels of the Bolivia SNT (>40 MeV, third panel, and >80 MeV, *fourth panel*). The moment of 17:36:40 UT is the *GEOTAIL* hard X-ray peak time. *Grey curves* show expected counts assuming a neutron flux derived from the Bolivia NM data

abundances because its nearness has made possible detailed optical, UV and X-ray spectroscopic analyses of its atmospheric radiation. The compositions of various regions of the solar atmosphere have been studied using a variety of techniques revealing significant abundance variations (*e.g.*, Meyer 1993).

Observations of solar energetic particles and the solar wind led Meyer (1985) to conclude that elements separate based on the level of their first ionization potential (FIP). Those elements with potential exceeding ~11 eV fall into the high-FIP category (*e.g.*, C, N, O, and Ne) and those below ~10 eV fall into the low-FIP category (*e.g.*, Mg, Si, and Fe). It has been shown (Meyer 1985) that the coronal abundances of elements with low FIP are enhanced relative to those with higher FIP, as compared to photospheric abundances. Spectroscopic measurements of various regions of the solar atmosphere have also shown considerable variation in the low-FIP enhancement (Meyer 1993).

Spectral observations of the solar flare of 4 June 1991 with the Oriented Scintillation Spectrometer Experiment (OSSE) on board the *CGRO* were used in abundance study by Murphy et al. (1996). They show that ambient elements with low FIPs appear to be enhanced relative to those with higher FIP (as compared to the photosphere) similar to the enhancement found previously for the *SMM* flare of 27 April 1981. They believe that such variability could possibly be due to time-dependent composition changes at the flare site or possibly due to the flare location changing with time, progressing from deeper in the chromosphere-photosphere toward the corona.

Similarly, flare-to-flare variations in composition could reflect different heights of emission for each flare. Possible evidence for this is discussed also by Share and Murphy (1995) and Share et al. (1996). These authors reviewed measurements of ten narrow gamma-ray lines (GRL) in 19 X-class solar flares observed by the *SMM*/GRS from 1980 to 1989. It was shown, in particular, that abundances of elements in the flare plasma are grouped with respect to their FIPs, indicated that both the Ne/O and C/O line ratios are dependent on the spectral index of accelerated particles, and suggested that the range in low-FIP/high-FIP line ratios is similar to that in comparing coronal and photospheric compositions. It means suggest that ions accelerated in different flares may interact at significantly different depths. In other words, gamma rays in flares may actually be produced in regions ranging from the upper photosphere to the corona (Share et al. 1996).

Trottet et al. (1996) have performed a time dependent analysis of the gamma-ray line spectra recorded by the PHEBUS instrument on board *GRANAT* during the giant flare of 1 June 1991. From the time behaviour of the ratio $R_F = \mathrm{F}(1.1\text{–}1.8)/\mathrm{F}(4.1\text{–}7.6)$ they obtained, in particular, a continuous enhancements of the abundances of the interacting heavy ions during the flare. In their opinion, the increase of R_F with time is most likely due to the combined effect of a steepening of the spectrum of the interacting particles with time and of relative enrichment of their composition in heavy nuclei (Ne, Mg, Si, S, and Fe) as the flare progresses. If one assumes that the acceleration region is located in the corona, such a change in the composition is most likely due to the acceleration process itself. Recall that the common origin of the γ-ray producing particles and the SEPs from impulsive flares was pointed out previously on the basis of both heavy element abundances and electron-to-proton ratios (Ramaty et al. 1993).

Recent paper of Meyer (1996) is primarily devoted to the heavy element composition anomalies, which are observed, as a rule, among the particles of ~MeV energies accelerated during the rather common impulsive, He-rich, solar energetic particle (SEP) event. Contrary to gradual SEP events, in which essentially solar wind material is being accelerated far away from the Sun in the interplanetary medium, impulsive events are believed to accelerate material very close to the solar surface, in the immediate environment of a flaring loop (see, *e.g.*, Reames et al. 1994).

The paper of Meyer (1996) summarizes the composition observations; describes the inferences on the source gas temperature that can be derived from these observations in a quasi-model-independent way; very briefly overviews the proposed models for selective heavy ion acceleration by plasma waves; presents and discusses a specific model for selective heavy ion acceleration in terms of wave damping by ^{4}He ions.

As a summary of all these points, Meyer (1996) notes, first of all, that in the energetic particles escaping impulsive SEP events, ^{3}He is commonly enhanced by huge factors of up to ~20,000, while C, N, O are not enhanced at all, Ne, Mg, Si are enhanced by factors of ~3, and Fe by factors of ~7.5 relative to ^{4}He. The trapped energetic particles seem to have a similar heavy element composition. The ^{3}He and heavy element enhancements are not correlated. The rarer species Na, Al, ^{22}Ne, ^{25}Mg and ^{6}Mg are enhanced by factors of ~2 relative to the neighbouring dominant ones. One observation (Luhn et al. 1987) has yielded surprisingly high charge state $Q \sim 20.5$ for energetic Fe ions; however, it requires confirmation.

Further, with the sole assumption that energetic particle composition is controlled by the ion-to-charge ratio Q/A in the source gas at equilibrium, this observed composition implies that the heavy elements are accelerated predominantly out of gases with temperatures in the ~2.5 to ~5 MK range, *i.e.*, with typical active region, not flaring loop temperature. This implies that both escaping and trapped particles are accelerated, either in the active region gas surrounding the flaring loop itself, or within the flaring loop before it got heated. These temperatures are totally inconsistent with the currently observed Fe charge states; either the observation is not valid, or the ions get further stripped after they have been extracted from the thermal pool.

As to the models for selective ion acceleration by plasma waves, Meyer (1996) divides them on two types. Models in terms of electron beam generated plasma waves may account simultaneously for the ^{3}He and heavy ion enhancements, as well as for the currently observed high ion charge states (late stripping by the beam electrons). However, they probably cannot accelerate the ions to ~MeV energies in one single step, so that another process is required to boost the ions to these energies. In turn, models in terms of the general ambient turbulence, and of its cascading, yielding a smooth wave frequency spectrum, deal with the heavy ion acceleration only. It implies that the huge, uncorrelated ^{3}He enhancements must then be produced by another process. They do not account for the currently observed high ion charge states, but they should be able to fully accelerate the ions in a single step.

In the context of this latter type of models, Steinaker et al. (1997) investigated in more detail the formation of the heavy element enhancements for the impulsive events in terms of damping of the electromagnetic He cyclotron waves associated with the general turbulence. It is suggested that the damping is due to interaction with the ions in the gas, and particularly with the abundant $^4He^{+2}$ ions. Proposed scenario is based on a general analysis of wave-particle interaction in the warm plasma surrounding an impulsive flare.

Steinaker et al. (1997) considered the damping of the waves which can accelerate heavy ions in >2.5 MK gases (*cf.* above), which all have values of $0.25 < Q_i/A_i < 0.50$ or, in terms of gyrofrequencies, $0.25 < \Omega_i/\Omega_p < 0.50$, where Ω_p is the proton gyrofrequency. In this range of frequencies, the wave damping is largely dominated by the energy transfer to the $^4He^{+2}$ ions, with a value of their ratio $Q/A = \Omega_i/\Omega_p = 0.50$. This effect is due to the comparatively huge ^4He abundance and of its low mass, which both tend to produce a very broad He damping region, or "Helium-Valley" of dearth of waves. The width of the He-Valley depends on the plasma temperature, and on the characteristic time scale for the He-Valley replenishment due to wave cascading (for more details see, *e.g.*, Meyer 1996, and references therein). The observed elemental composition, together with the ionization balance in the gas, imposes that some heavy ions remain unenhanced relative to ^4He, hence have their gyrofrequencies lie within the He-Valley, and that others be enhanced, and hence have their gyrofrequencies lie outside the He-Valley.

These requirements allow to impose constraints to the source gas temperature, $T \sim 2.4$ to 4.5 MK (which are very similar to those first obtained on more general grounds, see above), and on the rate of wave cascading into the He-Valley (Steinaker et al. 1997). This must suppress further broadening of the He-Valley after time scales somewhere in the range between $\Delta t \sim 0.1$ and $\sim 5 \times 10^{-3}$ s, which could be shifted by a factor of ~10 upward, depending on the density and field in the active region gas. According to Meyer (1996), the observed excess of the rarer species in the Ne to Al range might suggest that wave damping by the dominant species in this range is significant.

6.5 Particle Trapping and Transport in the Corona

6.5.1 Delayed Gamma-Rays and Particle Trapping

As it was mentioned above, during the flares of 11 and June 15 1991 (Kanbach et al. 1993; Akimov et al. 1991) gamma ray emission following the decay of pions has been recorded in the energy range about 50 MeV–2 GeV. In particular, the observation of Kanbach et al. (1993) for the flare of 11 June revealed, for the first time, the existence of pion radiation as late as 8 h after the impulsive phase.

Evidently, delayed gamma rays returns us to a well-known problem of time delay in release of solar cosmic rays from the corona (*e.g.,* Cliver et al. 1982). In particular, in the light of recent findings and new developments in the solar physics, some *dynamic* effects associated with the possible transport of SCR by *traveling* coronal structures (*e.g.*, by CMEs) and particle drifts from *expanding* magnetic bottles (loops) should be taken into account. Of special interest is an escape of the *first* relativistic protons recorded by ground-based and underground detectors.

On the other hand, there are certain evidences that electrons and protons are accelerated only during the impulsive phase of the flare and are subsequently mirror trapped in coronal magnetic loops. This poses the following dilemma (Mandzhavidze and Ramaty 1992): if the magnetic field lines in the loop are simple plane arches, the protons will drift across the cross section of the loop (in the X-direction) in seconds to minutes, rather than hours. For example, for a 1 GV proton in a typical loop of 10^9 cm height, the drift velocity is about 10^6–10^7 cm s^{-1}, so in a loop of 10^8 cm thickness in the X-direction at the mid-plane, a proton will be contained for seconds to a minute or two only. To solve the dilemma, several different approaches have been developed.

The production of GeV gamma rays as late as 8 h after the impulsive phase of a flare, according to the interpretation of Mandzhavidze and Ramaty (1992), could be due to either the continuous acceleration of particles to GeV energies or the trapping of such particles in closed magnetic structures. While the possibility of continuous acceleration is not ruled out by these authors, they provide two strong arguments that the bulk of the particles could be accelerated during the impulsive phase and subsequently trapped in closed magnetic structures, most likely loops (Ramaty and Mandzhavidze 1994a).

The first argument pertains to relativistic electrons. Gamma ray emission at energies >10 MeV was observed from many disk flares. This emission is mostly bremsstrahlung from ultrarelativistic electrons whose radiation pattern is highly collimated along the direction of motion of the electrons. Since it is much more likely that these electrons are accelerated in the corona rather than in the photosphere, in the absence of trapping the electrons would radiate predominantly downwards the photosphere because the amount of material above the acceleration site is negligible relative to the radiation length of relativistic electrons. In this case radiation would not be observed from disk flares. On the other hand, mirroring in convergent magnetic flux tubes, or pitch angle scattering by plasma turbulence, can reflect the particles and allow them to radiate in their way up in the atmosphere.

The other argument follows from the comparison of the number of interacting particles, as derived from the gamma ray observations, with the number of escaping particles from the same flare, obtained from interplanetary observations. This comparison shows that for electrons the ratio of the escaping to interacting particles (the escape ratio) is less than 1 for all the flares that were studied (Klecker et al. 1990; Daibog et al. 1990). For protons, the escape ratio can be both less than or greater than 1; but it is typically less than 1 for impulsive flares (Ramaty et al. 1993; Cliver et al. 1989), indicating that at least for these flares the bulk of the protons remain trapped at the Sun. In addition, Ramaty and Mandzhavidze (1994a) believed that long term trapping of particles in loops provides a natural explanation for the observation of high energy gamma ray emission hours after the impulsive phase of the flare.

6.5.1.1 Particle Drifts

To allow particles to be trapped for up to 8 h, the rate of pitch angle scattering must be much *lower* than that during the impulsive phase (Mandzhavidze and Ramaty 1992). On such long time scales the effects of particle drifts become important. For example, in a purely toroidal magnetic field the curvature of the coronal portion of the loop will produce drift velocities on the order of 10^6 cm s^{-1}, which could transport particles over distances on the order of the solar radius on time scales of hours.

As noted by Lau et al. (1993), the loop containment problem is very similar to that faced in magnetic fusion research with toroidal geometry (in a device called a *Tokamak*). In a torus having only toroidal magnetic field lines in horizontal plane, the electrons and ions drift vertically in opposite direction, giving rise to a vertical electric field E. The loop differs from the Tokamak in that Tokamak particles can circulate around the torus, while in the loop they are mirrored back and forth between the feet. Also, in the loop geometry the electric field would be shorted out by electrons moving along the field lines, which are connected to the photosphere. Based on guiding center theory, Lau et al. (1993) show that these differences are unimportant as far as the containment is concerned, and that if nothing else happens to the particles, there are surfaces in the loop on which particles will remain indefinitely. It is possible, however, if magnetic field lines have enough *twist*. Particles, however, can be removed from the loop by drift.

The effects of the drifts have been studied by Lau et al. (1993) employing a magnetic field model that satisfies the force-free equilibrium equation, $\nabla \times B = \lambda B$, and boundary conditions such that the photospheric magnetic field is concentrated in two spots separated by a distance L. The twist exhibited by the resulting loop-like structure is determined by the parameter λ. The particles can drift to the boundaries of the loop as well as into the loss cone. The presence of twist causes some of the particles to drift on closed paths, and these particles can remain trapped in the loop indefinitely. In the absence of twist ($\lambda = 0.1$), most of the high energy protons are removed from the loop after above 1 h. However, for $\lambda = 3.4$ (larger values lead to instabilities) a fraction (6 %) of these protons remain trapped indefinitely. On the other hand, because the time scale is proportional to L^2, if $L = 10^{10}$ cm, essentially all the high energy protons will remain trapped for at least 8 h independent of the amount of twist.

The particles, of course, lose energy by nuclear reactions and collisions as they bounce in the loop. By means of numerical calculations of drift orbits in twisted magnetic fields (within a class of force-free loop-like models) it was shown (Lau et al. 1993) that particles typically remain active for a time much longer than 8 h. However, this conclusion was obtained provided *ignoring* the effects of waves, *i.e.*, the particle scattering inside the loop. On the other hand, as shown some later by Ruffolo (1997a, b) from charge state data for interplanetary ions, any residence in coronal loops must be for <0.03 s, which rules out models of coronal transport (*e.g.*, birdcage model) in which escaping ions travel to distant solar longitude within coronal loops.

6.5.2 Prolonged Trapping or Continuous Acceleration?

Undoubtedly, high-energy gamma-ray data (Akimov et al. 1991; Kanbach et al. 1993) on two flares under consideration are of great interest. However, they do not cover, unfortunately, impulsive stages of the corresponding flares and are rather fragmentary. Therefore, one can not consider them as convincing evidences of continuous acceleration or prolonged trapping of the particles in the coronal magnetic loops. In spite of great enthusiasm of some workers (*e.g.*, Mandzhavidze and Ramaty 1992; Mandzhavidze et al. 1993; Ramaty and Mandzhavidze 1994a), they cannot suggest an unequivocal version of explanation for the gamma-ray data mentioned. Actually, their interpretation seems to be controversial.

For example, according to Mandzhavidze et al. (1993), the 15 June flare data are consistent, from one hand, with impulsive acceleration and subsequent trapping of the particles in magnetic loops and, from the other hand, with the assumption of continuous acceleration. The spectrum of accelerated protons over a broad energy range (E_p = 10 MeV − 5 GeV) can be represented, however, in both cases, in fact, by the same power law with differential indices of $\gamma = 3.8$ and $\gamma = 3.5$, respectively, with exponential cutoff at $E_0 = 2.7$ GeV. Later on, Ramaty and Mandzhavidze (1994b) and Mandzhavidze et al. (1996) have admitted the possibility that particles were accelerated during several discrete episodes and remained essentially trapped between them. Moreover, based on the similarity of time profiles of the pion decay and 2.223 MeV line emissions, it was shown (Mandzhavidze et al. 1996) that at least during the first 3 h of the 11 June flare, pure trapping cannot account for the observations. An argument against trapping, based on the similarity of the combined pion decay-nuclear line time profile and microwave time profiles, was also presented for the 15 June flare (Akimov et al. 1993; Kocharov et al. 1993).

The conclusion of Mandzhavidze and Ramaty (1994b) was re-examined by Mandzhavidze et al. (1996) using new 2.223 MeV and 4.44 MeV nuclear line data and 150–210 MeV continuum data. These data, together with microwave ones provide much more detailed description of the time profiles of the various gamma ray emissions than was hitherto available. The data indicate the existence of at least three distinct emission phases characterized by changes in the ion spectrum during transitions from phase to phase, with the spectrum probably remaining constant during the second and third phases. It is shown (Mandzhavidze et al. 1996) that ion spectrum hardened during the transition from the first to the second phase. The ion spectrum in the third phase is softer than that in the second phase. It is suggested, finally, that during the flare of 11 June the ions were accelerated episodically and subsequently trapped between the acceleration episodes.

6.5.3 Alternative Models

In the light of such an ambiguity, it is reasonable to apply to certain alternative approaches to the problem of time delay in particle release from the corona. One of them may be associated with a model of the energy dissipation in a vertical current sheet (CS) at the top of post-flare loops during the two-ribbon flare (Martens and Kuin 1989). Such a CS configuration gives rise to particle acceleration, in particular, along the solar surface.

On the other hand, this problem has been considered (Kahler 1994) taking into account that nearly all large solar proton events (SPE) are associated with fast ($u > 400$ km s^{-1}) coronal mass ejections (CMEs). It is suggested that the particles are accelerated in shocks driven by CMEs, and the ejection profiles of accelerated particles are functions of CME heights.

At last, Perez-Peraza et al. (1992) have suggested a model of two acceleration sources separated in time and located at different coronal heights. The first source (*I*) produces relativistic particles during the impulsive phase of a flare deeply inside the corona, and the second one (II) begins to operate $\leq$1 h later in the upper corona. The conditions for the particle escape from the source *II* turn out to be easier than those from the source *I*, so when observing by surface neutron monitors the effects of two sources undergo a *superposition* and experience a *re-arrangement* at time scale. Obviously, such a model is incompatible with a trapping of fast particles for a long time and does not need a CME-driven shock for particle acceleration. At the same time, it allows, in particular, to explain two-peak structure of intensity-time profiles observed during certain SPEs at relativistic energies (Miroshnichenko et al. 1996), for example, on 29 September 1989 (Vashenyuk et al. 1993; Miroshnichenko 1997; Miroshnichenko et al. 2000). Besides, it does not need the assumptions of continuous acceleration and/or prolonged trapping of relativistic particles in loop-like coronal structures.

Unfortunately, no one of mentioned three approaches does resolve the problem completely. In general, a scenario of particle acceleration and release remains unclear in some significant features, so the problem requires to search for more adequate approaches. In particular, it should be considered particle drifts from *expanding* magnetic bottles (loops), especially in the context of the problem of an escape of the *first* relativistic protons recorded by ground-based and underground detectors.

It seems to be pertinent also to use more widely the *source spectrum data* taking into account different conditions for particle propagation inside and outside the magnetic loop. Quantitative information on so-called "source" spectra of SCR (for protons) is available, by now, for 80 SPE in the period of 1949–1991. This information has been systematized in the form of Source Spectrum Catalogue

(Miroshnichenko et al. 1999). It contains, in practice, the first real grounds for development a *quantitative* model (scenario) of particle acceleration at the Sun. When constructing such a model, it is necessary, of course, to take into consideration some dynamic effects of particle transport in the corona. One of them is a coherent (convective) transport which is presumably realized either by flare shock or by expanding magnetic structure (bottle) associated with closed magnetic loops above the flare. Below we consider this issue in some detail relying upon the model of magnetic bottle developed by Schatten and Mullan (1977), Mullan and Schatten (1979), and Mullan (1983).

6.5.4 Gradient Drift from Expanding Bottle

According to estimations of Mullan (1983), a start of particle acceleration in flares is unknown with the accuracy of several minutes, the rest uncertainties are within the limits of 5–10 min. As an acceleration moment Mullan (1983) accepted the time $t_a > t_0 + \delta t$, where t_0 is the moment of impulsive phase of the flare and $\delta t < 1$ min – duration of acceleration. This estimation is confirmed by the findings obtained, in particular, from the measurements of flare gamma-rays. For example, based on gamma-ray data Forrest and Chupp (1983) have concluded that both electrons and ions must be accelerated together to relativistic energies and interact with matter in a time scale of *seconds* after the flash phase of the flare.

Hence, it appears that the time profile of SCR ejection is mainly determined by the dynamics of the bottle expansion and destruction, it being expected the escape onset of the *first* (relativistic) particles and the release of the *great bulk* of SCR are governed by *different* factors. As to relativistic protons, their escape must start in essence since the moment of generation. In application to coronal magnetic trap such a hard assumption needs a serious substantiation.

The most probable mechanism of escape seems to be a gradient drift in the inhomogeneous magnetic field which decreases gradually as the height increases. If so, an equation of particle transport in the corona may be written in the form:

$$\partial n/\partial t + V_d \partial n/\partial \mathrm{r} = Q\,(r,\ t) \tag{6.3}$$

where a drift velocity, V_d, depends on particle velocity v (in units of $\beta = v/c$) and rigidity R as $V_d \sim \beta^2 \sim R^2$. The source function $Q\,(r,t) = D_0 R^{-\gamma} \delta\,(r - r_0)\,\delta\,(t - t_0)$ describes an instantaneous generation of particles in the point r_0 with the spectrum of $D_s(R) = D_0 R^{-\gamma}$. The Equation (6.3) is linear but non-uniform one, and its solution can not be obtained in explicit form by the separation of the variables. Therefore, we use a simplifying method replacing (6.3) by a uniform equation

$$\partial n/\partial t + V_d \partial n/\partial \mathrm{r} \; = \; 0 \tag{6.4}$$

with suitable initial condition

$$n(r_0,t) = D_0R^{-\gamma}\delta(r - r_0)\delta(t - t_0) \tag{6.5}$$

The solution of (6.3) may be written in general form

$$n(R,r,t) = D_0R^{-\gamma}\exp\left[\lambda(r - r_0/V_d - \lambda(t - t_0)\right] \tag{6.6}$$

where λ is an arbitrarily constant which may be apparently determined only by observational data. The relation (6.6) was obtained for a differential spectrum "line", that is for the number density of particles with a given rigidity R. By fixing an escape point r_1 one can estimate a drift time $t_{dr} \sim r_1/R^2$. Hence, it appears that the time distribution of particle arrival to the point r_1 will be quadratic in the rigidity, the resulting curve of $\varphi(t) = n(R, r_1, t)$ being consist of individual "lines" with the intensities depending on the form of generation spectrum: $\varphi(t_1, R_1) = D_0R_1^{-\gamma}$, where $t_1 \sim r_1/R_1^2$, $t_2 \sim r_2/R_2^2$, and so on. The softer (steeper) this spectrum the larger will be a rate of the curve $\varphi(t)$ growth, and the contributions of particles with the different rigidities will *not be intermixed*.

As a result we get a "spectrum reversal" (or "inverse spectrum") on *time* scale. However, neither such a phenomenological scheme nor the relation (6.6) do not comprise the possibility of that the function $\varphi(t)$ may be formed as a *curve with the maximum*, at least, in the case of instantaneous point-like source (6.5). Obviously, to satisfy this requirement it is necessary to hypothesize that the initial distribution was not instantaneous but some extended in time. Anyhow, if the drift is a basic process of SCR transport from the generation point to ejection one (within the frame of bottle model) then ejection spectrum being integrated over total drift time should be identical to generation one.

It should be noted that the drift effects have been calculated by Mullan and Schatten (1979) to simulate particle motion *outside* the boundary of the magnetic bottle, in the *static* corona, using a 9th order harmonic expansion of the coronal field. Their results are therefore applicable only to particle transport in the quiet corona – due to perpendicular diffusion and drift, with gradual escape of particles to interplanetary space. In other words, those results are important to understand the passive role played by coronal fields prior to particle release. It is interesting that electron drifts due to curvature and gradients in the coronal fields are entirely negligible in comparison with proton drifts, due to large ratio of their masses m_p/m_e. It may be understood if we note that curvature and gradient drifts velocities are given by relation (*e.g.*, Delcroix 1965)

$$V_d = -(\gamma mc)/qBL)\left[v_1 + (1/2)v_2\right] \tag{6.7}$$

where L is the length scale of field variation, and the other symbols have their usual meaning: γ – Lorentz-factor, m and q are the particle mass and charge, respectively, v_1 and v_2 are velocity components along and across magnetic field B.

On the other hand, Schatten and Mullan (1977) have made some estimates in order to determine relative significance of different processes in release mechanism. It is suggested that the Rayleigh-Taylor phase of the bottle expansion begins with the interchange of open and closed field lines due to field line reconnection on a time scale $t_r = D/fV_A$, where D is the transverse dimension of the reconnection region, V_A is the Alfvén speed, and $0.1 < f < 1.0$. The field lines which are involved in the Rayleigh-Taylor interchange are (1) closed field lines near the inner boundary of the bottle and (2) open field lines near the outer boundary. When the interchange begins, the overall diameter of the bottle is of the order 1.0 solar radius r_s (Schatten and Mullan 1977). Therefore, the value of D in the present case is expected to be only a fraction of r_s; *i.e.*, D is of the order of 10^{10} cm.

To estimate V_A, we recall that the field at the top of the bottle is a few Gauss at an altitude of $\sim r_s$ above the photosphere (*e.g.*, Dulk et al. 1976). With a density of the order 4×10^7 cm^{-3} in the compressed material swept up by the bottle at that altitude (Dulk et al. 1976), V_A is therefore of the order of 10^8 cm s^{-1}, and so the reconnection time in the present case is $t_r \sim 10^2$–10^3 s. These times are sufficiently short to allow a large fraction of previously closed field lines to be reconnected to open field lines within a 1-h period, thereby releasing trapped particles into interplanetary space.

Schatten and Mullan (1977) do not exclude that there may be other processes contributing to the release of particles from inside the bottle to interplanetary space once the Rayleigh-Taylor interchange sets in. For example, when flux tubes become intertwined, particle drifts will contribute to the transfer of particles from closed to open field lines. If flux tubes are intertwined with a characteristic distance L between their axes, the drift speed for particles of energy E_k (MeV) is $V_d = 10^4 E_k/BL$ cm s^{-1}, where B is the mean field in gauss and L is expressed in units of 10^{10} cm. The time scale required for particles to drift from one flux tube to another is therefore of the order of $t_d = 10^6 BL^2/E_k$ s. Using $B = 3$ G and $L \sim 1$, we see that drifts contribute to escape of particles at a rate comparable to that caused by reconnection only in the case of particles with energies of ~3 GeV or greater. For particles of lower energies, drifts in the present context do not appreciably enhance the efficiency of particle escape by means of reconnection. Schatten and Mullan (1977) conclude that during a 1-h period following a flare, either reconnection alone or reconnection in association with particle drifts will indeed permit the efficient transfer of particles from closed to open lines as a result of the Rayleigh-Taylor interchange.

6.5.5 Particle Energy Losses in Expanding Bottle

A potential problem with the present expanding bottle model is the possible loss of particle energy as the volume V of the bottle increases. Under adiabatic conditions, the particles are accelerated at a time when the volume is V_2, and they are released when the volume is V_1. In such a case, conventional gas dynamics suggests that the particles will reduce their energy by a factor $(V_2/V_1)^{2/3}$ as a result of the expansion. Hence, if the acceleration were to occur solely during the very early stages of the bottle lifetime (*e.g.*, $V_1 << V_2$), the particles upon release would have lost a large fraction of their energy.

In considering this problem, we must recall that inside the magnetic bottle there is a two-component gas: the hot flare plasma itself and the cosmic ray particles. The reason for the existence of the cosmic rays is that an acceleration mechanism (so far unspecified) has been at work as a result of the physical processes involved in the flare. The acceleration need not to be confined to the early stages of bottle expansion, so it is not necessarily true that V_1 is in general much less than V_2. Particles accelerated early in the lifetime of a bottle may be subject to rather severe energy losses due to expansion, but such losses are expected to be small for particles accelerated at later stages of the bottle lifetime.

It is known that essentially all flares which produce solar cosmic rays are associated with type II radio bursts. The latter are generated by the MHD shock waves propagating through the corona. Our current knowledge of physical conditions in the corona (especially following a flare) are certainly not adequate to allow us to discuss in detail all of the processes which result in gains or losses of particle energy. Therefore, it is worth only to summarize briefly two independent approaches (Schatten and Mullan 1977) which indicate that, as a result of the turbulence induced in the corona by a flare, it is possible for the cosmic ray energy gain to outweigh the losses due to bottle expansion.

From one hand, it is suggested (Schatten and Mullan 1977) that energy losses due to bottle expansion can be offset by particle acceleration in the second-order Fermi processes behind the expanding shock. Furthermore, the acceleration mechanism can continue to operate as long as the expansion front retains the character of a shock and as long as the fast particles are trapped inside the bottle. In the other words, particle acceleration inside a magnetic bottle needs not to be instantaneous process at the very earliest stages of the bottle lifetime. It may be prolonged for a substantial fraction of the bottle lifetime, until either the shock disappears or trapping of the particles is no longer effective.

Recent models for ion and relativistic electron transport in solar magnetic loops and some problems of transport of accelerated particles in the corona (adiabatic motion, pitch angle scattering by plasma turbulence, drifts, and angular distributions) were reviewed in detail by Ramaty et al. (1990), Ramaty and Mandzhavidze (1994a). Acceleration of solar cosmic rays in extended coronal structures is analyzed in Chap. 7.

6.6 Physical Implications of Gamma Ray and Neutron Data

Measurements of accelerated charged particles near the Earth clearly indicate that such particles are produced in energetic phenomena at/near the Sun. These particles consist of both electrons and nuclei. But until the advent of solar gamma-ray astronomy, observations in the radio and X-ray bands had revealed only the existence of the electronic component in the flare region itself. In the hopes of finding the properties of accelerated protons and heavier nuclei in flares, a variety of theoretical studies of the possible nuclear reactions of such particles in the flare region have been made (for early references see, *e.g.*, Ramaty et al. 1975; Kocharov 1980; Kuzhevskij 1985).

One of the most dramatic manifestations of those reactions is the solar neutral emission (gamma rays and neutrons) produced by accelerated ions interacting with the ambient solar atmosphere. The main components of gamma-ray emission are: electron bremsstrahlung which dominates at energies of the photons of $\leq$1 MeV, and at energies of ~10–50 MeV; nuclear gamma-ray line (GRL) emission (of ~1–10 MeV) and pion decay emission (>50 MeV). The experiments on *SMM*, *Yohkoh*, *GRANAT*, *Compton Gamma Ray Observatory* (*CGRO*), *RHESSI*, *CORONAS-F* and *INTEGRAL* allowed to accumulate copious data on solar gamma-rays in different energy range, in particular, on annihilation 0.511 MeV line, neutron capture line at 2.223 MeV, nuclear GRL emission of 4–7 MeV, pion decay emission above 50 MeV. There are even some evidences of gamma-ray production at the Sun at energies above 1 GeV.

Notice that an Atlas of all the flares observed in 1980–1989 by the *SMM/GRS* has been published (Vestrand et al. 1999). About 10 years after, a number of spacecraft (*RHESSI*, *CORONAS-F* and *INTEGRAL*) with their large set of detectors have registered several recent energetic solar phenomena, in particular, the flares of 23 July 2002 (*e.g.*, Smith et al. 2003), 28 October and 2 November 2003 (*e.g.*, Arkhangelskaja et al. 2006; Kiener et al. 2006; Kuznetsov et al. 2011), and 20 January 2005 (*e.g.*, Arkhangelskaja et al. 2006, 2009a, b, c).

6.6.1 Recent Progress in Solar Gamma-Ray Astronomy

Gamma rays provide important information on many aspects of the Sun's physics, including the fundamental problem of particle acceleration in the solar atmosphere. Papers by Chupp (1996), Share and Murphy (2000), Murphy and Share (2005), Miroshnichenko and Gan (2012) give an extended view of the history of the field, its development and its current status, including some physical implications of the gamma-ray data.

In spite of some limitations, the experiments on *SMM*, *Yohkoh*, *GRANAT* and the *Compton Gamma Ray Observatory* (*CGRO*) have already provided data for fundamental discoveries over the past decades relating to particle acceleration, transport and energetics in flares and to the ambient abundance of the corona and chromosphere. These include (*e.g.*, Share and Murphy 2000): (1) enhancements in the concentration of low FIP elements where accelerated particles interact; (2) a new line ratio for deriving the spectra of accelerated particles at $\leq$10 MeV; (3) energies in accelerated ions that exceed those in electrons for some flares; (4) a highly variable ion to electron ratio during flares; (5) concentration of ^{3}He in flare-accelerated particles enhanced by a factor of $\geq$1,000 over its possible photospheric

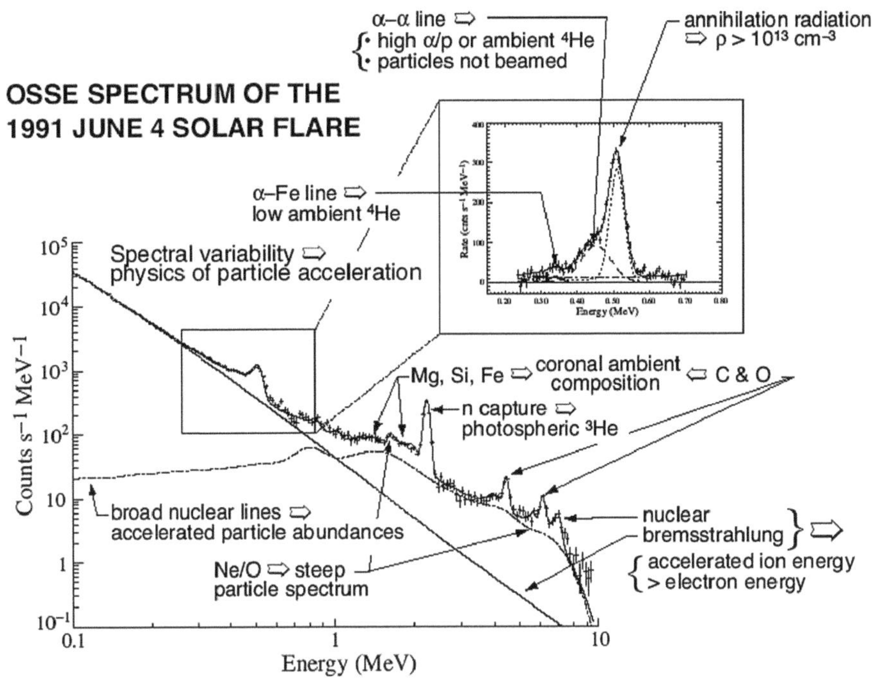

Fig. 6.10 Energy spectrum of the 4 June 1991 solar flare observed by the *CGRO*/OSSE instrument, with a summary of the physics to be revealed by gamma-ray spectroscopy (Share and Murphy 2000)

value; (6) an accelerated α/p ratio >0.1 in several flares and evidence for high ambient ^{4}He in some flares; (7) measurement of the positronium fraction and a temperature-broadened 511 keV line width; (8) new information on the directionality of electrons, protons, and heavy ions and/or on the homogeneity of the interaction region; and (9) the spectrum of broadened gamma-ray lines emitted by accelerated heavy ions that indicates Fe enhancements consistent with those observed in solar energetic particles. In addition to these important findings, Kuzhevskij et al. (1998, 2005a) have developed a new method for investigation of the solar flare plasma density based on the analysis of the 2.223 MeV gamma-line time profiles (for some details see Chap. 12).

Share and Murphy (2000) summarized some past findings and highlight recent discoveries based primarily on measurements made by *SMM*/*GRS* and *CGRO*/*OSSE* instruments. The state of our knowledge of high-energy flare emissions is visually demonstrated in Fig. 6.10 that shows the gamma-ray spectrum of the 4 June 1991 flare observed by the OSSE/*CGRO* experiment. As one can see, from the gamma ray data may be derived important information about energy spectrum, elemental abundances and other features of accelerated particles, as well as about the properties of the solar atmosphere. Later on, high spectral-resolution measurements of nuclear de-excitation lines and the 2.223 MeV neutron capture line have been carried out with the *Reuven Ramaty High Energy Spectroscopic Imager* (*RHESSI*) during the flare of 23 July 2002 (Smith et al. 2003; Murphy et al. 2003). The data on the time history of the narrow deuterium line at 2.223 MeV turned out to be very informative, in particular, to derive photospheric ^{3}He/H ratio during solar flares.

Mandzhavidze and Ramaty (2000) reviewed the results of gamma-ray investigations that provide information on the solar flare accelerated α/p and ^{3}He/^{4}He ratios, on the ambient He/H, Mg/O, Si/O and Fe/O in subcoronal regions of the solar atmosphere, and on the photospheric ^{3}He/^{4}He ratio. The data on the 2.223 MeV line from five more flares considered here confirms their previous conclusion that the ^{3}He/^{4}He ratio in the photosphere is lower than it is in the corona. These findings have major implications on the understanding of solar atmospheric dynamics, solar wind and solar flare particle acceleration and galactic chemical evolution.

6.6.2 Photospheric ^{3}He Abundance

Gamma-ray lines, elemental abundances and charge states of solar energetic particles are very important sources of astrophysical information. In particular, ^{3}He are thought to be primarily produced by nucleosynthesis in the early Universe, and its abundance is used to place a constraint on cosmological model. Since the photospheric ^{3}He abundance can not be determined spectroscopically, observations of the neutron capture line at 2.223 MeV from solar flares provide a direct means of determining the photospheric ^{3}He abundance.

Neutrons which are produced simultaneously with prompt gamma-ray lines by interactions of accelerated ions diffuse into the photosphere where the 2.223 MeV line are emitted by neutron capture on hydrogen (see above). Because of the time required for neutrons to slow down and be captured, the 2.223 MeV line is produced about 100 s after the production of the neutrons. The competing capture reaction ^{3}He(n, p)^{3}H affects the delay of the 2.223 MeV line emission.

The 2.223 MeV line flux from instantaneous production of neutron is assumed to fall exponentially in time with a time constant τ given by $1/\tau = 1/\tau_H + 1/\tau_{He} + 1/\tau_d$. Here τ_H is the time constant for capture on H, τ_{He} is the time constant for capture on ^{3}He and τ_d is the neutron decay time (918 s). The values of τ_H and τ_{He} are approximated by $1.4 \times 10^{19}/n_H$ s and $8.5 \times 10^{14}/n_{He}$ s, respectively, where n_H and n_{He} are the number densities of hydrogen and ^{3}He. In a case of the simplified approach (for details see, *e.g.*, Yoshimori et al. 1995a, b) the time profile of the 2.223 MeV line emission $F(t)$ is expressed by

$$F(t) = A\int_{t_0}^{t} \left[S\left(t'\right)/\tau\right]\exp\left[-\left(t - t'\right)/\tau\right]dt' \qquad (6.8)$$

Table 6.4 Data of photospheric $^{3}He/^{1}H$ ratio

$^{3}He/^{1}H$ ($\times 10^{-5}$)	Flare	Satellite/detector	Reference
<3.8	03 June 1982	*SMM*/GRS	Prince et al. (1983)
2.3 ± 1.2	03 June 1982	*SMM*/GRS	Hua and Lingenfelter (1987a)
2–5	11 June 1991	*GRANAT*/PHEBUS	Trottet et al. (1993)
2.3	04 June 1991	*CGRO*/OSSE	Murphy et al. (1997)
4.5	04 June 1991	*CGRO*/OSSE	Murphy et al. (2007)
2.3 ± 1.4	06 Nov 1997	*YOHKOH*/GRS	Yoshimori et al. (2000)
$11 \div 22$	20 Jan 2005	*CORONAS*/AVS-F	Arkhangelskaja et al. (2009c)

Table 6.5 Ratios of gamma ray fluxes from ^{12}C to that from other nuclei in the flare of 23 July 2002

Ratio	Experiment	Calculations	Calculations
$^{12}C/^{24}Mg$	1.01, max 1.90	0.40, variant "a"	4.40, variant "b"
$^{12}C/^{20}Ne$	1.34, max 2.47	2.65, variant "a"	1.53, variant "b"
$^{12}C/^{28}Si$	1.67, max 3.31	31.0, variant "a"	12.0, variant "b"
$^{12}C/^{56}Fe$	3. 81, max 8.02	30.0, variant "a"	2.60, variant "b"

where A is the constant, t_0 is the time when the gamma-ray lines are observed and $S(t')$ is the time profile of the neutron production. Temporal dependence of $S(t')$ is assumed to be similar to that of the C + O line emission. Using this formula, we can obtain τ which gives the best fit for the observed time profile of the 2.223 MeV line emission. The $^{3}He/H$ ratio is determined from this best fit τ, if n_H is assumed.

Yoshimori et al. (1999) and Miroshnichenko (2011) have summarized a few data of photospheric $^{3}He/H$ ratio obtained by different research groups from the gamma-ray line spectroscopy (Table 6.4). As they noted, in order to advance the understanding of the $^{3}He/H$ problem, we need more precise gamma-ray observations. Moreover, Share and Murphy (1997) suggested the procedure for determining the photospheric $^{4}He/H$ from the product of the solar wind $^{4}He/^{3}H$ and the photospheric $^{3}He/H$ ratio. The $^{3}He/H$ ratio is related to the $^{4}He/H$ ratio that is an important parameter for studies of stellar evolution and solar neutrino production.

One of the recent attempts to derive the photospheric $^{3}He/He$ ratio was undertaken by Arkhangelskaja et al. (2009c) by the data of *CORONAS*/AVS-F for the flarc of 20 January 2005. The authors have shown that this extraordinary solar event may be satisfactorily treated provided that strongly enhanced abundance of ^{3}He nuclei took place in the solar photosphere.

Above estimates of the ratio $^{3}He/^{1}H = (1.1 \div 2.2) \times 10^{-4}$ may be also considered as an independent confirmation of enhanced content of ^{3}He in the process of the 20 January 2005 flare based on the analysis of gamma-line emission of 2.223, 4.44 and 6.12 MeV. Earlier, Arkhangelskaja et al. (2009c) made a similar preliminary conclusion by the analysis of gamma-ray lines of 20.58 MeV and 0.937 MeV. Besides the data summarized in Table 6.5, we note the attempt by Murphy et al. (2003) to derive the photospheric $^{3}He/He$ ratio by the RHESSI data on the measured time history of the 2.223 MeV neutron capture line during the flare of 23 July 2002. This ratio, however, was not well-constrained, primarily due to uncertainties of the measured nuclear de-excitation-line flux used to represent the neutron-production time history.

6.6.3 Imaging and Mapping of Gamma-Ray Flares

Due to high resolution spectroscopy of gamma-ray lines from the flare of 23 July 2003 the first gamma-ray images of a solar flare were obtained by *RHESSI* detectors (Hurford et al. 2003). Two rotating collimators (with $35''$ and $183''$ resolution) were used to obtain images for the same time interval in four energy bands: the narrow deuterium line at 2.223 MeV formed by the thermalization and capture of neutrons produced in the collisions; the 3.25–6.5 MeV band that includes the prompt de-excitation lines of C and O; and the 0.3–0.5 and 0.7–1.4 MeV bands that are dominated by electron bremsstrahlung. The centroid of the 2.223 MeV image was found to be displaced by $20'' \pm 6''$ from that of the 0.3–0.5 MeV image, implying a difference in acceleration and/or propagation between the accelerated electron and ion populations near the Sun. Note, however, that this discovery is related to the regions of interactions (and emission production) of the electron and protons, respectively, but not to the sources (sites) of their acceleration. Also, we should keep in mind that the 2.223 MeV neutron capture line emission delays for about 100 s relatively to the production of proton above 10 MeV, and this time may be comparable with total duration of the particle acceleration.

Some later, Hurford et al. (2006) have presented the results of *RHESSI* imaging of three flares (2003 October 28 and 29 and November 2) in the 2.223 MeV neutron-capture gamma-ray line with angular resolution as high as $35''$. Comparisons of imaged and spatially integrated fluences show that in all cases most, if not all, of the emission was confined to compact sources with size scales of tens of arc seconds or smaller that are located within the flare active region. Thus, the ions producing gamma-rays appear to be accelerated by the flare process and not by a widespread shock driven by a fast coronal mass ejection. The 28 October event yielded the first such image to show double-footpoint gamma-ray line sources. These footpoint sources straddled the flaring loop arcade but were displaced from the corresponding 0.2–0.3 MeV electron-bremsstrahlung emission footpoints by $14''$ and $17'' \pm 5''$. As with the previously studied 2002 July 23 event (Hurford et al. 2003), this implies spatial differences in acceleration and/or propagation between the flare-accelerated ions and electrons (Fig. 6.11).

INTEGRAL/SPI observations of the 28 October 2003 flare (Kiener et al. 2006) show that the time profiles of the prompt C and O deexcitation lines (4.4 and 6.1 MeV) and hard X-rays (1,150 keV) are closely similar, while the 2.223 MeV profile is delayed by the characteristic ~100 s neutron thermalization time. The *RHESSI* observations (Hurford et al. 2006) begun about 4 min after event onset, and they displayed a smooth exponential decay lasting at least five ~280 s *e*-folding

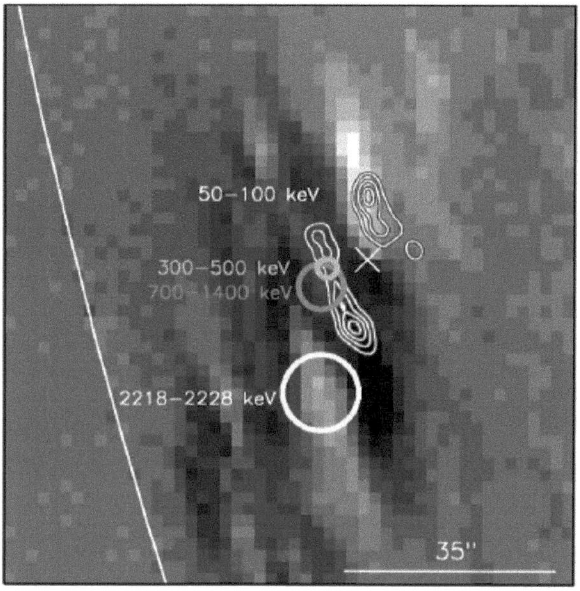

Fig. 6.11 Location of the gamma-ray sources of the flare of 23 July 2002 (Hurford et al. 2003). The *thick circles* represent the 1σ error for the 300–500 keV (*light gray*), 700–1,400 keV (*dark gray*), and 2,218–2,228 keV (*white*) maps made with identical parameters. The 35$''$ FWHM angular resolution is shown in the lower right. The white contours show the high-resolution 50–100 keV map made with 3$''$ resolution. The cross shows the centroid of the 50–100 keV emission made with the same lower resolution as the gamma-ray maps. The background image is a *SOHO*/MDI magnetogram acquired at 00:12 UT, 15 min prior to the flare

times for the 2.223 MeV line, while the 0.2–0.3 MeV band, dominated by electron bremsstrahlung, had a time profile similar to the 1,150 keV *INTEGRAL*/SPI hard X-ray profile. *RHESSI*'s 35$''$ resolution maps at 0.2–0.3 MeV and at 2.223 MeV (Fig. 6.12) are both dominated by two compact sources of comparable intensity (fluence ratio of 1.0 ± 0.4) separated by ~80$''$ and straddling the arcade of loops in the *TRACE* image.

With integration times of 1540s, the images time-average over any footpoint motions. Because of the exponential decay, however, emission in the first few minutes of the interval dominates. For the 0.2–0.3 MeV band, representative locations of the peaks of footpoint locations obtained with shorter time resolutions are indicated by the red plus signs in Fig. 6.12. The general trend was for the east and west footpoints to increase their separation, moving ~15$''$ toward the south and west, respectively. Only an upper limit of ~30$''$ could be placed on potential motions of the centroids of the 2.223 MeV footpoints.

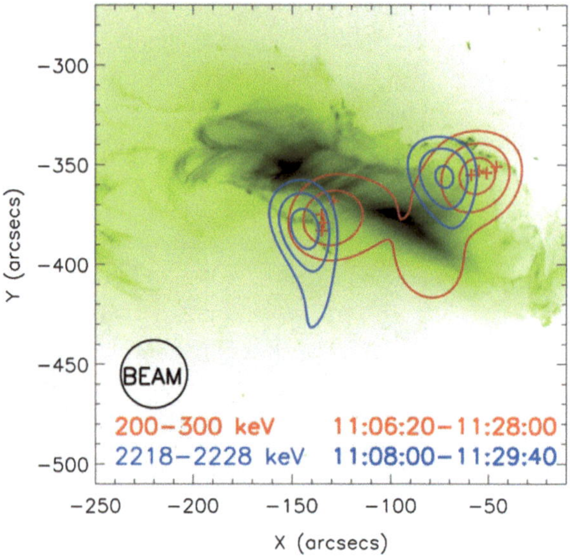

Fig. 6.12 Gamma-ray burst of October 28, 2003 (Hurford et al. 2006). Overlay of the 50, 70, and 90 % contours of 35″ resolution gamma-ray images made with a *TRACE* 195 Å image. The red plus signs indicate 200–300 keV footpoint locations for successive adjacent intervals of 100, 120, 180, and 240 s beginning at 11:06:20 UT. The *X* and *Y* heliographic offsets are positive west and north of Sun center

6.6.4 Heavy-Heavy Interactions of Accelerated Particles

As well-known, nuclear de-excitation lines result mainly from bombardment of ambient carbon and oxygen in the solar photosphere by the accelerated protons and alpha-particles (*e.g.*, Ramaty et al. 1975). Dramatic extensions of experimental possibilities (spacecraft *RHESS*I, *CORONAS-F, INTEGRAL* and others) in solar gamma-ray astronomy call for more detailed consideration of a set of physical problems related to the production of gamma-radiation in the processes of interactions of energetic (accelerated) heavy and middle nuclei with that of elements of the solar atmosphere (so-called heavy-heavy or *ij*-interactions). According to some estimates (Kuzhevskij et al. 2005b), a contribution of these interactions between accelerated and background nuclei in the gamma-ray production in the solar atmosphere may be more important that it was thought earlier. This conclusion is confirmed by a comparison of theoretical estimates (Kuzhevskij et al. 2005b) and observations during the *RHESSI* flare of 23 July 2002 (Smith et al. 2003).

As an example, we note that observed fluences of ^{12}C and ^{24}Mg nuclei (Smith et al. 2003) turned out to be the same ones (about 28 photons/cm^2), *i.e.*, their ratio is about 1.0 (maximum 1.9). In contrast to this, the calculations for two versions of abundances of these elements in the solar atmosphere ("a" and "b" correspond to Aller (1961) and Cameron (1973), respectively), give some evidence of that the ratios of fluxes of gamma rays from ^{12}C and from other nuclei in this experiment should be as those given in (Table 6.5), if we take into account only the processes of *pk*- and *αk*-interactions (for more details see Kuzhevskij et al. 2005b).

As it follows from Table 6.5, to explain the observed ratios we must assume that nuclei of ^{24}Mg and ^{28}Si have been effectively created in the solar active region during the flare due to *ij*-interactions (for example, the interactions between the nuclei ^{12}C and ^{16}O and ^{16}O with ^{16}O). As for the nucleus of ^{20}Ne, one can see that the contribution of interaction of nuclei C and O into its generation is small, because the initial concentration of this element in the solar atmosphere in amount is comparatively high. The abundance of ^{20}Ne in the solar atmosphere is five to seven times greater than the abundance of Mg and Si (Aller 1961; Cameron 1973). As for the ^{56}Fe, this nucleus cannot be created in the process of C-O and/or O-O interactions mentioned above.

As it has been shown several decades ago, nuclear de-excitation lines result mainly from bombardment of ambient carbon and oxygen nuclei *k* in the solar atmosphere by accelerated protons *p* and α-particles (*e.g.*, Ramaty et al. 1975). Dramatic extensions of experimental possibilities in solar gamma-ray astronomy (spacecraft *SMM, GRANAT, Yohkoh, Compton/GRO, RHESSI, CORONAS-F, INTEGRAL*), to our opinion, call for more detailed consideration of a set of physical problems related to the production of gamma-radiation in the processes of interactions of energetic (accelerated) heavy and middle nuclei *i* with that of elements of the solar atmosphere *j* (so-called heavy-heavy or *ij*-interactions).

In particular, due to observations of the *RHESSI* flare of 23 July 2002 (Smith et al. 2003) a need arose to revise the role of *ij*-interactions between accelerated and background nuclei in the production of gamma-rays in the solar atmosphere. According to recent calculations by Kuzhevskij et al. (2005b), the contribution of *ij*-interactions depends on the ambient medium composition and may be more important than it was thought earlier. These authors have noticed that observed fluences of gamma-rays from excited ^{12}C and ^{24}Mg nuclei (Smith et al. 2003) turned out to be the same ones (about 28 photons/cm^2), *i.e.*, their ratio is about 1.0 (maximum 1.9).

In contrast to this, the calculations for two versions of abundances of these elements in the solar atmosphere, provided that we take into account the processes of p-k and α-k interactions only, give some evidence of that ratios of gamma-ray fluences from excited ion of ^{12}C to that of other nuclei in this experiment should be quite different from observed ones. Results of calculations are presented in Table 6.5 where "a" and "b" correspond to standard solar composition by Aller (1961) and by Cameron (1973), respectively. We have never seen any discussions or clear explanations of the discrepancies in the fluence ratios of different gamma-lines noted by Kuzhevskij et al. (2005b).

To explain the observed ratios we must assume that nuclei of ^{24}Mg and ^{28}Si have been effectively created in the solar active region during the flare of 23 July 2002 due to *ij*-interactions (for example, between the nuclei ^{12}C and ^{16}O and/or ^{16}O with ^{16}O). As for the nucleus of ^{20}Ne, one can see that the contribution of interactions of the C and O nuclei into its generation is small, because the initial concentration of this element in the solar atmosphere in amount is comparatively high. The abundance of ^{20}Ne in the solar atmosphere is about five to seven times of the content of Mg and Si (Aller 1961; Cameron 1973). As to ^{56}Fe, this nucleus cannot be created in the process of C-O and/or O-O interactions mentioned above.

As one can see from Table 6.4, the results of estimates of calculated fluence ratios considerably depend on the model of standard solar composition. In one of the reviews for the standard solar composition, Grevesse and Sauval (1998) gave for C/Mg and O/Mg the ratios that differ from the values used above for about 20–25 % and about two times, respectively, in comparison with the data by Cameron (1973) and Aller (1961). Nevertheless, this does not change our main conclusion about important role of *ij*-interactions in the production of gamma line emission from ^{24}Mg in the flare of 23 July 2002. As to the ratios of C/Si and O/Si from Grevesse and Sauval (1998), they are even closer to the versions considered by Kuzhevskij et al. (2005b). Moreover, the conclusions on the contribution of *ij*-interactions into the creation of Ne and Fe nuclei become stronger due to the new solar composition results by Grevesse and Sauval (1998).

If we try to treat the results of above calculations in terms of thin-target and thick-target interaction models (*e.g.*, Ramaty et al. 1975), the conclusion is that in the case of thin-target model, with continuous compensation of energy losses by acceleration process, *ij*-interactions provide substantial contribution into the production of excited nuclei (Kuzhevskij 1985). In the case of thick-target model the role of *ij*-interactions is 10–35 times less because of ionization losses. Nevertheless, the *ij*-processes still remain important for $s \geq 4$. In the meanwhile, it is necessary to take into account that Kuzhevskij et al. (2005b) considered rather small energy interval for *ij*-interactions.

The results by Kuzhevskij et al. (2005b) have been criticized by Tatischeff et al. (2006) and Murphy et al. (2007). They evaluated the significance of *ij*-reactions by using the universal parameterization of total reaction cross sections given by Tripathi et al. (1996). Assuming a thick target interaction model, a power-law source spectrum for the fast ions (index s) and standard compositions (Kozlovsky et al., 2004) for the ambient and accelerated nuclei, they found that the heavy ion collisions should contribute less than a few percent of the total radioisotope production and can therefore be safely neglected. Thick-target radioisotope yields have been calculated for $s = 3.5$, 2 and 5. The first value is close to the mean of spectral index distribution as measured from analyses of gamma-ray line ratios (*e.g.*, Ramaty et al. 1996), whereas the two other values are extreme cases to illustrate the dependence of the radioisotope production on the spectral hardness. Note that the power-law index of energetic protons for the flare of 23 July 2002 was estimated to be about 3.52 (Gan 2004).

Murphy et al. (2007) have confirmed that for the narrow 1.634 MeV ^{20}Ne and 1.369 MeV ^{24}Mg lines produced by interactions of ^{12}C and ^{16}O, the yields are negligible due to both the lower abundances of both species and the higher energy loss of heavy particles. Accelerated ion interactions also create radioactive nuclei, which produce delayed gamma-ray lines that can be observed when the prompt emission is negligible (Tatischeff et al. 2006).

Of course, the most of cross sections applied in the calculations by Kuzhevskij et al. (2005b) now become out of date. On the other hand, estimates and calculations cited above are strongly dependent of the ambient medium composition and average composition of solar energetic particles (SEP). Besides, measurements of the energy spectra of SEP events over a broad energy range (~0.1 to 100 MeV/nuc) show that all large SEP events have spectral breaks organized by the charge-to-mass ratio (Q/M) of the ions (*e.g.*, Mewaldt et al. 2005b). This effect of spectrum softening at high energies may be important just for the *ij*-interactions at the Sun. There are also some evidences of particle acceleration in the flare of 23 July 2002 by specific mechanism (Kichigin et al. 2010, 2014) that provides rather soft power-law spectrum ($s \geq 4$).

Although significant efforts in the past decades have been devoted to measure many reaction cross sections, experimental data only covers a minute fraction of the entire data set required for some important nuclear physics applications. On the other hand, the nuclear physics community has developed specific tools which can shed light on the many approximations in nuclear astrophysical applications. One of these tools is the modern reaction code called TALYS (*e.g.*, Koning et al. 2008). This code is software for the simulation of nuclear reactions. This code includes many state-of-the-art nuclear models to cover all main reaction mechanisms encountered in light particle-induced nuclear reactions. TALYS provides a complete description of all reaction channels and observables and, in particular, takes into account all types of direct, pre-equilibrium, and compound mechanisms to estimate the total reaction probability as well as the competition between the various open channels. The code is optimized for incident projectile energies, ranging from 1 keV up to 200 MeV on target nuclei with mass numbers between 12 and 339. It includes photon, neutron, proton, deuteron, triton, ^{3}He, and α-particles as both projectiles and ejectiles, and single-particle as well as multi-particle emissions and fission.

The TALYS code was designed to calculate total and partial cross sections, residual and isomer production cross sections, discrete and continuum gamma-ray production cross sections, energy spectra, angular distributions, double-differential spectra, as well as recoil cross sections. Recently, Goriely et al. (2008) have updated TALYS to estimate reaction rates of particular relevance to astrophysics. In contrast to other codes developed for astrophysical applications, TALYS avoids many of the approximations mentioned above; it therefore provides a unique opportunity to test the robustness of these alternative codes. In general, the TALYS code provides a new tool to estimate all nuclear reaction rates of relevance to astrophysics with improved accuracy and reliability (Goriely et al. 2008).

Recently, a number of researchers (*e.g.*, Murphy et al. 2009; Chen and Gan 2011) have applied TALYS code to calculate solar flare gamma-ray spectrum in the whole. As it was noted by Share (2009) at the IX *RHESSI* Workshop, the TALYS code is most reliable for particle production reactions involving heavy nuclei; less so for gamma-ray production reactions involving light nuclei. It was found that the best approach is to use the energy dependence of the cross section supplied by TALYS code, but to normalize the cross section whenever possible to measurements. Unfortunately, not all light-element branching ratios are correct: there are some excited states that decay primarily via particles but TALYS treats them as a gamma-ray transition. So, the appropriate TALYS library files should be corrected (Murphy et al. 2009).

About 10 years ago, Murphy and Share (2005) present the gamma-ray line-production and loop transport models used in the calculations of high-energy emission. They discussed in detail the calculated interaction time history, the depth distribution, the interacting-particle angular distribution, and fluence ratios of the narrow gamma-ray line. It was shown that the pitch-angle distribution (PAD) of accelerated particles in the loop model is very important to estimate the GRL fluence. As to the calculations of the fluences for nuclear de-excitation lines and continuum from accelerated-particle interactions in solar flares, the situation remains rather vague one (Murphy et al. 2009). While laboratory measurements of the cross sections for production of the strongest lines seen in flare spectra are available, in fact, these measurements often only cover a limited range of projectile energies. In addition, the bulk of the gamma-ray emission arises from the numerous weaker lines for which there are no measurements. Even the problem of *ij*-interactions is not exhausted yet because of definite uncertainties with cross sections of different nuclear reactions and poor-known initial conditions of the gamma-ray production (shape of spectrum of accelerated particles, ambient nuclei abundances, density models of interaction region *etc.*).

As to the calculations of the fluences for nuclear de-excitation lines and continuum from accelerated-particle interactions in solar flares, the situation remains rather vague one (Murphy et al. 2009). While laboratory measurements of the cross sections for production of the strongest lines seen in flare spectra are available, in fact, these measurements often only cover a limited range of projectile energies. In addition, the bulk of the gamma-ray emission arises from the numerous weaker lines for which there are no measurements. Even the problem of *ij*-interactions is not exhausted yet because of definite uncertainties with cross sections of different nuclear reactions and poor-known initial conditions of the gamma-ray production (shape of spectrum of accelerated particles, ambient nuclei abundances, density models of interaction region *etc*).

We believe that real progress in this field as a whole may be achieved only by combination of gamma-ray data in different energy ranges with multi-wave and energetic particle observations during the same event. Such kind of the studies should be complemented of modeling of self-consistent physical and time scenario of the event.

Chapter 7
Acceleration and Release of Particles from the Corona

From above consideration it appears that existing models of particle acceleration at the Sun are still far from complete and self-consistent. In particular, a question remains if there are the continuous acceleration and/or long-time trapping of energetic particles in the extended coronal structures? In this context, a new paradigm raised of particle acceleration in different sources at or near the Sun – in solar flares, coronal mass ejections (CMEs), coronal and/or interplanetary shocks, *etc*. (as the modern reviews see, *e.g*., Miroshnichenko 2001; Miroshnichenko and Perez-Peraza 2008). This idea has been extensively developed and argued in a series of comprehensive studies (see, *e.g*., Cliver 1996; Kahler 1996; Reames 1996, 1999, 2013; Berezhko and Taneev 2003 and references therein). It turned out to be very fruitful for understanding many energetic solar phenomena. Nevertheless, a number of important unanswered questions remain, and the keenest of them is: Where and how solar particles are accelerated to very high energies up to ~100 GeV? Therefore, it is necessary to discuss some new approaches to the problem, in particular, a possibility of multi-step (multi-source) particle acceleration in the extended coronal structures (loops, streamers, CMEs, *etc*.).

Below we consider several important aspects of the problem, namely: release of the first relativistic particles and interpretation of characteristic injection time; inverse problem of particle transport and reconstruction of the pitch-angle distributions and ejection profiles of escaping particles; nature, location and main features of the behind-the-limb sources; acceleration of solar cosmic rays in the extended coronal structures; two-source model of acceleration and possible general scenario of large solar particle event.

L. Miroshnichenko, *Solar Cosmic Rays*, Astrophysics and Space Science Library 405, DOI 10.1007/978-3-319-09429-8_7

7.1 Release of the First Accelerated Particles

For a long time a persistent problem of SCR research was a lack of observations bearing on the timing and conditions in which protons that escape to the interplanetary medium are first accelerated in the corona. In contrast to the electrons, proton bremsstrahlung and gyrosynchrotron emission are negligible, and the observed gamma-ray line emission, directly attributable to the presence of energetic (E_p >10 MeV) protons (*e.g.*, Ramaty et al. 1975, 1979), may be unrelated to the solar protons observed in space at all (Chambon et al. 1981; Von Rosenvinge et al. 1981; Pesses et al. 1981).

To compensate this difficulty, inferences about coronal acceleration processes drawn from observations of solar protons at 1 AU are generally compromised by the effects of proton scattering in the interplanetary medium. One can, however, hope to learn something about the onset of proton acceleration and/or injection into space from observations of the earliest arriving protons for which the scattering effects should be minimized (*e.g.*, Cliver et al. 1982; Toptygin 1985; Perez-Peraza et al. 1992). It is especially interesting that the first relativistic (>500 MeV) solar protons observed in Ground-Level Events (GLEs) have rather short interplanetary travel times (≈11 min), and this makes the GLEs a unique data source from which to infer the timings of proton injection onsets in large solar flares.

7.1.1 Release of Relativistic Particles

Earlier attempts to infer the initiation of proton ejection from solar flares using the GLE onsets were made by Carmichael (1962) and Kodama et al. (1977). Carmichael (1962) first called attention to what he termed the "transit time anomaly" for solar cosmic-ray events. The transit time anomaly, T_A, is defined as follows:

$$T_A = T_n - 11\,\text{min}, \tag{7.1}$$

where T_n is the deduced Sun-Earth transit time for the *first* arriving relativistic protons and 11 min is the nominal transit time for ~2 GeV proton traversing a 1.3 AU without scattering along the Archimedean spiral path in the IMF. By making the assumption that protons are accelerated to GeV energies at the start of the flare-associated microwave burst, Carmichael (1962) found T_n values of 7–19 min for a small sample of so-called well-connected (20–90°W) flares. Using the same assumption, Kodama et al. (1977) reported a systematic minimum $T_n = 9$ min for all 26 GLEs observed through 1973 (see Table 2.1), independent of the longitude of the parent flare.

In the experiments carried out in 1966–1977 on board the *Proton* and *Prognoz* satellites using the Cherenkov and scintillation counters, the first arrivals of protons with the energies of >100 MeV and >500 MeV were measured for seven large solar flares (Volodichev and Savenko 1981): 7 July 1966 (class 2B), 22 July 1972 (behind-the-limb flare), 4 and 7 August 1972 (3B), 29 April 1973 (2B), 24 September (behind-the-limb flare) and 22 November 1977 (2B). It was assumed that the particles of the above mentioned energies are simultaneously injected into the interplanetary medium and cover the same path to the Earth. There were obtained the delays of escape of the protons between of 4–40 min with respect to the impulsive phase of the corresponding flare, namely, <20; 0–12; 17 ± 4; ~6; ~40; 4–8; and 24 ± 5 min.

Because of their high velocities, energetic electrons also have short interplanetary travel times. The injection onsets of both the non-relativistic and relativistic electrons have been discussed at length in the literature (*e.g.*, Volodichev and Savenko 1981; Cliver et al. 1982, and references therein). However, the ejections of these species relative to energetic protons have been discussed for only a very few events. In an effort to obtain more extensive learning about the onset of particle injection and/or acceleration in large solar flares, Cliver et al. (1982) reviewed the ~2 GeV proton onsets for the 32 GLEs observed between 1942 and 1978. For the GLEs since 1966, they also examined the observed onsets of the non-relativistic (~100 keV) and relativistic (~1 MeV) electrons.

For each event they listed six candidate times (T_i) for the proton injection onset: T_1, Hα onset; T_2, radio main onset; T_3, first significant ≥9 GHz peak; T_4, earliest reported metric type II onset; T_5 ≥9 GHz maximum; and T_6, Hα maximum. For each available time parameter Cliver et al. (1982) determined T_n (and hence T_A) by the relation:

$$T_n = t_n - (T_i - 8\,\text{min}) \tag{7.2}$$

where t_n is the earliest reported onset time of the GLE and (T_i – 8 min) is the time of the particular flare phase under consideration adjusted for the Sun-Earth propagation time of electromagnetic waves. The "radio main onset" time, T_2, used by Kodama et al. (1977) and Carmichael (1962) in their determination of T_A, is defined as the onset of the initial dramatic increase in the microwave flux-density profile during the flare. The "first significant microwave peak" time, T_3, refers to the initial ~500 solar flux units (1 $sfu = 10^{-22}\,W\,m^{-2}\,Hz^{-1}$) peak or plateau in the microwave temporal flux-density profile, with preference given to the higher (≥9 GHz) frequencies. For the 32 events under consideration, T_3 followed T_2 by 2.8 ± 1.8 min. In Fig. 7.1, the flare phases, T_i ($i = 1$–6), are indicated on the microwave burst profile for the flare associated with the GLE of 7 August 1972 (for more details of this GLE see, *e.g.*, Rao 1976).

The GLE onset times were obtained (Cliver et al. 1982) by careful inspection of the neutron monitor records from the world-wide observing network. As a result, the onset times of some events turned out to be slightly differ from those reported earlier in the literature. For example, for two of the events, February 23, 1956 and January 24, 1971, the onset time was moved earlier by 2 min upon re-examination of original and published data. Figure 7.2 contains histograms of T_A for the six flare-timing parameters under consideration. Only those events having T_A values with

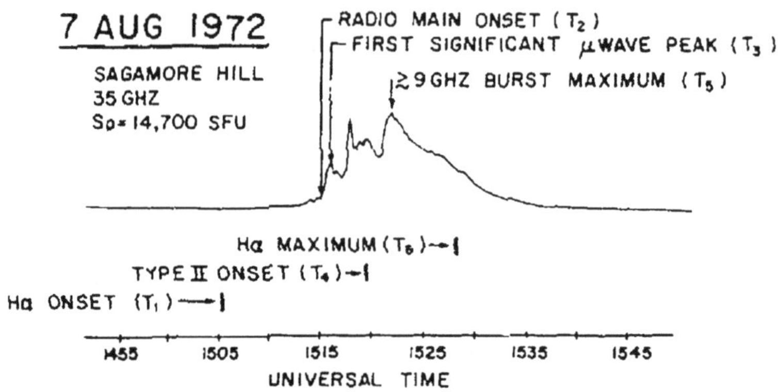

Fig. 7.1 Flux-density profile of the 35 GHz radio emission by the data of the Sagamora Hill Observatory for the GLE-associated flare of 7 August 1972. The flare phases T_i ($i = 1$–6) that were considered as possible candidates for proton acceleration/injection onset are indicated (After Cliver et al. (1982))

uncertainties $\leq$5 min were included in this figure. Contrary to previous investigations, Cliver et al. (1982) found no compelling evidence for a systematic delay in GLE onset times. The most likely time of GeV proton injection onset in these large flares appears to be near the maximum of the first significant microwave peak (FSMP). They noted that GLEs with long delays to onset tend to be small in size.

In addition, the data indicate a systematic phase relationship among the injection onsets of the three particle species considered, with the low-energy electron onset times preceding those of the relativistic protons by $\leq$5 min, and the relativistic electrons following the GeV protons by $\geq$5 min (Fig. 7.3). This phase relationship holds even when the inferred injection times of all three species follow the flare flash phase by >20 min. To account for these observations, Cliver et al. (1982) suggested a picture in which the earliest observed particles are injected when an outward moving acceleration region at a shock front intersects the open field lines connecting to the Earth.

7.1.2 CMEs, two Classes of Flares and Release of SEPs

Numerous studies over the last 25 years have shown that the protons and ions of solar energetic particle (SEP) events are injected from the Sun into interplanetary medium in two distinct phases (*e.g.*, Reames 1990a, b, 1993, 1996). The first phase of injection is closely associated with type III bursts (Cane and Reames 1990) of solar flares and appears to consist of coronal ions accelerated in a gyroresonant processes (Miller and Viñas 1993). These ions are accompanied by high fluxes of electrons and show dramatic enhancements in abundance ratios such as He/He and Fe/O.

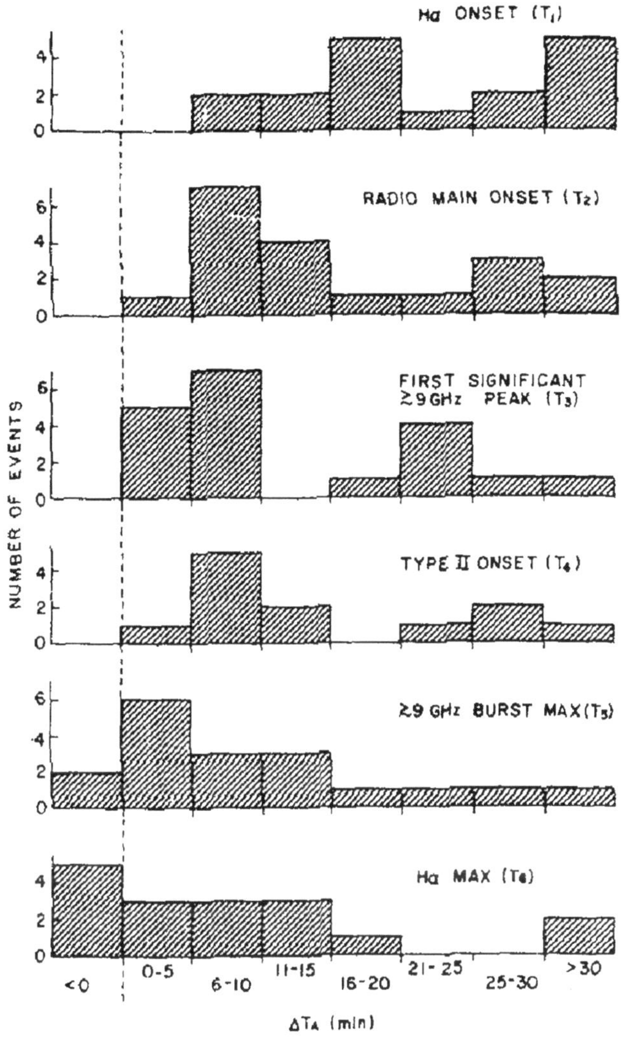

Fig. 7.2 Histograms of T_A, the transit-time anomaly, for each of the six flare phases considered as candidates for the injection onset of GeV protons. Only those GLEs with uncertainties in T_A <5 min were used (After Cliver et al. (1982))

The direct injection of $E > 10$ MeV/nucleon ions during the flare impulsive phase has been observed, in particular, on the *Helios* 1 spacecraft for a sequence of flares on 28 May 1980 (Kallenrode and Wibberenz 1991) and has been inferred for other events from the absence of associated coronal shocks (Cane and Reames 1990). These SEP events usually have relatively small peak fluxes and can be observed only within a longitudinal range of 20° from the solar sources (Reames et al. 1991). When the large SEP events are magnetically well connected to the

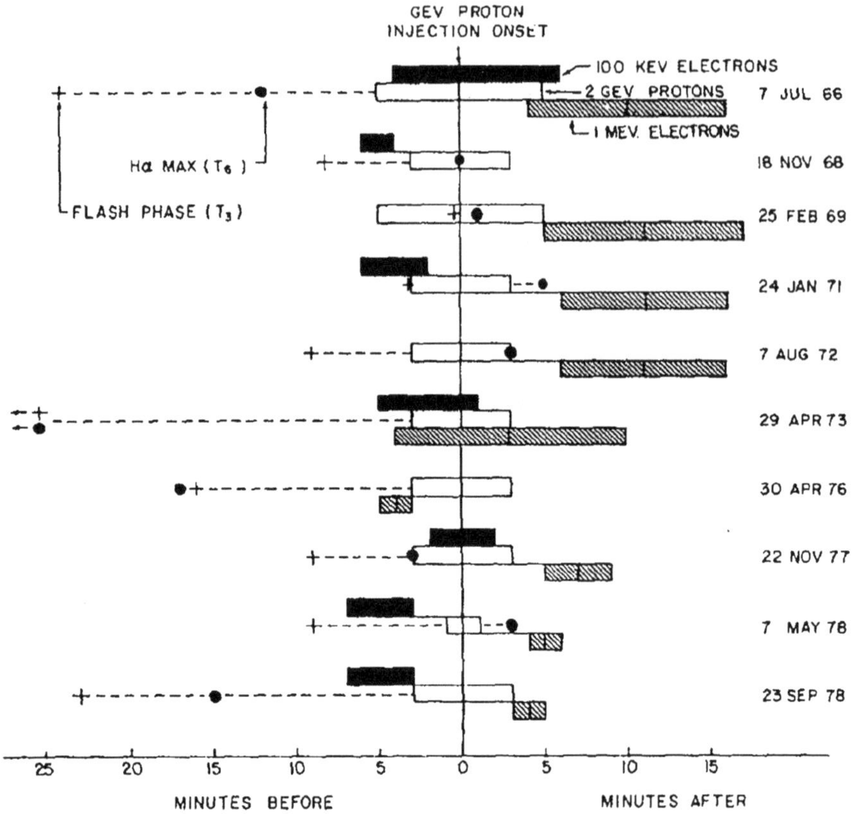

Fig. 7.3 Inferred injection onsets of the non-relativistic and relativistic electrons in GLE events relative to the inferred injection onsets of GeV protons. Only well-connected (20–90°W) GLEs with small onset time uncertainties (<5 min) were considered (After Cliver et al. (1982))

source flare region, Fe-rich material is sometimes seen early in the event, revealing the presence of the impulsive component (Reames 1990a, b; Cane et al. 1991).

Nearly all large-flux SEP events are produced during the gradual injection phase, which is associated with the occurrence of fast CMEs (Kahler 1993, 1996). These events consist of ions with abundances and ionization temperatures characteristic of the ambient coronal material (Mason et al. 1984), probably accelerated by coronal shocks. Large events have time scales of days and are sometimes known as gradual events. They can be associated with flares or CMEs from a broad range of solar longitudes (Cane et al. 1988). The duration of the associated flare X-ray events are usually several hours (Cane et al. 1986).

Apparently, if we can deduce the SEP injection profiles at the Sun relative to the flare impulsive phase and to the appearance of the CME, we can begin to understand the roles of the impulsive phase and coronal shocks in producing SEP events. Injection profiles are most easily inferred for SEPs of highest energy, which propagate to 1 AU with the longest mean free paths. Kahler et al. (1990) examined

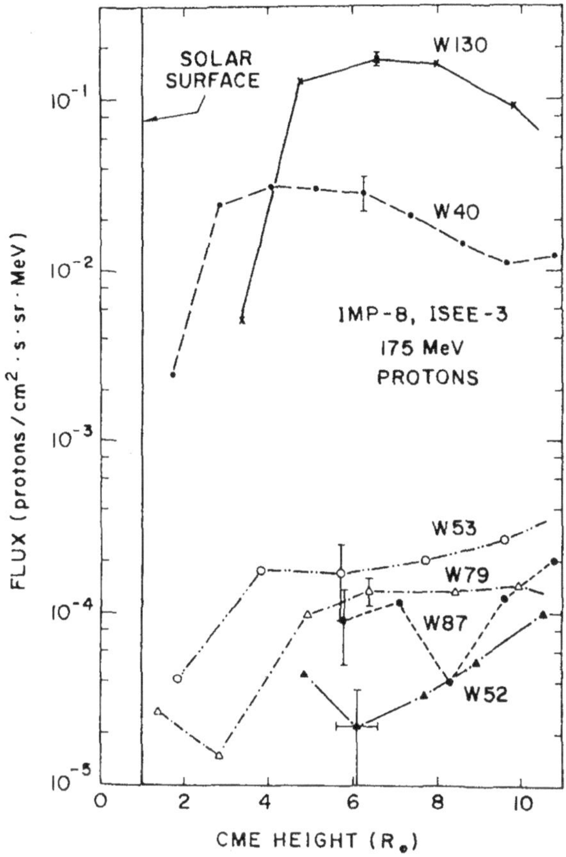

Fig. 7.4 Profiles of injected 175 MeV proton fluxes versus the heights of the leading edges of the associated CMEs observed by the *Solwind* coronograph. Longitudes of associated Hα flares are given for each SEP event (Kahler et al. 1990; Kahler 1994)

SEP injection profiles at the Sun as a function of the height of the associated CMEs observed with the *Solwind* coronograph. They found that the profiles were increasing when the CMEs passed through heights of 2–10 solar radii, R_s, as shown in Fig. 7.4 for 175 MeV protons. From these results they suggested that SEPs were accelerated in shocks driven by the CMEs, but a more detailed examination of such high-energy SEP profiles can help to resolve two major questions about SEP injections (*e.g.*, Kahler 1994). The first question is whether ions accelerated in gamma-ray flares contribute to SEP events at 1 AU. The second is whether gradual phase SEP acceleration is carried out in a single shock or in two shocks, one in the corona and one in the interplanetary medium.

7.1.3 Ion Injection from the Flare Impulsive Phase

It is known from the gamma-ray observations that protons can be accelerated to energies of 1 GeV (*e.g.*, Chupp 1990, 1996) or higher (Mandzhavidze et al. 1993) during the flare impulsive phase. The appearance of high-energy neutrons from these events shows that they are produced in regions of high densities, and the events are conventionally modeled in terms of ions in a confined magnetic loop (*e.g.*, Ramaty et al. 1990; Ryan and Lee 1991). Because of high energies involved in these events, one may ask: Whether GLEs, which result from ions of energies of $E > 1$ GeV, are due to the same energetic particles produced in the impulsive phase gamma-ray events or result only from acceleration in the gradual phase coronal shocks? The association of impulsive soft X-ray events with some GLEs has suggested a possible role for the impulsive phase as the source of GeV protons (Kahler et al. 1991). In their survey of GLE onsets, Cliver et al. (1982) concluded that GeV proton injection could begin near the first significant microwave maximum (FSMP), but, in general, the inferred injection onsets were delayed from those maxima by up to 30 min (see Fig. 7.2).

Thus far, the only particles observed in GLEs and clearly associated with impulsive phases of flares are neutrons (see, *e.g.*, Chupp 1984, 1990, 1996) or neutron-decay protons (*e.g.*, Shea et al. 1991a). For example, in a well-connected GLE on 24 May 1990 a pulse of neutrons associated with a gamma-ray flare (Kocharov et al. 1993, 1996a) preceded the onset of the main phase of the GLE (Shea et al. 1991b). In Chap. 6 we have already discussed in detail a neutron nature of this event, nevertheless, it is timely to mention here its new aspects. As assumed by Shea et al. (1991b), the neutrons were produced at the flare site by the same population of protons which have been seen later in the GLE. However, the 15 min delay between the onset of the neutrons and that of protons, both of energies up to ~8 GeV and speeds of ~0.95c, shows that the proton injection was delayed by >10 min from that of the neutrons. This may imply that the GeV ions of the flare impulsive phase remained trapped in a coronal shock, as proposed by Debrunner et al. (1993). Further evidence that the high-energy SEPs of GLEs arise in the gradual phase was provided by the GLE of 21 August 1979, which was associated with a flare with a weak impulsive phase (Cliver et al. 1983).

There is also evidence that low-energy (E_p ~10 MeV) proton injection in the impulsive phase is not important for large-flux SEP events. In particular, Cliver et al. (1989) found a large scatter of four orders of magnitude in the ratios of flare gamma-ray line (GRL) fluences to peak interplanetary proton fluxes at ~10 MeV. Since the GRL emission is predominantly an impulsive phase phenomenon, the poor correlation suggests that impulsive phase injection plays a minor role in large SEP events. Furthermore, in several *Helios* 1 scatter-free large SEP events associated with gradual flares, the ~10 MeV proton injections were delayed by 10–20 min from the 0.5 MeV electron injections (Bieber et al. 1980; Kallenrode and Wibberenz 1991).

7.1.4 Two-Component Gradual-Phase Injection

A second problem in understanding SEP injection is to determine whether the gradual phase events consist of one or two component. To make the case for a two-component model, Evenson et al. (1982) examined the flux-time profiles of three large-flux SEP events, two of which were accompanied by interplanetary shocks. The first component is directly injected at the Sun and usually results in a peak or plateau in the flux-time profile at 1 AU within about a day. The second component is produced by an interplanetary shock and usually results in a peak flux at 1 AU 2 or 4 days later, coincidental with the shock arrival. Cane et al. (1988) assumed the two-component model and stressed the importance of the interplanetary-shock component in determining the resulting flux-time profiles of intense SEP events at 1 AU.

On the other hand, Lee and Ryan (1986) have shown that one can model the observed profiles at 1 AU with a constant rate of particle injection at a single shock propagating out to 1 AU from the corona. With large particle diffusion coefficients, the first component dominates the subsequent shock-associated component, while for small diffusion coefficients the roles are reversed. Kallenrode (1993a) has suggested a single-shock model in which the particle acceleration efficiency of the shock decreases with distance from the Sun. Numerical modeling by Kallenrode and Wibberenz (1993) suggests that the acceleration efficiency may be substantially stronger in the corona than in interplanetary space.

Cane et al. (1990) examined a number of large-flux events observed by *Helios* 1 spacecraft which were associated with both CMEs and interplanetary shocks. They argued for the two-component model, with the first SEP component somehow associated with the CME and second component due to interplanetary shock. However, nearly all fast interplanetary shocks are driven by fast CMEs (Cane et al. 1987), so the CME must therefore be associated with both SEP components. If SEPs comprising the first flux component are accelerated and injected only when the CME is high in the corona and the interplanetary shock is already formed, this would argue for the single acceleration model (Lee and Ryan 1986; Kallenrode and Wibberenz 1993) and against the two-component model. The SEP injection profiles of Fig. 7.4 (Kahler et al. 1990) appear to support this view. To test this result, Kahler (1994) examined additional SEP profiles and estimated the timing corrections to the profiles due to effects of interplanetary scattering.

Previous studies with *Skylab* and *Solwind* CMEs have shown that nearly all large E_p >10 MeV SEP events are associated with fast CMEs (u >400 km s^{-1}). Kahler (1994) compared heights of CMEs observed on the SMM spacecraft with the flux-time profiles of five associated SEP events (March 25, 1988; March 17, August 16, September 29, and October 24, 1989) observed by the *GOES* spacecraft, including three events observed as GLEs by neutron monitors. The SEP injection profiles as functions of the CME heights were estimated from the "solar release times" and the effects of interplanetary scattering. The proton solar release times were calculated by subtracting

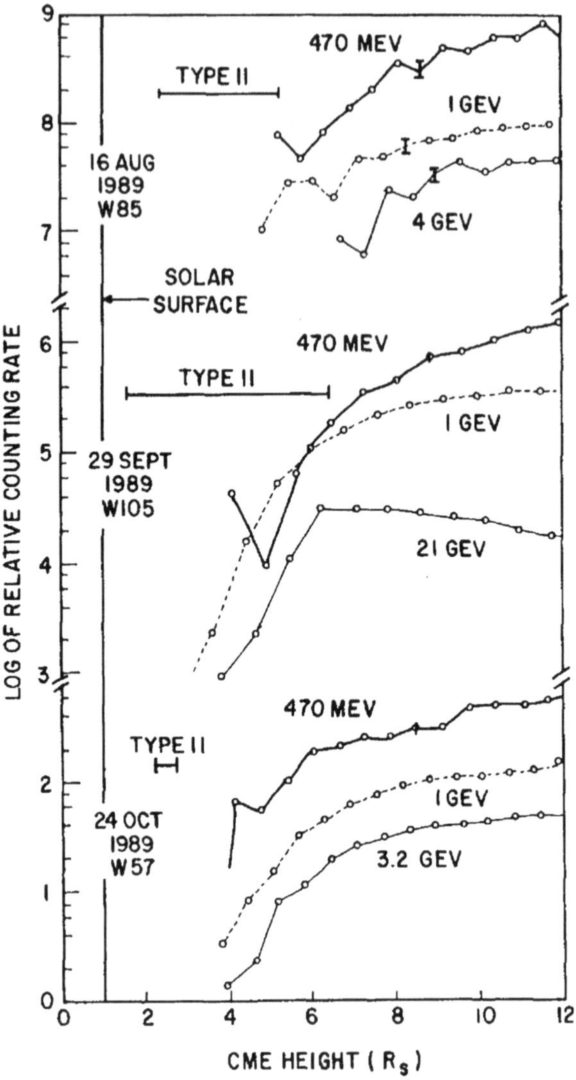

Fig. 7.5 Proton injection profiles (Kahler 1994) for the three GLEs (relative counting rates are given after background subtraction). Sample error bars are shown when they are larger than the data symbols. The profiles were plotted by the *GOES* HEPAD data for the >470 MeV protons and by the NM data at Kerguelen (1 GeV)-Climax (4 GeV), Calgary (1 GeV)-Mt. Norikura (21 GeV), and Kerguelen (1 GeV)-Moscow (3.2 GeV) for the GLEs of August 16, September 29, and October 24, 1989, respectively

$$\Delta t = 1.3\,\mathrm{AU}/v - 8.3\,\mathrm{min} \tag{7.3}$$

from the observed times at 1 AU. It was assumed that the SEPs travel 1.3 AU along the spiral field lines at a speed v with no scattering (*e.g.*, Toptygin 1985; Perez-Peraza et al. 1992).

The proton injection profiles as functions of CME heights are shown in Fig. 7.5 (Kahler 1994). For each event the plots show the background correcting counting rates from the *GOES* HEPAD P9 channel, which has an energy range of $430 < E < 505$ MeV (in average 470 MeV). Also are shown counting rate increases from two neutron monitors stations for each event. It was postulated (see Kahler 1994) that the average (effective) proton energy at each station to be twice the calculated cutoff energy: 1 GeV (Kerguelen, Calgary); 3.2 GeV (Moscow); 4 GeV (Climax); 21 GeV (Mt. Norikura). The duration of metric type II bursts (indicating the presence of the coronal shocks) are shown as reported in Solar-Geophysical Data (1989).

It was found that the peaks of the 470 MeV to 4 GeV injection profiles of the GLEs occur when CME heights reach ~(5–15)R_s or greater and that the onsets occur no earlier than the maxima of the flare impulsive phases. According to Kahler (1994), in those events SEP injection appears to result only from single CME-driven shock and not from the flare impulsive phase or from separate coronal and interplanetary shocks. In one small SEP event (on March 25, 1988) the impulsive flux-time profiles of >60 and >140 MeV protons are consistent with injection during the flare impulsive phase but could also be due to injection from a coronal shock over a limited time.

7.2 Reconstruction of Ejection Parameters

The observed time profiles of the intensity and anisotropy of SCR potentially provide a rich source of information about their ejection, as well as some transport parameters (*e.g.*, Toptygin 1973). In particular, as it was shown in a series of publications of Miroshnichenko and Sorokin (1985, 1986, 1987a, b, 1989), in some cases it becomes possible to reconstruct the energy spectrum of escaping particles, intensity-time ejection profile, and pitch angle distribution (PAD) of SCR near the Sun. This procedure of deconvolution of the source/ejection functions is carried out by a numerical solution of the inverse problem.

7.2.1 *The Inverse Problem in the SCR Studies*

The solution of the inverse problem is commonly reduced to the solution of multiple direct problems under different initial conditions, and the subsequent comparison with the observational data. In the cosmic ray variations, so far as we know, the only example of the direct solution of the inverse problem was provided by the method of the coupling coefficients or yield functions (see Chap. 10). The difficulties of direct solution of the inverse problems arise due to the fact that they are related to the class of *incorrect problems*. In order to that the problem be considered correct, its solution should satisfy three conditions: existence, uniqueness, and stability. It is commonly considered that the first condition is automatically fulfilled: if a measured effect is perceived then there is a reason for its existence. To prove the uniqueness and stability of the solution turned out to be not so trivial.

For these reasons until the early of 1960s it was thought that there is no sense to solving such problems. However, due to the efforts of Soviet mathematicians it has been shown (see, *e.g.*, Tikhonov and Arsenin 1979) that existing difficulties may be overcome if one use the a priori information on the characteristics of the expected solution (for example, the monotony or smoothness of searched function). In practical calculations the incorrectness is manifested due to the methodical uncertainties and/or statistical errors of s, which are inevitable in measurements of the particle fluxes, energies, *etc*. The method that enable to prove, within the error limits, the stability and uniqueness of the solution, was termed the *regularization method* (Tikhonov and Arsenin 1979).

If the transformation of the SCR flux in space is a linear process, then the solution of a boundary problem (or a Cauchy problem for a linear equation) can be written as follows

$$\int G(x, x_0) f(x_0) dx_0 = \Psi(x) \tag{7.4}$$

where $f(x_0)$ is the particle distribution function in the source for any parameters (initial or boundary conditions); $G(x, x_0)$ is the Green function characterizing the transformation of the SCR flux at its passage from the source to the point of observation; $\Psi(x)$ is the particle intensity measured, for example, at the Earth's orbit. It should be noted that x and x_0 in the general case represent any parameters ("coordinates") of the process $x = (x_1, x_2, x_3, \ldots, x_n)$, $x_0 = (x_{01}, x_{02}, x_{03}, \ldots, x_{0n})$, for example, the energy, time, space coordinates, pitch angle, *etc*. In other words, $f(x_0)$ can be a function of several variables, for example, $f(x_0) = f(E, t, \theta) = \varphi(E)\,\varphi(t)\,\varphi(\theta)$.

The expression (7.4) can be regarded as an integral equation of the first kind for $f(x_0)$, either the Fredholm equation (if the limits for the integration are constant), or the Volterra equation (if the upper limit is a variable quantity). One of the main difficulties in practical calculations is the reasonable choice of the Green function (a nuclei of the integral equation) which should describe adequately the particle transport in interplanetary space. A possibility to derive some parameters of accelerated particles by numerical solution of the inverse problem, for the first time, was considered by Toptygin (1973). He suggested to reconstruct the initial spectrum $f(E)$ based on the measured particle spectrum at the Earth's orbit, $D(E, r_E)$, and given (known) diffusion coefficient, $\kappa(E, r)$, in interplanetary space. The Green function obtained by Toptygin (1973), however, refereed to the case of quasi-stationary ejection of particle with energies of ~1–10 MeV from the solar atmosphere.

Taking into account the non-stationary nature of the SPE, it has been proposed (Miroshnichenko and Sorokin 1985) to reconstruct the SCR spectrum near the Sun, $D_s(E)$, from the data on intensity-time profile of accelerated particles near the Earth, $I(r_E, t)$. In this case it is necessary to use the time-dependent Green function, $G(r, t, E)$. So far, only in a few cases of a specific nature this function has been derived strictly. For example, Webb (1981) obtained the time-dependent Green functions for two cases: (1) $\kappa = \kappa_E = const$; (2) $\kappa = \kappa_E r$. In order to simplify the calculations, in many cases as a Green function one can use the solution of the diffusion equation in different modifications, for example, the solution by Krimigis (1965), where $\kappa(E,\ r) = \kappa_E(E)\,(r/r^b)$ and $0 < b\ < 1.0$. On the basis of described method Miroshnichenko and Sorokin (1985) attempted the first to reconstruct the ejection spectra for five proton events: September 28, 1961; February 5, 1965; May 28, 1972; August 7, 1972; and March 7, 1973. The choice of these events was dictated mainly by the existence pertinent data on the intensity-time profiles of protons with $E_p \geq 25$ MeV. Later it turned out to be possible to estimate in a similar manner the particle ejection spectra of relativistic protons (magnetic rigidity $R \geq 1$ GV) for three SPEs: November 19, 1949; December 7, 1982; and February 16, 1984 (see for details Miroshnichenko and Sorokin 1985, 1987a, b, 1989; Miroshnichenko et al. 1999, and references therein). Below we concentrate mainly on numerical calculations of the ejection functions and pitch angle distribution of escaping particles near the Sun by using the same integral equation (7.4).

7.2.2 Ejection Intensity-Time Profiles

If the SCR distribution law in the interplanetary is known, then the following integral equation, instead of (7.4), can be written

$$\int_0^t \varphi(t_0,\ E)\,G(t - t_0, r, E)\ \mathrm{d}E\ \mathrm{d}t_0 = I(r, t) \tag{7.5}$$

where the function $\varphi(t_0, E)$ describes the particle distribution in the source in terms of the time and energies; $I(r, t)$ is the observed intensity-time profile at a distance r (for example, near the Earth's orbit, $r = r_E$). For rather narrow energy intervals $(E_1–E_2)$ it can be assumed that the ejection time profile is about the same for particles of all energies, i.e., $\varphi(t_0, E) = \varphi(t_0)E^{-\gamma}$, and the equation (7.5) acquires the following form

$$\int_0^t \Psi(t_0) \int_{E_2}^{E_1} G(t - t_0, r, E)\,E^{-\gamma}\,\mathrm{d}E\mathrm{d}t_0 = I(r, t) \tag{7.6}$$

This expression can be regarded as the integral Volterra equation of the 1st kind in relation to ejection function $\Psi(t_0)$. The test computations showed (Miroshnichenko and Sorokin 1986) that the accuracy is best warranted if the right part in (7.6) is specified in the time interval of $0.1t_m \leq t \leq 1.5t_m$, where t_m is the observed peak time near the Earth. To reconstruct the function $\Psi(t_0)$ for a real proton event, namely, for the well-known GLE of February 23, 1956, the authors used observed profiles of neutron component at the stations Greenwich (the cutoff rigidity $R_C = 2.7$ GV) and

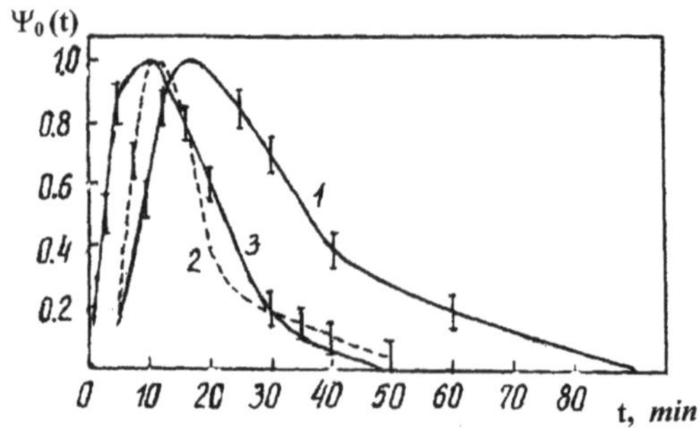

Fig. 7.6 Ejection intensity-time profiles of relativistic protons (Miroshnichenko and Sorokin 1986, 1989) derived by ground observation data for the GLEs of November 19, 1949; December 7, 1982; and February 23, 1956 (curves *1*, *2*, and *3*, respectively)

Berkeley ($R_c = 4.58$ GV). It was shown that the two reconstructed ejection profiles are closely similar, and within the error limits match each other. Their most distinctive feature is that the front of the increase of the number of ejected particles, or ejection function, $N_s(t)$, was found to be quite outstretched: the maximum occurs about 10 min after the beginning of the particle generation. The method based on the equation (7.6) enabled also to reconstruct the ejection profiles of relativistic protons for the GLEs of November 19, 1949 (by the data of the ionization chamber in Yakutsk, $R_c = 1.65$ GV) and December 7, 1982 (by the NM data at the Apatity station, $R_c = 0.61$ GV). As a nuclei of the integral equation in this case Miroshnichenko and Sorokin (1986, 1989) used the same Green function as it was used to reconstruct the source energy spectra for some SPEs.

The profiles obtained are depicted in Fig. 7.6 (curves 1 and 2, respectively). For a comparison the injection function for the GLE of February 23, 1956 (curve 3) is shown (Miroshnichenko and Sorokin 1989). The profiles are normalized to 1.0 at the moment of the peak ejection flux. Judging by the shape of the profiles, the processes that remove the relativistic particles from the solar atmosphere are of non-diffusive nature (at least for the three events indicated). Common features of the profiles are the rapid increase to the maximum, the asymmetry relative to the moment of ejection peak flux, and exponential decay beyond the maximum. A singular characteristic of the profile is its FWHM (Full Width Half Maximum): 25, 19, and 12 min., for the event of November 19, 1949; February 23, 1956; and December 7, 1982, respectively. Notice that, in fact, these profiles correspond to quite different effective energy of escaping particles, namely, ~15–20, ~5–7, and ~1 GeV, respectively.

A more detailed analysis of the particle ejection profiles should be made using the data on the observed time profiles of the flare photon fluxes at high energies. Such data are available, for example, for the event of December 7, 1982. In this particular case, the *SMM*/GRS recorded strong emissions in the nuclear energy band of 4–8 MeV and above 10 MeV (*e.g.*, Rieger et al. 1987; Vestrand et al. 1999). In addition, Rieger et al. (1987) estimated the flux above 25 MeV (resulting from pion decay) expected in the case if the spectral form of the directly measured particles, $\sim E^{-\gamma}$, and of the particles which caused the gamma rays, was the same. Recall that the main sources of the gamma rays with energies of 4–8 and >25 MeV are, respectively, the CNO nuclei excited by accelerated ~5–50 MeV protons, and neutral pions produced in the >180 MeV proton collisions with the H and He nuclei in the solar atmosphere.

Figure 7.7 shows the time history of the flare of December 7, 1982 in hard X- and gamma-rays, as well as the intensity-time profile of the GLE as it was recorded by the neutron monitor at Kerguelen (Rieger et al. 1987). One can see that the photon emission below 10 MeV lasts for more than 30 min. There are three distinct pulses at the profiles of hard X-rays and in the nuclear 4–8 MeV band, their peak values being grown subsequently (the event belongs to the class named gradual). The emission observed above 25 MeV, however, does not follow the time history of the nuclear GR lines. Note that the significant (though moderate) increase of the counting rate at the Kerguelen neutron monitor was observed only in coincidence with the third (largest) pulse, and its halfwidth was about 7–8 min.

Rieger et al. (1987) calculated the >25 MeV photon flux under the assumptions that: (1) the distribution of accelerated protons in the source was isotropic, and (2) their spectral index was $\gamma = 2.8$ in the entire range of energies $E_p = 5$ MeV – 5 GeV. It was found that the calculated time profile contains three distinct peaks also for the >25 MeV photons. This does not consistent with observed time picture in Fig. 7.7 (bottom panel), where significant peaks, in practice, are indiscernible at the background of large fluctuations. This discrepancy is very likely due to the flux of accelerated >180 MeV protons was not isotropic. Moreover, their spectral index in relativistic range (E_p >500 MeV), contrary to the above suggestion ($\gamma = 2.8$), seems to be noticeably larger, $\gamma = 4.5$ (Miroshnichenko and Sorokin 1989). At any rate, the observed time profile of the >25 MeV photon flux has a likeness neither with the observed NM intensity profile, nor with the reconstructed ejection profile of relativistic protons. In turn, an absence of significant photon flux at >25 MeV is an indirect evidence of the source spectrum softening at high proton energy.

7.2.3 Angular Distribution of Escaping Particles

Using the method described in Sect. 7.2.1, it also turned out to be possible to reconstruct the pitch angle distributions (PADs) of escaping particles based on the observed PADs for some SPEs. It should be noted that the observational data on the PADs near the Earth are rather limited, so the PADs near the Sun have been reconstructed only for several GLEs: May 7, 1978, February 16, 1984, September 29 and October 22, 1989 (Miroshnichenko and Sorokin 1986, 1987a, b; Miroshnichenko et al. 1997, 1998). In this case the integral equation (7.4) has the form

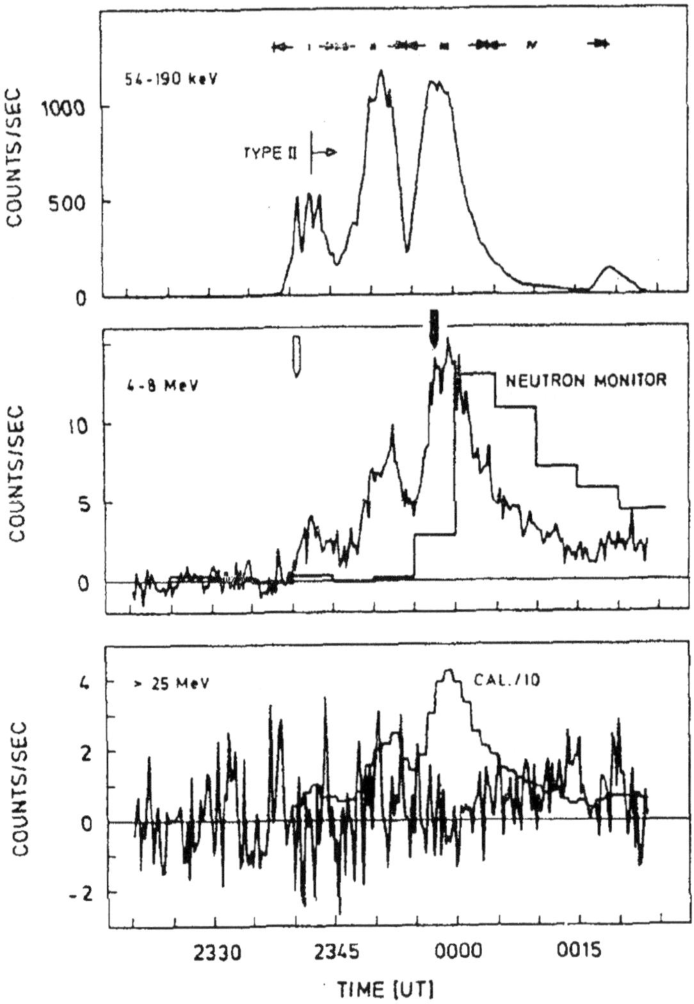

Fig. 7.7 Time history of the event of December 7, 1982 in hard X- and gamma-rays (Rieger et al. 1987). The histogram in the middle panel is a short period of the Kerguelen neutron monitor record. The full arrow indicates the release time at the Sun (+8.33 min) of a 1 GeV protons; the open one corresponds to the time of the first significant microwave peak at 17 GHz at the Nobeyama Observatory

$$\int_{-\pi/2}^{\pi/2} f(\theta)\, G(z, \theta; z_0, \theta_0)\, d\theta_0 = \varphi(\theta) \tag{7.7}$$

where and θ_0 and θ are, respectively, the initial and current pitch angles of the particle in the interplanetary magnetic field (IMF); the functions $\varphi(\theta)$ and $f(\theta)$ describe the observed and reconstructed PADs, respectively; z_0 and z are,

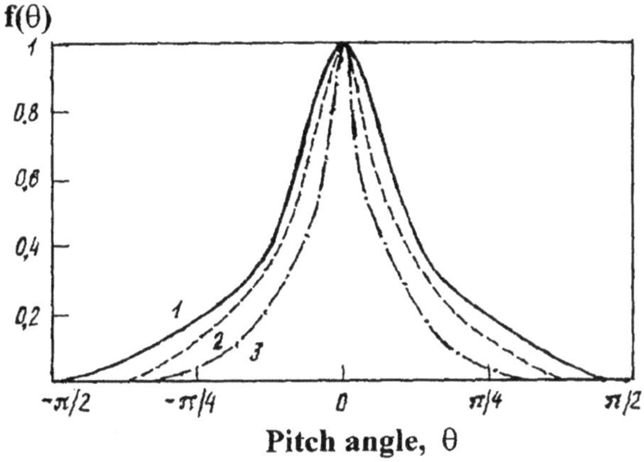

Fig. 7.8 Pitch angle distributions of solar relativistic protons near the Earth in the event of February 16, 1984 (curve *1*, Bieber et al. 1986) and reconstructed ones (Miroshnichenko and Sorokin 1987b) for the two values of the mean square angle $<\theta^2> = 0.07$ and 0.3 (curves *2* and *3*, respectively)

respectively, the position of the particle source and the distance from the Sun along the guiding line of force of the IMF. In application to the problem under consideration it seems more feasible to describe the SCR distribution on the basis of solution of the kinetic equation for anisotropic phase of the SPEs (*e.g.*, Toptygin 1985), taking into account the conservation of the adiabatic invariant and particle scattering on the IMF inhomogeneities. Therefore, as a nuclei of the integral equation (7.7) Miroshnichenko and Sorokin (1986, 1987a, b) used the Green function obtained from the kinetic equation by Dorman et al. (1973). When analyzing the PAD data for the GLEs of September 29 and October 22, 1989, Miroshnichenko et al. (1998) relied upon the solution of the transport equation (Earl 1995) as a nucleus of the integral equation (7.7).

An important parameter in the computations of $f(\theta)$ using the equation (7.7), is the mean square pitch angle $<\theta^2>$in the Green function. Some uncertainty in the choice of $<\theta^2>$ does not enable the function $f(\theta)$ near the Sun to be unambiguously defined; however, according to the estimates by Toptygin (1985), it can be assumed that $<\theta^2> <1.0$, and probably $<\theta^2> <0.5$. In Fig. 7.8 we show the results of reconstruction of the PAD near the Sun, $f(\theta)$, for the GLE of February 16, 1984, together with the data obtained by Bieber et al. (1986) for the PAD near the Earth (curve 1). The results of computations are given for two values of $<\theta^2>$: 0.07 and 0.3 (curves 2 and 3, respectively). It can be seen that in both cases the particle distribution near the Sun is more narrower than at the Earth. This can be explained (Miroshnichenko and Sorokin 1987b) by the prevalence of the scattering over the adiabatic focusing as the IMF decreases with the distance from the Sun. At the beginning of the region of transformation of the $f(\theta)$ function (near the source) the picture may be opposite.

In conclusion of this discussion, we note that the inverse problem method was applied by Guglenko et al. (1990b) to estimate the number of neutron produced at the Sun in the flare of 3 June 1982. Some later, Ruffolo et al. (1997) developed a new deconvolution technique to determine ejection profiles and spectra of SCRs assuming that the interplanetary transport is mainly parallel to an Archimedean magnetic field, with a constant radial mean free path, Λ. It is important to emphasize that this automated fitting technique yields the flux of particles ejected onto the observer's magnetic line near the Sun as a function of time and particle energy. In spite of some simplifying assumptions, the technique have yielded promising results in the non-relativistic energy range of 27–147 MeV for several SPEs (July 20, 1981, January 2, 1982, and others).

7.3 Relativistic Particles in Extended Coronal Structures

From the very beginning of ground-based observations of solar cosmic rays (SCR) it became clear (see Chaps. 2 and 4) that solar particles of relativistic energies (above 500 MeV for protons) provides an unique opportunity to obtain new information of great interest about particle acceleration processes in space plasma and to make clear some characteristics of the solar accelerator (short acceleration time, upper intensity and energy limits for accelerated particles, etc.) under extreme astrophysical conditions. A high accuracy of ground observations by neutron monitors makes possible, in particular, to study fine temporal structure GLEs and to estimate a number of important parameters of the SCR sources. Moreover, during two last decades there were discovered some GLE peculiarities which may significantly change traditional interpretation of SCR generation and transport (see, *e.g.*, Borovkov et al. 1987; Smart et al. 1987a, 1991; Torsti et al. 1991, 1992; Perez-Peraza et al. 1992; Vashenyuk et al. 1993, 1995, 1997; Cramp et al. 1995a, b, c, 1997; Lovell et al. 1998; Miroshnichenko et al. 1995a, b, 1996, 1998, 2000; Shea and Smart 1996b, 1997a; Miroshnichenko 1997; Karpov et al. 1998).

7.3.1 Unusual Features of Intensity-Time Profiles

The form of intensity-time profile contains important information about the duration of SCR ejection and their azimuthal propagation through the solar corona. Based on the ground observation data, Borovkov et al. (1987), for the first time, pointed out to a possible existence of two separate relativistic components in certain GLEs. Figure 7.9 shows four typical profiles for GLEs registered in 1981–1984 by neutron monitor at the Apatity station (geomagnetic cutoff rigidity $R_c = 0.57$ GV). One can easily see a systematic difference in the form of SPE profiles, namely, these four events may separated on two groups – "prompt" and "delayed" ones with the narrow (sharp) and broad (smooth) profiles, respectively. Such a difference was interpreted by Borovkov et al. (1987) as possible manifestation of two populations of relativistic particles – prompt and delayed components (PC and DC) – in certain SPEs.

As mentioned above, the events of solar cycle 22 also demonstrated several peculiarities which need to be interpreted on the new concept base. In particular, the

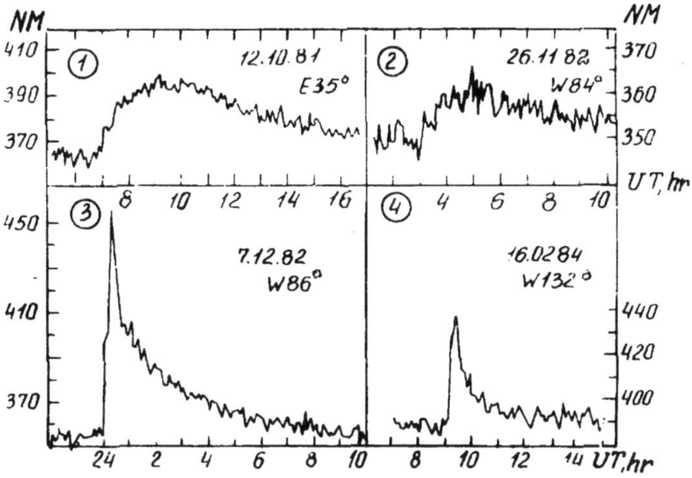

Fig. 7.9 Intensity-time profiles of four GLEs registered in 1981–1984 by neutron monitor at the Apatity station ($R_c = 0.57$ GV)

Fig. 7.10 Intensity-time profiles of the May 24, 1990 event by the NM data in Hobart ($R_c = 1.84$ GV) and Tixie Bay ($R_c = 0.48$ GV)

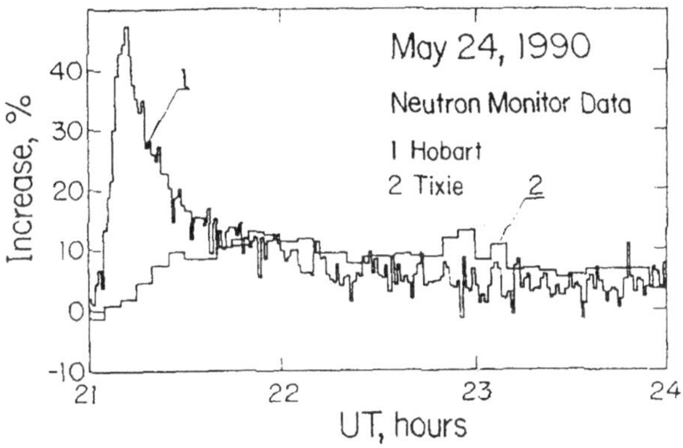

shape of intensity-time profile for a number of events displays some features that possibly imply the presence of two SCR components. For example, Fig. 7.10 illustrates the counting rate profiles obtained by neutron monitors at Tixie Bay ($R_c = 0.48$ GV) and Hobart ($R_c = 1.84$ GV) during May 24, 1990. This event is classified as a prompt one, and the time profiles in Fig. 7.10 show a distinct two-peak structure. The first (sharp) peak, in our opinion, corresponds to the arrival of the PC, and the second (smooth) peak is probably due to the DC.

Similar effect was recorded on October 22, 1989. In Fig. 7.11 we present the intensity-time profiles observed by NMs at the Antarctic station South Pole ($R_c = 0.09$ GV) (Bieber et al. 1990) and in the Northern hemisphere, Oulu ($R_c = 0.78$ GV) and Apatity ($R_c = 0.57$ GV). It is interesting to note that in the Northern hemisphere (for example, at Thule, Greenland, $R_c = 0.00$ GV) the time

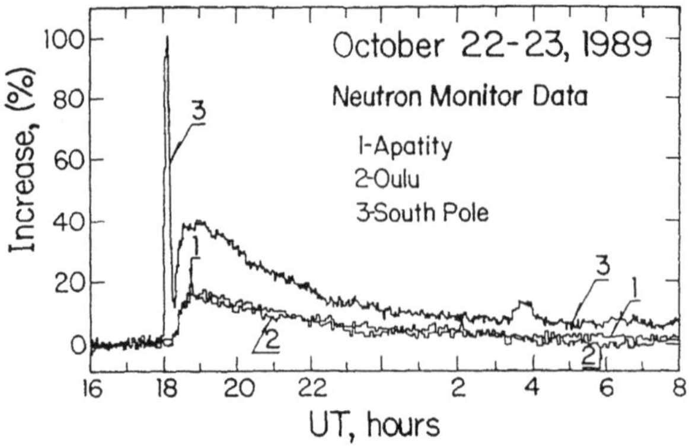

Fig. 7.11 Intensity-time profiles of the October 22, 1989 event by the NM data at three different stations: *1* Apatity ($R_c = 0.57$ GV), *2* Oulu ($R_c = 0.78$ GV) and *3* South Pole ($R_c = 0.09$ GV)

profile of this event was rather smooth, and according to the Apatity NM data the event was not classified as a prompt one. A less distinct but noticeable two-peak structure was observed at Apatity during the events of May 21 and May 26, 1990.

Retrospective analysis of several GLEs observed in the 19th solar cycle (Shea and Smart 1996b) also exhibit an unusual initial anisotropic spike, as it was found, in particular, for the event of November 15, 1960. These structures were typically recorded by polar stations having narrow asymptotic cones of acceptance (see Fig. 1.8) presumably viewing in the direction of the initial incoming particle flux. In general, as noted by Shea and Smart (1996a, b), the identification of the "spike-like" structures in several SPEs of the 22nd solar cycle, due to increased time resolution, indicates that these type of events may be more common than originally thought.

7.3.2 *Evidence of Two-Phase Structure*

The peculiarities of many GLEs from Table 2.1 were analyzed in some details by Vashenyuk et al. (1993, 1994, 2008), Miroshnichenko et al. (1995a, b, c, 1996), Perez-Peraza et al. (2008). In addition to a visual intercomparison of the intensity-time profiles measured by different NMs for the same event, there were used so-called vT_m-technique (v is the particle velocity and T_m – the time of maximum increase at 1 AU) proposed by Reinhard and Wibberenz (1973, 1974), and the distribution of GLEs on a specific parameters $T_{1/2}$ – the width of intensity-time profile at its half height (Vashenyuk et al. 1993). Besides, the data on SCR anisotropy and energy spectra at different phase of SPEs have been taken into account.

The probable existence of two components of relativistic SCRs may be demonstrated by the data of the GLE of December 7, 1982. In Fig. 7.12 we show observed intensity-time profiles at two cosmic ray stations (Deep River and Kerguelen) which have almost identical cutoff rigidities ($R_c = 1.14$ GV), but located in opposite

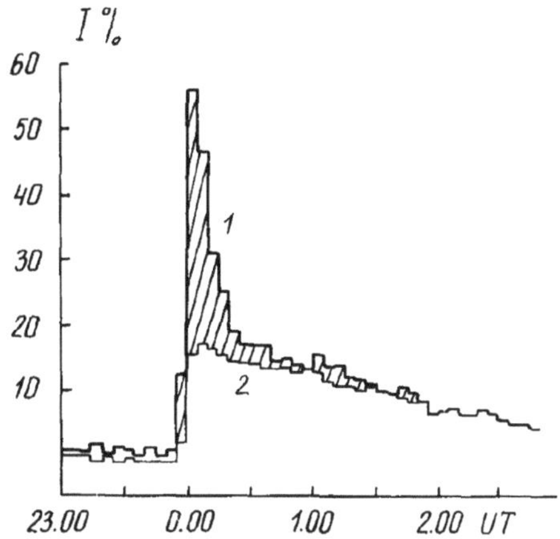

Fig. 7.12 Intensity-time profiles of the event of December 7, 1982 by the data of two neutron monitors of Kerguelen (*1*) and Deep River (*2*), located in the opposite hemispheres of the Earth. The cross-hatched difference between two profiles corresponds to the prompt component contribution (Perez-Peraza et al. 1992)

hemispheres of the Earth. According to Smart et al. (1987a), all stations whose asymptotic cones of acceptance were viewing the "forward" flux propagating along the IMF direction recorded an impulsive increase, with a maximum of 56 % by the 5-min data at the neutron monitor of Kerguelen Island.

Neutron monitors whose asymptotic cones of acceptance were viewing primarily the "reverse" particle flux opposite to the IMF direction (for example, the NM at Deep River), observed a gradual increase over about 30 min. It is remarkable that a magnitude of "reverse" flux was about 1/4 of the maximum increase observed by "forward-viewing" neutron monitors at an equivalent cutoff rigidity. The spectrum of the high amplitude flux turned out to be harder than that of "reverse" flux: with a simple power law in rigidity, $D(R) = D_0R^{-\gamma}$, the difference in the γ values was about 1.0. In addition, during the entire event there was a persistent bi-directional anisotropy in the relativistic proton flux. Thus, the observations of this event at high rigidity certainly indicate a two-phase structure (Smart et al. 1987a).

At the same time, the scattering of particles in interplanetary space near the Earth seemed to be relatively efficient. In fact, the stations with asymptotic cones viewing in the "reverse" direction (for example, neutron monitor at Deep River) started to record the event within 5 min after the stations viewing the "forward" propagating flux. Therefore, it is rather difficult to separate the contribution of impulsive peak at the intensity-time profiles of the neutron monitors viewing in the "reverse" direction.

Comparing two profiles observed at the NM Kerguelen (*1*) and Deep River (*2*), ascribing the cross-hatched region of profile *1* to the prompt component and calculating the difference in areas occupied by profiles *1* and *2*, one can to obtain that the PC amounts to about 25 % of the integrated flux of relativistic particles in this particular event (Perez-Peraza et al. 1992), *i.e.*, the ratio PC/DC = 1/3. If the DC does dominate, it imposes certain limits on the parameters of the PC generation in two-component events (see below Sect. 7.6).

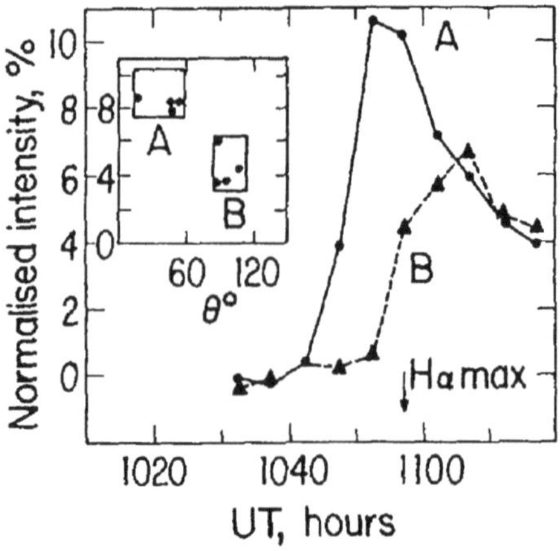

Fig. 7.13 Prompt (*A*) and delayed (*B*) components of solar cosmic rays in the GLE of November 18, 1968 (Duggal 1979). The angle $\theta = 0°$ corresponds to average IMF direction

7.3.3 Anisotropy Data

The existence of two components in certain GLEs is evidently proved by the anisotropy data. Figure 7.13 shows time profiles of anisotropic, *A*, and scattered, *B*, components in the GLE of November 18, 1968 (Duggal 1979). These profiles are averaged on the NM stations shown by points located in the boxes *A* and *B* at the diagram (upper left) displaying the dependence of the increase at given station on the angle distance of their mean asymptotic direction from the anisotropy axis $(\theta = 0°)$.

It is seen that *A* and *B* profiles are similar to the profiles of prompt and delayed SPE (*cf.* Fig. 7.9), respectively. A shift in time between their onsets, in this particular case, is about 20 min. So, it seems likely (Miroshnichenko et al. 1995a, b, c) that profiles *A* and *B* in Fig. 7.13 are the intensity-time profiles of the PC and DC in the November 18, 1968 event. The anisotropy characteristics of both components may reflect the properties of their sources at the Sun. Obviously, the PC must have an anisotropic source located in the region with open field lines, probably rather high in the corona. As to the DC, it must have principally different source of particles, a "diffusive" or isotropic one, associated, for example, with the expanding magnetic bottle (Schatten and Mullan 1977; Mullan 1983).

If an equilibrium exists between the processes of scattering and focusing in the interplanetary medium (Earl 1976), *i.e.*, if the propagation conditions do not change significantly during the time of SPE development, then the pitch angle distribution (PAD) of solar particles will be retained in course of their transport from the Sun to the Earth. The situation shown in Fig. 7.13 is typical for many SPEs (*e.g.*, Shea and Smart 1996a, b), and in all these cases one can not always find any suitable shock or any other ideally reflecting boundary behind the Earth's orbit producing the scattered, or "reversed", component as it is often supposed (*e.g.*, Shea and Smart 1982). So, PC is evidently ejected from the corona in an anisotropic manner, as confirmed also (see Fig. 7.8) by calculations of Miroshnichenko and Sorokin (1987b), and DC seems to have an isotropic source.

7.3.4 Spectral Differences

The PC and DC can be distinguished also due to differences in the slope of their rigidity (energy) spectra. Long ago by the data of February 23, 1956 event it was already noted (Pfotzer 1958) that the momentum spectrum of "direct" fraction of radiation observed in the "impact zones" (anisotropic component) was harder than the spectrum of "deflected" (isotropic) one (see, *e.g.*, Fig. 4.4). At present level of our understanding, this spectral difference may be treated (Vashenyuk et al. 2008) as a manifestation of two-component nature of the GLE05. Abnormally hard spectra were also recorded in two similar GLEs of May 7, 1978 (Shea and Smart 1982) and February 16, 1984 (Bieber et al. 1986; Miroshnichenko and Sorokin 1987a). The event of February 16, 1984 is of special interest because it was recorded in the PC form only (Miroshnichenko et al. 1990).

As it was shown (*e.g.*, Cramp et al. 1997; Dvornikov and Sdobnov 1998), the October 22, 1989 GLE has an extremely anisotropic onset (see Fig. 7.11). In particular, none of the neutron monitors with narrow asymptotic cones looking outwards of the Sun recorded any significant increase until about 1815 UT (Cramp et al. 1997). Between 1815 and 1820 UT, there was a significant increase in the count rates of these monitors. Further evolution of the PAD implies a bidirectional flow of particles along the local IMF line. It is of great interest that the initial sharp "spike" in intensity was also observed as an increase in the spin-averaged proton flux at energies between 36 and 550 MeV by particle detectors at geosynchronous orbit (Nemzek et al. 1994; SGD I, No.543, p.14, 1989).

The picture of the spectrum behaviour in time was also rather complicated. The rigidity spectra of relativistic protons derived by Cramp et al. (1997) at the Earth for 18:05, 18:20, 19:00, and 19:20 UT are shown in Fig. 7.14. It is seen that at the event onset (first spike-like peak in Fig. 7.11 at 18:05 UT) the spectrum was considerably harder than about ~1 h later (second maximum at the NM South Pole at 19:00 UT), though at 18:05 UT the spectrum turned out to be slightly steeper (softer) than at 18:20 UT. Cramp et al. (1997) used the form and evolution of those spectra to give some evidence of shock acceleration mechanism. They modified a theoretical form of shock acceleration spectrum by Ellison and Ramaty (1985) to obtain its rapid steepening at high rigidity. A comparison of the modified form with the derived spectrum at 18:40 UT (as well as at other times nearby the second peak in Fig. 7.11) indicates a close agreement of two spectral forms.

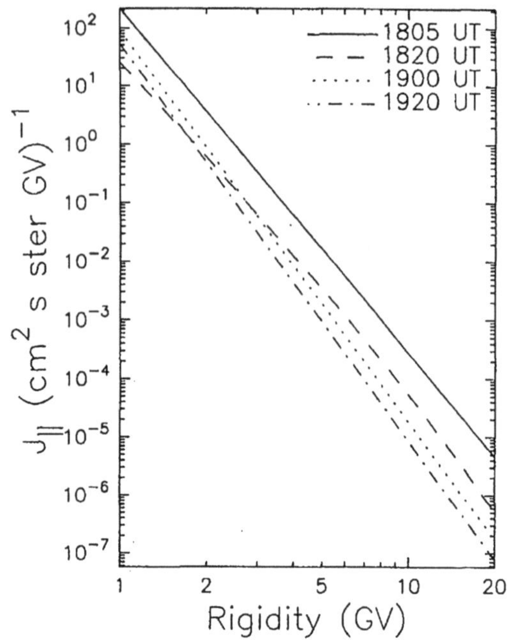

Fig. 7.14 Rigidity spectra of relativistic protons for several time intervals during the GLE of October 22, 1989 (Cramp et al. 1997)

7.3.5 Width of Intensity-Time Profile

To derive new additional information from observational GLE data, Vashenyuk et al. (1993) proposed to use a specific parameter $T_{1/2}$ – the half-width of the intensity-time profile as measured at the level of half of maximum intensity. This parameter seems to be a measure of the time the main bulk of relativistic particles spend in the corona after acceleration. The time spent by relativistic protons in interplanetary space may be hardly as large as several hours.

Figure 7.15 shows the GLE half-width as a function of heliolongitude of the proper flare (SCR source). Unfortunately, from total statistics of 54 GLEs detected before 1993 (Shea and Smart 1993a, b) only 43 events have distinct time profiles suitable to the analysis, 31 of them being recorded by the Apatity NM since 1969. For some earlier events occurred in 1956–1968, Vashenyuk et al. (1993, 1994) used the data from other high-latitude NMs with the same cutoff rigidity (~0.5 GV). If the event has a spike-like profile with $T_{1/2}$ <1 h, it is considered as a "prompt" one; in the case of $T_{1/2}$ >1 h, the event is called "delayed".

From the point distribution plotted in Fig. 7.15 one can separate two distinct event groups. One of them, at $T_{1/2}$ >1 h, has nearly V-shaped form with the minimum at W50°, *i.e.*, nearby the nominal longitude of the footpoint of the Sun-Earth conjunction IMF line. Such a distribution is similar to that of spectral index for solar protons in the range of 20–80 MeV (van Hollebeke et al. 1975). All the events of this group have as a rule more or less smooth ("diffusive") profile, with low anisotropy near the time of maximum intensity. Notice that the "eastern"

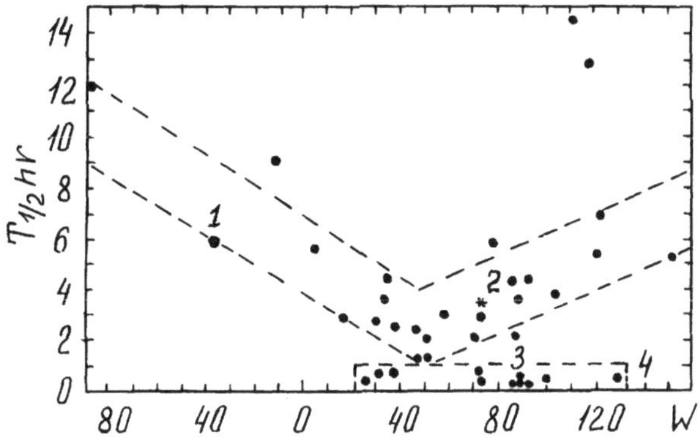

Fig. 7.15 Heliolongitude distribution of the 43 GLEs (*solid circles*) on the parameter T1/2 – the half-width of their intensity-time profile (Vashenyuk et al. 1993, 1994). The numbers *1*, *2*, *3*, and *4* denote the events of October 12, 1981; November 26, 1982; December 7, 1982; and February 16, 1984. The event of September 29, 1989 is marked out by an asterisk

event of October 12, 1981 and "western" one of November 26, 1982 (in Fig. 7.15 the points *1* and *2*, respectively) turned out to be the events of the same class but belonging to different wings of the *V*-shaped distribution.

The events of second group (points inside the box) are characterized by short duration, $T_{1/2} < 1$ h, and high degree of anisotropy. Most of them were caused by the flares occurred near the west limb of the Sun or behind it. Typical events of this group are GLEs of December 7, 1982 and February 16, 1984 (in Fig. 7.15 the points *3* and *4*, respectively). There is no heliolongitude dependence on the event duration, but, on the other hand, there is a slight tendency to avoid by them the optimal heliolongitudinal interval of 50–70° W. Notice that some events may be not pure "prompt" or "delayed", but "combined", or "mixed", ones. The events of this kind are characterized by initial spike and complex two-peak structure of the entire profile, as it was observed, for example, on September 29, 1989 (this event is marked by an asterisk in Fig. 7.15).

It is timely to present a complete list of the GLEs with well-defined prompt component. As the main features of PC we consider the pulse-like time profile (or pronounced initial pulse) and very hard spectrum. Proceeding from this definition, the following 16 events were found to contain the prompt component with proton energy $E_p > 3$ GeV (Vashenyuk et al. 1993, 1994, 2011; Shea and Smart 1996b): November 19, 1949; February 23, 1956; May 4, 1960; November 15, 1960; July 20, 1961; November 18, 1968; November 22, 1977; May 5, 1978; December 7, 1982; February 16, 1984; September 29, October 22 and November 15, 1989; May 21, May 24 and May 26, 1990. In the case of September 29, 1989 event the PC at the energies $E_p < 3$ GeV was distinguished, too. From this point of view, some other GLEs since 1942 (in particular, the events of July 17, 1959; February 25, 1969; August 7, 1972, and April 10, 1981) should be additionally examined (Shea and Smart 1996b). More extended analysis of all 70 GLEs (Table 2.1) have been carried out quite recently (Vashenyuk et al. 2011), and summary data on their energy spectra for 35 GLEs from 71 ones (Table 2.1) are presented in Table 9.4 (Chap. 9).

7.3.6 Data Analysis by vT_m-Technique

Probable existence of two separate SCR components is revealed very visually by the method proposed by Reinhard and Wibberenz (1973, 1974), followed and developed by other researchers (*e.g.*, van Hollebeke et al. 1975; Ma Sung et al. 1975; Bazilevskaya and Vashenyuk 1979). As follows from those works, for the most SPEs the following relations take place:

$$vT_n = A_n + B_n v \tag{7.8}$$

$$vT_m = A_m + B_m v \tag{7.9}$$

where v is the velocity of the particle, T_n and T_m are the times of the SPE onset and maximum, respectively. The parameters A_n and A_m have a simple physical meaning – they are the summary interplanetary paths of the first particles and the main bulk of them, B_n and B_m are the times spent by particles of the respective population in the corona, all these parameters constants for a given SPE. The products vT_n and vT_m are, respectively, the total distances traveled by first particles and main bulk of them from the moment of generation to the arrival at the detector. A region of applicability (linearity) of the relations (7.8) and (7.9), however, turned out to be limited: they are valid at low and moderate energies of solar protons, but are violated at the energies of E_p >100 MeV (Bazilevskaya and Vashenyuk 1979). Just this interesting fact was used to separate the prompt component in a series of SPEs (Vashenyuk et al. 1993, 1994, 1997; Miroshnichenko et al. 1990, 1995a, b, c, 1996; Vashenyuk and Miroshnichenko 1998; Karpov et al. 1998).

The reliability of the results obtained by the vT_m-technique should be apparently dependent on a correct choice of the time of generation for the particles with a given energy. As it was discussed above (Sect. 7.1.1), when analyzing the times of ejection of the ~2 GeV protons and ~100 keV and ~1 MeV electrons, Cliver et al. (1982) found two the most appropriate time of generation of those particles – the occurrence time of the "first significant microwave peak" (FSMP) at the frequencies of >9 GHz and the onset of type II radio burst. In their study, Cliver et al. (1982) have made the choice in favour of the FSMP. It should be pointed out, however, that the FSMP precedes in time the type II onset within 1–3 min only, and using the last one instead of FSMP will not influence significantly the results of further vT_m-analysis. We prefer the onset of type II radio burst as more reliable signature of the particle ejection. In addition, this onset is near the moment of flash phase of a flare, and it is thought to correspond to the moment when the particle acceleration and generation of a shock wave are originating (*e.g.*, Mullan 1983). In some cases, however, the type II radio bursts were not recorded before the GLEs (*e.g.*, Kahler et al. 1991). As a moment of particle generation in the event of this kind one can choose a peak time of the soft X-ray burst (1–12 Å) since this time is also near to the flare phase when the type II burst starts.

With this information in mind and based on the relations (7.8) and (7.9), a number of GLEs have been analyzed (October 12, 1981; November 26, 1982;

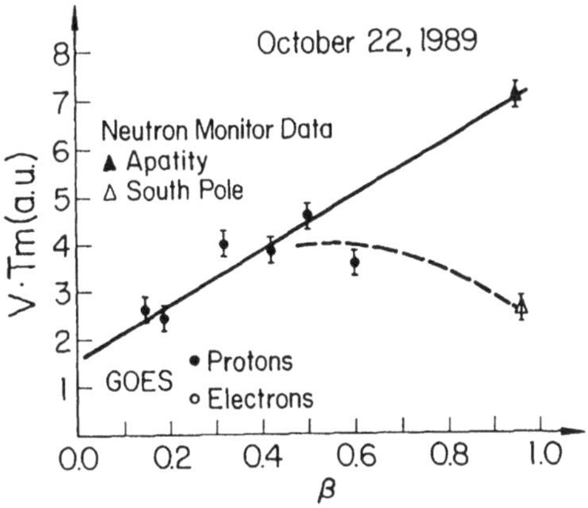

Fig. 7.16 Results of vT_m-analysis of intensity-time profiles for the October 22, 1989 event (Miroshnichenko et al. 1995a, b, c, 1996). Two different components of solar protons, DC (*straight line*) and PC (*curve*) are clearly separated up to the energies as low as 200 MeV

December 7, 1982; February 16, 1984; September 29 and October 22, 1989; May 24, 1990). In addition to the NM data, there were used also the data of stratospheric measurements of solar particles and the results of riometric observations, as well as the satellite data from *Meteor* (Avdyushin et al. 1984), *ICE* (Bieber et al. 1986) and *GOES* (Solar-Geophysical Data, 1983–1990), the particle velocity v being normalized to the speed of light c (for the proton with the energy of $E_p = 1$ GeV a value $v/c = 0.87$). It was found that the delayed events of October 12, 1981 and November 26, 1982 consist of the delayed component, with $B_m \approx 2.1$ and 1.8 h, respectively. The prompt event of December 7, 1982 contains the delayed component including low-energy protons and electrons of >2 MeV as well as the prompt component for which the storage time in the corona is close to zero ($B_m \approx 0$). It was evidenced again that the February 16, 1984 event seemed to consist of the PC alone ($B_m \approx 0$).

Figure 7.16 demonstrates the results of vT_m-analysis for the GLE observed on October 22, 1989 (Miroshnichenko et al. 1995a, b, c, 1996). It is seen, in particular, that this event had both components, the PC being registered by NMs (above 500 MeV) as well as by proton detector on board the *GOES* (above 200 MeV), meanwhile the DC was registered in the entire range of SCR energies, at least, starting from 30 MeV up to several GeV. The same is true for the May 24, 1990 event, the DC being present also in relativistic electron population (>2 MeV).

One can discover two components (PC and DC) in the population of the first particles, too. Lack of the data on the "first" particles does not allow us to construct the plots like that shown in Fig. 7.16. However, based on the relations (7.1) and (7.2), it is easily to evaluate the coronal storage time, $T_A = B_n$, for relativistic protons by subtracting from the arrival time of the first particles, T_n, a minimal time of particle transport from the Sun to the Earth, ~11 min. Setting $t_g = T_i$ as a time of particle generation and (t_g – 8 min) as a time of certain flare phase (for example, the onset of the type II radio burst), we revised all available estimates of the B_n value.

Figure 7.17 shows the distribution of 39 GLEs from Table 2.1 on the parameter B_n, as it was derived by Miroshnichenko et al. (1990, 1995a, b, c, 1996) using the

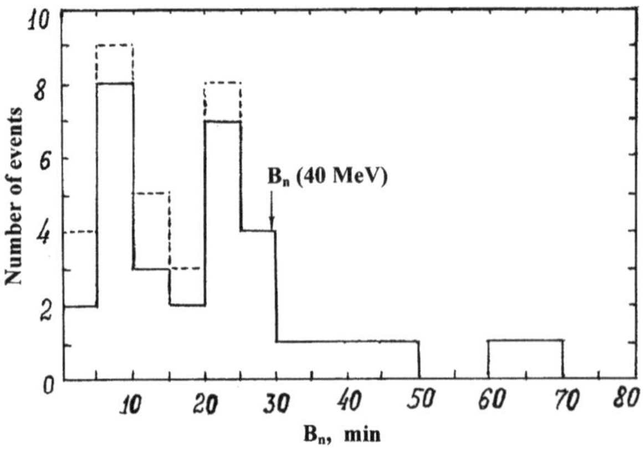

Fig. 7.17 Distribution of 39 GLEs from Table 2.1 on the parameter B_n derived by Miroshnichenko et al. (1990, 1995a, b, 1996) for the period 1979–1992 (*solid line*) and complemented by the data of Cliver et al. (1982) for the GLEs through 1978 (*dashed line*). *Arrow* denotes the value $B_n = 29$ min obtained by Bazilevskaya and Sladkova (1986) for proton with the mean energy of 40 MeV

NM Apatity data of 1979–1992 and complemented by the estimates of Cliver et al. (1982) for the GLEs through 1978. As a rule, the time of particle generation, t_g, was chosen in coincidence with the onset of the type II radio burst which is determined more reliable in comparison with the FSMP. In several cases when the type II radio bursts were absent, the times of the FSMP or SHR maximum have been used as a proxy of t_g.

As one can see, the distribution in Fig. 7.17 has two distinct maxima (a similar tendency may be traced also in Fig. 7.2 taken from Cliver et al. 1982). For the most of SPEs a release of the first particles from the corona starts after the time $B_n = 8$ min from the beginning of type II radio burst (first group, PC), for the rest of the events $B_n = 25$ min (second group, DC), and it is near to the value of $B_n = 29$ min found by Bazilevskaya and Sladkova (1986) for protons with the mean energy of ~40 MeV, *i.e.*, the value of B_n does not depend on the particle energy. So, two components seem to exist in the population of first particles, too, B_n and B_m being energy independent for the DC (Bazilevskaya and Vashenyuk 1979).

7.4 Two Components in the GLE of September 29, 1989

A well-known SPE of September 29, 1989 (see Fig. 1.9) was the largest GLE from February 23, 1956 and have been widely discussed (for a review see Miroshnichenko et al. 2000). The event was supposedly caused by a behind-the-limb-flare which manifested itself in X-rays (X9.8, maximum at 10.47 UT) and a very strong radio emission. In addition to the microwave bursts, there were also

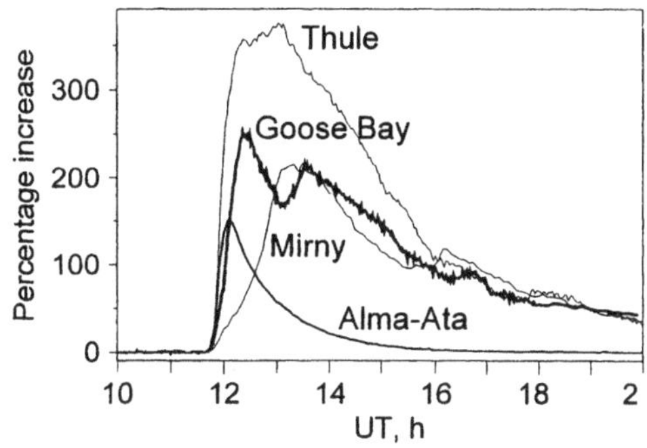

Fig. 7.18 Intensity-time profiles of the September 29, 1989 event by the NM data at four NMs with different cutoff rigidities (R_c): Alma-Ata (6.61 GV), Mirny (0.03 GV), Goose Bay (0.64 GV) and Thule (0.00 GV)

several type II and type IV radio bursts associated evidently with a CME. The CME was observed by the coronograph/polarimeter on board the Solar Maximum Mission (*SMM*) satellite (Burkepile and St.Cyr 1993). As reported by Bhatnagar et al. (1996), post-flare Hα loops were observed at the Udaipur Observatory from 1119 UT onward. Also, several eruptions were observed. In addition, a limb brightening was visible in Hα at 1230 UT, and a spectacular loop structure was distinctly observed at the Sacramento Peak Observatory for more than 10 h (Smart and Shea 1990a, b; Swinson and Shea 1990) from 1326 UT.

7.4.1 Intensity-Time Profiles

The GLE of September 29, 1989 was not only large in intensity but remarkable also by very complicated intensity-time profiles as measured on different cosmic-ray station. As follows from the data of the Apatity, Deep River, Calgary and many other stations (*e.g*., Ahluwalia et al. 1991; Smart et al. 1991; Miroshnichenko et al. 1995a, b, c; Vashenyuk et al. 1997), the event of September 29, 1989 belongs to delayed or rather to combined (mixed) type. This is easily seen from the four NM intensity-time profiles given in Fig. 7.18. Namely, the impulse-like profile of the NM at Alma-Ata shows the first (prompt and hard) ejection of relativistic particles, and the Mirny station evidences the second (delayed and soft) ejection. Goose Bay station shows two maxima, one of which seems to be due to the first ejection because it nearly coincides in time with the Alma-Ata profile. The second peak at the Goose Bay station seems to be formed by the second ejection because it coincides with the delayed profile of the Mirny neutron monitor. The flat maximum at the Thule profile is probably a result of summation of decreasing prompt and increasing delayed ejections (Vashenyuk and Miroshnichenko 1998).

Relying upon their analysis of the NM Oulu and Lomnicky Stit data, Torsti et al. (1991, 1992) believe that these peculiarities of the time profiles are indicative of the two-fold ejection of SCR; Miroshnichenko (1997) suggested to explain them based on a two-source acceleration scenario. In Fig. 7.19 a proposed reconstruction of the two-fold ejection process based on the Oulu station data (Torsti et al. 1992) is

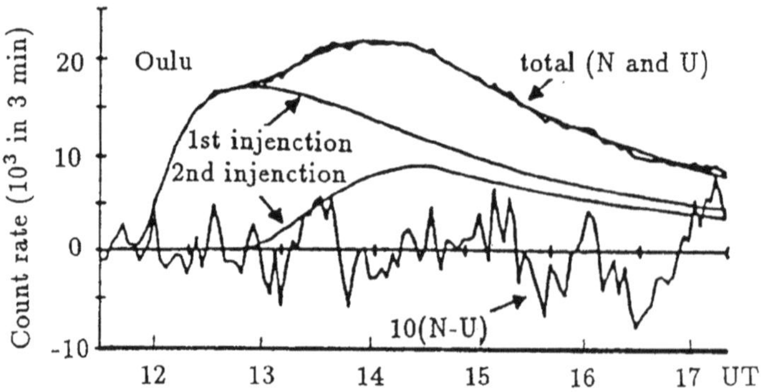

Fig. 7.19 Residual variation of the observed, N, and calculated, U, intensities during the September 29, 1989, event at the NM Oulu under the assumption two-fold ejection of accelerated particles (Torsti et al. 1991, 1992)

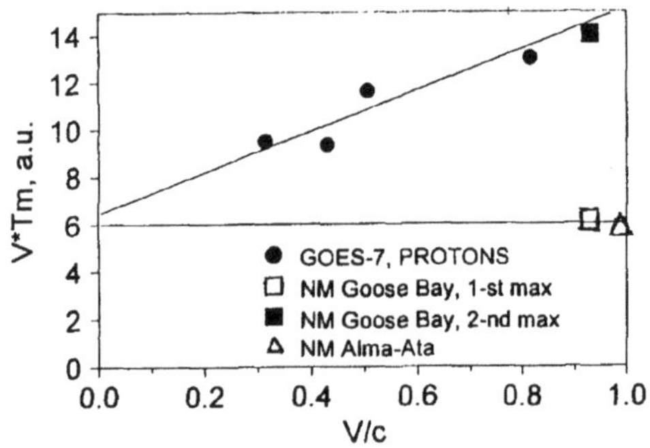

Fig. 7.20 Separation of two relativistic components in the event of September 29, 1989 by the vT_m technique (Vashenyuk et al. 1997)

shown. It should be noted that the asymptotic cones of both stations were directed "vertically" to the IMF lines during the first hours of the GLE. Nevertheless, both NMs turned out to be able to record the two ejection processes in the same time intervals as the forward viewing stations.

7.4.2 Specific Features of Particle Release

Two SCR components in the event of September 29, 1989 can be distinctively separated also by the vT_m-technique (7.8) and (7.9). In Fig. 7.20 we present the vT_m-diagrams constructed by the data of Fig. 7.18 (Vashenyuk et al. 1997). Because of absence of direct data on the time of SCR generation, Vashenyuk et al. (1997) referred the beginning of the particle release to the onset of the type II radio burst (1126 UT). To construct the dependence of vT_m on the particle velocity, v/c, they used, in addition to the NM data from Fig. 7.18, also the *GOES*-7 data from four energy intervals.

It is seen that observational points in Fig. 7.20 form two linear dependencies of type (7.9). One of them, with great inclination, unites data of non-relativistic solar protons measured by the *GOES*-7 spacecraft and the second maximum recorded the NM in Goose Bay. All these particles apparently belonged to the same population (DC) which was delayed in the corona and then released simultaneously through the same time $B_m \approx 2$ h. Another possibility is a simultaneous acceleration of the DC particles at the post-eruption stage of the flare (*e.g.*, Chertok 1995). The second straight line nearly parallel to the horizontal axis ($B_m \approx 0$) is drawn through the points corresponding to the intensity maximum of the Alma-Ata NM (PC) and the first maximum at the NM Goose Bay. Thus, the PC left the Sun without any delay and it is represented by the relativistic protons only.

The straight line which provides the best fitting of the DC data in Fig. 7.20 crosses the ordinate axis in the point $A_m = 6.3$ AU. The PC data are located, in practice, at the same level. It implies that a coronal time B_m for the prompt component is about zero, i.e., those particles left the Sun without any delay, immediately after their acceleration. On the other hand, according to the simple model of anisotropic diffusion in the interplanetary space, the parameter A_m is related to the transport path of protons in radial direction as $\Lambda_r = r_E/2A_m$ (Reinhard and Wibberenz 1974) where r_E is the radius of the Earth's orbit. Hence, we get $\Lambda_r = 0.08$ AU which is equivalent to the mean free path along the IMF $\Lambda_{\|} = 2\Lambda_r = 0.16$ AU. This value is within the limits of admissible magnitudes of the mean free path (0.08–0.3 AU) for the protons in the rigidity range $R = 0.5$ MV – 5.0 GV (*e.g.*, Miroshnichenko and Petrov 1985; Bieber et al. 1994). The corresponding estimate of the parameter A_m for the event of October 22, 1989 (Miroshnichenko et al. 1996) is $A_m \approx 2.0$ AU, hence it follows that the value of $\Lambda_{\|} \approx 0.5$ AU is considerably larger than that for the event of September 29, 1989. A detailed analysis by Vashenyuk et al. (1993, 1994), however, clearly demonstrated a limitation of traditional diffusion approach to the study of the intensity-time profiles in such a complicated SPE as the GLE of September 29, 1989.

7.4.3 Temporal Evolution of Rigidity Spectrum

The separation of two relativistic components in the event of September 29, 1989 is confirmed also by specific form of rigidity spectra derived by Cramp et al. (1993a) and revised recently by Lovell et al. (1998) for three different intervals of observations: 1215–1220, 1325–1330, and 1600–1605 UT. The spectrum corresponding to the first peak at the intensity-time profiles in Fig. 7.18 (at about 1217 UT) was significantly harder than that of the second one. Independently, based on the same observational data, Dvornikov and Sdobnov (1995a, 1997) calculated differential spectra of the relative variations of cosmic rays intensity $\Delta D_s/D_g(R)$ for several time intervals (Fig. 7.21) during the event (symbols "*s*" and "*g*" correspond to solar and galactic particles, respectively). Dashed lines in this figure show the extrapolation of the spectra to the low-energy range. The asterisks, open triangles, open

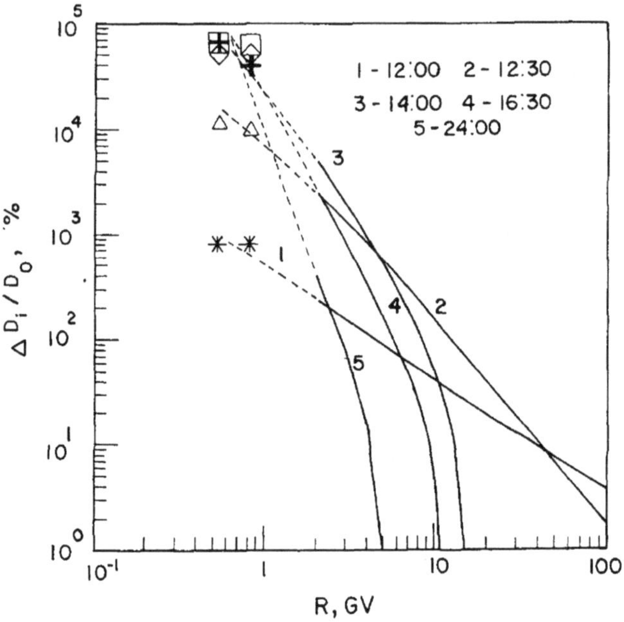

Fig. 7.21 Differential rigidity spectra of SCR intensity variations relatively to GCR background during the GLE of September 29, 1989 for five time intervals (UT): *1* 11:45–12:00, *2* 12:15–12:30, *3* 13:45–14:00, *4* 16:15–16:30, and *5* 23:45–24:00 (Dvornikov and Sdobnov 1997). The *curves* demonstrate a very hard spectrum in the early phase and a softening of the proton spectrum at the late phase of the event

diamonds, open squares, and plus signs correspond to *GOES*-7 measurements (SGD, 1989, 1990) as a percentage of background values for particles with $R \sim 0.54$ GV and $R \sim 0.82$ GV for the corresponding time intervals. The time profiles of the SCR intensity with $R \sim 1.4$ GV measured by the satellite *GOES*-6 agree with those obtained by these authors from ground-based data, with the correlation coefficient ~0.986. The curves demonstrate a very hard spectrum in the early phase and a softening of the proton spectrum during the late phase of the event.

In the early phase the rigidity spectrum of variations in Fig. 7.21 is approximated quite well by the power law $\Delta D_s/D_g(R) \sim R^{-\delta}$ with the exponent $\delta \approx 1.0$ at 1200 UT and $\delta \approx 1.7$ at 1230 UT. An increase in intensity is traceable right up to R ~100 GV. This may imply that the highest energy of the flare-accelerated particles is not lower than 100 GeV (Dvornikov and Sdobnov 1995a, 1997). A similar conclusion was drawn by Krymsky et al. (1990), Karpov et al. (1998) and some other researchers (see details in Chap. 4). The declining phase is remarkable for a continual softening of the variation spectrum, and by the end of the day the spectral index increased to $\delta \approx 5$.

As to the anisotropy of relativistic protons, the results obtained by different research groups are rather controversial (for details see Miroshnichenko et al. 2000). Nevertheless, we are inclined to think that there was more than one source direction (*e.g.*, Smart et al. 1991; de Koning and Bland 1995; de Koning and Mathews 1995). The primary source direction was in the northern hemisphere, and the second one in the southern hemisphere; both of them were west of the garden-horse field line. The second source direction was only observed in the low rigidity data ($\leq$2 GV); it is the source of the second peak observed by many NMs. A third source direction might exist in the anti-solar direction, but it was only seen at high rigidities (>4 GV). The recent findings by Vashenyuk et al. (1997) and Vashenyuk and Miroshnichenko (1998) seem to support the proposed bi-directional picture. All these results evidence again that relativistic SPEs require some new descriptive and analytical approaches.

As known, the giant GLE69 of 2005 January 20 was the second largest on record (and largest since 1956), with up to 4,200 % count rate enhancement (NM Terre Adélie, 1-min data) at sea level. Bieber et al. (2013a) have analyzed data from the "Spaceship Earth" network, supplemented to comprise 13 polar NM stations with distinct asymptotic viewing directions and Polar Bare neutron counters at South Pole, to determine the time evolution of the relativistic proton density, energy spectrum, and three-dimensional directional distribution. Similarly to our findings (Miroshnichenko et al. 2000) for the GLE42 (29 September 1989), Bieber et al. (2013a) have identified two energy-dispersive peaks, indicating *two solar injections*. It was also found that the relativistic solar protons were initially strongly beamed, with a peak maximum-to-minimum anisotropy ratio over 1,000:1.

7.5 Source and Acceleration Models

As mentioned above, the observational data on the September 29, 1989 event are susceptible of several different interpretations. In fact, three possible scenarios have been used in describing the main features of this GLE: (1) acceleration by a CME-driven coronal shock; (2) post-eruption particle acceleration in the corona; and (3) a combined two-source acceleration.

7.5.1 Acceleration by a Coronal Shock

There are some evidence (e.g., Cliver et al. 1993b; Bhatnagar et al. 1996) that indicated the presence of a CME-driven coronal/interplanetary shock in the event of September 29, 1989. In addition, in spite of the location of the originating flare behind the limb, a strong 2.223 MeV gamma-ray line, which is normally limb-darkened, was observed (Vestrand and Forrest 1993). It implies that accelerated particles were interacting well onto (~30°) the visible disk. To explain this spatially extended GRL emission, Vestrand and Forrest (1993) first postulated that the interacting particles may diffuse from flare loops or precipitate from a coronal shock. The latter scenario was then developed by Cliver et al. (1993b) and Cliver

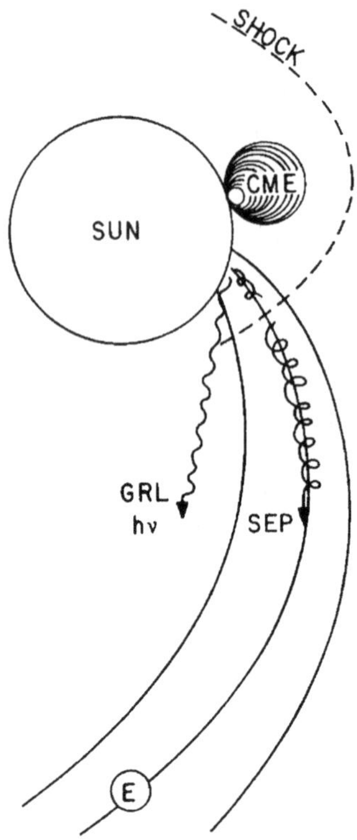

Fig. 7.22 Proposed scenario for shock acceleration of GRL-producing protons from a behind-the-limb flare on September 29, 1989 (Cliver et al. 1993b)

(1996) in some detail. They suggested that the CME/shock ensemble was responsible for the acceleration of the particles that caused the observed front-side gamma-ray line emission, as depicted in Fig. 7.22. Cliver et al. (1993b) and Cliver (1996) find this scenario appealing because of its simplicity: particles accelerated on open field lines can either escape to be observed as solar energetic particles (SEPs) or precipitate to give rise to GRL emission. They point out that in both cases fast "transport" of energetic particles is accomplished by widespread shock acceleration.

Cliver et al. (1993b) mention that, spatially, the CME/shock ensemble should be broad enough, based on the CME latitudinal extent, to encompass the front-side regions from where the 2.223 MeV emission originated. They admit that, temporally, the onset of GRL emission is only *marginally* consistent with the presence of high-energy particles in the corona; meanwhile, from ~3 to ~30 % of the protons accelerated at a coronal shock would need to precipitate to the Sun to produce the observed 4–7 MeV emission. A similar "precipitating-shock" model has been proposed by Ramaty and Murphy (1987) to account for the pion-rich phase of gamma-ray emission observed in the flare of June 3, 1982. As Cliver et al. (1993b) emphasized, it is an open question, however, whether the spatially extended GRL emission on September 29, 1989 and high-energy gamma ray emission on June 3, 1982 were the same.

As one can estimate, for a source ~10° behind the west limb the occultation height would be ~7,000 km. Such a source will contribute to prompt GRL emission, but this would only tend to lower the ratio of the 2.223/4–7 MeV fluences (Cliver 1996). Recall that the 2.223-MeV neutron-capture deuterium line is not generally seen from limb flares since it is produced by the flare neutrons deep in the photosphere. The neutrons must be thermalized before they can be captured by protons to produce deuterium nuclei in an excited state. These nuclei decay then to produce (with some delay) the 2.223 MeV line. However, because of the large absorption, gamma rays cannot escape in the direction of an observer tangential to the solar surface.

Unfortunately, the early impulsive phase of the flare was not observed because the *SMM* was in the South Magnetic Anomaly at that time and the Gamma Ray Spectrometer was not collecting data. Consequently, as emphasized by Dennis (1996), a correction must be applied to account for the delay in the production of the 2.223 MeV line. If this is taken into account, one can estimate the corrected intensity which is normal, relative to the flux in other lines, for a *disk flare* (see Dennis 1996, and references therein). The observation of the deuterium line in this event can only mean that accelerated ions must produce the neutrons on the visible disk. But how did they get there from the flare site?

One possibility is that the ions were accelerated in the shock of the associated CME as it expanded outwards. If this were the case, then the gamma-ray spectrum would provide direct information on the shock-accelerated particles. However, recently Somov (1996) argued that shock accelerated ions can not produce the delayed component of gamma-ray emission since the shock is already too high in the corona by the time this component appears. Some earlier, citing timing and composition inconsistencies in other events, Mandzhavidze and Ramaty (1993) and Ramaty et al. (1993) have also expressed doubt about the role of shocks in accelerating the particles that produce the observed gamma rays. In their opinion, for most flares the gamma ray production is due to particles from impulsive flare acceleration (Ramaty and Mandzhavidze 1996).

Mandzhavidze (1994) has criticized the proposed scenario of Fig. 7.22, because the gamma-ray spectrum of this event was similar in some aspects to the spectra of electron-dominated events, whereas shocks are not thought to be efficient accelerators of electrons. Indeed, gradual flares in which shock acceleration is believed to play a dominant role have low e/p ratios, while particles that interact and produce gamma rays are always electron rich. She suggested that the shock picture might apply if the spectrum were pion-dominated. Namely, if the spectrum of accelerated ions was very hard (harder than ~E^{-2}), the 1–10 MeV energy range could be dominated by bremsstrahlung of the secondary electrons and positrons from charged pion decay. This would be also consistent with the unusually hard proton spectrum derived from the NM data during the early stage of the GLE (see Sect. 7.4). Thus, as noted by Ramaty and Mandzhavidze (1996), in this particular case, gamma-ray observations may be reconciled with a shock acceleration scenario.

Alternatively, as it was proposed by Vestrand and Forrest (1993), the accelerated particles could be transported in a large-scale magnetic loop that connected the invisible acceleration source (impulsive flare) to the visible hemisphere of the Sun. In that case, the particles might not be related to the CME shock at all. One way to separate these two explanations in the future will come from imaging gamma ray observations. For shock acceleration, the 2.223 MeV line should come from a large diffuse area whereas with the magnetic loop hypothesis, the gamma ray source should be much more compact (Dennis 1996; Ramaty and Mandzhavidze 1996). In fact, Dennis (1996) presented evidence for that the X-ray spectrum is hardening during the time of the GRL emission, this effect being a signature of the post-flare loops above the limb (see also Cliver 1996).

On the other hand, Cliver (1996) noted that the densities of ~10^{11} cm^{-3} of an elevated source would be too low to support either 2.223 MeV line emission or the pion radiation (with any efficiency). Hence, he concludes that the shock picture remains viable. At any rate, he believes that the source must be *diffuse*, because a GRL fluence ~10 photon/cm^2 observed in this case should be accompanied by an Hα flare. However, no frontside emission was reported except for an eastern hemisphere subflare. In terms of his Expanded SEP Classification System (see Table 2.7), Cliver (1996) considers this event as a reminder that just as flare-associated particles may interact at the Sun or escape to the interplanetary medium, shock accelerated particles may also appear in both venues. He proposes to investigate precipitation from a shock as an alternative (or contributing) mechanism relative to acceleration via reconnection in post-flare loop systems responsible for the long-lasting gamma-ray emission. The latter possibility is discussed in the next Sect. 7.5.2.

Apparently, detailed modeling efforts will be required to determine whether a coronal shock in the proposed scenario (Fig. 7.21) can precipitate up to 30 % of its protons with energies >30 MeV and still efficiently accelerate protons to energies ≥20 GeV. In Sect. 7.5.3 we propose another scenario for interpreting the gamma-ray data, relying upon the concept of two acceleration sources in the corona.

7.5.2 Post-eruption Acceleration

When analyzing and treating SEP events, usually only two acceleration processes are considered: (1) an impulsive (primary) flare energy release in the upper chromosphere or lower corona; and (2) a gradual acceleration in coronal/interplanetary shocks driven by large and fast CMEs. Moreover, a point of view exists (see Kahler 1994, 1996; Reames 1996, and references therein), that even relativistic protons in the GLEs (up to ≥20 GeV) and energetic Fe ions (up to 200–600 MeV/nucleon) are produced in shocks only, but not in a region of the primary energy release. Meanwhile, there is another plausible source of the particle acceleration at the Sun – a post-eruption (PE), or secondary, energy release – also closely associated with CMEs (Chertok 1995, 1996, 1997a, b). Below we explore this suggestion in

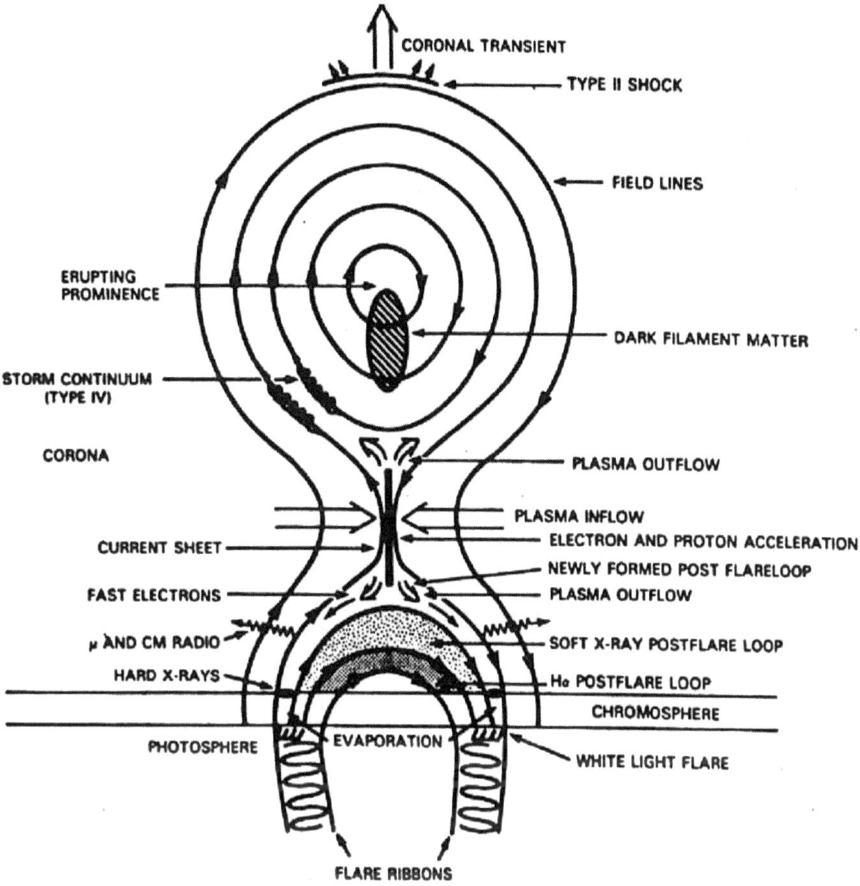

Fig. 7.23 The global structure of the two-ribbon flare and the location of the major observed processes of energy conversion, viewed in a cross section along neutral line (Martens and Kuin 1989)

connection with post-eruption acceleration in the corona as outlined by Bhatnagar et al. (1996) and Akimov et al. (1996).

According to Bhatnagar et al. (1996), available data on the event of September 29, 1989 allow to suggest the following general scenario of the event involving the CME and PE energy release. The CME erupts under the influence either of an instability of the large-scale coronal structures, or of the primary (flare) energy release. The latter manifests itself by the first component of radio emission observed at all frequencies. There is no doubt that when passing through the corona, the CME disturbs the coronal magnetic field over a large area. After the CME has passed, the disturbed magnetic field will relax to its initial state via magnetic field reconnection. This process may proceed in a quasivertical reconnecting current sheet (RCS), in the configuration of Fig. 7.23 proposed by Martens and Kuin

(1989). The relaxation is assumed to accompany by a prolonged (secondary) energy release and effective particle acceleration (Chertok 1995; Bhatnagar et al. 1996). Bhatnagar et al. (1996) believe that available radio burst data allow to associate these two ejections with the impulsive and PE acceleration, respectively.

The reconstruction of the coronal magnetic field leads to the forming of new, higher loops giving rise to the second (delayed) burst component. Spatial and temporal relations between the CME eruption, impulsive flare energy release and PE energy release have been studied by Chertok (1997a, b). It was found that the microwave burst of September 29, 1989 included the impulsive and post-eruption components. As it is seen from Fig. 7.19, the observed GLE also seems to reveal two subsequent particle ejections from the Sun (Torsti et al. 1992), with a time difference of several tens of minutes.

Bhatnagar et al. (1996) consider the PE energy release as a spatially extended source of the broad band emission. In particular, the observed gamma-rays (Vestrand and Forrest 1993) may be due to particle precipitation through large-scale magnetic field loops that connected the source of the PE energy release above the AR 5698 behind the limb with the AR 5703 located on the visible disk at (24°S, 62°W). According to Bhatnagar et al. (1996), this interpretation is more adequate than the suggestion of Cliver et al. (1993b) that a CME-driven shock is a plausible source of energetic protons producing gamma-rays on the visible disk.

The post-eruption acceleration appears also to be responsible for the long duration increase of the low-energy (tens of MeV) proton fluxes in the interplanetary space. In addition, Bhatnagar et al. (1996) suppose that the particles accelerated in a process of the PE energy release may be trapped inside the propagating CME and then leak out gradually from the trap. It would give a considerable contribution to the 10–30 MeV proton fluxes observed near the Earth's orbit, as well as to the GLEs with a complicated intensity-time profile.

According to Chertok (1997a, b), such a scenario is corroborated, in particular, by the analysis of recent measurements of prolonged and high-energy gamma-ray and neutron flare emissions in six very large homologous flares occurred on 1, 4, 6, 9, 11, and 15 June, 1991. For example, the time profiles of microwaves in these flares reveal two main components – very strong impulsive one and the relative less intense delayed component separated by a time interval of 15–70 min. The delayed component manifests some features allowing to identify it with the PE energy release. A notable similarity takes place in those flares between the time histories of different emissions (in particular, for the delayed components). This implies that the various emissions observed well after the impulsive phase, appear to be initiated by the prolonged acceleration of electrons and ions at the late phase of the flares, rather than by a long-term trapping of particles accelerated at the flare onset. Theoretical studies (e.g., Litvinenko and Somov 1995) show that a direct electric field in the reconnecting current sheet (see Fig. 7.23) can indeed result in the prolonged particle acceleration up to GeV energies (for protons). Thus, the PE energy release following a CME seems to be important for various energetic phenomena on the Sun, including the production of solar cosmic rays, alongside with the impulsive and shock acceleration.

Unfortunately, this concept was not developed quantitatively for the September 29, 1989 event. Instead, it has been considered in detail for the GLE of June 15, 1991 (Akimov et al. 1996) to substantiate the long-duration emission (>2 h) of high-energy gamma rays (>2 GeV). By the way, the two-stage energy release has displayed also in the much shorter and less powerful gamma flare of March 26, 1991, although no pronounced CME was observed in this case (Akimov et al. 1996). It means that the preflare magnetic structures may be disturbed not only by a large CME, but by other factors such as rapidly expanding and evolving coronal loops.

7.5.3 Two-Source Model

As shown above, evidence exists for two separate components of SCR in the event of September 29, 1989, the so-called prompt (PC) and delayed (DC) components. According to Perez-Peraza et al. (1992), increases of the SCR flux in events with a PC are of impulsive nature and have an anomalously hard spectrum, which may indicate the specific mechanism of fast acceleration. A magnetic loop or bottle (Mullan 1983) is evidently a possible source of the DC. Presumably, the PC is generated in the region of reconnection of magnetic loops under the driving action of an expanding magnetic bottle. We believe that some of the peculiarities of this GLE may be explained by a model with two separate sources of acceleration (Vashenyuk et al. 1993, 1994, 1997, 1998a, b; Miroshnichenko et al. 1995a, b, 1996; Miroshnichenko 1997; Vashenyuk and Miroshnichenko 1998).

A magnetic bottle was chosen by Perez-Peraza et al. (1992), as the basis for the interpretation of two-peak SPEs, since this model seems to be well substantiated physically and elaborated numerically (Schatten and Mullan 1977; Mullan and Schatten 1979; Mullan 1983). Besides, as noted by Mullan and Schatten (1979), this model does not contradict the CME concept. Moreover, within the uncertainties of a few minutes, the data on SCR release from the corona compiled by Cliver et al. (1982) are found to be reconcilable with the predictions of the bottle model (Mullan 1983), contrary to the conclusions of Cliver et al. (1982). At the present level of our knowledge of solar flare physics, we can identify a magnetic bottle with an extended coronal structure as shown in Fig. 7.23.

In the two-source acceleration model a flare is assumed to develop at coronal heights $h \approx (0.07\text{–}0.14)R_s$, in accordance with the scenario of Mullan (1983). When expanding, the flare-generated magnetic bottle gets in touch with the neighbouring magnetic arcade at heights $h \approx (0.5\text{–}1.0)R_s$, where a current sheet (CS) with length L may be formed due to the process of magnetic reconnection between the lines of opposite polarity. Local particles in the non-adiabatic region of the current sheet may be accelerated by the intense impulsive electric fields produced by the magnetic emerging process. According to Priest (1982), data on coronal transients indicates that the magnitude of the magnetic field, B, amounts to a few units or tens of Gauss and the plasma density, n, in the upper part of the magnetic bottle can

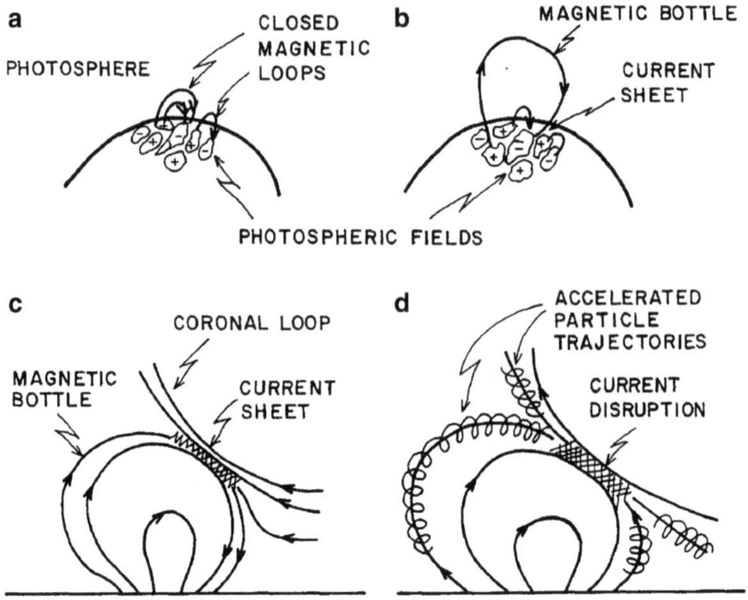

Fig. 7.24 Source model for the prompt component of SCR in the corona (Perez-Peraza et al. 1992): (**a**) active region with complex magnetic structure; (**b**) formation of a magnetic bottle and start of its interaction with the adjacent magnetic arch; (**c**) formation of a current sheet; (**d**) acceleration and escape of particles (Perez-Peraza et al. 1992)

be several times higher than in the surrounding corona; for example, at heights ~ (0.5–1.0)R_s the plasma density may reach the values n ~10^6–10^7 cm^{-3}. The evolution of the proposed magnetic configuration (Perez-Peraza et al. 1992) is shown in Fig. 7.24.

As has been found earlier (Perez-Peraza 1986), the maximum of the additional flux of accelerated particles (PC) in the magnetic configuration of Fig. 7.24 should be observed before the maximum of the delayed component. Later on, Perez-Peraza et al. (1992) showed the applicability of this scenario in fitting the source spectra at the early stage of a number of GLEs. The relations (5.11, 5.12, and 5.13) were used in describing the source spectrum formation under the action of electric fields in the reconnecting current sheet (RCS).

The theoretical source spectrum (5.11) of the PC of three events (February 23, 1956, December 7, 1982, and February 16, 1984) may be adequately fitted to observed spectra provided the source parameters for the three GLEs are: $B = 30$, 20 and 20 G; $n = 2 \times 10^7$, 2×10^6 and $5 \times 10^6\, cm^{-3}$; $L = 10^{10}$, 2×10^{10} and 2×10^{10} cm, respectively (Perez-Peraza et al. 1992). These values correspond to generation heights $\geq 0.5R_s$. The accelerating electric fields are in the range of ~10^{-2}–10^{-1} V cm^{-1}, which provides multi-GeV proton production (up to 250 GeV in the case of February 23, 1956 flare).

As to the event of September 29, 1989, very preliminary estimates of the DC spectrum have been obtained under a rather conventional scenario (Miroshnichenko et al. 1995b). According to the proposed scenario, the bulk of the energetic particles are generated in the flare volume or its vicinity. It was assumed that the acceleration of this component was carried out by magnetosonic turbulence, with initial particle energy around E_0 and monoenergetic injection into the resonant stochastic process. The accelerated particles are subsequently trapped in an expanding magnetic bottle. As a result of the Rayleigh-Taylor instability, the bottle is destroyed at a height $\leq 0.9r_s$, and the energetic particles are released into interplanetary space approximately 0.5–1.0 h after the flare (Mullan 1983). The calculated rigidity spectrum for the DC (Miroshnichenko et al. 1995b) was compared with the observed spectrum for the second intensity peak at 1325 UT (Cramp et al. 1993a; Lovell et al. 1998). Assuming monoenergetic injection at $E_0 = 0.5$ MeV and a mean confinement time of particles in the acceleration region of $t \approx 1$ s, the best fit was obtained with an acceleration efficiency of $\alpha = 0.04\ \text{s}^{-1}$. It should be noted that the fitting was carried out without taking into account possible interplanetary modulation of the observed spectrum.

Recently, Vashenyuk and Pchelkin (1998) estimated the parameters for rigidity spectra outside the magnetosphere at different phases of this GLE in the framework of two working hypothesis: (1) unidirectional anisotropy during the first peak, and (2) bi-directional anisotropy during the second peak. It was found, in particular, that early in the event (at 12:25 UT) the spectrum near the Earth can be described by Equation (1.9), where $D_0 = 1.94$ particles $(\text{cm}^2\ \text{s sr GV})^{-1}$ and $\gamma = 1.08$ for $R < 2$ GV, the value of γ being increased by $\Delta\gamma = 0.13$ per 1 GV for $R > 2$ GV.

The source energy spectrum for the PC was then estimated by Vashenyuk et al. (2000) who transformed the near-Earth spectrum into energy scale and recalculated it at the Sun using a simple empirical procedure (Perez-Peraza et al. 1992), under the assumption of scatter-free interplanetary propagation. Fitting the spectrum calculated by Eq. (5.11) to that estimated from the experimental data, by the parameter optimization procedure, Vashenyuk et al. (2000) obtained the following source parameters: $B = 91$ G; $n = 1.2 \times 10^7\ cm^{-3}$; $L = 10^9$ cm. Such values of B and n are characteristic for the trailing part of the coronal transient (behind an eruptive filament) at coronal heights of several tenth of solar radius, and the value of L is of the order of the filament length. As far as we know, the above theoretical determinations of the source spectra, calculated using the two-source model, give the only numerical estimates of B, n, and L for the event of September 29, 1989 available in the literature. The two other models – CME-driven shock and post-eruption acceleration – do not yet have any similar estimates either for the source spectrum or for the source parameters in this particular event.

7.5.4 General Scenario of the Event

As one can see from the above evidences and estimates, the model of two SCR sources separated in time and space (Perez-Peraza et al. 1992) in application to the event of September 29, 1989 led to reasonable results (Vashenyuk et al. 1993, 1994, 1997, 1998a, b; Miroshnichenko 1997; Miroshnichenko et al. 1997, 1998, 2000; Vashenyuk and Miroshnichenko 1998). On the other hand, as far as we know, two previous approaches (CME-driven shock and PE acceleration) can not suggest, at present, a single self-consistent scenario of the event based on the totality of the observational data. Therefore, at present, only the two-source model allows us to outline a possible general scenario of the generation, release and transport of SCR in this event, though our scenario, of course, is far from irreproachable.

Initial stage. The initial stage of the SCR generation is proposed to be associated with a "classic" flare that occurred in the lower corona behind the western limb of the Sun. The particles are accelerated during the impulsive phase and are transported then into the upper corona in an expanding magnetic bottle (source I, delayed component, DC). When expanding, the flare-generated magnetic bottle (loop) gets in touch with a neighbouring magnetic arcade (system of long-lived coronal loops). Then, at the height $h \geq 0.5r_s$ and heliolongitudinal distance below $\pm 50°$ from the flare site θ_f (Reinhard and Wibberenz 1973; Perez-Peraza 1986), a current sheet (CS) may be formed between the magnetic bottle and the extended coronal structure due to the process of magnetic reconnection of lines of opposite polarity, with the subsequent acceleration of particles (source II, prompt component, PC). Note that the protons with energies $\geq$100 MeV, according to some estimates (Schatten and Mullan 1977; Perez-Peraza 1986), may occasionally escape from the trap by gradient and curvature drifts from the very beginning of the bottle expansion; however, the bulk of DC particles is convected inside the bottle up to its destruction.

On the other hand, the generation and escape of the PC particles occur shortly before the bottle opening; the lifetime of the bottle is estimated to be of ~5–50 min (Mullan 1983). Soon after, these particles come to open lines of force of the IMF and easily reach the Earth in the form of a beam with a strong anisotropy (the first maximum of SCR increase). When opening at the height $\leq 0.9r_s$, the magnetic bottle may have a heliolongitudinal extension of $\theta < \theta_f \pm 50°$, so the release of the trapped particles does not only proceed with some delay, but over a wide range of heliolongitudes as well. As a result, a second SCR maximum with a weak anisotropy is observed at the Earth. The details of this scenario depend on the lifetime of the magnetic bottle, on the geometry of extended coronal structures, and on the mutual position of the originating flare and the Earth.

Event development. For the purpose of timing the event development, it is worth restating the key observational points in condensed form. From the detailed data of Bhatnagar et al. (1996) it follows that the enhancements of microwave (3.1 and 5.2 GHz) as well as of soft X-ray (1–8 Å) emissions started almost simultaneously (about 1045–1047 UT). The radio data reveal that at least two phases of energy release occurred during this flare. In particular, the first significant microwave burst (FSMB) at the frequency of 3.1 GHz was recorded within the interval 1120–1126 UT (Chertok 1995; Bhatnagar et al. 1996). The more prolonged second component became visible after 1126 UT in microwaves (below 19.6 GHz), and somewhat later in the decimetric range.

It has been estimated (*e.g*., Perez-Peraza et al. 1992; Kahler 1994) that energetic solar particles travel about 1.2–1.3 AU along the spiral field lines of the IMF at a speed v with no scattering. Hence, for relativistic solar protons ($v \approx c$) the travel time from the Sun to the Earth will be about 11 min (the travel time of the electromagnetic waves is about 8.33 min). As estimated by Cliver et al. (1993b) and Kahler (1994), the first relativistic protons (~21 GeV) began to arrive at the Earth at 11:35–11:40 UT. This means that they were ejected not later than at 11:24–11:29 UT, and were generated earlier. The latter follows from the fact that the GRL emission started not later than at 11:24 UT. If we assume, then, that the moment of the PC generation coincides with the onset of the FSMB at about 11:20 UT (as proposed earlier by Cliver et al. (1982) for other events), we get that source II in the upper corona started at about 11:11 ± 05 UT. The uncertainty ascribed to this value (±05 min) is due to a slight discrepancy in different estimates of the time of the GLE onset. For example, according to the 1-min NM data from Mt. Norikura, Japan (R_c =11.48 GV), the main increase commenced at 1145 ± 0001 UT (Takahashi et al. 1990).

With this in mind, it is easily to construct a *kinematic scheme* based on a two-source model (Miroshnichenko 1997). In accordance with the findings and estimates of Schatten and Mullan (1977) and Mullan (1983) obtained for the parameters of magnetic bottle, let us assume that the source of the DC (source I, or magnetic bottle) starts at a height of $0.1R_s$ above the photosphere, with an expansion velocity of ~300 $km\ s^{-1}$, at the moment of ~30 min earlier than the source II, *i.e*, about 10:41 UT. Then, with an expansion velocity of ~300 $km\ s^{-1}$, at 11:11 UT the top of the bottle will reach a height of $0.877R_{s.}$ This height is almost equal to the characteristic height of $0.9r_s$, where the bottle should start to disintegrate (Mullan 1983). Evidently, it is just the moment (about 11:20 UT) that has to be taken for the onset of the CME movement (Cliver et al. 1993b; Chertok 1995; Bhatnagar et al. 1996). This moment also coincides with the FSMB onset at 3.1 GHz.

Locations of SCR sources. A geometric sketch of the locations of the two proposed sources at the Sun may be depicted as follows (Fig. 7.25). Source I (point *A*) starts at a height ~$0.1R_s$ above the photosphere, where the magnetic bottle forms. The top of the bottle raises to a height ~$0.9R_s$, where it is destroyed. Further, the bottle comes into contact with long-lived coronal loop (extended magnetic structure) at point *B*, giving rise to source II; the latter, according to the data on the GRL source (Vestrand and Forrest 1993; Cliver et al. 1993b), is viewed from point *A* at an angle of ~30° relative to the line connecting source I and the top of the bottle top. Since the geometry of the sources is given (Fig. 7.25), it is not difficult to estimate that source II is located at a height ~$0.7R_s$. This does not contradict the suggestion (Perez-Peraza et al. 1992) that the magnetic bottle interacts with an adjacent magnetic structure at heights ≥(0.5–1.0)R_s. It is worth noting that the separation in two ejection times, according to an independent estimation by Torsti et al. (1992), is of the order of 30–50 min.

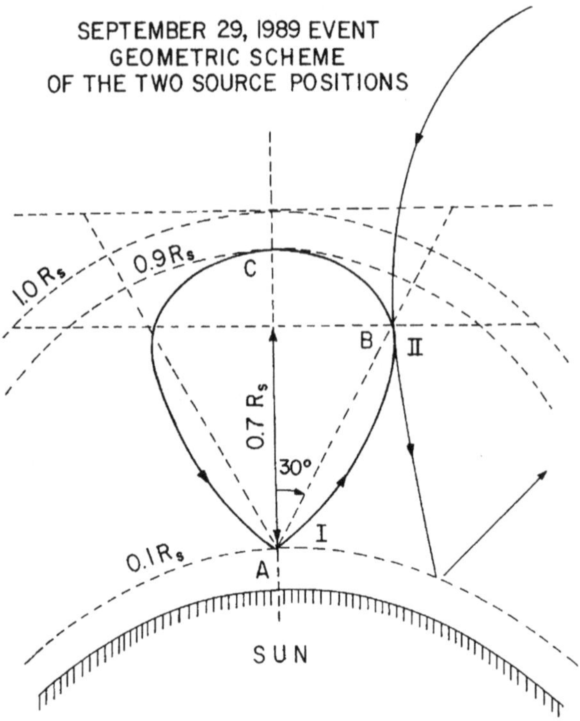

Fig. 7.25 Geometric scheme of suggested positions for two sources of relativistic protons in the event of 29 September 1989 (Miroshnichenko 1997)

Observations of extended coronal structures before and during the event (see Sect. 7.4) satisfy one of the main requirements for the application of the two-source model. It is important to note that geomagnetic activity during this event was low; therefore, one can assume that the IMF was quiet and had an Archimedean spiral formation. If the solar wind speed was ~350 km s^{-1} (see details in Miroshnichenko et al. 2000), then the garden-hose field line of the IMF originated at heliolongitude $\theta_E \approx 66°$W; this is ~35–40° (along the arch of a great circle) from the estimated site of the originating flare. The Earth's heliolatitude at the same moment was about 7.25°N (see Figs. 1.5 and 3.13). The details of proposed temporal scenario are summarized in Table 7.1.

As mentioned in Chap. 2, the GRL flare was remarkable for the observed high (~0.2) ratio of the 2.223 to 4–7 MeV emission. Because of the large attenuation of the 2.223 MeV line near the limb, this ratio implies that a significant fraction of the GRL emission originated on the visible disk, as far as ~25–30° from the flare centroid (Vestrand and Forrest 1993; Cliver et al. 1993b). Hence, taking into account the estimates of the mutual position of the originating flare and the time of the hard SCR ejection (see Table 7.1), Miroshnichenko (1997) concluded that the flare of September 29, 1989 provided the first evidence of a prompt component generation in the corona (around ~0.7R_s), rather than of a spatially extended component of GRL emission from solar flares.

Table 7.1 Temporal scenario of the September 29, 1989 event: two-source model

Time (UT)	Observed object/parameter	Data sources/ comments
	Pre-flare situation	
05:00–05:40	Behind-the-limb ejection (optics)	Lomnicky Stit, Slovakia, Sept. 28
07:05–07:57	Filament eruption (radio, 5.2 cm)	Badary data, Russia, Sept. 29
10:22	Last of the eruptions (optics) before a CME start	Hα observations
10:30–10:40	Onsets of the soft X-ray and microwave bursts	*GOES*-7 data
10:41 ± 05	Originating flare behind the W-limb, formation of magnetic bottle (source I)	Estimated for the delayed component
	Event development	
11:11 ± 05	Current sheet formation around ~0.7r_s (source II)	Estimated for the prompt component
11:19–01:19, Sept.29–30	Loop-like structure (optics)	Udaipur (India), Sacramento Peak (USA)
11:21	Start time of the CME	Estimated by *SMM* data
11:24–11:28	Metric type III emission	Weissenau, Germany
11:25–11:27	First significant microwave burst	IZMIRAN, Russia
11:26	Type II onset	Weissenau, Germany
11:27	First CME observation at ~1.0r_s above the photosphere	*SMM* data
11:31–11:33	Soft X-ray maximum	*GOES*-7 data
11:33–11:50	GRL records at 2.223 and 4–7 MeV	*SMM* data
11:41	Flare 1B at the W-limb (optics)	Hα (24°S, 90°W)
11:43	CME leading edge at ~4.5r_s	*SMM* data
	Relativistic protons near the Earth	
11:45 ± 00:01	Arrival of the first relativistic protons at the Earth	NMs at Mt.Norikura and Tokyo
12:00–13:00	Flux maximum of ≥20 GeV protons	Hourly average data at UMT Embudo
12:14 ± 00:02	Flux maximum of muons at detector "Carpet" (BNO – Baksan Neutrino Observatory)	First peak of the GLE at the Earth's surface
12:17 ± 00:02	Flux maximum at NM Inuvik	First peak of the GLE
13:15 ± 00:02	Flux maximum at NM Inuvik	Second peak of the GLE
13:30–13:45	Underground muon burst at BNO	BUST observations

This conclusion seems to avoid some of the difficulties of the model (Cliver et al. 1993b) based on CME-driven coronal shock. The main problem is that a shock should efficiently accelerate protons to energies >20 GeV and provide their fast (practically free) escape into interplanetary space, giving rise to a smooth increase of the relativistic particle flux at the Earth for a rather long time. In addition, the time to accelerate high-energy protons should be longer than that for low-energy protons. However, in the event of September 29, 1989, as it follows from the recent analysis by Kahler (1994), the situation turned out to be quite different (Fig. 7.5). The intensity profile of the >21 GeV protons derived from the data of NM at Mt. Norikura ($R_c = 11.48$ GV) displays a break (maximum) at the moment when the CME was at the distance about $6R_s$. At the same time, the >1 GV proton profile (estimated from the data of NM at Mt. Calgary, $R_c = 1.08$ GV) continued to rise very smoothly and reached maximum when the shock was at $12R_s$ or more (see Reames 1996). Moreover, the background corrected counting rate from the *GOES*-7 HEPAD P9 channel, which has an energy range of 430–505 MeV (or effective energy of 470 MeV), demonstrates also a very peculiar "trough". Meanwhile, observed intensity-time profile at the NM Calgary (as well as at several other NMs) shows apparent two-peak structure (see, *e.g.*, Smart et al. 1991), in accordance with the two-source scenario, but in contradiction to the CME-shock model.

It is important to note that the average proton energy indicated in Fig. 7.5 was taken (see Kahler 1994) to be twice the calculated cutoff energy at Mt. Norikura ($R_c = 11.48$ GV). Besides, the ejection profile of >21 GeV protons in Fig. 7.5 was derived under the assumption that there was no scattering in the interplanetary medium. Any scattering would evidently lengthen the effective travel distance of relativistic protons to the Earth, and thus the onset of their ejection would be moved to an earlier time. This would result, in particular, in approaching the time of the first detection of GRL emission on board the SMM satellite.

7.6 Magnetic Reconnection in Acceleration Scenario

From the above considerations it follows that two of the three proposed models of the event under study are based on the concept of magnetic reconnection in the solar corona. In order to complete the foregoing scenario for the event (see Table 7.1), we estimate the time, t_f, required for the formation of the reconnecting current sheet (RCS) in the region of source II, and the time for acceleration of protons by an electric field, t_{ac}, to energies ≥10–100 GeV. First we will introduce corresponding estimates of Litvinenko and Somov (1995) for RCS which is supposed to form during the rise of a CME at the post-eruptive stage of the flare.

A typical CME velocity of upward motion equals the Alfvén speed in the corona $V_A \cong 1{,}000$ km s^{-1} under characteristic values of the coronal magnetic field $B \cong 100$ G and plasma density $n \cong 10^{11}$ cm^{-3}. Assuming the speed of plasma inflow into the RCS to be $u = 0.1V_A$ (fast reconnection under high, but finite conductivity) we obtain $t_f = L/u = 10^2$–10^3 s, where $L = 10^9$–10^{10} cm is the characteristic scale for width and length of the sheet. Further, it should take into account the effect of transverse electric field outside the RCS. It was shown (Litvinenko and Somov 1995) that this field efficiently locks non-thermal ions inside the sheet. Such a confinement allows the particles to be accelerated with a characteristic time $t_{ac} \cong 0.03$ (E_p/1 GeV) s. It follows the proton requires only 3 s to be accelerated up to energy E_p ~100 GeV (Litvinenko and Somov 1995; Akimov et al. 1996; Somov 1996).

On the other hand, under derived conditions for the PC generation at the source II in Fig. 7.25 for the event of 23 February 1956 ($B = 30$ G, $n = 2 \times 10^7$ cm^{-3}, $L = 10^{10}$ cm; Perez-Peraza et al. 1992), one can estimate the Alfvén speed $V_A = 1.5 \times 10^9$ cm s^{-1}. If we take $u = 0.1V_A$, then the time for formation of the RCS will be $t_f \cong 66.7$ s. This is close to the lower estimate of Litvinenko and Somov (1995). For the event of 29 September 1989 ($B = 91$ G, $n = 1.2 \times 10^7$ cm^{-3}, $L = 10^9$ cm; Vashenyuk et al. 2000), the time for formation of the RCS is considerably less, $t_f \cong 1.74$ s. However, we should bear in mind that if the magnetic bottle (with an expansion velocity V_c ~300 cm s^{-1}) interacts with a coronal arch, there will probably be stimulated (explosive) reconnection. As shown by Yokoyama and Shibata (1994), its rate is determined not only by the parameters of the stimulating (driving) process, but also strongly depends on the plasma resistivity (uniform or anomalous) near the neutral point. It appears that the formation of magnetic islands (plasmoids) and their subsequent ejection from the current sheet is a key physical process leading to fast reconnection (Yokoyama and Shibata 1994). Anyway, and this is important, the problems of magnetic reconnection and coronal mass ejections are closely related (Somov 1991, 1992).

Overall, it is fair to say that the two-source model is consistent with modern theories of magnetic reconnection in the solar corona, including the possible acceleration of protons to energies ~10–100 GeV. We note that if the reconnection speed is $u = 0.1V_A$, instead of accepted earlier $u = V_A/18$ (Priest 1982), the calculated number of accelerated particles changes considerably (Perez-Peraza et al. 1992). For example, for $E_p = 25$ MeV, the number of accelerated protons, according to Eq. (5.11), increases by a factor of 2.4.

From these estimates it is concluded that the acceleration of the prompt component of relativistic protons in the September 29, 1989 event may be understood in the framework of reconnection models of Martens and Kuin (1989) and Litvinenko and Somov (1995). Here the particle acceleration proceeds in the electric field that is produced between reconnecting magnetic field lines in the trailing part of coronal transient behind the eruptive filament. On the other hand, while gaining energy in the electric field, particles may accomplish an azimuthal drift in the neutral sheet carrying them to the visible side of the Sun from the-behind-the-limb flare. So, the prompt arrival of particles and gamma-ray emission from the behind-limb flare (Vestrand and Forrest 1993) may be easily explained as well.

However, the two-source model can not yet answer, of course, all the questions involved. At least, three important problems remain unresolved theoretically, namely, the drift effects of relativistic particles in expanding bottle (loop), possible adiabatic loss of particle energy as the volume of the bottle increases, and maximum rigidity of accelerated particles. Though the first two problems were treated in several works (*e.*

g., Mullan and Schatten 1979; Mandzhavidze and Ramaty 1992; Ramaty and Mandzhavidze 1994), many questions remain unclear (for example, the escape of the *first* relativistic protons from expanding magnetic structures).

As to maximum rigidity of accelerated particles, available acceleration models do not exclude large values of R_m (or E_m), and the problem seems to reduce to the search for adequate magnetic configurations (structures) in the solar corona. For example, the model of two SCR sources (Perez-Peraza et al. 1992) gives a value of E_m ~250 GeV for the flare of February 23, 1956 type; in the electromagnetic model of solar flare (Podgorny and Podgorny 1990) maximum proton energy may be as large as 10^6 GeV. On the whole, however, all such estimations depend heavily on the choice of acceleration model. Moreover, to compare the estimated values with observational results it is not only important to calculate E_m, but also to resolve a more difficult problem, namely, to determine the SCR spectrum shape at the source and the number of accelerated particles of extremely high energy. In this respect, the results of the generalization of the SCR spectrum data (Miroshnichenko 1994, 1996) for the most powerful SPEs impose certain upper limitations. In the range of energies from several units to several tens GeV, the data point to a steepening behaviour of the SCR spectrum (*e.g.*, Miroshnichenko et al. 2000). At any rate, they do not give convincing grounds for its extrapolation (Kolomeets et al. 1993, 1995) by the power-law function with the same (unchanging) slope to the higher energies.

As to the BUST muon burst during the event of September 29, 1989 (see Sect. 5.3), it is difficult to explain, first of all, its delay for a time >1 h relative to the first intensity peak at the surface muon telescopes. At the same time, it is obviously impossible to accept a hypothesis about the trapping and prolonged containment of relativistic protons in magnetic loops of the solar corona during certain SPEs (*e.g.*, Mandzhavidze and Ramaty 1992). The presence of source II high in the corona (Vashenyuk et al. 1993, 1994, 1997; Miroshnichenko 1997) would be a possible explanation of above fact. Such a suggestion, however, comes in collision with the fact that the proton intensity corresponding to the BUST burst does not agree with the spectrum of relativistic protons at the early stage of this GLE (Karpov et al. 1997b, c, 1998). It becomes clear that in application to the BUST burst the existing two-source model must be modified to take into account either possible additional acceleration of solar particles at the shock front far from the site of the proper flare, or eventual modulation of galactic cosmic rays at the energies above 500 GeV (Karpov et al. 1997b, c, 1998).

At this stage of our knowledge about energetic solar processes, in particular, for the case of the September 29, 1989 event, it seems to be reasonable to incorporate all three approaches as *contributing* to the comprehensive spatial-temporal scenario of the September 29, 1989 event rather than *competing* or even *mutually excluding* one another.

As to predictive capacities of the solar and cosmic ray communities in this particular case, we quote with agreement Peggy Shea (1990): "*There are times when nature puts to a severe test man's presumed knowledge and technology. The event of 29 September is one of these times*".

Chapter 8
Solar Cosmic Rays in the Interplanetary Space

Transport of energetic particles of galactic and solar origin in the interplanetary medium is an important topic of research in space physics over several decades. As well-known, the first observations of the interplanetary magnetic field (IMF) established the anisotropic nature of the interplanetary medium, and particles become excellent probes to study the structure of the fields and the particle interactions with interplanetary plasma irregularities.

Unfortunately, there is no way to observe the complete trajectory of an individual energetic charged particle from its source to the point of detection. What is measurable is the intensity of charged particles of a given type as a function of time, energy and direction of incidence relative to the local magnetic field (*i.e.*, pitch angle). To relate these observations to the characteristics of the medium a theoretical treatment has to take into account its known properties and to make some assumptions. Other approaches to the problem (empirical and numerical) can also be pursued (*e.g.*, Gombosi and Owens 1981; Valdes-Galicia et al. 1988).

This very extended and diverse area of investigations includes a number of fundamental problems of particle interactions with the plasma turbulence in space conditions, and the most of them are out of the scope of our considerations (see, *e.g.*, Parker 1963; Dorman and Miroshnichenko 1968; Toptygin 1985; Berezinsky et al. 1990; Dröge 1994a, b). It is the purpose of this Chapter to summarize and discuss briefly recent studies, preferably concerning the transport of solar cosmic rays and their interactions with the turbulent magnetic fields (pitch angle scattering), magnetic clouds, interplanetary shocks, corotating interaction regions, *etc.*

Since the largest SPE of February 23, 1956 it was proposed to treat the time history of solar cosmic rays as a result of a fundamental diffusive process in the interplanetary space (Meyer et al. 1956). Initially, scattering by magnetic irregularities was appealed as a physical mechanism responsible for diffusion, and numerous efforts were made to infer information on this scattering process from observed intensity-time profiles of solar cosmic rays (as the reviews see, *e.g.*, Dorman and Miroshnichenko 1968; Sakurai 1974; Palmer 1982; Bieber et al. 1994). Later it was noted that the diffusion could equally well be occurring

L. Miroshnichenko, *Solar Cosmic Rays*, Astrophysics and Space Science Library 405, DOI 10.1007/978-3-319-09429-8_8

in the solar corona (*e.g.*, Reid 1964; Axford 1965). These developments presented opportunity for using observations of SEPs to study physical processes at/near the Sun; at the same time, the interpretation of solar particle events became also greatly complicated. More recent studies have resorted to complex numerical procedures that allow to model the time histories of both the intensity and the anisotropy of solar cosmic rays (*e.g.*, Bieber et al. 1986, and references therein). Some new developments in the understanding of the interplanetary transport of solar cosmic rays have been reviewed by Dröge (1994a, b).

As it was discussed in Chap. 3, a new class of observations has revealed two distinct populations of SEPs, with completely different origins, based upon the abundances, ionization states and time profiles of the particles as well as the longitude distribution and the radio, optical, X-ray and gamma-ray associations of the event (see, *e.g.*, Reames 1993, 1999). Moreover, those observations gave rise to a new paradigm of particle acceleration in different sources at/near the Sun (flares, CME-driven shocks, *etc*.). In fact, these new findings resulted in declining the concept of coronal diffusion of particles across magnetic field lines. The consequences of this demise, however, are not fully appreciated by some authors (see, *e.g.*, the reviews by Reames 1996, 1999 and references therein).

8.1 Theory of Particle Transport

The behaviour of SCRs in the interplanetary space is determined by magnetic fields frozen into the solar wind plasma. Electrical fields of the solar wind, in the first approximation, can be neglected, and collisions of SCR with particles of the solar wind are insignificant. As the Sun rotates with an angular velocity Ω_s, the mean magnetic field, <*B*>, in the plane of the solar equator has a spiral form. The angle between the field <*B*> and radius-vector from the Sun is $\varphi = \mathrm{arctg}\ (\Omega_s\ r/u)$, where u is the solar wind velocity. In the heliocentric system of coordinates the B vector may be represented as a sum of three component $B = B_x + B_y + B_z$, respectively, radial, azimuthal and normal to the ecliptic plane. On that topology are superimposed fluctuations of the solar wind parameters (magnetic fields, plasma density and velocity, *etc*.) in a wide range of time scales τ (or wave numbers k, or frequencies f) – from ~ 27 days (period of the Sun's rotation) to hours, minutes and seconds (*e.g.*, Alfvén waves, rotational and tangential discontinuities). It has been established by direct and indirect measurements that there exists a broad spectrum of irregularities in the interplanetary medium that is typical of a developed turbulence. Observed spectrum of turbulence can be presented in the power-law form (*e.g.*, Hedgecock 1975):

$$P(k) = dh^2/dk = A/(k_0 + k)^q = A/(f_0 + f)^q \tag{8.1}$$

where h is the random field, A is the normalization constant, k (or f) is the wave number (or frequency) of fluctuations, $q = 1.5 \pm 0.1$, and k_0 (or f_0) corresponds to so-called main scale of turbulence L_0. The expression (8.1) has a simple physical sense: it describes approximately the distribution of density of magnetic energy between fluctuations of various scales. Depending on the spectrum of turbulence, characteristics of particle transport (mean free path, anisotropy, characteristic time of propagation, *etc.*) also change.

Strictly speaking, transport of solar particles in the IMF should be considered by the methods of kinetic theory (*e.g.*, Toptygin 1985). Such an approach is based on consideration of accelerated particles as a highly energetic component of interplanetary plasma. Here the trajectory of an individual particle can not be calculated because of the stochastic nature of the small-scale magnetic fields. In such a field the motion of particles is of random nature and the system of particles can be described conveniently by distribution function, $f(r, p, t)$, satisfying the kinetic equation. The latter should be averaged over possible values of the random components of the magnetic and electric fields.

Owing to the topology of the solar wind as a divergent supersonic flow of plasma both kinematical and geometric effects should be observed in the behaviour of SEPs, namely: convection; adiabatic deceleration or acceleration; adiabatic focusing; mirror reflection; stochastic acceleration; gradient drift (due to the gradient of the regular part of the IMF); drift of the curvature of magnetic lines of force; pitch angle scattering on the magnetic irregularities (plasma turbulence). In practice, however, the distribution of magnetic fields in the interplanetary space is so complicated and variable that it is very difficult to present a complete and strict description of the behaviour of particles in most cases, in spite of the simplicity of the physical basis of the process (Lorentz and electric forces in the moving magnetic field). In such a situation it is necessary first to construct a model of IMF on the basis of observation data and then to deduce a transport equation. For comparison with the observed characteristics of SCR (flux, time profile, anisotropy, *etc.*) solution of the transport equation is usually simplified to the diffusion limit, i.e., it reduced to the diffusion approximation (diffusion coefficient, κ; mean free path, Λ, *etc.*).

The equation describing the dynamics of energetic particle propagation in moving weakly turbulent plasma has the form (*e.g.*, Tverskoi 1981):

$$\partial f/\partial t + u\nabla f - (1/3)\mathrm{div}\, up(\partial f/\partial p) = (\partial/\partial x_i)\kappa_{ij}\ (\partial f/\partial x_i) + \left(1/p^2\right)(\partial/\partial p)D(p)p^2(\partial f/\partial p) \quad (8.2)$$

where $f(r, p, t)$ is the distribution function of particles on their directions r, momentum p and time t; u is the solar wind velocity; x_i is the current spatial coordinate; κ_{ij} is the tensor of spatial diffusion; $D(p)$ is the coefficient of diffusion in the momentum (phase) space. The second term on the left-hand side of (8.2) describes the convection of particles, the third one corresponds to the adiabatic effects (deceleration and acceleration); the second term on the right-hand side refers to the stochastic acceleration in the phase space, the meaning of the other terms is obvious. When deriving Eq. (8.2), it was assumed that the mean free path is small,

$\Lambda \ll L_c$, as compared to the characteristic scales, L_c, of the system (for example, $\Lambda \ll 1.0$ AU), and the distribution function is near isotropic. In certain cases some of the term may be neglected. For instance, the last term on the right-hand side describing the stochastic acceleration is important only when acceleration in the interplanetary medium is considered.

Equation (8.2) serves as the common theoretical basis for interpreting many phenomena: 11-year variation and Forbush-decreases of galactic cosmic rays, transport of fast particles from the flares, acceleration in the interplanetary medium, *etc*. Specific forms of this equation depend on the IMF model. Usually, the IMF is represented as a sum of regular and random components: $B = B_0 + \delta B$, where $<B> = B_0$; $<\delta B> = 0$; $|\delta B/B| \ll 1.0$. The averaged component B_0 (regular or background field) determines a zero order of the particle motion, and the fluctuations δB act as the disturbances that cause a resonant scattering of the particles (*e.g*., Jokipii 1966; Jokipii 1971b; Völk 1975; Toptygin 1985). In such a model, relative contribution of the regular and random components into the particle motion depends on the particle energy (momentum). Resonant scattering takes place under the following condition:

$$k_{res} \cong 1/\rho = eB/cp = B/R \tag{8.3}$$

Thus, Eq. (8.3) allows visually to link the particle energy (or momentum p, or rigidity R) and Larmor radius ρ with the spectrum of turbulence (8.1).

The well-defined overall structure of the IMF, namely the Archimedean spiral, gives energetic particles a preferential direction of propagation, while on the other hand, irregularities present in the field make the particle scatter in pitch angle. As a consequence, the motion of the particles has two components, an approximately adiabatic motion along a smooth field and random walk in a pitch angle space. The main equation for the distribution function $f(m, z, t)$ of particles traveling parallel to inhomogeneous regular field one can derive from the general kinetic equation (8.2) by passing along the Fokker-Plank approximation under some simplifying assumptions (*e.g*., Roelof 1969; Earl 1976a, b; Toptygin 1985; Bieber et al. 1986; Valdes-Galicia 1991):

$$\begin{aligned} \partial f/\partial t + \mu v(\partial f/\partial z) = {} & (1/2)\,(\partial/\partial\mu)\; D(\mu)\;(\partial f/\partial\mu) \\ & - (v/2L)\left(1 - \mu^2\right)(\partial f/\partial\mu) \end{aligned} \tag{8.4}$$

where z is distance along the mean magnetic field B_0 (positive outward from the Sun); μ is the cosine of the particle pitch angle, θ ($\mu = \cos\theta$); L is the characteristic length of variations of the mean (smoothed) magnetic field, or so-called focusing length, and $D(\mu)$ is the Fokker-Plank coefficient for pitch angle scattering (or the coefficient of particle diffusion in the pitch angle space). In fact, this equation is the Boltzmann equation for cosmic ray transport in the presence of adiabatic focusing and pitch angle scattering. According to Jokipii (Jokipii 1971b), the coefficient of the pitch angle scattering $D(\mu)$ may be written as

$$D(\mu) = < (\Delta\mu)^2 > /\Delta t = A_\mu |\mu|^{q-1}(1 - \mu^2) \quad (8.5)$$

where A_μ is a parameter related to the particle energy and strength of magnetic fluctuations; q is the spectral index of the power spectrum of magnetic fluctuations (8.1). The parameter q determines the dependence of the scattering rate on pitch angle, with $q = 1$ corresponding to isotropic scattering. Though a validity of the expression (8.5) has been questioned (*e.g.*, Duggal 1979) because of the lack of strict theory of pitch angle scattering, it was accepted as a working formula that characterizes visually the amplitude and anisotropy of scattering. In particular, it may be seem that scattering is absent at $\mu = 0$, i.e., at $\theta = 90°$ (effect of mirror reflection).

The effect of adiabatic focusing due to the radially decreasing magnitude of the IMF is described by the second term on the right-hand side of (8.4). The focusing length L is defined by

$$(1/L) = (-1/B)\,(\partial B/\partial z) \quad (8.6)$$

For positions near the ecliptic plane in a Parker spiral field, L may be conveniently calculated from the acute spiral angle ψ (angle between the magnetic field line and the radial direction):

$$L = r/\left[\cos\psi\,\left(1 + \cos^2\psi\right)\right] \quad (8.7)$$

where r is radial distance from the Sun. Thus, at the Earth's orbit ($r = 1.0$ AU), where the average value of ψ is 45°, we get from (8.7) the average focusing length $L \approx 0.94$ AU.

If the values A_μ, q, and v are known, the scattering mean free path Λ may be calculated according to

$$\Lambda = \left(3v/\mathrm{A}\;A_\mu\right)/[(2 - q)(4 - q)] \quad (8.8)$$

Alternatively, the scattering strength may be characterized by a diffusion coefficient, κ, in the coordinate space

$$\kappa = \Lambda v/3 \quad (8.9)$$

In this context, it should be noted that the phenomenon of diffusion which occupies a central place in the problem of particle transport, initially has been discovered heuristically in the middle of 1950s. Only about 10 years after the diffusion equation has been deduced directly from the equation of particle motion in the inhomogeneous magnetic field (see, *e.g.*, Toptygin 1985).

In order to trace the evolution of distribution function, Earl (1976a) considered a complete transport equation (8.4) in some details. It was shown that diffusion solutions are the lowest order ones of the equation (8.4). If one takes into account

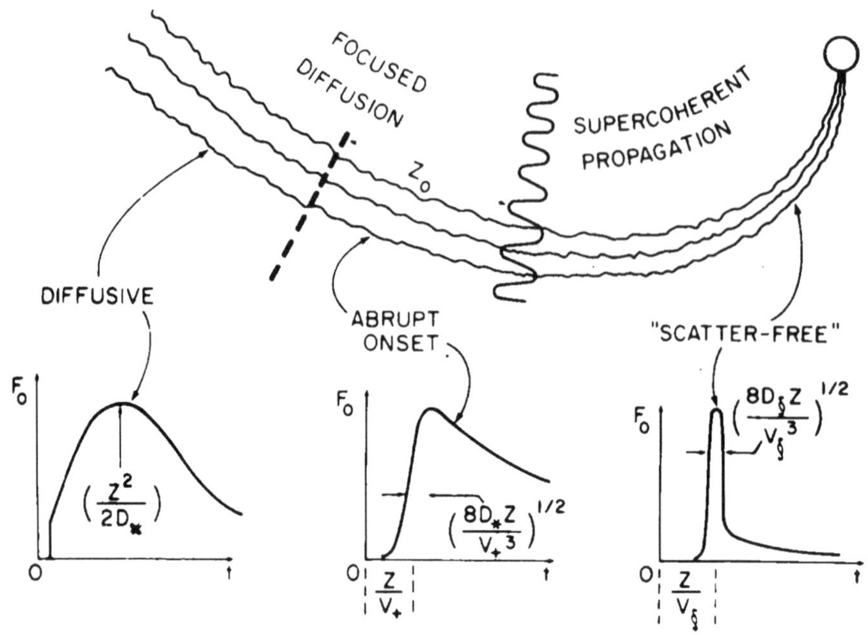

Fig. 8.1 Schematic diagram of super-coherent propagation and focused diffusion regions near the Sun (Earl 1976a). These regions are characterized by three types of intensity-time profiles of SEPs. The location of the Earth, in relation to the wiggly and dashed dividing lines, depends on the strength of the IMF fluctuations and also on the velocity and rigidity of the particles

the higher order terms, a number of non-diffusive effects become important, such as coherent propagation, particle velocity dispersion, and exponential decay with a focused diffusion. The focusing of the particles, i.e., the diminishing of their pitch angles, develops due to the divergence and decrease of the interplanetary magnetic field, $B(r)$, the first adiabatic invariant being preserved, $\sin^2\theta/B = \text{const}$.

Since it is not possible to get a strict solution of (8.4) in the closed form, Earl (1976a) suggested to use an extension of the distribution function on the eigenvalues of the combined operator of focusing and scattering. Such approach allows to understand at the qualitative level a great variety of the SEP profiles. The results of this study are shown schematically in Fig. 8.1. It seems that near the Sun a coherent propagation takes place up to a certain boundary z_0, where a region of adiabatic focusing begins. In turn, this region may be divided on two parts, with the prompt (gradual) increase onset and exponential (power-law) decay. In the coherent region the intensity-time profile has a spike-like form. The Earth's position relatively to these regions (and observed form of the profile as well) will be determined by the spectrum of the IMF fluctuations and particle rigidity. Numerous aspects of non-diffusive propagation of energetic solar particles are described in more details elsewhere (*e.g.*, Toptygin 1985; Dröge 1994a, b; Ruffolo 1995).

8.2 Change of Average Energy and Spectrum Transformation

In the process of particle transport in interplanetary space the distribution function of the SCR experiences significant changes. They are evident in the observations in the form of space-time variations of the angular distribution and the energy spectrum of the particles. Observations near the Earth's orbit indicate, in particular, an increase in the hardness (flatness) of the spectrum of solar protons in the energy range $E_p \leq 10$ MeV (*e.g.*, Miroshnichenko and Petrov 1985). This tendency is exceptionally important for estimates of the energetics of the SCR and of the flare as a whole (Miroshnichenko 1987). No unambiguous explanation of this effect yet exists.

At present, two possibilities are discussed; (1) formation of the spectrum with a variable slope at the source itself; (2) adiabatic deceleration of the particle in the interplanetary medium. As reviewed by Miroshnichenko (1987), acceleration models based on the concept of a magnetic reconnection indicate the variable form of particle spectrum at the source. Further, typical proton and alpha particle spectra observed near the Earth may be fitted by Bessel function following from a stochastic acceleration model (see Fig. 5.1). Predicted proton energy spectra from acceleration at turbulent shocks (Fig. 5.2) also are consistent with a spectral flattening at low energy. As it was demonstrated in Chap. 5, theoretical problems of particle acceleration at the Sun are still far from complete resolving. Therefore, it is of interest now to determine quantitatively what contribution adiabatic deceleration can make to the deformation of the spectrum.

It is customary to assume that the third term on the left-hand side of (8.2) corresponds to the adiabatic cooling, or to the anti-Fermi mechanism of particle deceleration in a spherically diverging solar wind. Estimates and studies of this effect have been made by many authors. In particular, Dorman et al. (1979) noted that the exchange of energy between the SCR and the solar wind plasma does not reduce to a simple adiabatic slowing, but depends on the strength of the redistribution of the particles with different energy in space due to diffusion. These authors were the first to point this out in solving the problem of GCR modulation: it was shown that, in general, the mechanism of energy exchange between the cosmic rays and the interplanetary medium is determined by the specific form of the particle distribution function and is incompatible with the traditional concept of deceleration based on intuitive thermodynamic considerations.

The complete transport equation (8.2) must be solved in order to calculate the change in the average energy and the corresponding deformation of the proton spectrum in interplanetary space. Practical estimates of the quantity dE/dt by this way, as far as we know, have not been made, and most investigators have used a simplified approach. If diffusion and convection are ignored, then the change of the energy due to adiabatic slowing is described by the usual formula (*e.g.*, Parker 1965; Toptygin 1985)

$$dE_k/dt = -(pv/3)\ \nabla u\,(\alpha E_k/3)\,(2u/r) \tag{8.10}$$

where $\alpha = 2$ in the non-relativistic ($pv = 2E_k$) and $\alpha = 1$ in the relativistic ($pv = pc = E_k$) cases, and the expansion of the solar wind is assumed to be strictly radial. The solution of (8.10) is

$$E_k(\mathrm{t}) = E_0\ \exp(-t/\tau_{\mathrm{ad}}) = E_0\ \exp(-4ut/3r) \tag{8.11}$$

where τ_{ad} is the characteristic time of adiabatic cooling (see, *e.g.*, Toptygin 1985). For $u = 400$ km s^{-1} at the Earth's orbit we obtain $\tau_{\mathrm{ad}} = 78 \pm 4$ h. The results of observations during the SPE of June 7, 1969 (Murray et al. 1971) were considered (Jokipii 1971a; Dorman 1972; Toptygin 1985) as the first direct evidence of the change of the solar proton energy in the interplanetary medium. By comparing the observed proton spectra at different times, Murray et al. (1971) obtained the characteristic time of energy change, $\tau_{\mathrm{n}} = 210 \pm 10$ h with an exponential approximation of the $E_p(t)$ dependence of the form (8.11). The considerable difference between τ_{n} and τ_{ad} is not surprising since the time constant τ_{ad} corresponds to ignoring the diffusion process.

Daibog et al. (1981, 1984) have made an interesting attempt to take diffusion mixing into account based on a simplifying model in which the diffusion maximum of the SCR intensity, $t_m = r^2/6\kappa$, moves with the velocity $u = dr/dt = 3\kappa/r$, and the diffusion coefficient is assumed to depend on the energy $\kappa(E_k) = \kappa_0 E^a{}_k$. In this case the expression

$$E_k(t) = [E^a{}_0 - (4/9)a\,(ur/\kappa_0)]^a \tag{8.12}$$

may be used instead of (8.11). Using the data for the SPE of November 22, 1977 (Daibog et al. 1981), let us compare the estimates of the change in the proton energy $\Delta E_k/E_k$ (in %) during the time to reach a maximum near the Earth, t_m (1 MeV) = 20 h and t_m (10 MeV) = 6.3 h, for $u = 300$ km s^{-1}, $\kappa_0 = 5 \times 10^{20}$ cm^2 s^{-1}, and $a = 0.6$. The results of calculations of $\Delta E_k/E_k$ from Eqs. (8.11) and (8.12) for $\tau_{\mathrm{n}} = 210$ h and $\tau_{\mathrm{ad}} = 78$ h are listed in Table 8.1. The obvious discrepancies between the expected and observed values of $\Delta E_k/E_k$ are seen here.

For a non-contradictory interpretation of the results of Murray et al. (1971) and Daibog et al. (1981) it is interesting to consider, in addition to diffusion, also the role of convection besides diffusion for $u = u(r)$ and a possible acceleration of the particles in the interplanetary space. At a sufficiently small energy the diffusion of the particles can be ignored compared with their convection ($\kappa \nabla n \ll un$), and the transport equation can be integrated (*e.g.*, Jokipii 1971a). In practice, it is of interest the case when $u = u(r) = u_r + u_\theta$, *i.e.*, when the solar wind velocity $u \neq$ const and has a transverse component u_θ, with $u_\theta/\ u_r \leq 0.1$, according to the data of direct measurements. Hence, one can write instead of (8.10) the equation (Toptygin 1985):

Table 8.1 Change of solar proton energy in the interplanetary medium

Estimation formula	(8.11)	(8.11)	(8.12)
$\Delta E_k/E_k$ $\Delta E_k/E_k$, (1 MeV MeV), %	9.1	23	35
$\Delta E_k/E_k$, (10 MeV), %	3.0	7.75	10

$$dE_k/dt = -(\alpha/3)\,[2u_r/r - (\partial u_r/\partial r)]\;E_k \qquad (8.13)$$

During the observations by Murray et al. (1971) the quantity u_r varied within the limits of 377–455 km s^{-1}. This gives, according to Eq. (8.13), a correction of only ± 10 % to $\tau_{ad} = 78$ h, which is within the error limits of the estimate. If it is assumed that an acceleration of the particles also occurs, besides cooling, then the characteristic acceleration time (regardless of the assumption about the specific mechanism) can be obtained by the relation $\tau^{-1}{}_{n} = \tau^{-1}{}_{ad} - \tau^{-1}{}_{a}$, from which $\tau_a = 125 \pm 10$ h (Murray et al. 1971).

For the case of stochastic Fermi acceleration Jokipii (1971a, b) estimated the acceleration rate, $dE_k/dt \approx (8V^2{}_A/3\kappa)E_k$. Hence, for $\kappa \sim 10^{20}$ cm^2 s^{-1}, $E_k \geq 1$ MeV, $B \sim 7 \times 10^{-5}$ G and $n \sim 2$–7 cm^{-3} at the Earth's orbit, one can get $\tau_a = 120$ h which agrees with the estimate of Murray et al. (1971). Because of the quadratic dependence τ_a on V_A, the estimate (Jokipii 1971a), however, is very unstable with respect to variations in the quantities B and n. If the average values $B \sim 5 \times 10^{-5}$ G and $n \sim 5$ cm^{-3} are used, then we obtain $\tau_a \approx 417$ h. According to theoretical calculations by Toptygin (1985), in the presence of a developed large-scale turbulence in the interplanetary medium the characteristic acceleration time for protons with an energy $E_p = 1$ MeV amounts to $\tau_a \approx 1110$ h, and for acceleration of particle by intense small-scale MHD waves $\tau_a \approx 7000$ h. These times are 1–2 orders of magnitude greater than the adiabatic deceleration time τ_{ad} τ_{ad} at the Earth's orbit. Nevertheless, under extreme conditions (for example, after the arrival of a shock wave from the 4 August 1972 flare) the time τ_a, conversely, can be 1–2 orders of magnitude smaller, than τ_{ad}; this entails a strong turbulent acceleration of the particles (Toptygin 1985).

The observations show that the fluctuations (irregularities) of the IMF are statistically anisotropic, *i.e.*, the interplanetary medium is gyrotropic. There are some evidences (*e.g.*, Matthaeus and Goldstein 1981) that such a medium also has a nonzero value of a helicity of the magnetic field, $H_m = \langle B \text{ curl} B \rangle \neq 0$ (this pseudoscalar measures the departure of B from mirror symmetry). Under this condition, the average electric field E_h can arise, being directed along the large-scale magnetic field. The process of particle acceleration in such electric fields was first considered by Kichatinov (1983). The propagation of particles in turbulent statistically anisotropic electromagnetic fields was investigated by Dorman et al. (1988) when a change in particle energy due to the stochastic Fermi acceleration mechanism occurs in addition to an acceleration of the particles in the average electric field. The corresponding transport equation in the phase space is

$$(\partial f/\partial t) = \nabla_\alpha \kappa_{\alpha\beta} \nabla_\beta f - (u_0 \nabla) f + (u_0 \nabla)(p/3)(\partial f/\partial p) - (2/3)p(w\nabla)(\partial f/\partial p) + (1/p^3)(\partial/\partial p)D(\partial f/\partial p) \quad (8.14)$$

where u_0 and w are the average and effective solar wind velocity, and D is the diffusion coefficient in the phase space. The first three terms on the right-hand side of Eq. (8.14) are analogous to the corresponding terms in Eq. (8.2); the forth term reflects the change in particle energy due to the expansion of the solar wind with an effective velocity w, which depends on the value of H_m; the last term describes the acceleration processes with the parameter H_m and the non-monotonic rigidity dependence of the transport mean free path of the particles, $\Lambda(R)$, taken into account (*e.g.*, Miroshnichenko 1980; Palmer 1982; Miroshnichenko and Petrov 1985; Bieber et al. 1994).

The relative efficiency of the acceleration due to action of the average electric field E_h and Fermi acceleration is determined by the relationship $\eta = w^2/\langle u^2 \rangle$, where $\langle u^2 \rangle$ is the mean-square random component of the solar wind velocity. The calculations show (Dorman et al. 1988) that the particle acceleration by the average electric field is most effective at small energies. If the quantity H_m in interplanetary space is sufficiently large, then the acceleration by the field E_h greatly exceeds the acceleration due to the action of the Fermi mechanism. The joint work of the two mechanisms should lead to a decrease in the characteristic acceleration time τ_a toward a value of $\tau_a = 125 \pm 10$ h, which follows from the observations (Murray et al. 1971). However, calculations of the quantity τ_a, the change in the proton energy in interplanetary space and the possible deformation of the proton spectrum were not made by Dorman et al. (1988).

Chebakova et al. (1985) and Daibog et al. (1986) have carried out the calculations of proton spectrum transformation on the basis of a numerical solution of the transport equation (8.2). However, it is impossible to accept their results as a convincing proof of effective adiabatic cooling of the particles. Actually, the spectrum hardening at the source obtained by these authors at small energies is the result of the combined action of diffusion, convection and slowing, with the typical dependence on particle energy and distance to the Sun, namely, $\kappa(E_k, r) = \kappa_0 E_k^{0.6} r$, being assumed for the diffusion coefficient (Daibog et al. 1984). Using the solution of the corresponding diffusion equation (ignoring convection and slowing), it is easy to show (Miroshnichenko 1992b) that the calculated change in the power-law index from g (at the source) to $\gamma \sim 3.0 \pm 0.2$ (at the Earth) in the energy range $E_p = 1$–100 MeV can be completely explained by the effect of diffusion (within the error limits of the observations). Notice that the accuracy of the measurements on spacecraft in this energy interval amounts to tens of percent for the hourly intensity values (Murray et al. 1971; Bengin et al. 1985).

The controversy over the role of adiabatic deceleration is of fundamental character: if the point of view of Daibog et al. (1981, 1986) and Chebakova et al. (1985) is adopted then the steepening of the spectrum of ejected protons at low energies increases the estimates of their total energy by a factor of 5–10; this is extremely important for estimates of the energetics of the flare as a whole and for the identification of the acceleration mechanisms (*e.g.*, Miroshnichenko 1981a). The above discussion emphasizes again that detailed calculations by the kinetic equation (8.2) are still necessary to obtain strict numerical estimations of the contribution of the effect under consideration into the transformation of observed spectrum in different energy intervals.

In the framework of focused transport theory, comprehensive analytical description of adiabatic deceleration of cosmic ray particles was given recently by Ruffolo (1995). If one does not neglect the solar wind effects, two reference frames should be considered: the fixed frame (where the Sun is always in origin) and the local solar wind frame (co-moving with the solar wind velocity at a given point). Because the large-scale structure of the magnetic field is taken to be stationary in the fixed frame, the process of focusing conserves the absolute value of the particle velocity here, *i.e.*, it is a natural frame to simulate the focusing. The small-scale irregularities in the magnetic field can be considered to be frozen in the solar wind. For this reason, the process of scattering should be simulated in the solar wind frame. Similar approach was earlier employed for large-angle scattering (see Toptygin 1985, Sect. 13.3 therein).

Figure 8.2 shows the results obtained by Ruffolo (1995) for the 2 MeV protons at 1 AU with (*1*) no solar wind effects, (*2*) convection only, (*3*) deceleration only, and (*4*) all solar wind effects. The main effect of convection is to speed up the pulse, so that protons start to arrive sooner. At later times, convection leads to a slightly steeper decay slope, because the peak in the distribution is convected farther from the point of observation. Deceleration creates a widening deficit in the intensity as a function of z, and accounts for most of the change in the decay slope. Finally, the simulation including all effects exhibits both an earlier rise and a steeper decay at late times. These results justify the consideration of adiabatic deceleration in focused transport models. It seems, however, to be erroneous to think that the effect of deceleration estimated from the equation of the (8.11) type could be simply added to that due to the other processes (*e.g.*, Murray et al. 1971). In fact, Ruffolo (1995) has shown that the effect of including deceleration can be much greater than the simple estimate, because the effect is greatly amplified by an interaction with diffusive processes. The magnitude of amplification depends on the cumulative effect of deceleration on the shape of the distribution function f, which in turn depends on the detailed history of the distribution function. Thus, as emphasized by Ruffolo (1995), to accurately evaluate the effect of deceleration on the transport of solar cosmic rays, it is necessary to incorporate the process fully into a numerical code for focused transport.

Adiabatic deceleration of SEPs as deduced from Monte Carlo simulation of the focused transport was considered by Kocharov (1997). Monte Carlo simulations of interplanetary transport are employed to deduce adiabatic energy losses of solar protons. The model includes both small-angle anisotropic scattering in solar wind frame and focusing. The author uses this code to deduce intensity-time profiles, which were expected to be close to the profiles hitherto calculated by means of a finite-difference method. Adiabatic deceleration directly calculated from the energy of individual arriving particles is also compared with the

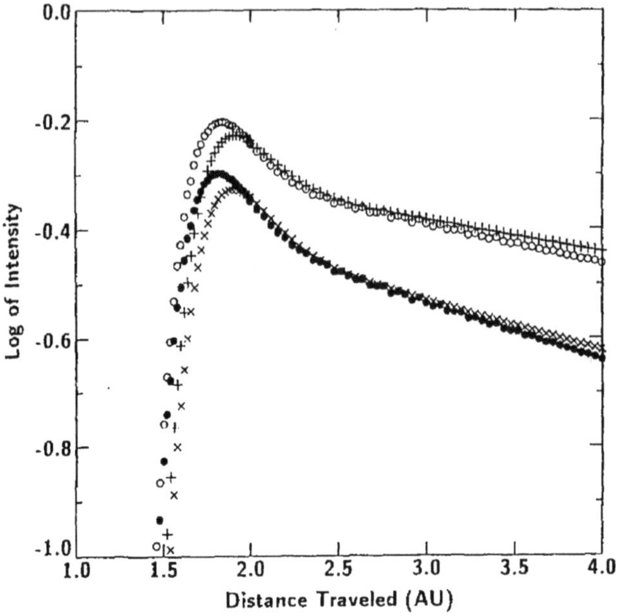

Fig. 8.2 Logarithm of the intensity of 2 MeV protons *versus* distance traveled, *s*, for simulations that included no solar wind effects (*plus signs*), convection only (*open circles*), deceleration only (*crosses*), and all solar wind effect (*filled circles*), for a radius of 1 AU (Ruffolo 1995). Note that convection results in an earlier arrival of protons, and deceleration causes a lower intensity and a faster decay after the peak

traditional estimations based on the complete transport equation (8.2). Kocharov (1997) proposes an impulsive and isotropic ejection of protons near the Sun (at $r = 0.01$ AU). Because the effect of focusing of SEPs is always essential as compared with scattering, no matter how small the value of the mean free path may be (Kocharov et al. 1996c).

The results of the simulation confirm the conclusion by Ruffolo (1995) that the focused transport model should be used for the consideration of adiabatic deceleration at least when the mean free path is not very small. Hatzky et al. (1997) compared different schemes of numerical solution for the expanded transport equation (geometric interpolation, flux-corrected transport and flux-limiter methods). They found that the flux limiter method gives the better results. The results of numerical calculations are used by Hatzky et al. (1997) to analyze some peculiarities of angular distribution of solar energetic particles. Kota and Jokipii (1997) suggested to generalize the model of Ruffolo (1995) taking into consideration a non-uniform, non-steady solar wind and non-spiral magnetic fields.

8.3 Rigidity Dependence of Transport Path

Understanding the mechanism by which energetic charged particles scatter in turbulent magnetic fields continues to be one of the foremost and difficult problems of modern cosmic-ray physics and space plasma physics. Fitting cosmic ray observations to diffusion model, especially in SEP events, has allowed us to determine representative values of the scattering mean free path, Λ, and has given some indication of how this key parameter of particle propagation varies with particle rigidity (*e.g.*, Palmer 1982; Miroshnichenko and Petrov 1985). It should be noted that the rigidity dependence of Λ has been derived from the SEP observations in the middle of 60's (see, *e.g.*, Dorman and Miroshnichenko 1968), but Palmer (1982) was first who compiled most of observational estimates of Λ for protons and electrons accumulated before 1982. However, up to now a satisfactory theory for deriving the observed Λ values from the known properties of turbulence in interplanetary space is still lacking (*e.g.*, Toptygin 1985; Bieber et al. 1994). Meanwhile, a comprehensive analysis of data on the energy (rigidity) dependence of Λ, especially in low-energy region, is of great interest to ascertain the limits of applicability of different approximation of kinetic theory.

8.3.1 Long-Standing Problem

The consideration is usually made under assumption that different types of IMF inhomogeneities (8.1) give different contributions into particle scattering. From different points of view, it is also important to keep in mind an energy dependence of SEP flux anisotropy. In particular, a non-monotonous dependence of parallel mean free path, Λ, on proton energy E_p should be taken into consideration. According to Toptygin (1985), so-called "normal" value of the mean free path may be presented as

$$\Lambda_{\|} \cong \left(B^2{}_0/ < B^2{}_{\rm st} >\right) (L_0/\rho)^{q-1}\rho \sim \rho^{-q+1}\rho \sim \rho^{-q+2} \sim R^{-q+2} \qquad (8.15)$$

where B_0 and B_{st} are regular and stochastic components of IMF, respectively, L_0 is a correlation length of interplanetary turbulence, and $\rho = cp/ZeB_0 = R/B_0$ is a gyroradius of particle with the momentum p (rigidity R) in regular magnetic field B_0. If this field is rather strong, the magnitude of $\Lambda_{\|}$ under some conditions may exceed largely its "normal" value (8.15). It is due to an anisotropy of particle scattering in strong magnetic field as well as due to an anisotropy in the distribution of wave vectors of turbulent pulsations.

Early treatments of scattering theory employed quasi-linear approximation in a magnetostatic "slab" model of the IMF (Jokipii 1966; Jokipii 1971a, b). In such a quasi-linear theory (QLT) the wave vectors of the turbulent fluctuations are aligned with the mean magnetic field. In his comprehensive review paper, Palmer (1982) compared the predictions of standard QLT for magnetostatic, dissipationless turbulence with slab geometry (Jokipii 1966) with the Λ estimates derived from numerous observations. This comparison displayed two major discrepancies. First, the observed mean free paths turned out to be generally larger than the theoretical prediction, with a typical discrepancy amounting to nearly a factor of 10 for 10 MeV protons (so-called "magnitude problem"). Second, the observations are broadly consistent with a rigidity-independent mean free path in the range of 0.5 MV–5.0 GV, whereas the QLT predicts that the mean free path should increase with increasing rigidity (so-called "flatness problem"). Those two discrepancies are considered at length in Sect. 8.3.3.

Since the middle of 60's, many theorists contributed to the problem, considering particle scattering in geometries other than the axisymmetric slab model (see Toptygin 1985; Bieber et al. 1994, and references therein). There were investigated, for example, isotropic geometry and oblique wave models. Within the context of resonant magnetostatic theory, such geometries generally yield reduced rates of scattering (compared with the slab model) and increased Λ values. A special attention of many researchers was paid to the mechanism by which particles scatter through 90° pitch angle. Among proposed mechanisms are "mirroring" by fluctuations of the magnetic field magnitude, a variety of nonlinear extensions of the theory, wave propagation effects, and effects of dynamic turbulence. These efforts emphasized again the importance of the problem and its gravity as well. Below we describe briefly some new aspects of the problem developed by different research groups within two last decades.

Kurganov and Ostryakov (1992) proposed a new treatment of propagation of decay protons from the flare neutrons. They have obtained a numerical solution of general kinetic equation taking into account the anisotropy of secondary protons. The calculations were accomplished by well-known Monte Carlo method. When comparing the calculated intensity-time profiles of protons with observational data for the neutron flares of June 21, 1980, June 3, 1982, and April 24, 1984 the values of transport path for protons of energy 30–40 MeV were derived about 0.5, 0.1–0.2 and 0.3 AU, respectively. In addition, the angular distributions of secondary protons were calculated for different time intervals during the same events. It should be noted that a numerical Monte Carlo technique in application to the scattering of solar protons in interplanetary space with the purpose of determining of the transport path value have been first used by Gorchakov et al. (1975).

As mentioned above, the QLT has been undergoing development for many years and it remains a valuable tool for modeling solar particle transport. However, the QLT has a persistent (and embarrassing) resonance gap at 90° ($\mu = 0$). This problem was addressed, in particular, by Ng and Reames (1995) who pointed out that any proton with energies >25 keV will resonate with hydromagnetic waves present in the interplanetary medium. They suggested modifications to QLT which involve expanding the representation of the interplanetary turbulence. Bieber et al. (1995) demonstrated that allowing 2D turbulence in the formulation generates a much better fit between the theoretical results and the observations.

The general problem of focused transport in the inner heliosphere and resulting pitch angle distributions of solar protons in the MeV energy range were considered by Hatzky et al. (1995) and Hatzky and Wibberenz (1995). They have developed a fundamental theorem which may deserve a special attention of the specialists in this area, in connection with conflicting determinations of the mean free path existing for the last 30 years (see, *e.g.*, Kunow et al. 1990; Valdes-Galicia 1993). In their formulation, it is suggested the following concept (Hatzky et al. 1995; Hatzky and Wibberenz 1995): while there is a "global mean free path" appropriate for overall modeling, the local mean free path will dominate in the local plasma domain. These local changes may be observed by various observers, as the particles propagate in the interplanetary space. The results obtained by those authors also show that the anisotropy is a function of mean free path. They further suggest that the pitch angle of ejection is not an important parameter because the focusing length and scattering down the propagation path will remove this information.

There were a number of efforts to fit new developments of transport theory to observations. It is well known that the angle averaged intensities of protons can be fitted with various combinations of ejection and interplanetary transport functions (*e.g.*, Kallenrode 1993a, b, c). Hence, Vainio et al. (1995b) suggested the method of determination of both spatial and energy dependencies of the mean free path based on comparison of Monte Carlo simulated transport of particles with the observations of *GOES* satellites and ground-based neutron monitors. They studied the events of May 24, 1990 and of October 19, 22, and 24, 1989. As a result of simulations, a set of interplanetary transport Green functions was found for various values of particle energy E_p and path length Λ. These functions may be very useful to reconstruct the probable ejection profiles (see Sect. 7.2).

8.3.2 *Shift in the Transport Paradigm*

Recently Bieber et al. (1994) presented new theoretical and observational evidence suggesting that "consensus" ideas about cosmic-ray mean free path may require drastic revision. It is proposed, specifically, that proton and electron may be fundamentally different at low to intermediate rigidities (<50 MV). Notice in this context that for the effective scattering of electrons the resonant sizes of the inhomogeneities are different from that for protons, as it follows from the relation (8.3).

Figure 8.3 shows the Palmer (1982) consensus plot re-created end extended by Bieber et al. (1994). Solid and open symbols denote mean free paths derived from electron and proton observations, respectively. Circles in Fig. 8.3 corresponds to actual determinations of the mean free path, while upward-pointing triangles reflects lower-limit values. Bieber et al. (1994) noted that in most cases the limit values were estimated from studies that neglected possible extended ejection near the Sun and as a result underestimated the true mean free path. Palmer (1982) proposed that the empirical mean free paths could be characterized by a

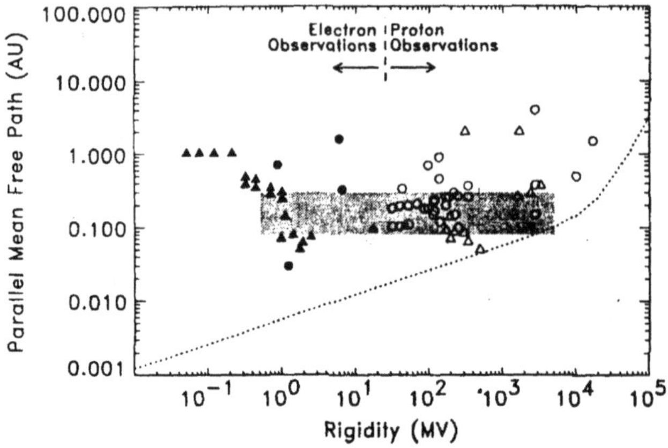

Fig. 8.3 Rigidity dependence of cosmic-ray parallel mean free path, $\Lambda_{\|}$ (After Bieber et al. 1994). *Filled and open symbols* denote estimates derived from electron and proton observations, respectively. *Circles* and upward-pointing triangles denote actual values and lower-limit values, respectively. The shaded band is the observational consensus enunciated by Palmer (1982). The *dotted line* represents the prediction of standard quasi-linear theory (QLT) for magnetostatic, dissipationless turbulence with slab geometry (Jokipii 1966)

"consensus" band (shaded in Fig. 8.3) which extends from 0.08 to 0.3 AU and spans a rigidity range of 0.5 MV–5.0 GV. In total, Fig. 8.3 contains the 68 data points.

In putting Fig. 8.3 together, Bieber et al. (1994) have consulted the original sources cited by Palmer (1982) (in all 17 papers). They also have added some new Λ determinations reported in eight other papers published since 1982. Figure 8.3 displays two well-known discrepancies between the scattering theory and cosmic-ray observation: the mean free paths predicted by the QLT are "too small" and have the "wrong" rigidity dependence. It is interesting to note that no one of the 68 data points falls on or below the theoretical prediction shown by the dotted line. The energy dependence exhibited by theoretical curve in Fig. 8.3 is related to the slope of the reduced power spectrum of the turbulence (8.1) and is determined by the relation (8.15). Bieber et al. (1994) used the model spectrum with a Kolmogorov ($\sim k^{-5/3}$) inertial range, so the predicted mean free path varies as $R^{1/3}$ for cosmic ray particles resonant with inertial range fluctuations. At very low wave numbers, the spectrum (8.1) turns over (*i.e.*, q decreases), which causes the rigidity dependence to steepen above ~10 GV.

In contrast, the observed mean free paths in Fig. 8.3 show slight tendency to decrease with decreasing rigidity. One can see also a great deal of variation among individual events, and a substantial number of the data points are only lower-limit values. In the whole, the observations taken with no distinction between proton and electron seem to lend no support to the rigidity dependence predicted by QLT. In addition, Fig. 8.3 highlights one aspect of the Palmer (1982) consensus that is usually overlooked: electron and proton measurements contribute to entirely distinct rigidity ranges that do not (at present) overlap. Mean free paths below 25 MV

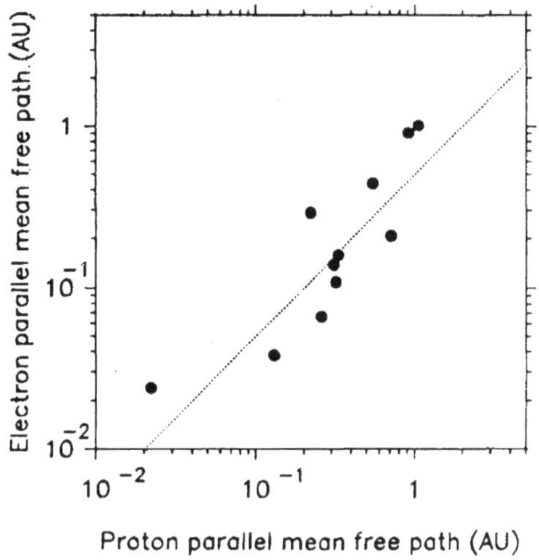

Fig. 8.4 Comparison of parallel mean free paths of 1.4 MV electrons and those of 187 MV protons (Bieber et al. 1994). The dotted line corresponds to an electron mean free path that is exactly half the proton mean free path

are derived exclusively from electron observations, whereas those above 25 MV are derived exclusively from proton observations.

As noted by Bieber et al. (1994), in the past it has been a commonplace to ignore the distinction between electron and proton data, because in resonant magnetostatic scattering theory rigidity is the only property of the particle that influences, according to (8.15), the mean free path. Pure rigidity dependence of mean free path, however, is closely tied to the magnetostatic approximation (8.15). Meanwhile, recent studies have relaxed this approximation by considering finite wave propagation speeds (*e.g.*, Schlickeiser 1989a, b; Achatz et al. 1993) and by introducing a fully dynamical representation of the scattering turbulence (Bieber and Matthaeus 1991, 1992). Both approaches predict that the mean free path has explicit speed dependence, such that electrons have a different mean free path from protons of the same rigidity. The difference is largest at the lower rigidities, where such speed differences are most pronounced, and becomes vanishingly small at high rigidities, where both species of particles are relativistic.

Bieber et al. (1994) have computed particle scattering for protons and electrons based on dynamical model of turbulence (Bieber and Matthaeus 1991, 1992). It was shown that electrons will have a larger mean free path than protons of the same rigidity. The experimental confirmation of the calculations was obtained from the *Helios* observations of solar energetic particles in 1978–1981. In Fig. 8.4 we present the results of Λ estimates obtained by Bieber et al. (1994) by the data on 9 proton events. One can see that the mean free path of 1.4 MV electrons is often similar to that of 187 MV protons, even though proton mean free paths continue to decrease comparatively rapidly with decreasing rigidity down to the lowest channels (~100 MV) observed. In light of these new results, "consensus" ideas about cosmic-ray mean free paths may require serious revision.

8.3.3 Modern Treatment of the Problem

The need for a correct quantitative treatment of the interactions between cosmic rays and turbulent magnetic fields continues to be one of the fundamental problems of modern astrophysics. The deficiencies of the first, pioneering scattering theories of Jokipii (1966) and Hasselmann and Wibberenz (1968), using a quasi-linear approach and assuming a magnetostatic approximation for the fluctuations (standard QLT) have been partly overcome by Bieber and co-workers (1994) introducing the concept of dynamical turbulence and resonance broadening, and by Schlickeiser (1989a, b) and Achatz et al. (1993) who interpret the magnetic fluctuations as waves in a hot, disperse plasma and investigated their interaction with energetic particles (for a recent review see, *e.g*., Dröge 1994a, b).

However, mean free paths Λ derived from those models with observed fluctuation spectra are still considerably too small unless somewhat arbitrary assumption are made. Thus, the question remains: what is the three-dimensional structure of the fluctuations? An important feature of the above, improved models which can be tested by studying solar particle events is that, in contrast with standard QLT, they predict different values of Λ for electrons and protons below some 100 MV (see, *e.g*., Schmidt and Droge 1997).

In particular, the functional form of rigidity dependence of electron mean free paths can give information about the spectral shape of the magnetic fluctuations the electrons interact with and the parameters controlling the resonance broadening. In the range of 0.1–100 MV this dependence is a key parameter to distinguish between different models for the nature of interplanetary magnetic turbulence. Because the level of fluctuations, and of scattering mean free paths, derived from fits to particle events, can vary by more than an order of magnitude from one event to the next, it is important to investigate the variation of Λ with respect to rigidity over a large range in rigidity on an event by event basis.

First results of such a study were presented recently by Dröge et al. (1997). The rigidity dependence of solar electron mean free paths was investigated for several events in the range of 0.3–20 MV. The results of this study are shown in Fig. 8.5. One can see from Fig. 8.5 that between 0.1 and 1 MV this dependence has a power-law form with slope – 0.2, and in the range from 1 to 20 MV they got a constant Λ. At higher rigidities electrons from solar flares usually have fluxes too low to derive meaningful values of Λ, but it is interesting to note that the behaviour of electron mean free path models well proton observations from the same event (for example, SPE of November 22, 1977) – usually a power-law rigidity dependence with slope ~0.3 from 30 to 300 MV. This indicates that there is a close connection between electron and proton scattering at ~50 MV.

These results, however, do not remove the apparent discrepancy between the scattering length Λ_q, calculated with QLT, and the length Λ_f, obtained by observational data (Palmer 1982; Wanner and Wibberenz 1993; Dröge 1994a, b; Bieber et al. 1994), and it remains to be a long-standing problem of cosmic ray physics. As has been pointed out by Palmer (1982) the discrepancy consists of two parts: first,

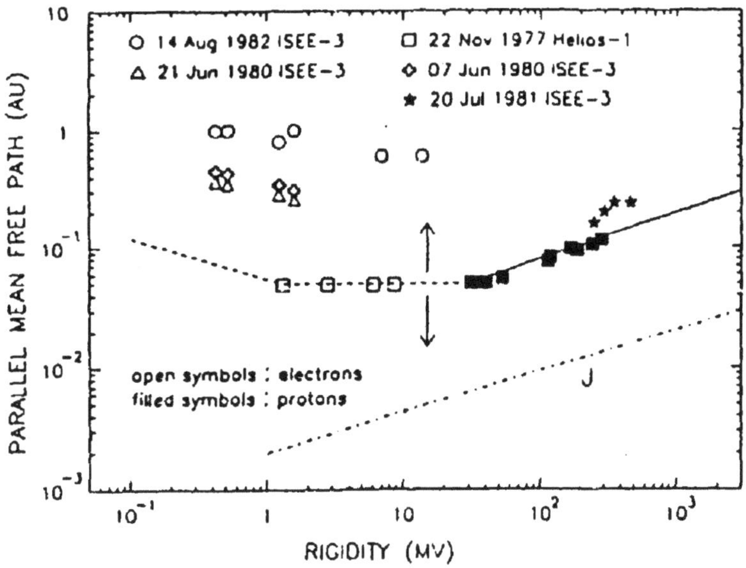

Fig. 8.5 Parallel mean free path at 1 AU *versus* particle rigidity for selected solar particle events (Dröge et al. 1997). The form of the rigidity dependence as indicated by the upper curve seems to be consistent with observations from any given event, only the absolute height of the curve varies. The lower curve represents the predictions of standard QLT based on typically solar wind conditions

the fitted lengths are typically an order of magnitude larger than the theoretical ones (referred to as "magnitude problem"), and second, the observations are broadly consistent with a rigidity-dependent Λ from 0.5 MV to 5.0 GV, while according to the QLT the mean free path should increase with increasing rigidity (referred to as "flatness problem").

As it is well-known, within quasi-linear theory the scattering length results from the pitch-angle average of the inverse of the pitch angle Fokker-Plank coefficient $D_{\mu\mu}$ as

$$\Lambda - (3v/8) \int_{-1}^{1} d\mu (1 - \mu^2)/D_{\mu\mu} \tag{8.16}$$

Fisk (1979) has made the important observation that because of the enhanced scattering near zero pitch angle ($\mu = \cos\theta = 0$), the coefficient $D_{\mu\mu}$ becomes larger, so that according to relation (8.16) nonlinear corrections would reduce the value Λ_q further, and thus worsen the discrepancy. Hence, as noted by Schlickeiser and Miller (1997), the resolution of the discrepancy can only be achieved by improving on the assumptions underlying the QLT calculations of mean free paths. More recent proposals to resolve the discrepancy, based on changing the nature of the

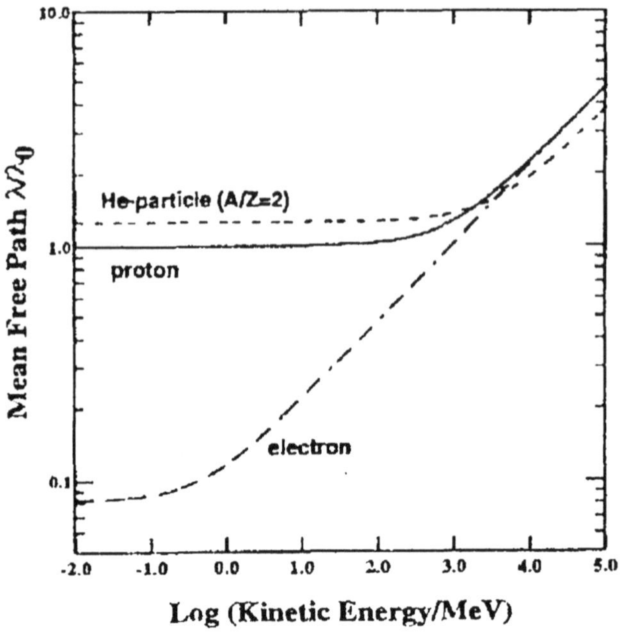

Fig. 8.6 The mean free paths as a function of kinetic energy for three different cosmic-ray particle species in the case of an admixture of slab Alfvén waves and isotropic fast mode waves (Schlickeiser and Miller 1997). A power-law spectral density of index $q = 5/3$ is assumed

scattering centers (see, *e.g.*, Bieber et al. 1994; Dröge 1994a, b, and references therein), solve the magnitude problem, but they offer no solution to the flatness problem. If interplanetary turbulence consists of a mixture of slab Alfvén waves and isotropic fast mode waves, as proposed by Schlickeiser and Miller (1997), then both the magnitude and flatness problems can be resolved within QLT, due to the presence of transit time damping of fast magnetosonic waves. Figure 8.6 shows the mean free paths as a function of kinetic energy for three different cosmic-ray particle species (electrons, protons and alpha particles) calculated by Schlickeiser and Miller (1997), who assumed equal spectral densities and neglecting any cut-off effects. One can see that the mean free path is constant for protons and alpha particles in the range of non-relativistic energy. The neglect of cut-off effects is problematic for electrons with energies below 10 MeV, and probably leads to a severe underestimation of the electron mean free path. Nevertheless, they conclude that these results may account for the legendary discrepancy between Λ_f and Λ_q.

In Sect. 8.2 we have already noted that some time ago it was argued theoretically an influence of magnetic helicity, H_m, on cosmic ray scattering (*e.g.*, Hasselmann and Wibberenz 1968; Goldstein and Matthaeus 1981; Matthaeus and Goldstein 1982). Acting in concert with adiabatic focusing, magnetic helicity can alter the parallel mean free path in the IMF in a manner dependent upon particle charge sign and the magnetic polarity of the Sun (*e.g.*, Bieber et al. 1987; Bieber and Burger 1990). In particular, the scattering of solar energetic protons in the IMF is enhanced when a negative polarity field has a negative helicity or a positive polarity field has a positive helicity. Such helicity effects will be most pronounced for particles resonant in the regime of steady helicity below 10^{-5} Hz. For typical conditions at 1 AU, this corresponds to particle rigidities ~10 GV and up. Otaola and Valdes-Galicia (1995) attempted to estimate the effects of helicity on the propagation characteristics of eight SEP events observed with the *Helios* spacecraft. They found that the helicity effects were negligible, but their helicities were computed for a wavenumber range of 10^{5}–10^{-3} km^{-1}, a range over which the values show substantial fluctuations (Smith and Bieber 1993).

The sign of helicity is steady only for time scales of about a day or more, so one would expect that if helicity plays a significant role in the propagation of SEPs, it would be observable during onset phases (Earl 1992) only for particles with very large gyroradii. Therefore, Kahler and Shea (1997) looked for this effect in the rise phases of GLEs, for which the rigidities generally exceed 1 GV. They assumed that GLEs with increased scattering will appear statistically to have longer rise times to peak flux. They studied 18 events of 1978–1992 during the different epochs of the Sun's magnetic polarity and found that the large range of rise-time scales of GLEs are not ordered by the helicity effects of the IMF. This range is more likely due to variations in the acceleration and ejection conditions at shocks or other phenomena.

8.4 Anisotropy and Spike Structure of Proton Events

In Chap. 7 we have already described some spectacular peculiarities observed in the intensity-time profiles of certain GLEs. In particular, the GLE of October 22, 1989 displayed an extremely anisotropic onset, with an initial sharp "spike" in intensity (see Fig. 7.11). At present, there is no widely accepted model which could convincingly explain all features of this event. At the same time, it provides a good test for the different models of particle acceleration and propagation.

One hypothesis which may explain the two-phase nature of this event is that the spike was caused by solar neutrons or the products of solar neutron decay. In their initial discussion Bieber et al. (1990) noted that such a spike may be ascribed to the decay protons. At the same time, these authors (see also Bieber and Evenson 1991) also stated that this explanation was not very likely, primarily because solar neutrons would have to be vastly more numerous and emitted from the Sun more anisotropically than in previously observed events. In fact, there is no evidence for direct solar neutrons from any ground-based neutron monitor near the subsolar point, in particular, in Huancayo which was the best situated for this. Moreover, the size of the spike at McMurdo NM precludes solar neutron decay in the absence of direct solar neutron observations (Cramp et al. 1997).

On the other hand, a number of theorists (*e.g.*, Fedorov and Shakhov 1993; Earl 1995; Fedorov et al. 1995; Ruffolo and Khumlumlert 1995, and others) have stated that the scattering conditions for solar particles prior to the establishment of steady state conditions are such that anisotropic spike may be expected early in some events. These are followed by more isotropic particle distributions as the diffusive mode takes over from the coherent mode. As noted by Cramp et al. (1997), this scenario is inconsistent with the strong anisotropy of the forward pitch angle distribution (PAD) which persisted until quite late in the event of October 22, 1989. It is also expected that such a mechanism would not produce the depression in intensity seen at some stations between the spike and the later enhancement (see Fig. 7.11).

An alternative explanation is that there were two individual particle ejections, as it was proposed by Torsti et al. (1991, 1992) in their interpretation of the event of September 29, 1989. For the event of October 22, 1989, however, there is no evidence of two phases in the metric radio emission. The soft X-ray emission also exhibits only a single peak. Although there are signatures of structure in the 10-cm radio emission, Cramp et al. (1997) found no compelling evidence of that a two-phase source existed at the Sun consistent with the intensity-time profile of relativistic proton flux at the Earth.

In principle, the observed two-peak profiles could have arisen if the particles followed two different paths through the IMF. One possible scenario would be that the magnetic field connection between the particle source and the Earth changed between the time of spike and the subsequent enhancement. This speculative argument (Cramp et al. 1997) could explain the abrupt decrease in intensity from the forward direction between 1805 and 1820 UT. It might also account for the change in apparent particle arrival direction between 1805 and 1820 UT (see Fig. 7.11). However, it is not compatible with an interpretation of the bidirectional particle flow (Cramp et al. 1997), as the reverse propagating particles could no longer be reflections of the original spike. As it turned out, the stations which viewed the reverse propagating particles saw a signature of the reflected spike. This was evidenced, for example, by a small but significant ($>$10 s) spike at the Deep River neutron monitor, coincident with a sharp rise at the Mawson station. Evidence of the reverse propagating particles is found for all stations having an appreciable portion of their asymptotic cones viewing in the “reverse” direction. Therefore, Cramp et al. (1997) conclude that the available evidence does not support particle transport along two different IMF paths.

In their opinion, the most logical explanation for this particular event appears to be an impulsive particle ejection followed by continuous shock acceleration over an extended period of time, in agreement with conclusions by Torsti et al. (1995). Earlier, other authors have reached similar conclusions for different events (Reames et al. 1990; van Hollebeke et al. 1990) The changes in the apparent particle arrival direction must be due to changes in the direction of the local IMF line. Unfortunately, there are no measurements of field direction with which one can compare the derived arrival directions during the event of October 22, 1989. However, data from preceding and following days indicate that changes in the IMF direction of the same order (*e.g.*, approximately 20°) were present on days either side of this relativistic solar proton event.

The effects of anisotropy of relativistic SCR during the GLE of September 29, 1989 were studied by Vashenyuk et al. (1993) on the basis of the hypothesis of the two-component ejection of the particles from the solar atmosphere. The first component was manifested at the Earth in the single maximum increase at the low latitude cosmic ray stations, high degree of anisotropy and very hard energy spectrum (see Chap. 7). The axis of the anisotropy in this increase passed through the asymptotic cone of the Thule station, Greenland (Vashenyuk et al. 1995).

The second component in this event of displayed as a second intensity maximum at many high altitude stations (see Fig. 7.18). Very significant temporal variations during the second maximum could be described if one assumes that a large-scale magnetic structure was passing through the Earth at this time and the anisotropy axis was not strongly changing its direction in space during all the event. Such a possibility has been studied in some details by Vashenyuk et al. 1997 (see also Vashenyuk and Miroshnichenko 1998), and the results obtained are described in Sect. 8.6.

As noted by Smart (1996), the computed position of the maximum flux directions often do not correspond to the quiet time Archimedean-spiral direction. It was found, in particular, for the GLEs of the 22nd solar cycle, perhaps, because many of these events occurred near the solar-activity maximum. There is often dramatic evolution of the maximum flux direction as the event evolves, for example, during the GLE of May 24, 1990 (Morishita et al. 1995). The major events have sufficient statistics, so that flux contours in space can be derived, along with spectral evolution and rigidity-dependent pitch angle distributions (*e.g.*, Cramp et al. 1995a, b, c, 1997; de Koning and Bland 1995; de Koning and Mathews 1995, 1996; Dvornikov and Sdobnov 1995a, b, 1997, 1998).

As to the general problem of a coherent pulse of solar cosmic rays, the consensus of the solar particle theorists seems to be that this pulse is a natural feature of solar particle propagation. This feature should be expected in the inner heliosphere whenever there are long mean free paths involved. The modeling work of Ruffolo and Khumlumlert (1995) indicates, in particular, that diffusion is not really effective at propagation distances less than two mean free paths from the ejection position. Whenever the focusing length, L, dominates the scattering length, $L \gg \Lambda$, at distances not too far from the ejection site, then these coherent pulses (or "flash phase" in the Earl (1995) terminology) should be expected at the beginning of an event. The computations by Fedorov et al. (1995) shows that the time profiles observed during a GLE will depend on the neutron monitor asymptotic viewing direction in space with respect to the particle propagation direction.

On the basis of the Boltzmann kinetic equation, Fedorov (1995, 1997) has calculated the particle time profiles, spatial and pitch angle distributions at different regimes of particle ejection from the Sun (anisotropic initial distribution, instantaneous or prolonged ejections, etc.). Such a kinetic approach was applied to several GLEs to estimate the half-width of corresponding ejection time profiles, ΔT, and mean transport lengths, Λ. According to estimates by Fedorov (1997), the pairs of these parameters were $\Delta T = 8$ min and $\Lambda = 0.7$ AU, and $\Delta T = 19$ min and $\Lambda = 0.3$ AU, for the events of February 16, 1984 and September 29, 1989, respectively. A similar approach was used by Fedorov et al. (1997) to the GLE of May 24, 1990 which displayed a large anisotropy at the event onset and some signatures of two-fold ejection of relativistic protons (see, for example, Miroshnichenko et al. 1995a, 1996). It was postulated a prolonged, energy dependent escape of accelerated particles into interplanetary space. Fedorov et al. (1997) found that the observed intensity-time profile at the Hobart station corresponds to the ejection profile with a half-width $\Delta T = 19$ min at the value of $\Lambda = 0.6$ AU. Such an approach, however, seems to be insufficient to explain a great time delay between anisotropic peak at several NM stations and a smooth isotropic maximum at the others, until one assumes a second ejection. In general, the underlying physical circumstances leading to the initial spikes and two-peak structures in some GLEs are not presently understood. Thus, taking into account the results of Chap. 7 (see especially Sect. 7.3), we do not believe that the above hypothesis of "an interplanetary origin" of the features mentioned can resolve alone the problem of relativistic proton events.

8.5 Energy Density and Flux Instability of Solar Protons

For a typical IMF value $B = 5 \times 10^{-5}$ G its energy density is $w_B \sim 10^{-10}$ erg cm^{-3}. The energy density of directed motion of the solar wind is equal to $w_k = n_p m_p u/2 \sim 0.7 \times 10^{-8}$ erg cm^{-3} (for $n_p = 5$ cm^{-3} and $u = 400$ km s^{-1}), whereas the energy density of chaotic (thermal) motions $w_t \sim 2 \times 10^{-10}$ erg cm^{-3}, if the electron and proton temperatures are $T_e = 1.5 \times 10^5$ K and $T_p = 5 \times 10^4$ K, respectively. Thus, in interplanetary space the kinetic energy of the solar wind prevails over the thermal energy, whereas the magnetic and thermal energies are comparable. It is usually assumed that the ratio of the kinetic pressure of the interplanetary plasma to the magnetic pressure $\beta = 8\pi n(T_e + T_p)/B^2 \sim 1.0$. On the other hand, as shown by Miroshnichenko (1984), in the largest SPE the energy density of non-relativistic protons can reach values comparable to the energy density of the magnetic fields in interplanetary space and in remote portions of the magnetosphere (for example, in the geomagnetic tail). Thus, during the events of May 7, 1978, November 22, 1977, and August 4, 1972, the proton energy density w_p amounted to 1.7×10^{-11}, 4.8×10^{-11}, and 6.5×10^{-9} erg cm^{-3}, for the 5–90, 15–90, and 10–60 MeV energy intervals, respectively. The fact that the ratio $w_p \sim w_B$ is satisfied in certain cases, may indicate that the accelerated solar particles can play a considerable dynamic role in the nearest portion of the heliosphere, including a collective effect on the external geomagnetosphere.

The rigorous solution of this problem requires a self-consistent approach since the fluxes (or beams) of accelerated particles in the plasma, as is known, can alter the properties of the medium noticeably. In terms of empirical estimates, this problem was reviewed in some details elsewhere (Miroshnichenko 1992a), so we describe below only several characteristic results. In conjunction with accumulation of data on flare electrons Yukhimuk (1982) considered the problem of instabilities in the solar wind produced by anisotropic fluxes of electrons with energies >40 keV. It is assumed that these fluxes appear in interplanetary space as a result of their impulse ejection from the solar atmosphere. Because of beam instability, these electrons can be the cause of the formation of a non-uniform interplanetary plasma structure, *i.e.*, the appearance of a local plasma (or Langmuir) turbulence and, possibly, MHD turbulence. Actually, an approximate analysis showed (Yukhimuk 1982) that for typical values of the density of flare electrons in the solar wind plasma low-frequency disturbances can be generated with characteristic dimensions of $1.2 \times 10^{10} < l < 2 \times 10^{11}$ cm, which were actually observed by means of space probes as well as by radio astronomy methods (see, *e.g.*, Toptygin 1985). A scattering of solar protons occurs at inhomogeneities of such a scale, and it is especially effective in the energy range $E_p = 30$–300 MeV (see Fig. 8.3). However, the relative contribution of such low-frequency waves to the overall turbulence spectrum in the interplanetary medium (see, *e.g.*, Hedgecock 1975) has not yet been estimated at all.

On the other hand, by analyzing the intensity-time profiles of electrons with E_e >30 keV for the 1972–1974 period measured at the Earth's orbit, Kurt et al. (1976) found a two-component structure of increases. The prompt component was observed only near a line of force connected with the flare site, within the narrow interval of heliolongitudes, $|\Delta\theta| < 10°$. The delayed component was detected usually within a wider interval of heliolongitudes, $|\Delta\theta| < 100°$, and it was described quite well by the diffusion equation with the diffusion coefficient κ decreasing with an increase in the intensity of the event. Such a behaviour of κ implies that the scattering of non-relativistic electrons occurs not at permanently existing inhomogeneities of the IMF, but at plasma oscillations excited by the electron fluxes themselves (Kurt et al. 1976). The difficulty of this attractive hypothesis, however, lies in the fact that for the effective scattering of electrons the resonant sizes of the inhomogeneities, unlike the estimates of Yukhimuk (1982), must be $l \leq 10^7$ cm, and the turbulence level, W_l, necessary for isotropization of the electrons was not estimated at all. Nevertheless, as it was shown by Bespalov and Trakhtengerz (1974), a separation of the high-energy electrons ($V_e \gg V_{Te}$) into a component escaping along the magnetic field, and a diffusion component is possible, in principle. In this case, it must be expected that $\kappa \sim 1/W_l$, where W_l in turn depends on the intensity of the electron beam. Let us return, however, to the flare protons.

In the absence of any magnetic field inhomogeneities, shock fronts, *etc.*, the SCR protons will move essentially in the form of a beam along the IMF, with preservation of the adiabatic invariant, $\sin^2\theta/B = \text{const}$. But such an idealized case is the classical example when beam instability develops (*e.g.*, Berezinsky et al. 1990). The problem of beam stability is very important for an interpretation of the time behaviour of the anisotropy of SCR at different energies. One can calculate the "plasma" frequency of the SCR beam from the formula $\omega_s = (4\pi e^2 n_p/m_p)^{1/2}$, where n_p is the density of SCR protons and m_p *is* the proton mass. For specific estimates of ω_s we use the n_p values for protons with energies of 10 MeV and 1.0 GeV.

According to SPE Catalogue data (Dodson et al. 1975; Akinyan et al. 1983; Bazilevskaya et al. 1986, 1990a; Sladkova et al. 1998; Logachev et al. 2014), one can assume $I_p(>10\ \text{MeV}) \geq 10^4\ \text{cm}^{-2}\ \text{s}^{-1}\ \text{sr}^{-1}$ and $I_p\ (>1\ \text{GeV}) \geq 1.0\ \text{cm}^{-2}\ \text{s}^{-1}\ \text{sr}^{-1}$ (*e.g.*, Miroshnichenko 1994, 1996) as the upper values of the integral intensity (see Fig. 4.4). Hence, it follows that $n_p(>10\ \text{MeV}) = 2.9 \times 10^{-5}\ \text{cm}^{-3}$ and $n_p(>1\ \text{GeV}) = 4.8 \times 10^{-9}\ \text{cm}^{-3}$. From the formula given above we now obtain $\omega_s(>10\ \text{MeV}) \approx 72\ \text{s}^{-1} \gg \omega_B$ and $\omega_s(>1\ \text{GeV}) \approx 9.1 \times 10^{-9}\text{s}^{-1} \ll \omega_B$, where $\omega_B = 0.5\ \text{s}^{-1}$ $\omega_B = 0.5\ \text{s}^{-1}$ is the cyclotron frequency in the IMF with an average intensity of $B = 5 \times 10^{-5}$ G.

Such ratio between the frequencies ω_s and ω_B indicates that the non-relativistic solar protons can excite certain plasma effects in the interplanetary medium, whereas in the case of relativistic SCR they are excluded in practice. In order to prove this, let us estimate the increments (or characteristic development times) τ_s^{-1} of the instabilities. In the case under consideration, in order to estimate the increment, we can use the formula $\tau_s^{-1} \sim \omega_B(n_p/n)(V/V_A - 1)$, obtained from the analysis of the flux instability of galactic cosmic rays in interstellar space (Kulsrud and Cesarsky 1971; see also Berezinsky et al. 1990, in Chap. 9 therein). Here V is the velocity of the proton beam as a whole with respect to the interplanetary medium, and V_A is the Alfvén velocity in the interplanetary plasma for $n = 5\ \text{cm}^{-3}$ and $B = 5 \times 10^{-5}$ G. As shown by Kulsrud and Cesarsky (1971), the value of V is restricted on the high side to a value of $V \cong 0.3c$ (for relativistic protons); for non-relativistic particles one can assume $V \leq 0.1c$. Since $V_A \cong 5 \times 10^6\ \text{cm s}^{-1}$, then the relation $V_A \ll V$ is known to be satisfied. Under the given estimated values of the parameters, we obtain $\tau_s^{-1}(>10\ \text{MeV}) \sim 3.3 \times 10^{-4}\text{s}^{-1}$ and $\tau_s^{-1}(>1\ \text{GeV}) \sim 10^{-6}\text{s}^{-1}$, or $\tau_s(\geq 10\ \text{MeV}) \sim 0.9$ h and $\tau_s(\geq 1\ \text{GeV}) \sim 12.5$ days. Typical propagation times t_m of protons from the Sun to the Earth amount to $\sim 10^4$ s and 10^3 s, respectively. It is seen from this that if a beam of protons with an energy $E_p \geq 10$ MeV has a sufficiently large intensity, it starts to break up through scattering at a distance ≥ 0.6 AU from the Sun, whereas a beam of relativistic protons apparently retains stability and a high level of anisotropy at least to the Earth's orbit, if isotropization due to scattering of the protons by the permanently existing inhomogeneities of the IMF is ignored.

Let us now estimate at what proton density the proton beam is at the stability limit for $B = 5 \times 10^{-5}$ G, *i.e.*, when the condition $\omega_s = \omega_B = 0.5\ \text{s}^{-1}$ is satisfied. It is easy to prove that the density $n \sim 1.5 \times 10^{-7}\ \text{cm}^{-3}$ satisfies this condition. For protons with $E_p = 1$ GeV such a density corresponds to the intensity of I_p $\sim 3.1 \times 10^2\ \text{cm}^{-2}\ \text{s}^{-1}\ \text{sr}^{-1}$. This value clearly exceeds the limit of the capabilities of the solar accelerator (Miroshnichenko 1994, 1996). However, even for $E_p = 100$ MeV the fulfillment of the condition $\omega_s \sim \omega_B$ becomes attainable. In principle, on the basis of the existing SPE Catalogue data one can calculate at what energies the beam of SCR becomes unstable (for a given intensity). Estimates of the proton energy density for the largest SPEs and upper limit spectrum of protons (see Fig. 4.4) can play an auxiliary role in such calculations. Rigorous estimates of the isotropization time of the SCR beam, of course, must be made with the pitch angle distribution of the particles at the source and their multiple scattering in the non-uniform IMF taken into account.

In principle, non-linear effects of particle propagation may be considered on the basis of an equation of non-linear heat conduction with the source and sink. This equation has a general form:

$$U_t = (U^\sigma U_x)_x + U^{\sigma+1} - U; \mathrm{t} > 0; \quad x \in R; \quad \sigma > 0 \tag{8.17}$$

$$U_t^i = D^{ik} \; U^k{}_{xx} + F^i(U, U_x, \gamma) \tag{8.18}$$

where U^i (x, t) is a multi-component internal parameter of system under consideration; D^{ik} is a constant diffusion matrix; F^i describes the nonlinear properties of the system; γ is a set of controlling parameters which characterize the medium state.

Non-linearity of particle transport in interplanetary space may be expressed by simple relation

$$\kappa \sim \Lambda \sim n^{-\sigma}, \quad \sigma > 0 \tag{8.19}$$

where n is a number density of particles. A decrease in the spatial diffusion coefficient is possible behind the shock where the solar plasma turbulence increases, and the particle scattering enhances, as it was assumed in a number of studies (*e.g.*, Chih and Lee 1986; Wibberenz et al. 1997; Cane 1998; Wibberenz 1998; Miroshnichenko et al. 1998). If the conditions (8.19) are valid, then the variables in (8.17) and (8.18) may be separated, and the solution may be obtained analytically.

A suitable example seems to be the interplanetary transport of flare electrons. As mentioned above, this process sometimes is probably following by additional excitation of plasma turbulence and, as a result, by significant changes of scattering properties of interplanetary medium (Kurt et al. 1976). In other words, the level of turbulence depends on the flux of solar electrons, and this effect results in the Λ decrease and in enhancement of electron scattering. Kallenrode et al. (1992b), however, did not find similar effect in their study of a number of ~0.5 MeV electrons. It should be noted also that some theorists (*e.g.*, Toptygin 1985) do not believe in non-linear properties of interplanetary medium at all: in their opinion, this medium changes its parameters slowly.

A new aspect of the problem under consideration rose recently in connection with a hypothesis that large fluxes of solar protons may produce their own turbulence, or so-called self-generated waves, SGW (*e.g.*, Reames 1989; Ng and Reames 1994). Since wave-particle interaction is generally accepted to be the cause of the scattering, these newly created waves will in turn modify the particle transport conditions. An important consequence of this situation, from theoretical point of view, would be that the mean free path of energetic particles will be a time-dependent parameter. Reames (1989) suggested that the protons may excite interplanetary Alfvén waves significantly enough to affect their own propagation. More recently, Ng and Reames (1994) developed a model of focused transport of

1 MeV solar protons through interplanetary Alfvén waves in the presence of wave evolution (amplifying or damping). They concluded that the protons may produce their own turbulence and have a significant effect in the scattering process in large events. This in turn modifies the coefficient of pitch-angle diffusion and the time profiles of the particle intensity and anisotropy. A key prediction of the model is the increase magnetic fluctuations in association with the arrival of intense anisotropic fluxes of SEPs. Are there any observations in support of this prediction?

According to the model by Ng and Reames (1994), the chances of observing SEP-driven wave growth increase with event size and decrease strongly with distance from the Sun. The >10 MeV protons, however, contribute negligibly to wave growth because of their relatively small number. Beeck et al. (1990) reported an IMF power spectral density that increased by more than an order of magnitude over 1 day for the event of 20 July 1981 observed on *ISEE* 3 at 1 AU. Wanner and Wibberenz (1993) presented a time series of a total power of IMF fluctuations in the wavenumber range 10^{-9}–10^{-8} cm^{-1} for the event of March 28, 1976 observed on board *Helios* 2 at 0.5 AU. The total power increased by a factor of 3 early in the event, in agreement with the model (Ng and Reames 1994). Nevertheless, the correlation between the arrival of the SEP and the growth of the IMF fluctuations in the observations of Beeck et al. (1990) and Wanner and Wibberenz (1993) may be fortuitous (Ng and Reames 1994).

A study of Valdes-Galicia and Alexander (1997) is different from that of Beeck et al. (1990) and Wanner and Wibberenz (1993) as it includes the separated contributions of the directional fluctuations in the three spatial directions, not only fluctuations perpendicular to the average IMF. They also added helicity spectra not considered previously and used a technique that permits to see more clearly the temporal evolution of the spectra. In order to represent a variety of physical circumstances regarding the proton fluxes and IMF conditions, there were chosen the eight SEP events observed at proton energies of 4–13 MeV by *Helios* 1 and 2 in the range of 0.31–0.93 AU (Valdes-Galicia et al. 1995). The time evolution of the directional power and helicity spectra show increases which may be associated with SEP self-generated waves in two of the analyzed events, namely, 28 March 1976 and 11 December 1978. In three other events the signatures of the effect under consideration are uncertain, and the remaining three events show no evidence of SGW.

Because of great importance of this issue for interplanetary physics, in the next paper Alexander and Valdes-Galicia (1998) returned to the study of three most relevant proton events (see above). They incorporated into their analyses the new information, namely both IMF and plasma data that allowed, in particular, to determine the energetic content of inward and outward propagating waves. Unlike to the results of previous work, they have not found any clear evidence of SGW due to solar energetic protons, even in those cases which approach the optimum conditions stated by the theoretical model (Ng and Reames 1994). Therefore, to their opinion, it is not clear whether the effect, if it exists, is too small to be detected, or if it becomes permanently masked by other phenomena also present in the interplanetary medium.

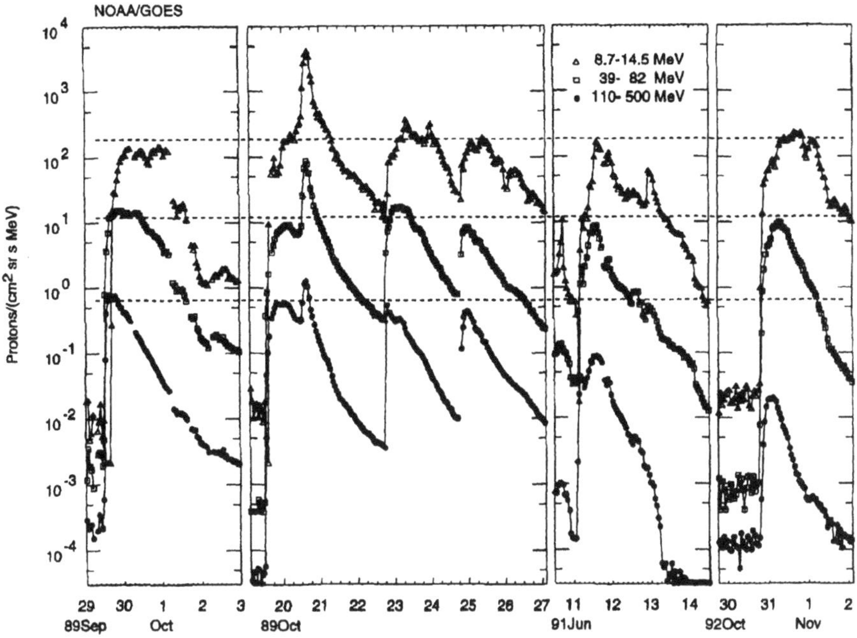

Fig. 8.7 Intensity-time profiles of protons in three energy channels for six large SEP events as detected on the *GOES* spacecraft in 1989–1992 (Reames and Ng 1998). Streaming-limited intensity values are shown as dashed lines

Meanwhile, this discussion seems to be continued, in particular, due to the difference in the distributions between the flare- and SSC-associated events in Fig. 5.10 that may evidence their different origins. As we mentioned in Sect. 5.8, there is an obvious interplanetary effect to explain, at least, the change in slope at the 10^3 pfu value for shock-associated events. That is so-called streaming-limited saturation of SEP events (Reames and Ng 1998, and references therein). In proposed nonlinear scenario, protons streaming outward from an intense source near the Sun reach a maximum-intensity plateau (Ng and Reames 1994) due to particle scattering by self-generated Alfvén waves. According to Reames and Ng (1998), at energies of a few MeV the limiting intensity is attained for a dozen or more events per solar cycle. In Fig. 8.7 intensity-time profiles of protons in three energy intervals (8.7–14.5, 39–82, and 110–500 MeV) as measured on board the *GOES* spacecraft in 1989–1992 are shown for six large SEP events, together with the probable streaming-limited intensity values. The data for the first two intervals demonstrate that those events peak at or near the nominal limits. For the 110–500 MeV interval the limiting intensity was only attained for the four large proton events in September-October 1989 (Reames and Ng 1998).

Such a streaming limit may be closely linked with the particle acceleration at interplanetary shock waves that give rise to the large gradual SEP events. It is especially important in view of distinct manifestation of the slope break in the

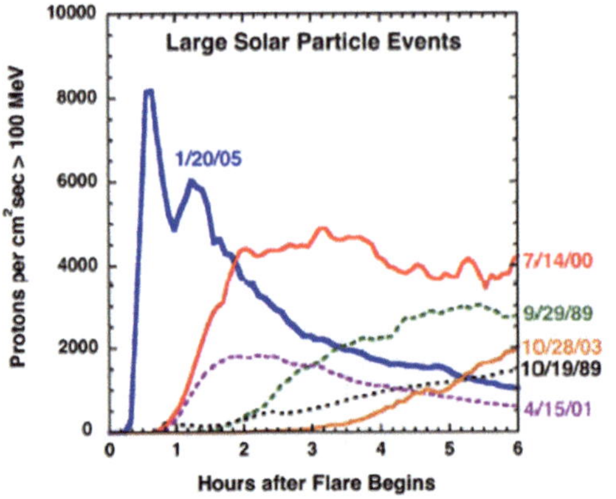

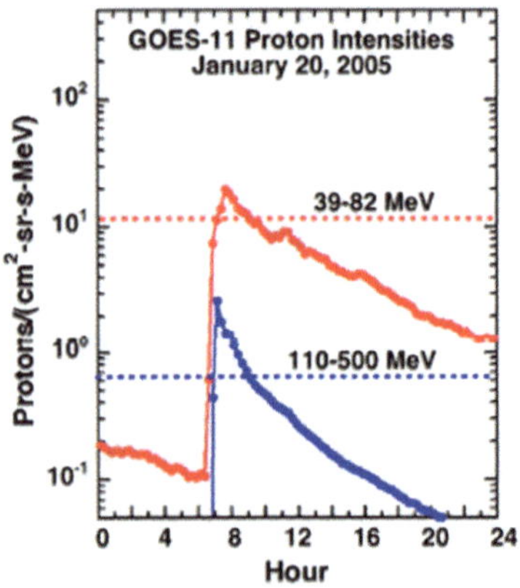

Fig. 8.8 *Left*: Intensity-time profiles of >100 MeV protons for six large SEP events as detected on the *GOES* spacecraft in 1989–2005 (Mewaldt et al. 2007). *Right*: The 110–500 MeV proton intensity measured by *GOES-11* during the 20 January 2005 event exceeded the "streaming limit" (*dotted line*) of Reames and Ng (1998) by a factor of ~4

distribution of the shock-associated >10 MeV events (Fig. 5.10). It is evidenced that the high-intensity break effect may be real and physically meaningful for the shock-associated proton events. If so, an effect of streaming-limited saturation of SEP events seems to be very important, especially for the large "gradual" proton events. At this point we agree with Valdes-Galicia and Alexander (1997) that further research for Alfvén wave generation by SEPs are worth pursuing with refined criteria for the SEP events to be used (see Sect. 2.7). In particular, the data must be clean of any perturbing interplanetary magnetic feature such as shock waves or sector crossing.

As well-known, SEPs constitute a radiation hazard to both humans and hardware in space, to astronauts (cosmonauts) and spacecraft systems (*e.g.*, Miroshnichenko 2003a). It is essential to determine the highest particle intensities reached in SEP events, especially at the energies that pose serious risk to human health and spacecraft performance. Therefore, at present, with the development of extensive plans of space exploration by different countries, there is renewed interest in understanding and forecasting large SEP events (*e.g.*, Mewaldt et al. 2007; Lario et al. 2008, 2009). Over the past few decades there have been significant advances in our knowledge of the composition and energy spectra of SEP events, leading to new insights into the conditions that contribute to the largest events. In this context, the interplanetary conditions that affect the intensity of largest events are of special interest (*e.g.*, Lario et al. 2008, 2009).

In this context, the event of 20 January 2005 served as a wake-up call when the >100 MeV protons intensity near the Earth reached maximum within ~30 min of the onset of the X-ray flare, resulting also in the largest GLE since 23 February 1956 (see Table 2.2). This event had a much faster rise time than other large SEP event (Fig. 8.8, left), and it also exceeded the 110-MeV streaming limit established by

Table 8.2 Events with prompt component above the previously determined streaming limit

Event	Solar flare
14 July 2000	X5/3B, N22W07
8 November 2000	M7/3 F, N10W77
28 October 2003	X17/4B, S16E08
20 January 2005	X7/2B, N12W61

Reames and Ng (1998) by a factor ~4 (Fig. 8.8, right) (Mewaldt et al. 2007). It is possible that streaming limits depend on the level of pre-existing turbulence, or on the proton energy spectrum. In fact, this event had a very hard spectrum (*e.g.*, Mewaldt et al. 2005a, b). In particular, by *GOES*-11 data between 60 and 300 MeV, a spectral index of differential proton spectrum near the Earth was estimated to be about 3.00 (Miroshnichenko and Gan 2012).

Lario et al. (2008) have argued that the highest particle intensities measured during large SEP events occur in association with the passage of CME-driven shocks and generation of the particle population historically termed "energetic storm particle" (ESP) component. Furthermore, it has been argued that the intensities measured early in the SEP events (known as the prompt SEP component) are bounded by a maximum intensity plateau that results from wave-particle interactions that restrict the free streaming of particles (also called the "streaming limit"). They analyzed proton intensities measured by the GOES spacecraft at the energy channels P5 (~39–82 MeV) and P7 (~110–150 MeV) during solar cycles 22 and 23 and examined whether the highest intensities were measured during the prompt or the ESP components of the SEP events.

The authors found three (one) SEP events in which the highest proton intensities measured during the prompt component at the energy channel P5 (P7) exceeded by a factor of 4 or more the previously determined "streaming limit" (Table 8.2). One of the scenarios proposed to explain intensities that exceed this limit in these events invokes the existence of transient plasma structures beyond 1 AU able to confine and/or mirror energetic particles (Lario et al. 2008).

With the same scenario, Lario et al. (2009) have studied whether other particle events with prompt-component intensities close to the previously determined streaming limit are similarly affected by the presence of interplanetary structures. Whereas such structures were observed in four out of the nine events studied by Lario et al. (2009), the authors concluded that only the events on 22 October 1989, 29 October 2003, and 17 January 2005 show interplanetary structures that can have modified the transport conditions in a way similar to those events with prompt components exceeding the previously determined streaming limit (see Table 8.3). The other six events with prompt components close to the previously determined streaming limit were characterized by either a low level of pre-event solar activity and/or the absence of transient interplanetary structures able to modify the transport of energetic particles.

As one can see from Table 8.3, a modern treatment of the streaming limit problem is rather ambiguous, some observations (*e.g.*, 22 October 1989 and 29 October 2003) contradict to the scenario proposed by Reames and Ng (1998)

Table 8.3 SEP events with prompt components close to the streaming limit

Event	Solar flare	Pre-event IP structure beyond 1 AU	Consistency with the scenario by Lario et al. 2008	Consistency with the scenario by Reames and Ng 1998
Events with intense ESP component				
19 October 1989	X13/4B, S27E10	No	Yes	Yes[a]
4 November 2001	X1/3B, N06W18	No[b]	Yes	Yes[a]
Events without intense ESP component				
12 August 1989	X2/2B, S16W37	No	Yes	Yes
29 September 1989	X9/–, ≈S26W105	No	Yes	Yes
22 October 1989	X2/2B, S27W31	Yes	No	No[c]
15 April 2001	X14/2B, S20W85	Yes	Yes[d]	Yes[d]
22 November 2001	M3/2B, S25W67	No[b]	Yes	Yes
	M9/2N, S15W34			
29 October 2003	X10/2B, S15W02	Yes	No	No[c]
17 January 2005	X3/2F, N13W23	Yes	No	Yes

Notes:
[a]Transient structures around or in front of the CME-driven shock contributed to both form an intense ESP component at the arrival of the shock at 1 AU and reduce the prompt component of the SEP event
[b]Presence of an ICME beyond the observer at the time of the onset of the SEP event that presumably did not produce large solar-wind and magnetic-field enhancements able to reflect energetic particles back to 1 AU
[c]First injected SEPs propagated within ICME where there is a deficit of MHD waves to amplify, contrary to the scenario proposed by Reames and Ng (1998)
[d]Diffusive transport inferred from the analysis of NM observations favors the amplification of MHD waves (consistent with Reames and Ng 1998) and reduces the role of IP structures beyond 1 AU. Additionally, the presence of intervening structures implies a diminished role of the IP structures beyond 1 AU

because, in order to maintain particles below the streaming limit, it is necessary to generate enough waves to restrict the particle streaming. The event of 17 January 2005 occurred when multiple transient structures were present in the interplanetary medium. This contradicts the scenario proposed by Lario et al. (2008) to explain, in the context of the Reames and Ng (1998) theory, the exceeding of the streaming limit. They concluded (Lario et al. 2009) that in this event either the seed particle sources were not intense enough to produce particle intensities above the streaming limit, or the transient structures did not have an effect of confining and mirroring SEPs, and hence, the elevated intensities at or after their passage were not observed.

8.6 Particle Motion in the Large-Scale Magnetic Structures

The transport equation (8.4) in most cases gives a satisfactorily description of the anisotropic diffusion of SCR in IMF in the presence of a smooth spiral (or regular) component B_0 and turbulent (or random) component B_{st}. However, the conditions of the diffusion approximation ($\Lambda \ll r$, $t \geq r/v$) are knowingly not satisfied in the initial phase of the SPE and often break down near $t \sim t_m$. In some cases the impression appears that the solar particles are propagating almost without scattering ($\Lambda \geq 1.0$ AU). Actually, however, according to present-day concept (*e.g.*, Earl 1976a, b; Toptygin 1985), two competing processes are occurring in the interplanetary medium - a pitch angle scattering in the random field B_{st} and adiabatic focusing (or collimation of the particles) in the diverging regular field B_0. To describe such a complicated transport, the Eq. (8.4) derived in the guiding center approximation is used.

It is easy to show that for an arbitrarily pitch angle distribution in the source the particles are expected to focus by the magnetic field at a small distance from the source, and then they will propagate along B_0 in the form of a highly anisotropic beam. Large values of the transport path $\Lambda_{\|}$ have been obtained just in the attempt at a purely diffusion description of such transport.

The characteristic length of change of the magnetic field B_0 (or the focusing length L) which is defined by the expression (8.6) serves as a focusing parameter. A numerical simulation of such focused diffusion showed (Bazilevskaya and Golynskaya 1989, 1990) that for vt_m <1.5 AU the traditional diffusion model (ignoring focusing) is unsuitable for estimating $\Lambda_{\|}$, while for $vt_m = 2$–4 AU the model gives overestimated $\Lambda_{\|}$ values. The time dependence of the particle density and anisotropy in the growth and maximum phases of the event, within the range of parameters $z \leq 2.5$ AU, $\Lambda_{\|} = 1.5 - 3.2$ AU, and $\Lambda_{\|} = 0.15 - 2.3$ AU, depends primarily on the local value of $\Lambda_{\|}/L$ and is not very sensitive to the spatial variation of $L(z)$; the exponent q in the power density spectrum of B_{st} (8.1) is in the limits of 1.0–1.5. The effect of the latter is displayed primarily in the decay rate of the density n after the event maximum phase.

The significant role of particle drift in the IMF was first pointed out in a theoretical analysis of the modulation of galactic cosmic rays (Jokipii et al. 1977). Ten years later, Kolomeets and Sevostyanov (1988) attempted to take drift effects into account in order to explain some peculiarities of the time profiles of SPEs, the shape of spectrum and the time delays of the particles *versus* the angular distance between the source and the escape site of the SCR. It was proposed that the IMF has a two-sector structure, with B_0 being positive in the northern hemisphere of the Sun (see Figs. 3.10 and 3.11). Under certain simplifying assumptions, the transport equation was solved with drift included (Kolomeets and Sevostyanov 1988). It turned out that drift effects can be significantly dependent on the relative position of the source and observation point, the phase of the heliomagnetic cycle, and the geometry of the heliospheric current sheet (HCS). In particular, the delay in the arrival of the first protons with respect to the flash phase of the flare was explained by the slow drift of the particles perpendicular to the IMF. As we have already seen, for instance, in the description of the September 29, 1989 event (Sect. 7.4), such an interpretation of the delay is not the only possible one.

If the direction of B_0 is reversed (toward the Sun), then the antisymmetrical elements of the diffusion tensor $\kappa_{\alpha\beta}$, describing drift effects, change sign, whereas the diagonal components, corresponding to diffusion, remain unchanged. In other words, the known variations of the sector structure of the IMF exert no influence on $\Lambda_{\parallel}$ and $\Lambda_{\perp}$ – the basic characteristics of the SCR determined from observations (Smart and Shea 1991). In the strong-field approximation the diffusion becomes highly anisotropic since parallel diffusion coefficient $\kappa_{\parallel}$ is much larger than the other components of the tensor $\kappa_{\alpha\beta}$, *i.e.*, transverse diffusion and drift are strongly suppressed (*e.g.*, Toptygin 1985; Smart and Shea 1991). In our opinion, during propagation in the IMF the drift effects for the SCR cannot be significant because of the strong collimation of the particles. Moreover, the strong softening of the spectrum predicted by Kolomeets and Sevostyanov (1988) in the case of a point-like source is not observed, at least for the prompt component of the SCR (see Sect. 7.4), which is always characterized by a very hard spectrum (*e.g.*, Perez-Peraza et al. 1992).

As seen from the foregoing comments, the focusing and drift of SCR lend themselves to a rather simple analysis on the basis of a numerical solution of the known equations of kinetic theory. The propagation of particles in the presence of complex structures in the interplanetary medium - corotating high-speed solar wind streams, magnetic loops (or bottles), magnetic plugs (or mirrors), traps, clouds (plasmoids), *etc.* - is much more difficult to investigate. For some of these formations, namely, high-speed streams, magnetic loops and mirrors, there were used numerical methods of solving the transport equations in various modifications (*e.g.*, Kolomeets and Sevostyanov 1987; Kolomeets et al. 1988). The conclusion that protons with an energy of several MeV propagate just in the corotating high-speed solar wind streams, follows from an analysis of data on the velocity u and the value of $|B_0|$ within a period of $\pm$ 27 days with respect to the SPE of December 17, 1976 (Kolomeets and Sevostyanov 1987). It suggests that isolated channels for the propagation of fast particles apparently exist in the IMF. These channels are associated with long-living regions of the interaction of streams of different velocity, in which motion of the SCR occurs with weak scattering.

Earlier, a similar result was obtained by the method of spectrographic global survey in a study of the propagation of protons with energies of several GeV (Dvornikov et al. 1983, 1984). They have determined amplitude and angular characteristics of the cosmic ray pitch angle anisotropy in the interplanetary medium for some intervals preceding the SPEs of 24 January and 1 September 1971, 4 and 7 August 1972. Stable pairs of recurrent "bursts" of pitch angle anisotropy were found for three solar rotations before each of the above events. The duration of each pair of bursts was about 3 h, the time interval between the bursts in each pair was 36–48 h. These data are indicative of the existence of two narrow IMF structures in interplanetary space that lasted for four solar rotations, including the moments of SPE occurrence. The latter coincided with the moments which are likely to be a kind of a channel for SCR propagation. An analysis of the power density spectra of the IMF fluctuations also indicates the existence of channels in the interplanetary medium that provide for anisotropic "conductivity" for the cosmic ray "current" (*e.g.*, Morfill et al. 1979).

Kolomeets et al. (1988) used numerical solution of the kinetic equation in the drift approximation (Toptygin 1985) for appropriate boundary conditions in order to describe the propagation of SCR in the presence of loops and magnetic mirrors. Thus, in the case of a mirror it was assumed that a "concentration" of the lines of force of the IMF occurs at a certain distance from the Sun beyond the observation point. Then the distance to the concentration point and the plug mirror ratio for the IMF serve as the parameters of the model.

In a loop structure the particle motion was modeled in the following manner. At some distance from the Sun (behind the spacecraft) a particle makes a transition to another line of force and reverses its pitch angle, then it moves toward the Sun, is reflected into the enhanced IMF region, again reaches a transition point and, finally, is detected at the spacecraft from the antisolar direction. The parameters of such a model are the distance to the transition point (the vertex of the loop) and the characteristic escape time of the particles from a given loop structure because of the drift in the non-homogeneous field of the loop.

A comparison of the calculations with observations for protons with $E_p = 1$–5 MeV on the *Venera* 4 spacecraft during the August 1, 1967 event has showed (Kolomeets et al. 1988) that the presence of loop or magnetic plug type structures is actually required in order to describe the time profiles of the intensity and anisotropy of such protons. According to the reflection model of SCR transport proposed by Lubimov (1988), the presence of loops and plugs in the interplanetary medium is due to the existence of discrete arcades in the corona and with the release of these structures to the heliosphere. Simultaneous observations on two or more spacecraft separated in heliocentric distance could give an unambiguous answer.

The concept of magnetic clouds (*e.g.*, Klein and Burlaga 1982; Ivanov et al. 1989; Burlaga et al. 1990) is of considerable interest for the SCR transport problem. The possibility of their existence in interplanetary space was shown as far back as 1974 (see, *e.g.*, Burlaga et al. 1990). For example, an unusual, large-scale magnetic field configuration observed in the solar wind in December 1982, according to the data of *ISEE* 3 at 1 AU and that of *IMP* 8 in the Earth's magnetosheath, was identified as a magnetic cloud (*e.g.*, Lepping et al. 1991). Although similar configurations have been studied recently, this particular event is especially important because of its large size at 1 AU (heliocentric radial extent $\geq$0.4 AU) and because it was observed by two spacecraft in markedly different plasma regimes. The existence of magnetic clouds was confirmed by direct measurements on the *Vega* 1 and *Vega* 2 spacecraft in 1986 (*e.g.*, Ivanov et al. 1989).

There are a number of serious problems with regard to an identification of the clouds, their physical nature, modeling, *etc*. In particular, Klein and Burlaga (1982) suggested that interplanetary structures with the following characteristics can be considered as clouds: (1) a radius of ~0.25 AU; (2) a compressed magnetic field

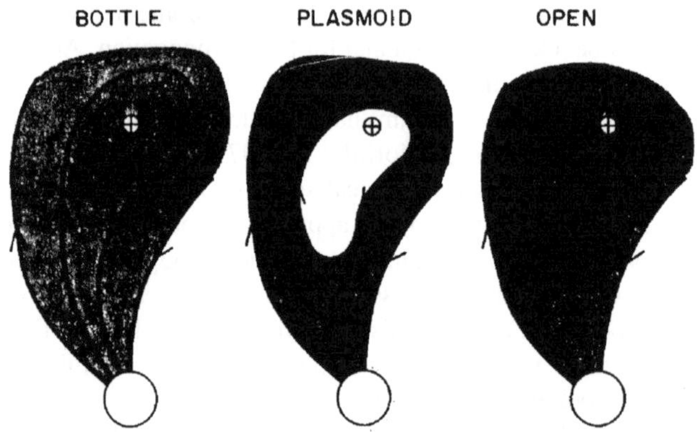

Fig. 8.9 Three proposed topologies of magnetic clouds, as projected onto the ecliptic plane (Kahler and Reames 1991). A *circle with the cross* is the Earth; the Sun is at the *bottom*. The *heavy solid lines* indicate magnetic field lines extending from the Sun into interplanetary space, and the shaded regions show the areas accessible to SEPs. In all cases a bidirectional particle flux could be observed at the Earth

>10 nT; (3) a loop-type field geometry. Based on these criteria, 64 such formations during the 1967–1982 period were identified, and many workers attempted to use solar energetic particles to probe the topology of those structures (see Kahler and Reames 1991, and references therein). It is important to note that in most early measurements, the bidirectional fluxes of protons and electrons were observed. These particles were associated with a solar flare which occurred earlier in the day and were therefore ejected at the Sun. As candidate topologies for the interplanetary magnetic field, three alternative formations were discussed - the magnetic bottle, plasmoid, and open field configuration. Those three kinds of magnetic topology that could explain the bidirectional particle events are shown in Fig. 8.9, as projected onto the ecliptic plane.

Kahler and Reames (1991) used a simple technique to determine whether the plasmoid structure shown in Fig. 8.8 is characteristic of bidirectional proton events. Their analysis was based on the following considerations. If solar energetic particles are injected into interplanetary space from the Sun while a plasmoid is present in space, those particles should be excluded from the region of the plasmoid. The subsequent passage of such a region past the Earth should therefore be accompanied by a substantial reduction in SEP fluxes. The absence of such reductions would imply that the plasmoid topology then can be ruled out. Similarly, the observation of promptly arriving SEPs at the Earth while the Earth is embedded within a region with bidirectional proton fluxes is inconsistent with the plasmoid topology. Kahler and Reames (1991) examined the corresponding measurements of the fluxes of protons with $E_p = 22$–27 MeV and electrons with $E_e = 0.2$–2.0 MeV for 36 SEP events, and it was found no cases that suggest that particles are excluded from a region of interplanetary space. So, these results argue strongly against the plasmoid topology, and, conversely, weighty proofs were obtained in favour of the bottle topology.

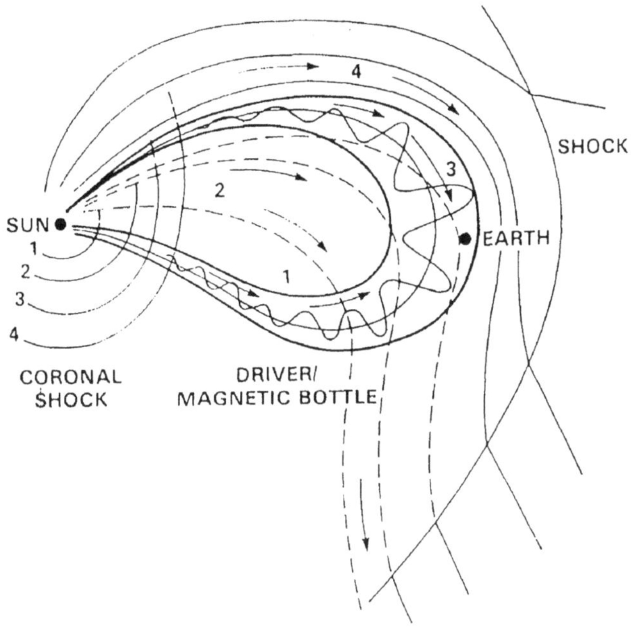

Fig. 8.10 Sketch of the proposed model for prompt eastern particle events showing a pre-existing bottle-like field region associated with the driver of an interplanetary shock, extending from the Sun (Richardson et al. 1991). Energetic particles are accelerated by an expanding coronal shock initiated by the eastern flare (for details see the text)

The rapid arrival of SEPs from far eastern flares also indicates the propagation of particles along the lines of force of a magnetic bottle extending to the Sun (Richardson et al. 1991) rather than the presence of a closed-loop plasmoid-like field region. Such a conclusion was obtained by the data on two "eastern" events, namely, of October 15, 1980 and October 12, 1981 which were measured in the energy ranges of $E_p = 4.5–45$ MeV, $E_e = 0.2–2.0$ MeV, and $E_p = 4.2–230$ MeV, $E_e = 3.6–19$ MeV, respectively. Moreover, during the GLE of October 12, 1981 also relativistic protons showed the bidirectional anisotropy (Cramp et al. 1995a, b). Proposed magnetic configuration for prompt eastern particle events is shown in Fig. 8.10.

It is suggested (Richardson et al. 1991) that energetic particles are accelerated by an expanding coronal shock, initiated by the eastern event, and are injected into the bottle and adjacent regions. Particle arrives first at the Earth, guided along the eastern leg of the bottle (*1*). As the shock expands, field lines draped over the driver (*2*), in the western leg of the bottle (*3*), and draped over its western edge (*4*) may be filled with energetic particles.

As could be deduced from the neutron monitor data, the well-known GLE of September 29, 1989 displays the certain signatures of bidirectional anisotropy, too, at the late stage of the event (Vashenyuk et al. 1997; Vashenyuk and Miroshnichenko 1998). Such an effect seems to exist, in spite of the western origin

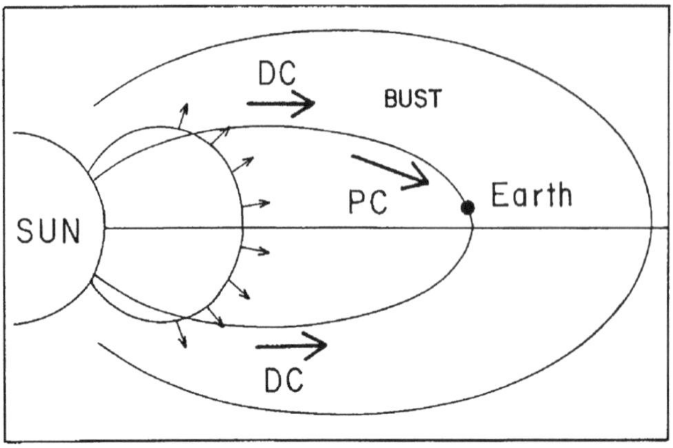

Fig. 8.11 Large-scale loop structure of the interplanetary magnetic field in the meridian plane as proposed source of bidirectional anisotropy of relativistic protons observed during the event of September 29, 1989 (Vashenyuk et al. 1997). *PC* and *DC* are the prompt and delayed relativistic components, respectively. The possible source of the *DC* in the solar corona is also shown

of relativistic particles and probable presence of the two relativistic components - prompt and delayed ones (see Sect. 7.4). This bidirectionality may be explained if one assumes that the IMF during the event had a shape of a giant loop, with its both legs rooted into the Sun (Fig. 8.11). The source of the delayed component (DC) presumably occupied an extended area as large as a CME. So, the particles of the DC could be injected into the both roots of the loop forming the bidirectional anisotropy. This looped structure of the IMF was possibly created by solar eruption preceding the GLE42 of September 29, 1989 (see as a review Miroshnichenko et al. 2000).

Undoubtedly, the bidirectional particle fluxes are an effective tool for examining the geometrical properties of the magnetic field, such as the field direction or divergence. At the same time, those fluxes apparently cannot determine the field topology, *i.e.*, the large-scale connections of the field lines (Kahler and Reames 1991). In this context one interesting problem arises. The fact that clouds are nearly transparent to SEPs indicates the open field structures. If we believe that the CME magnetic field topology must be closed, then bidirectional particle fluxes are not exclusively signatures of CMEs. On the other hand, if CMEs are characterized by open field topologies, a serious revision of modern view of magnetic fields in CMEs seems to be necessary (Kahler and Reames 1991).

The simulation of magnetic cloud propagation in the heliosphere (see, *e.g.*, Vandas et al. 1995, 1996, and references therein) have shown that propagating magnetic clouds (and, probably, other ejecta) widely modify the surrounding interplanetary medium. The simulations indicate a possibility that particles can be trapped not only inside the cloud but also outside on open magnetic field lines. Particles carried inside the cloud are subject to adiabatic deceleration during the expansion of the cloud, and possible scattering and diffusion lead to their losses. High-energy particles (*e.g.*, protons with $E_p \sim 10$ GeV) simply penetrate the cloud while trajectories of particles with slightly lower energy (*e.g.*, protons with E_p ~1 GeV) are bent, and they leave the cloud without being captured (Vandas et al. 1995). Low-energy particles (*e.g.*, protons with $E_p < 100$ MeV) can enter the cloud through its poles, and their density would be significantly higher near the poles and at the cloud's axis. This could account for some spatial increases of energetic particles inside some magnetic clouds (one example is treated by Vandas et al. 1995). The presence of strong turbulence, waves, or field inhomogeneities near the surface or inside the cloud may change the situation in favour of trapping. However, it is necessary to keep in mind that similar conditions also work in the opposite direction, causing enhanced losses of trapped particles. Possible locations of magnetic traps and their magnetic characteristics have been studied numerically by Vandas et al. (1996).

Chapter 9
Spectrum of Solar Cosmic Rays Near the Earth

As it was postulated in Chap. 1, an energy distribution (or a shape of the energy spectrum) of solar cosmic rays (SCR) is of great significance for the formulation of self-consistent model of particle acceleration at the Sun. In turn, the main problems of fundamental interest in the theory of particle acceleration at the Sun lie now at two boundary domains of SCR spectra, namely, in low-energy (non-relativistic) and high-energy (relativistic) ranges. The most important of them are: initial acceleration from the thermal background (see, *e.g.*, Vlahos et al. 1989; Simnett 1995; Miroshnichenko 1995; Miller et al. 1997) and final stage of acceleration to extremely high energies (see, *e.g.*, Miroshnichenko 1994, 1996; Karpov et al. 1995a, b, 1998).

9.1 Key Aspects of Spectrum Measurements

Since February 28, 1942 (an historical beginning of SCR observations) the generous data have been accumulated on the fluxes of SCR, and their spectra have been intensively studied in the energy range from ~1 MeV to ~10 GeV and even more. Now we have a set of ground-based data for 59 Ground Level Events (GLE) at the rigidity above 1 GV (see Table 2.1). However, spectral data at the rigidities above ~1 GV ($E_p = 435$ MeV) are fairly scarce and rather controversial (Miroshnichenko 1990, 1994, 1996).

Ground-based data on GLEs of the 22nd solar cycle (since September 1986) are of special interest due to unusually high occurrence rate and large energy content of the events (*e.g.*, Smart et al. 1991; Miroshnichenko 1992b, 1994). Another exciting finding of recent flare studies turned out to be a registration of high-energy (pion) gamma-rays (E_γ >1 GeV) in June 1991 (*e.g.*, Akimov et al. 1991; Kanbach et al. 1993). In particular, the observation of the flare of 11 June 1991 (Kanbach et al. 1993) revealed for the first time the existence of pion radiation as late as 8 h after the impulsive phase.

L. Miroshnichenko, *Solar Cosmic Rays*, Astrophysics and Space Science Library 405, DOI 10.1007/978-3-319-09429-8_9

The problem of SCR generation in relativistic range was strongly actualized about 10 years ago due to first confident observations of high-energy underground effects correlated with energetic solar phenomena (large solar flares, coronal mass ejections, etc.) and GLEs. In particular, significant increases of counting rate at several muon telescopes were recorded during GLE of September 29, 1989 (Krymsky et al. 1990; Swinson and Shea 1990; Filippov et al. 1991; Smart and Shea 1991), including one very peculiar muon burst (Alexeyev et al. 1992) at the Baksan Underground Scintillation Telescope (BUST).

Moreover, new significant bursts found by the BUST data for a set of GLEs since 1981 (Alexeyev and Karpov 1993, 1994; Karpov et al. 1995a, b, 1998) give a challenge to our present understanding of the upper capacities of particle accelerators at the Sun (see Chap. 4). In other words, all these results require that the traditional concepts of particle acceleration at the Sun to be considerably revised, as to the location of acceleration region(s), magnetic structure of the upper corona, time scenario and total duration of acceleration, maximum energy and absolute fluxes of accelerated particles, *etc*.

Relativistic solar protons are distinguished from lower energy particles (so-called solar energetic particles, or SEP) by the fact that the former are observed mostly at ground level. Sometimes they are also detected at the spacecraft; for example, one of the GOES-7 sensors has an energy window between 640 and 850 MeV (see Fig. 1.9), and the *Meteor* satellite measures solar protons with the energy E >600 MeV (*e.g*., Nazarova et al. 1992).

The lower energy solar particles are detected partly on airborne balloons in the stratosphere, and mostly on the Earth-orbiting satellites and spacecraft in outer space. These sub-relativistic particles (>10–100 MeV/nucleon) contain rich information concerning mechanisms of particle acceleration on the Sun and subsequent transport through turbulent interplanetary space. The lowest energy particles (<10 MeV) also reach the vicinity of the Earth, but they can only be detected in the upper atmosphere within the polar cap. Relativistic protons produce in the upper atmosphere a cascade of secondary particles. For a measurable number of secondary nucleons to penetrate the Earth's atmosphere the incident protons must have a rigidity greater than about 1 GV, corresponding to an energy approximately 435–450 MeV (very often this threshold is assumed to be 500 MeV). The occurrence rate of the events detected at the ground level, amounts on the average the average to about 1 per year, which is much less than the number of observations in the upper atmosphere or outer space (see Chap. 2).

Typical energy spectra of ions and electrons measured near the Earth's orbit have been demonstrated in Figs. 2.2 and 2.3. Here we concentrate mainly on the features of proton spectra. As an introduction to the following discussion, in Fig. 9.1 we show the energy spectrum of accelerated protons in the form proposed by Simnett (1991). This spectrum form was compiled taking into account existing observational uncertainties and theoretical restrictions imposed by the models of particle acceleration in the coronal current sheets. Simnett (1991) suggests that all accelerated spectra of non-relativistic protons may be represented by the general form shown in the hatched portion of Fig. 9.1.

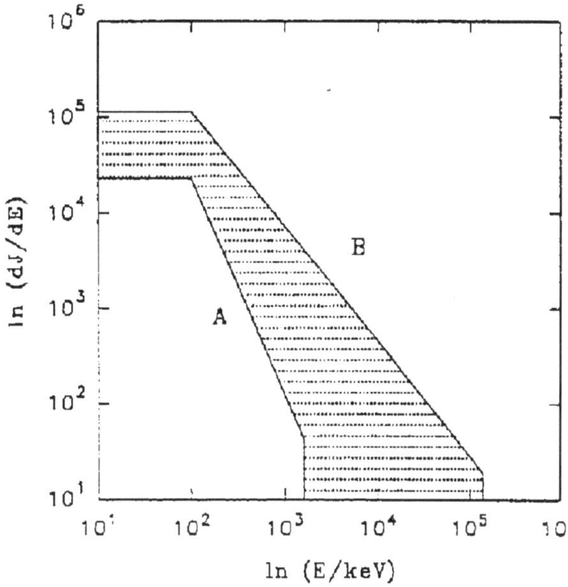

Fig. 9.1 Spectrum of accelerated protons proposed by Simnett (1991) to reconcile observations in space with acceleration of non-relativistic protons in a coronal current layer. Variations in the spectrum within the hatched area are due to the changes in the current layer parameters

Spectrum *A* represents particles accelerated in small events, while *B* corresponds to the spectrum generated by a major flare. Variations in the spectrum within the hatched region are produced by changes in the dimensions and other parameters of the current layer. In fact, it is a very difficult problem – to assess the true form of the accelerated spectrum. Coronal shocks and propagation effects, coupled with velocity dispersion, modify the observed spectrum to give the variety of spectral forms reported over the last four decades. Only the most energetic particles (>500 MeV) represent the true source, and even they may undergo propagation effects (Simnett 1991).

For lower energy (non-relativistic) particles (ions) the best estimate of propagation effects comes from observing the velocity dispersion in the arrival of prompt particles, *i.e.*, from the energy-dependent diffusion which resulted in the energy dependence of the time of maximum intensity, t_m. Such a time-of-maximum (TOM) technique was justified in detail by Forman et al. (1986) and was applied by many researchers (*e.g.*, Miroshnichenko et al. 1999). In particular, an extensive work has been carried out by the authors of several SPE Catalogues (Akinyan et al. 1983; Bazilevskaya et al. 1986, 1990a; Sladkova et al. 1998) who estimated the power-law indexes of proton energy spectra for the 334 events observed near the Earth over the period 1970–1996 with the threshold intensity of 1 pfu.

As concerns relativistic solar protons (RSP), the energy spectra for the most powerful SPEs have been compiled in Fig. 4.4, and comparing those data with the spectra in Fig. 9.1, one can see that, in fact, the both groups of data are characterized by the same scales of uncertainties. Thus, below we discuss in some detail the methods of spectrum determination by the ground observations.

9.2 Methods of Spectrum Data Analysis

The SCR spectrum in a large SPE may cover of 4–5 orders of the energy value (from ~1 MeV to >10 GeV), and differences in the intensity of particles at the ends of the spectrum (due to its great steepness in the range of high energies) may amount to 6–8 orders of magnitude (see, *e.g.*, Figs. 4.1, 4.2, 4.3, and 4.4) (Adams and Gelman 1984; Miroshnichenko 1994). This creates certain experimental and methodical difficulties in the measurements of SCR fluxes near the Earth (on the background of galactic cosmic rays) and interpretation of the data obtained. For determining the SCR spectrum in a wide interval of energies one has to compile, as a rule, the results of a few kind of measurements (on board the satellites, in the stratosphere, at the Earth's surface and so on), thus introducing additional uncertainties into derived values of spectral characteristics, as they are defined in Chap. 1, Eqs. (1.5), (1.6) and (1.7), namely, absolute flux D_0, power-law exponent γ, and characteristic rigidity R_0.

9.2.1 Measurement Limitations

The proton spectra for solar particle events are usually obtained from the data of any one group of detectors only within a narrow energy range. For instance, it is typical to use only spacecraft data (<500 MeV), or only the data from stratospheric balloons (hundreds of MeV), or ground level observations (>500 MeV). So, only a small part of the spectra can be studied. But information about the total spectrum within the wide energy range is usually necessary. In particular, we should know the complete particle spectrum over energy range >10 MeV to estimate the radiation dose in space during large proton events (see Chap. 11). The problem is to unify all the parts to obtain one complete spectrum, as it was attempted in Fig. 4.4. When doing this procedure, however, several serious methodical difficulties arise.

An attempt has been undertaken (Miroshnichenko 1983c) to analyze some possible reasons (sources) for methodical uncertainties and errors in determining SCR spectra near the Earth and in the source. In particular, errors in measurements of an absolute SCR flux in interplanetary space, in magnetosphere, at various heights in the atmosphere, on the Earth's surface were considered in detail. Also noted should be such factors as anisotropy of SCR, additional acceleration of particles in interplanetary space and possible existence of the upper limit rigidity, R_m, in the source spectrum. More recently, it was noted (Flückiger 1994) that some unusual intensity-time variations in the counting rate of neutron monitors still lack a satisfactory explanation.

It is true, in particular, for a number of "poor identified GLEs" which have been observed at single neutron monitor stations (for example, the GLE of 24 August 1998), as well as for other effects discovered by the application of new data analysis technique. In other words, some open questions in the analysis of GLEs by ground-based data remain, in spite of more than 50-year experience of SCR observations. A subject of special interest is the integral multiplicities of generation of secondary components of cosmic rays. As it will be shown below, available specific yield (and/or response) functions are still under question due to different approaches to their calculations, the discrepancies between different calculations being within one-two orders of magnitude. Meanwhile, the uncertainties in determining the SCR spectrum depend strongly on the accuracy of integral multiplicity values.

9.2.2 Analytical Techniques

To deduce the rigidity spectrum of relativistic solar protons at rigidities above ~1 GV, the world-wide network of neutron monitors are used. For determining the spectra, accurate specific yield functions, $m(R, h)$, are required, the latter give the number of secondary particles recorded by neutron monitors at given depths, h, in the atmosphere as a function of primary proton rigidity, R. The counting rate of a neutron monitor at a time t may be expressed as

$$I_i(> R_c, h, t) = \int_{R_c}^{\infty} m_i(R, h) D(R, t)\, dR \tag{9.1}$$

where I_i is the integral intensity of secondary particles of the i kind at the h depth in the atmosphere at the point with the geomagnetic threshold R_c (cutoff rigidity); $D(R)$ is the differential spectrum of galactic cosmic rays (GCR); m_i is the integral multiplicity of generation, *i.e.* the number of secondary particles of the i kind (*e.g.*, neutrons) generated by a primary particle with the rigidity R and registered at the h depth (*e.g.*, at the sea level). A gross yield function (or integral multiplicity function) $m(R, h)$ relates the counting rate of a neutron monitor to all the charge components of the primary cosmic-ray spectrum. Empirically, this function is taken as the ratio of the measured differential latitude variation in counting rate of a detector, $N(R, h, t) = dI(R, h, t)/dR$, at a specific atmospheric depth, to the measured differential intensity of primary rigidity spectrum. In other words, differentiating (9.1) with respect to R, we obtain the formula describing the latitude variation of GCR:

$$|dI_i/dR| = m_i(R, h) D(R) \tag{9.2}$$

Hence, the principal relationship for estimating integral multiplicities of generation follows:

$$m_i(R, h) = |dI_i/dR| D(R) \tag{9.3}$$

The values $m_i(R, h)$ may also be calculated theoretically taking into account nuclear-cascade processes in the atmosphere. If the values $m_i(R, h)$ are known, then the following formula for determining the SCR spectrum follows from (9.1), (9.2), and (9.3):

$$D_s(R) = (dI_i/dR)_s/m_i(R,\ h) \tag{9.4}$$

where the s index refers to solar particles, and $m_i(R,\ h)$ – to the total GCR flux, irrespectively to the content of the nuclei with different charge Ze. Meanwhile, the protons are the dominant part of SCR, their content being much more in respect to nuclei with $Z > 2$ in comparison with the GCR. It means that instead of empirical values of m_i derived from the latitude survey of GCR intensity (for example, for neutron component, m_n) an integral multiplicity for protons, m_p, should be used in determining the SCR spectrum by the (9.4). It causes a serious limitation in the accuracy of SCR spectrum determination by the formula (9.4). As it is well known, about one half of total amount of nucleons in the GCR flux, in fact, are free protons. Hence an important limitation follows:

$$m_p(R,\ h)\ < m_n(R,\ h)\ <\ 2m_p(R,\ h) \tag{9.5}$$

i.e., the integral multiplicities m_n calculated by observations of GCR (9.3) may be differ from the m_p values no more than by factor of <2 (for details see Dorman and Miroshnichenko 1966, 1968). Just the factor <2 imposes a principal limit on the accuracy of SCR spectrum estimation by the formula (9.4).

9.2.3 *Effective Rigidity*

Taking into account the non-linear pattern of relationship (9.3) between the primary and secondary fluxes of particles, it is possible to state that the intensity being measured refers, in fact, not to R_c, but to a certain effective rigidity R_e which may be determined from the formula:

$$1/R_e \int_{R_c}^{\infty} D(R)\, m(R)\, dR = \int_{R_c}^{\infty} R^{-1} D(R)\, m(R)\, dR \tag{9.6}$$

In an explicit form R_e may be obtained if one approximates $D(R)$ and $m(R)$ by simple analytical functions (*e.g.*, power-law or exponential):

$$D(R) = D_0 R^{-\gamma}; \quad m(R) = m_0 R^{\beta} \tag{9.7}$$

$$D(R) = D_0\ \exp(-R/R_0); \quad m(R) = m_0 R^{\beta} \tag{9.8}$$

Taking into account that m(R) is a monotonous non-decreasing function of R (i.e., $\beta > 0$) and doing the necessary computations we obtain from (9.6) in the simplest case (9.7) the expression

$$R_e = (\gamma - \beta)/(\gamma - \beta - 1)R_c \tag{9.9}$$

Under condition (9.8) expressions for R_e become rather cumbersome, so we write down them only for $\beta = 1, 2, 3$:

$$R_e = R_0 + R_c; \quad \beta = 1 \tag{9.10}$$

$$R_e = (R_0 + R_c) + R_0^2/(R_0 + R_c); \quad \beta = 2 \tag{9.11}$$

$$R_e = 3R_0 + R_c^2/\left[R_0^2 + (R_0 + R_c)^2\right]; \quad \beta = 3 \tag{9.12}$$

Thus, for the assigned R_c the effective value of rigidity of primary particles making the main contribution into the observed intensity on the Earth's surface, depends in a certain way on the shape of the SCR spectrum (upon the value γ or R_0) and on the value β. In the case of the neutron component b is within the limits $2 \leq \beta \leq 3$ (see, *e.g.*, Miroshnichenko and Petrov 1985). With this limitation the relevant restrictions for γ and R_e follow from above:

$$(\gamma - 3)/(\gamma - 4)R_c > R_e > (\gamma - 2)/(\gamma - 3)R_c; 2 \leq \beta \leq 3; \ \gamma > 3 \tag{9.13}$$

$$(R_0 + R_c) + R^2{}_0/(R_0 + R_c) < R_e < 3R_0 + R_c^3/\left[R_0^2 + (R_0 + R_c)^2\right] \tag{9.14}$$

For practical purposes it would be reasonable to carry out numerical calculations of R_e for different versions of the SCR spectrum and tabulated values of $m(R)$ for the neutron component (by now, there are no such computations).

9.3 Integral Multiplicities for Neutron Monitors

Integral multiplicities for the neutron component were calculated by many authors (see references, *e.g.*, in Dorman and Miroshnichenko 1966, 1968; Nagashima et al. 1989; Stoker 1995) on the basis of the latitude effect of GCR:

$$m(R) = (dI/dR)/D(R) \tag{9.15}$$

From (9.4) and (9.15) a simple relationship follows:

$$D_s(R) = (dF_s/dR)/(dF/dR)D(R) \tag{9.16}$$

If latitude curves for GCR and SCR were calibrated in the same way, then the magnitude of the calibration factor does not affect the $D_s(R)$ magnitude, which in this case is determined by a relative amplitude of increase $(dF_s/dR)/(dF/dR)$ and by an absolute flux of GCR (in proton flux units, *pfu*) for the moment of SPE.

As is seen from the previous section, the procedure of spectrum determination by ground-based observations requires precise data on the latitude effect and primary spectrum of GCR. Meanwhile, it is well known that the GCR spectrum changes substantially during an 11-year cycle of solar activity, in particular, in the range R <1 GV. Besides, in some periods of the solar cycle the GCR intensity displays certain anomalies associated, apparently, with the process of variation of polarity of the global magnetic field (GMF) of the Sun. One of such anomalies was observed at the end of 1971 – at the beginning of 1972. During the same period (October 1971-January 1972) on board the research ship "Academician Kurchatov" the latitude effect of the neutron component was measured (Bednazhevsky and Miroshnichenko 1982) at the entire course of the ship route from Kaliningrad (USSR) to the South Orkney Islands (Antarctica).

On the other hand, Charakhchyan et al. (1973) managed to restore the absolute GCR spectrum near the Earth's orbit for the end of 1971 taking into account the change of the GMF polarity in 1971–1972 (see Figs. 3.11 and 3.12). As it was shown by Stozhkov (1978), due to anomalous behaviour of solar activity the GCR intensity at R >0.5 GV already by the end of 1971 (just 2 years after the maximum of the cycle) had reached almost 95 % as regards the 1965 level. This created almost a unique opportunity for calculating integral generation multiplicities basing on the data on the latitude effect of GCR with a distinct reference to their primary spectrum.

The results of such estimations (Bednazhevsky and Miroshnichenko 1982) are shown in Fig. 9.2 (curve 1). The results of similar calculations based on the measurements on board the same research ship "Academician Kurchatov" in

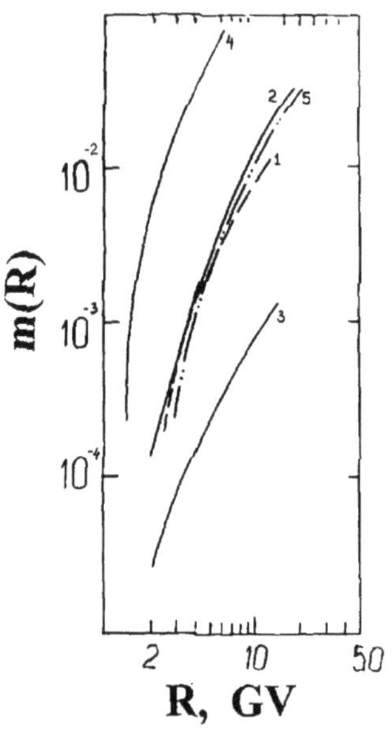

Fig. 9.2 Comparison of integral multiplicities of generation for the neutron component according to the data of different measurements of the latitude effect of cosmic rays (Bednazhevsky and Miroshnichenko 1982): *1* 1971–1972; *2* 1969; *3* 1954; *4* 1965; *5* estimates taking into account nuclear cascade processes in the Earth's atmosphere

1969 are given for comparison (curve 2). In addition, the $m(R)$ values are shown that were obtained (Ilencik et al. 1978) by the data of latitude measurements in 1954 and GCR spectrum measurements in 1965 (curve 3). A similar curve 4 constructed by the measurements of the primary spectrum and latitude effect of GCR for the summer of 1965 (Lockwood et al. 1974), and a theoretical curve 5 (Yanke 1980) are also given. It is seen that discrepancy between curves 1 and 3 is >10 and that between curves 3 and 4 is almost two orders of magnitude.

As it became clear not long ago, in estimations of Ilencik et al. (1978) there had been a methodical inaccuracy which led to underestimation of the $m(R)$ value by 4π times as compared with the computation results by Bednazhevsky and Miroshnichenko (1982). The discrepancy between the curves 3 and 4 is attributed to the fact that Ilencik et al. (1978) and Bednazhevsky and Miroshnichenko (1982) calculated absolute values of $m(R)$, whereas the $m(R)$ values was found by Lockwood et al. (1974) under the normalization the observed intensity of GCR (9.1) to their intensity on the equator, as it presented by equation (9.17) below.

In order to simplify the procedure of using the latitude curve of GCR during SPE it is reasonable to construct curves $A(R)$ of the latitude variation of a 1 % increase separately for the NM of the IGY type (this was done by Ilencik et al. 1978) and for the SNM-64 taking into account the intervals of averaging $\Delta t = 1$ h, 15, 5 min, *etc.* These curves will be parallel one another, but absolute values differing by approximately seven times. They will make possible to transfer to the latitude effect of SCR $F_s(>R)$ in absolute units, even if the maximum rigidity in their spectrum, R_m, slightly exceeds 1 GV; notice that when normalizing the $m(R)$ values at $R = 1$ GV (Lockwood et al. 1974) such an approach is impossible.

Apparently, in order to properly determine $m(R)$ on the basis of latitude measurements it is necessary to use data on the primary GCR spectrum for the same period of observations. The obtained $m(R)$ curves should be used with care for determining the absolute SCR spectrum bearing in mind the fact that the accuracy of estimating m(R) by the ground-based data does not exceed the factor 2. In addition, notice the authors Lockwood et al. (1974) when comparing their estimations with the results of other works had also discovered discrepancies in the $m(R)$ values of 50 % to one order of magnitude. Therefore, an error of determining the absolute SCR flux by the ground-based data (R >1 GV) apparently comprises >50 % and an error in a parameter γ may reach ± 1.0. As a rule, stratospheric and satellite measurements yield significantly smaller errors of the indicated paramcters of the spectrum (Miroshnichenko 1983c). Thus, a comparison of the estimated spectra with the data of direct SCR measurements in the range of R <1 GV may serve as an additional test for checking the accuracy of the $m(R)$ estimations.

Because of different detection efficiencies of neutron monitors of different design, different response functions of neutron monitors at different pressure levels, anisotropy of SCR and some other factors (*e.g.*, Miroshnichenko 1983c; Stoker 1995), the application of the discussed procedure may lead to essential inaccuracy in determining the SCR spectrum. However, at the same time, just that approach leads to a more comprehensive utilization of ground-based observation data (see Sect. 9.4).

As one can see from Figs. 2.2 and 4.3, the energy spectrum of solar cosmic rays is usually much steeper than that of galactic cosmic rays. This distinctive feature turned out to be very helpful for determining the SCR spectrum during significant GLEs. The point is that the super-imposing of two primary fluxes – galactic and solar – leads to a significant change of properties of secondary components.

This first of all manifests itself in a path change for the absorption of secondary components in the lower atmosphere, in a change in the barometric coefficient (β), which is especially important for the neutron component. The reciprocal of the barometric coefficient ($\lambda = 1/\beta$) is called absorption length in the atmosphere (this is a path on which particle flux decreases by a factor *e*). The procedure for taking into account the barometric effect in the presence of two components (galactic and solar) with different paths for absorption, λ_g and λ_s, respectively, was suggested more than 50 years ago by McCracken (1962a) and was subsequently developed in (Wilson et al. 1967; Kaminer 1967). Specifically, Wilson et al. (1967) proposed a method used to directly measure absorption length λ_s based on data from a pair of suitable NMs. For example, on 28 January 1967, GLE17 was registered with two Canadian NMs (Calgary and Sulphur) at altitudes of 1,128 and 2,283 m, respectively. These two stations have threshold rigidities and asymptotic cones of acceptance which are very similar, so that it was possible to determine the absorption length of the solar particles directly. This value was estimated to be $\lambda_s = 103 \pm 3$ g cm^{-2}. When the method of two absorption lengths is applied, the following values are as a rule accepted: $\lambda_s = 140$ g cm^{-2} for neutrons produced by GCRs and $\lambda_s = 100$ g cm^{-2} in the case of SCRs.

This method was applied by Ahluwalia and Xue (1993) to the GLE of September 29, 1989. They used the data of three representative pairs of neutron monitors: Tokyo-Mt. Norikura; Tbilisi – Alma Ata; and Durham – Mt. Washington. It was found that early in the event the computed mean values of λ_s are significantly larger than the consensus value of 105 ± 5 g cm^{-2} reported in the literature (*e.g.*, Duggal 1979). At later times, they approach the consensus value for the high latitude neutron monitors. Moreover, the mean values of λ_s exhibit a linear correlation with the effective cutoff rigidities in the range 1.1 GV $\leq R_c \leq$ 19 GV, as it shown in Fig. 9.3.

To find the values of spectral power-law index γ for the same event of 29 September 1989, de Koning (1994) applied the method of Wilson et al. (1967) to the data of three pairs of mountain-sea level NMs which have very similar threshold rigidities and asymptotic cones of acceptance: Mt. Washington ($R_c = 1.48$ GV) – Durham ($R_c = 1.64$ GV); Campo Imperatore ($R_c = 6.14$ GV) – Rome ($R_c = 6.32$ GV), and Mt. Norikura ($R_c = 11.48$ GV) – Tokyo ($R_c = 11.63$ GV). From the data of these pairs at 1230 UT there were found $\gamma = 4.90 \pm 0.05$; $\gamma = 6.8 \pm 0.2$; and $\gamma = 11 \pm 2$, respectively. Based on the details of the method, these values represent average values of γ for the rigidity ranges of $R > R_c$, namely, for $R >$1.56, >6.23, and >11.56 GV, respectively (*cf.*, however, Fig. 7.21).

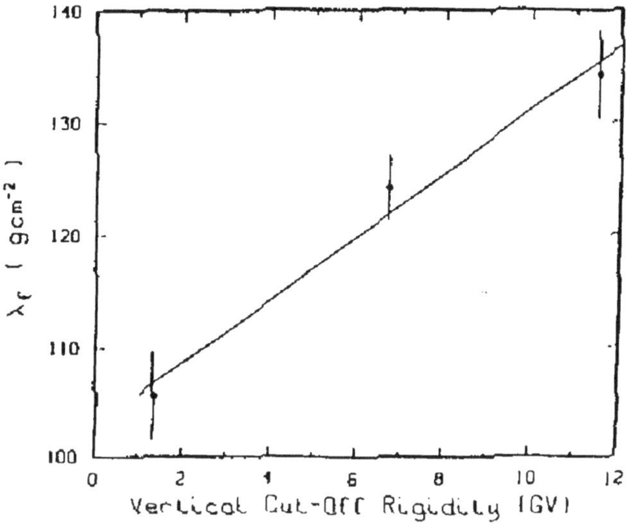

Fig. 9.3 The mean attenuation length λ_s (absorption length) for secondary particles during the GLE of September 29, 1989 *versus* the effective threshold rigidity R_c computed by the data of three pairs of neutron monitors (Ahluwalia and Xue 1993). A linear correlation between the two parameters is apparent; the correlation coefficient is 0.99

9.4 Rigidity Spectrum of Relativistic Protons

Below we compile and analyze available data on absolute fluxes of relativistic protons near the Earth, taking into account the flare importance and heliocoordinates, spectrum form (its slope, or power-law exponent), conditions of interplanetary propagation, acceptance cones of acceptance in the geomagnetic field, *etc*. As coupling coefficients, many researchers use the so-called specific yield function (Debrunner et al. 1982) which covers the range of rigidities from ~1 to ≥20 GV with an acceptable accuracy. In this context, it is important to remind that the SCR spectrum is usually much steeper (softer) than the GCR spectrum. This difference was very useful for determining the SCR spectrum during the GLEs since the superposition of two CR fluxes (galactic and solar) results in a substantial change in the properties of the secondary components.

From the total number of 71 GLEs by the present time absolute SCR spectra above 1 GV have been estimated by various researchers for 30 events of 1942–1991 (see Table 9.2). The traditional procedure of determining spectra was, in essence, identical and used formula (9.1). However, Dorman and Miroshnichenko (1966, 1968), Miroshnichenko (1970), Miroshnichenko et al. (1973), Ilencik (1979), Ilencik et al. (1978), Kepicova et al. (1982), and Miroshnichenko and Petrov (1985) have used absolute $m(R)$ values, while Lockwood et al. (1974), Bazilevskaya and Makhmutov (1983), Nagashima et al. (1989), Zusmanovich and Shvartsman (1989), Kohno (1991), Smart et al. (1991) and other researchers have preferred yield functions $S(R)$, normalized to the counting rate of the neutron monitor at the equator, N_e, in the minimum of solar activity (for example, notice the yield functions calculated by Debrunner et al. 1982). In this latter case the following relationship may be written instead of (9.1):

$$\Delta N(R_c,\, t)/N_e = \int_{R_c}^{R_m} S(R)\, D_s(R)\, dR \qquad (9.17)$$

where $\Delta N(R_c, t)$ is the increase in the counting rate of the neutron monitor at the point with a cutoff rigidity R_c at the moment t as a result of arrival of SCR with the spectrum $D_s(R)$ at the atmosphere boundary; R_m is the maximum rigidity of solar protons, and $S(R) = m(R)/N_e$.

Since the end of 1980s, some new properties of the GLEs have been found, a new concept of two-component GLE was born (*e.g.*, Miroshnichenko et al. 1990), and more sophisticated technique for SCR spectrum determination has been developed (see Sects. 9.8 and 9.9). As a result, the properties of different components of relativistic solar protons have been studied in detail (see Table 9.4), and finally we come to a new understanding of SCR generation and propagation.

9.4.1 Comparison of the Yield Functions

For the purpose of their mutual comparisons, in Table 9.1 we give integral multiplicities (specific yield functions) deduced by Ilencik et al. (1978), Bednazhevsky and Miroshnichenko (1982), and Debrunner et al. (1982), and in Fig. 9.4 the spectra determined by various researchers for several GLEs are represented.

9.4.2 Statistics of Spectral Data

It is of interest to note that in those cases when the rigidity intervals under consideration coincided or overlapped, the results obtained by various authors for spectra were the same (within the uncertainty limits of the technique). This refers, in particular, to the January 28, 1967 event which was analyzed independently by Miroshnichenko (1971), Lockwood et al. (1974), Ilencik (1979), Bazilevskaya and Makhmutov (1983). When estimating spectra for the GLEs of January 28, 1967, January 24, 1971, August 5, 1972, September 24, 1978, and October 12, 1981 (straight lines 2–4, 8, and 9, respectively, at the Fig. 9.4), Bazilevskaya and Makhmutov (1983) have used unpublished results of theoretical calculations of specific yield functions $S(R)$ performed by Debrunner et al. (1982). Due to reasons indicated above, errors in Fig. 9.4 are not shown. It should also be noted that the

Table 9.1 Comparison of integral multiplicities for neutron component

Points	R, GV	m(BM)	m(IDM)	S(DFL)
1	0.8	–	–	1.00×10^{-5}
2	1.0	–	–	2.50×10^{-4}
3	1.5	–	–	2.50×10^{-3}
4	2.0	–	–	5.00×10^{-3}
5	2.5	4.00×10^{-4}	5.00×10^{-5}	9.00×10^{-3}
6	3.5	1.79×10^{-3}	9.95×10^{-5}	2.25×10^{-2}
7	4.5	2.25×10^{-3}	1.65×10^{-4}	4.00×10^{-2}
8	5.5	3.64×10^{-3}	2.96×10^{-4}	6.40×10^{-2}
9	6.5	4.00×10^{-3}	3.91×10^{-4}	9.40×10^{-2}
10	7.5	5.33×10^{-3}	4.95×10^{-4}	1.25×10^{-1}
11	8.5	5.65×10^{-3}	6.02×10^{-4}	1.54×10^{-1}
12	9.5	6.59×10^{-3}	7.12×10^{-4}	1.81×10^{-1}
13	10.5	8.06×10^{-3}	8.26×10^{-4}	2.07×10^{-1}
14	11.5	8.28×10^{-3}	9.41×10^{-4}	2.32×10^{-1}
15	12.5	–	–	2.57×10^{-1}
16	13.5	–	–	2.82×10^{-1}

Notes: *BM* Bednazhevsky and Miroshnichenko (1982), *IDM* Ilencik et al. (1978), *DFL* Debrunner et al. (1982)

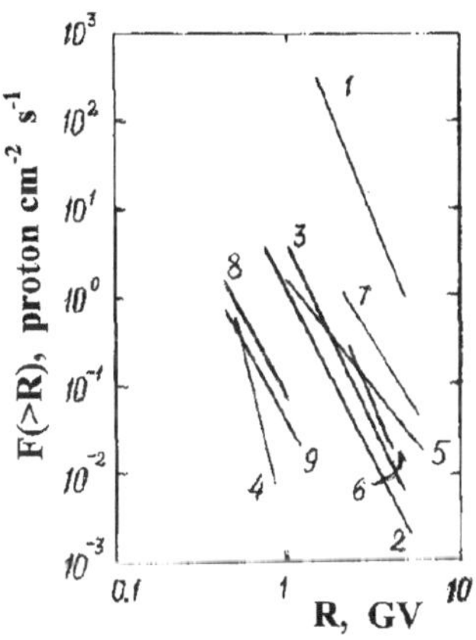

Fig. 9.4 Integral rigidity spectra of relativistic solar protons at the moment of the maximum flux near the Earth for a number of GLEs: *1* February 23, 1956; *2* January 28, 1967; *3* January 24, 1971; *4* August 5, 1972; *5* September 24, 1977; *6* November 22, 1977; *7* May 7, 1978; *8* September 23, 1978; *9* October 12, 1981 (Compiled by Miroshnichenko 1990)

procedure discussed yields satisfactory results only for the time moments when the anisotropy amplitude does not exceed 50 % (Bazilevskaya 1984).

Spectra in Fig. 9.4 are constructed mainly for the times of maximum increase, t_m, on the Earth's surface. It is seen that, as regards the slopes of spectra, the events differ very little from one to another. An exception is the August 5, 1972 event (straight line 4) when the spectrum appeared to be very soft. Though the increase was recorded on the Earth's surface (in particular, in Apatity its amplitude reached ~15 % by the 15-min NM data), it cannot be considered as ordinary GLE: it is very likely that this GLE was caused by protons accelerated between two interplanetary shock waves (Kuzmin et al. 1983; see also Smart and Shea 1989b). The same appears to be true also for the event of July 17, 1959 (see Table 2.1). A hard spectrum for the May 7, 1978 event, probably, was due to a strong anisotropy of SCR. Notice that differences in intensities between the events with similar parameters of the spectrum may amount to 2–3 orders of the magnitude.

Events with relatively small amplitudes of increase on the Earth's surface are of certain interest. Let us take as an example the SPEs of September 19 and 24, 1977 which were, apparently, the first SCR increases recorded on the Earth's surface in the 21st cycle of solar activity. According to the hourly data of the NM South Pole ($R_c = 0.11$ GV, the elevation is 2,820 m above sea level), the maximum amplitude of increase did not exceed 3.2 % and 11.8 % on September 19 and 24, respectively. Due to a small amplitude of effects and complicated helio-geophysical conditions in September 1977, the analysis of these events presents certain difficulties. But at the same time, as it was shown (Kepicova et al. 1982), certain quantitative information on the SCR spectrum may be obtained on the basis of the method of integral multiplicities.

Taking into account the effect of the atmospheric cutoff, one can assume that polar stations with $R_c \to 0$ actually record solar particles with the energies of E_p >435 MeV (R >1 GV), because the latitude curve for SCR at R <1 GV has a plateau. An additional flux of secondary neutrons $F_s(>R_c)$ is related to the SCR spectrum at the atmosphere boundary by the expression

$$F_s(> R_c) = \int_{R_c}^{\infty} D_s(R)\, m(R)\, dR \tag{9.18}$$

where the function $m(R)$, using the results of Bednazhevsky and Miroshnichenko (1982), may be approximated in the form

$$m(R) = 5.8 \times 10^{-5} R^{2.1} \tag{9.19}$$

which is valid with an accuracy to the factor ≤2, at least, within the rigidity interval $R = 1$–10 GV. Comparing the F_s values for the stations McMurdo ($R_c = 0.01$ GV, $A = 1.7$ %) and Swarthmore ($R_c = 1.92$ GV, $A = 1$ %), we get from (9.18) and (9.19) a crude estimate of the spectrum: $D_s(R) = 1.7\ R^{-4}\ \mathrm{cm}^{-2}\ \mathrm{s}^{-1}\ \mathrm{GV}^{-1}$. Within the accuracy of the method (factor ≤2) this estimate does not contradict to the shape of spectrum observed at the moment t below R <1 GV (Kepicova et al. 1982). The integral intensity of particles with R >1 GV at such a spectrum is $I_s(>1\ \mathrm{GV}) = 1.8 \times 10^{-1}\ \mathrm{cm}^{-2}\ \mathrm{s}^{-1}\ \mathrm{sr}^{-1}$, which, in fact, coincides with the value $1.73 \times 10^{-1}\ \mathrm{cm}^{-2}\ \mathrm{s}^{-1}\ \mathrm{sr}^{-1}$ obtained independently some later (Bazilevskaya et al. 1990a).

For the 24 September 1977 event, taking into consideration the equation (9.19), the spectrum is estimated to be $D_s(R) = 1.74 R^{-3.5}\ \mathrm{cm}^{-2}\ \mathrm{s}^{-1}\mathrm{GV}^{-1}$. The integral intensity of protons with R >1 GV is 2.2×10^{-1} pfu, and this value, within the uncertainties of the technique, agrees with the estimate by Bazilevskaya et al. (1990a).

A similar analysis was carried out by Miroshnichenko (1990) for the three other small events: 21 August 1979, 10 April 1981, and 10 May 1981. Also, Bazilevskaya et al. (1990a) have determined (or made precise) the spectra for large GLEs of 22 November 1977; 7 May 1978; 23 September 1978; 12 October 1981, 26 November 1982; 7–8 December 1982 and 16 February 1984. The most of derived differential proton spectra above >1 GV for the 31 GLEs of 1942–1991 have been summarized earlier (Miroshnichenko et al. 1995c). Since then, we extended the list of spectrum up to 35, and all those data have been revised and considerably specified in Sect. 4.7 (see Table 4.5). The estimates of spectral parameters for relativistic solar protons (RSP) observed after 1955 have been obtained with the data of several Catalogues of Solar Proton Events (Dodson et al. 1975; Akinyan et al. 1983; Bazilevskaya et al. 1986, 1990a; Sladkova et al. 1998). Also, some other sources of the GLE-RSP data have been taken into account (*e.g.*, Lockwood et al. (1974); Duggal 1979; Smart and Shea 1990b, 1991; Belov and Eroshenko 1996; direct *Meteor* satellite data by Nazarova et al. 1992, *etc.*). Table 9.2 comprises basic spectral data for 31 GLEs observed in 1942–1991, together with unique information about the spectrum evolution during certain events (for example, 23 February 1956, 28 January 1967, 7 May 1978, and some others).

In spite of essential and inevitable methodical errors and uncertainties, the obtained absolute spectra of the R >1 GV solar protons near the Earth present real quantitative information useful for further analysis and application. The data of Table 9.2 have been used, in particular, to compile Table 4.5 with estimates of absolute integral proton fluxes, F_m(>1 GV), for the moments of maximum increases. Also, those data were used in constructing the GLE size distributions in Chaps. 4 and 5 (see Figs. 4.8 and 5.11, respectively). To better understand the properties of the observed SCR spectrum, it seems necessary to perform a detailed comparison of computation results $m(R)$ and $S(R)$, to find out the cause and magnitude of discrepancy between them within different rigidity intervals, to determine absolute spectra for other GLEs from Table 2.1, and then to analyze the obtained set of data taking into account positions of relevant flares, conditions of SCR propagation in the interplanetary space, cones of acceptance in the geomagnetic field, *etc.*

As it was shown by Miroshnichenko (1971), the normalized values of $S(R) = m(R)/N_e$, which were used, *e.g.*, by Lockwood et al. (1974), are overestimated as compared with integral ones (Bednazhevsky and Miroshnichenko 1982) by about 2 orders of magnitude. This may affect, in particular, estimates of the supposed upper limit, R_m, of SCR spectrum when using the procedure suggested by Heristchi et al. (1976): a) by assessing at which cutoff rigidity R_c (*i.e.*, at which cosmic ray station) the SCR flux is compatible with the GCR background; b) basing on the ratio of the increase amplitudes A_1/A_2 at stations with similar cones of acceptance (iteration method for the isotropic stage); c) basing on the latitude effect of SCR with using the $m(R)$ values (an isotropic stage). In the latter case, it is of significance to know absolute $m(R)$ values, but not normalized ones. Analysis of the January 24, 1971 GLE demonstrates an accuracy of the R_m estimates (Heristchi et al. 1976):

Table 9.2 Differential spectra of relativistic solar protons observed in 1942–1991

GLE number	GLE date	Time, UT	Rigidity ΔR, GV	D_0, $(\text{cm s GV})^{-1}$	γ
GLE01	28 Feb 1942	13:00	>1.0	8.33×10^2	4–5
GLE02	07 Mar 1942	06:00	>1.0	1.04×10^3	4–5
GLE04	19 Nov 1949	12:00	>1.0	2.78×10^3	4–5
GLE05	23 Feb 1956	05:00	1.5–5.0	1.25×10^4	6.8
		06:00	1.5–10	6.40×10^3	7.0
		08:00	1.5–4.5	2.30×10^3	7.0
		10:00	1.5–4.0	1.00×10^3	7.0
		16:00	1.5–4.0	3.15×10^2	7.0
		19:30	0.4–1.0	7.08×10^1	2.0
		20:00	1.5–4.0	8.00×10^1	7.0
GLE08	04 May 1960	10:50	2.0–5.0	6.30×10^1	3.4
		11:20	2.0–5.0	1.85×10^1	3.8
GLE10	12 Nov 1960	20:00	0.98–3.45	1.70×10^2	5.2
GLE11	15 Nov 1960	04:00	1.5–4.0	1.55×10^2	5.0
		08:00	1.5–4.0	6.00×10^1	5.0
GLE16	28 Jan 1967	12:00	0.44–10	1.25×10^1	4.5
		12:00	1.02–4.35	3.10×10^1	4.9
		12:00	1.5–5.0	5.00×10^0	5.0
		12:00	1.6–5.3	2.01×10^1	5.0
		12:00	0.76–1.35	5.00×10^0	4.9
		14:00	1.5–5.0	4.10×10^0	4.9
		14:45	0.76–1.35	1.30×10^0	4.9
		16:00	1.5–5.0	3.20×10^0	4.9
GLE19	18 Nov 1968	11:00	1.6–5.0	1.57×10^1	5.0
		12:00	1.02–4.35	6.70×10^0	4.6
		12:00	0.5–10	2.50×10^1	5.0
GLE20	25 Feb 1969	10:00	1.02–4.35	9.50×10^0	4.1
		10:00	1.6–5.0	1.10×10^1	4.5
		10:00	0.5–10	1.10×10^1	4.5
GLE21	30 Mar 1969	14:00	1.0–3.0	2.45×10^0	4.0
		14:00	0.44–10	8.10×10^{-1}	4.0
GLE22	24 Jan 1970	24:00	1.0–5.0	1.66×10^1	5.0
		24:00	0.4–10	1.25×10^1	5.0
	25 Jan 1970	02:00	1.02–4.35	9.50×10^0	4.1
		08:30	0.435–1.0	8.42×10^{-2}	6.0
GLE23	01 Sep 1971	22:00	1.0–5.0	1.57×10^1	5.5
		22:00	0.5–10	1.10×10^1	4.5
		23:00	1.02–4.35	3.88×10^1	5.3
GLE24	04 Aug 1972	16:00	1.0–1.6	2.04×10^1	8.0
	05 Aug 1972	03:00	0.52–0.875	2.30×10^{-2}	9.5
GLE25	07 Aug 1972	17:00	1.02–3.0	7.00×10^0	4.0
		17:00	1.02–3.0	7.00×10^{-1}	5.0

(continued)

Table 9.2 (continued)

GLE number	GLE date	Time, UT	Rigidity ΔR, GV	D_0, (cm s GV)$^{-1}$	γ
GLE28	19 Sep 1977	14:00	>1.0	2.40×10^{-1}	4.0
GLE29	24 Sep 1977	10:00–12:00	1.0–6.3	4.00×10^{0}	3.4
GLE30	22 Nov 1977	12:00	2.29–4.0	5.0×10^{2}	5.5
GLE31	07 May 1978	03:45	2.15–6.20	4.10×10^{2}	4.1
		04:00	2.15–6.20	1.10×10^{2}	4.0
		04:15	2.15–6.20	5.34×10^{1}	4.2
		04:25	2.15–6.20	9.81×10^{1}	5.1
GLE32	23 Sep 1978	11:15	>1.0	1.88×10^{1}	4.8
		15:30	0.435–1.0	3.00×10^{-1}	4.5
GLE33	21 Aug 1979	07:00	>1.0	5.73×10^{0}	4.6
GLE34	10 Apr 1981	17:30	>1.0	1.72×10^{0}	4.5
GLE35	10 May 1981	10:00	>1.0	2.00×10^{0}	4.3
GLE36	12 Oct 1981	10:00	>1.0	1.37×10^{1}	4.4
		13:00	0.4–1.22	1.36×10^{-1}	4.6
GLE37	26 Nov 1982	04:55	>1.0	5.67×10^{0}	4.1
GLE38	07 Dec 1982	00:45	>1.0	8.62×10^{1}	5.5
GLE39	16 Feb 1984	09:15	>1.0	7.25×10^{0}	4.2
		09:15	>1.0	1.83×10^{2}	3.5
GLE42	29 Sep 1989	12:17	>1.0	9.32×10^{0}	2.9
		13:27	>1.0	1.52×10^{1}	3.0
GLE51	11 Jun 1991	01:56	1.0–4.0	1.55×10^{1}	5.5
GLE52	15 Jun 1991	08:10	1.0–4.0	6.20×10^{1}	6.0

$R_m = (3.5\text{–}4.0) \pm 0.6$ GV, $\gamma = (3.7\text{–}3.9) \pm 0.4$ according to the data for an isotropic stage; $R_m = (4\text{–}4.6) \pm 0.6$ GV, $\gamma = (4\text{–}4.6) \pm 0.4$ according to the data for an anisotropic stage. The existence and nature of the maximum rigidity, R_m, is of principal interest for physics of solar flares. The analysis of this problem started by Heristchi et al. (1976), has been continued by Bazilevskaya (1984), Bazilevskaya and Makhmutov (1988), Miroshnichenko (1990) and other workers (see Chaps. 3 and 4).

The worldwide NM network provides continuous ground-based recording of the hadronic component in atmospheric secondary radiation which is related to primary cosmic rays of galactic and solar origin. Long ago (Simpson 1948) it was discovered that the latitude variation of the secondary hadronic component was considerably larger than that of the muon component. It suggested that a neutron monitor is more sensitive to lower energies in the primary spectrum. Physical, experimental and methodological problems related to NM records, on different reasons, are still in the focus of attention of many researchers (*e.g.*, Struminsky and Belov 1997; Clem and Dorman 2000; Bhattacharya et al. 2013). Clem and Dorman (2000) have reviewed and discussed the different methods of determining the NM response function, including early results. These authors also provide new calculation results, including angle dependent yield functions for different neutron monitor types, in particular, for IGY and NM64 configurations using the standard $^1BF_3$ detectors and the new ^{3}He detectors to be used in the Spaceship Earth Project (Bieber et al. 1995, see Chap. 12). Statistical analysis was recently performed by Bhattacharya et al. (2013) to investigate the degree of response of different monitors towards cosmic ray counts. No significant difference was observed in statistical results if cosmic ray counts are normalized with respect to their mean counts in respective solar cycles.

9.5 Spectrum Compatibility in Different Energy Ranges

Recently, based mainly on the copious data on the solar proton events of the 22nd solar cycle, some new attempts have been undertaken to obtain compatible proton spectra in different energy ranges (*e.g.*, Humble et al. 1991a, b; Kohno 1991; Bieber and Evenson 1991; Belov and Eroshenko 1996; Miroshnichenko et al. 1999, 2000). In particular, when analyzing the event of September 29, 1989, Kohno (1991) plotted the relativistic proton intensity observed at the Tokyo NM, together with the GMS-3 proton spectrum in the range of 4–68 MeV. It was found that all these data can be fitted by a single power-law with an exponent about 2.6. Within a factor of 2 the spectrum of Kohno (1991) is consistent with the model spectra needed to fit the Australian NM data (Humble et al. 1991a, b), although the location of the NMs mentioned on the globe (*i.e.*, geomagnetic conditions) was quite different from each other.

In more extensive study, based on the observations of the 4 GLEs (29 September and 19 October 1989; 24 May 1990; 15 June 1991), Belov and Eroshenko (1996) have tried to obtain proton spectra near the Earth within a wide energy range (from 10 MeV to 10 GeV) directly from experimental data by applying an original empirical method as described briefly below. They used 1-h data from the spacecraft *IMP*-8 (Armstrong 1993) and 5-min data from the *GOES*-6 and 7 (Wilkinson 1992) as well as 5-min data from the world network of neutron monitors. This provided about 50 independent time sets of the data within the different energy ranges. These data were combined into several groups according to their energies. For example, the *IMP*-8 proton channels of >10, >30 and >50 MeV were unified into one group, and other groups were composed of the low, medium and high energy channels from the *GOES*-6 or 7. Neutron monitor data from the stations with cutoff rigidities R_c >1.4 GV were divided into separate groups according to the cutoff rigidities. Neutron monitor data for R_c < 1.4 GV have not been used because the response functions for low energies are doubtful.

The parameters of differential power-law energy spectra were estimated for each data group and 5-min intervals by the least squares method. The procedure can be easily done for spacecraft data, but it is more difficult task for neutron monitor data because the energy dependence of the neutron monitor response functions is rather complex. For a given group of the time data sets and a given time interval Belov and Eroshenko (1996) proposed to use so-called "optimum energy" E_0 that corresponds to a minimum of the root mean square error for the calculated proton flux. To define this optimum energy for spacecraft data one can write:

$$\ln E_0 = \Sigma c_i l_i / \Sigma c_i^2 \tag{9.20}$$

where i is the channel number with the energy $E_l < E_u$ (the lower and upper limits of energy for the channel); c_i and l_i are the usual and logarithmic coupling coefficients. In the case of a power-law energy spectrum with the index γ the parameters c_i and l_i are:

$$c_i = \left[1/(\gamma + 1)\, E_l^{\gamma+1} - E_u^{\gamma+1}\right] \tag{9.21}$$

$$l_i = \int_{E_l}^{E_u} E^{\gamma}\, lnE\, dE \tag{9.22}$$

It is more convenient to use the optimal rigidity R_0 instead of E0 for a set of neutron monitor data, or other ground level detectors data. So, if to replace E_0 with R_0 in equation (9.20), we get:

$$c_i = \int_{R_c}^{\infty} W_i(R_c, h_i, R) R^{\gamma} dR \tag{9.23}$$

$$l_i = \int_{R_c}^{\infty} W_i(R_c, h_i, R) R^{\gamma}\, lnR\, dR \tag{9.24}$$

where $W_i(R_c, h_i, R)$ is the neutron monitor response function at the atmospheric depth h and the geomagnetic cutoff rigidity R_c.

A slope of the energy spectrum was estimated near this optimum energy E_0 for all existing data groups, which might be partly overlapped. This procedure can be done for any period. In particular, one can choose time periods when a maximum intensity is observed in each energy range to get the peak spectrum. The next step was to combine results into one general spectrum. This was done (Belov and Eroshenko 1996) using the regression analysis for every energy range. Only the data from energy ranges close to the optimum energy were taken, so that the changes of ln$E$ should not be more than 1. Figure 9.5 represents the peak proton spectra obtained by this technique for the selected four events. As we mentioned, all these events are GLEs. This fact already provides a limitation to the possible spectral variety, but, nevertheless, all these spectra differ from each other by their magnitudes and slopes. The distinctions are especially great for the higher energies, but there is one common feature to their behaviour: as the energy increases, all spectra become softer.

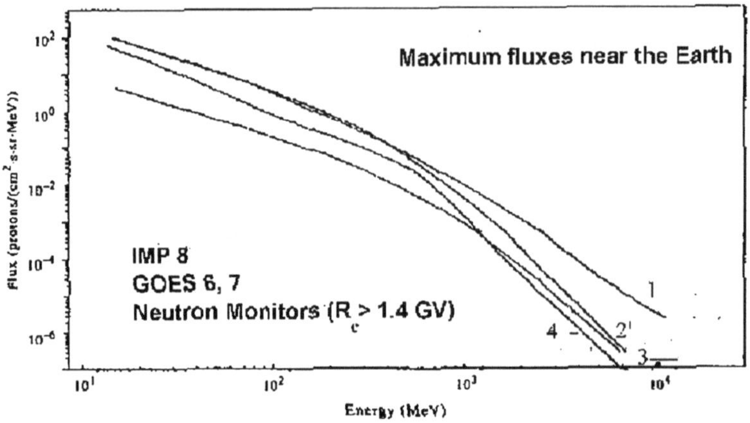

Fig. 9.5 The differential proton spectra for several proton events obtained within the wide energy range by spacecraft and ground-level observation data (Belov and Eroshenko 1996): *1* September 29, 1989; *2* October 19, 1989; *3* May 24, 1990; *4* June 15, 1991

However, the softening may be different. In some case it is gradual (as in the GLEs of 29 September 1989 and 24 May 1990), and sometimes there is a sharp break near the energy of about 1 GeV (as in the GLE of 15 June 1991). Of course, this method allows to obtain spectra for any other given time besides the peak spectrum. Evidently, it is an efficient way to study the acceleration processes in flares, to use the peak spectra found from the observations to model the ejection spectra (see Chap. 4).

9.6 Efficiency of Different Techniques in Ground Data Fitting

From previous considerations, it appears that the Earth, with its world-wide NM and MT network, serves as an unique omnidirectional detector to study the distribution function of relativistic solar protons outside the magnetosphere. All techniques available to date (*e.g.*, Kuzmin 1968; Krymsky et al. 1981; Shea and Smart 1982; Dvornikov et al. 1983; Baisultanova et al. 1987; Smart et al. 1991) make simplifying assumptions about the isotropic character of the distribution function, or about the total separation of angular and energy variables, or about the separation of these variables for each of the spherical harmonics of the distribution function. In addition, the rigidity dependence of the anisotropy is considered to be independent of time and is specified on the basis of some simplified theoretical models for the interplanetary propagation of particles.

In this context, we should mentioned the results of Cramp et al. (1993a) and Lovell et al. (1998) who have performed an extended analysis of the spectral and angular characteristics of relativistic protons during the GLE of 29 September 1989, relying upon the data from 42 NM stations from around the world, over a wide range of geomagnetic cutoff rigidities, from essentially zero at polar stations to R_c ~14.09 GV at Darwin (Australia). To fit the data, Cramp et al. (1993a) used the standard technique (Shea and Smart 1982; Smart et al. 1991), which is an attempt to reproduce the intensity response observed by all the NMs, through the NM yield functions and the asymptotic directions of approach (or asymptotic cones of acceptance). In the past, these "viewing cones" (see Fig. 1.7) were calculated only for undisturbed geomagnetic conditions. At present, due to recent improvements in the understanding of the structure of the disturbed geomagnetic field, one can take into account these deformations (*e.g.*, Flückiger and Kobel 1990). As a result, the accuracy of the viewing cone calculations has been dramatically improved, and the fitting quality of the model has increased at most of the 42 stations (Cramp et al. 1993a). Nevertheless, they noted a difficulty in fitting the model to the NM data at low rigidity, because the yield functions at the lowest rigidities are not as reliable as at higher rigidities. Similar difficulties have been encountered in previous analyses.

In contrast to this traditional approach, Dvornikov and Sdobnov (1995a, b, 1997) used the method of *spectrographic global survey* (Dvornikov and Sdobnov 1993) when analyzing the event of September 29, 1989, which is free from many of usual assumptions mentioned above. Instead, the anisotropy in the distribution of incoming particles is attributed to a dependence of their intensity on the pitch-angle in the IMF and to a density gradient over the Larmor radius of the particles. Dvornikov and Sdobnov (1985a, b, 1997) have analyzed data from 31 NMs, the Sayan Spectrographic Complex (Dvornikov et al. 1983), and the Irkutsk muon telescope, dealing with the angular distribution of particles of different energies, their intensity-time profiles and the time delay with respect to the onset of the soft X-ray burst. When developing their method, they used more than ten representations of the distribution function that reflect different models for the propagation of particles in the heliosphere. The best fit to ground-based and satellite data was obtained for a distribution function which combines two anisotropic components with an isotropic one, the rigidity spectra of all of them being approximated by the power-law series of inverse degree of rigidity (Dvornikov and Sdobnov 1993).

The relative efficiency of two techniques is demonstrated in Fig. 9.6 by comparing the results of Dvornikov and Sdobnov (1995a, b, 1997) with those of Cramp et al. (1993a). The diagrams demonstrate a considerable difference between them in fitting to the data, especially in the range of large amplitudes. Dvornikov and Sdobnov (1997) state that the reason for the difference is that their rigidity spectrum is significantly harder early in the event (see Fig. 7.12) than that derived by Cramp et al. (1993a). In later phases of the event the difference between the expected and observed amplitudes found by Cramp et al. (1993a) is also significantly higher than that found with the technique of Dvornikov and Sdobnov (1997).

One distinctive feature of their model should be pointed out; namely, for later times in the event the high-energy particle distribution in the sky includes regions with a decreased (with respect to background) intensity. To explain the results

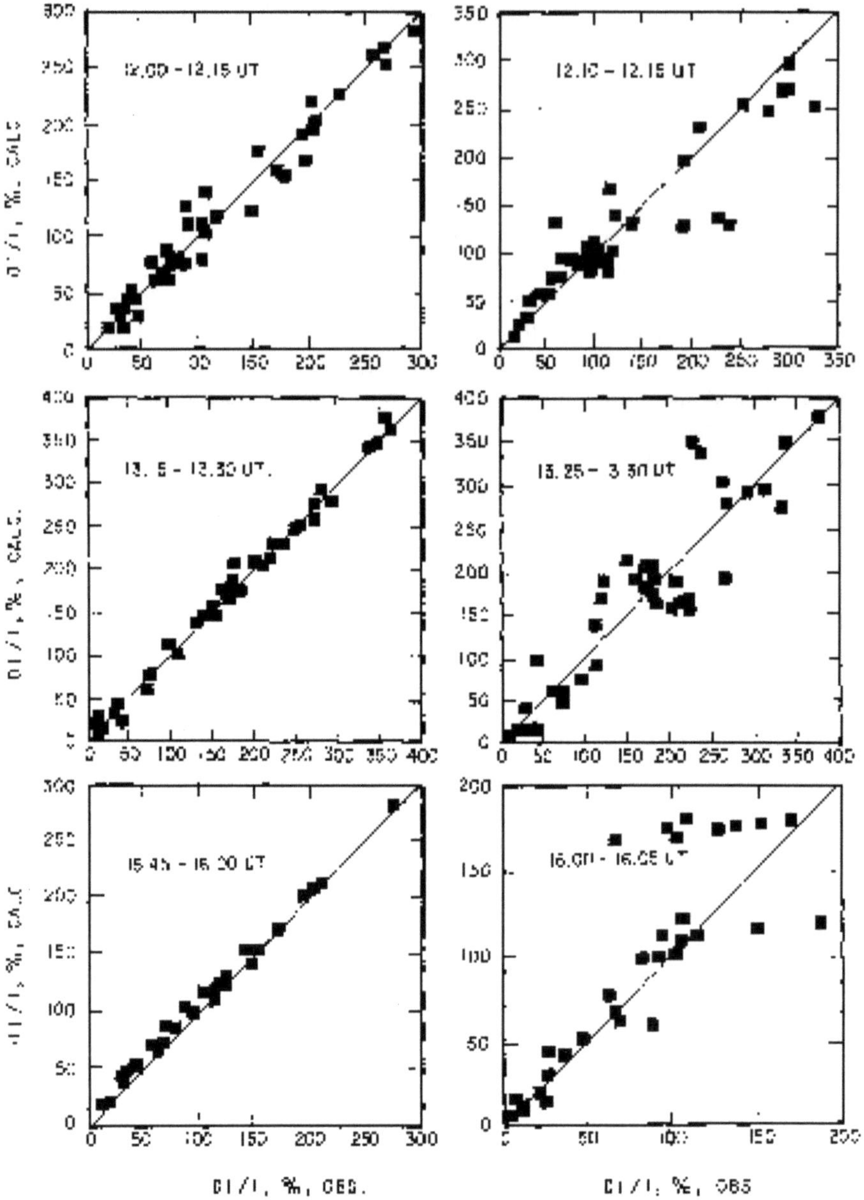

Fig. 9.6 Results of comparison of calculated variation amplitudes with observed amplitudes in the analysis of the GLE of 29 September 1989, when using (*left*) a technique developed by Dvornikov and Sdobnov (1993, 1995c, 1997) and (*right*) a technique reported by Shea and Smart (1982) and applied by Cramp et al. (1993a). The diagrams demonstrate a considerable difference between two techniques in fitting to the data, especially in the range of large amplitudes

obtained, it is speculated (Dvornikov and Sdobnov 1997) that a particle beam, accelerated in the flare, propagated along the heliospheric current sheet (HCS). As a result, polarization electric fields are produced in interplanetary space, through which an energy exchange occurred between SCR and background particles of solar wind plasma and galactic cosmic rays. In the context of this scenario, they propose to treat the process of SCR propagation in the heliosphere not in the terms of the propagation of particles in the external electromagnetic field, but with due regard for the self-consistency of fields with the particle distribution function.

Although this non-traditional method seems to be fruitful, the lack of quantitative statistical analysis, such as a χ^2-estimate for each method, reduces the value of the comparison published by Dvornikov and Sdobnov (1997). It is possible that a better fit was reached simply because they used more unknown parameters (15 in all). Therefore, we can not conclude that the model of Dvornikov and Sdobnov (1995a, b, 1997) is better than any of the other models that have been used and reviewed. Also, for the present, any quantitative estimates of the concept, within the framework of proposed general scenario, are absent.

Nevertheless, the approach developed by Dvornikov and Sdobnov (1995a, b, 1997) has yielded an improved fit to the data. Again, it called the researchers' attention to the complexity of the event of September 29, 1989, to the significant scatter in the results, and to a broad spectrum of interpretations of the physics behind the event (for a review see Miroshnichenko et al. 2000). This was clearly seen above in the results related to the observed rigidity spectrum at the Earth, the arrival directions of SCR at the Earth, and the scattering mean free path in the IMF. The variety and uncertainties in results and interpretations can be attributed to the different techniques of surface data analysis, the uncertainties in the asymptotic directions of acceptance, the pitch angle distribution functions, the primary rigidity spectrum, and the yield functions.

The same technique was applied to the GLE data of 22 October 1989 (Dvornikov and Sdobnov 1998). Using data from 34 neutron monitors, they shown that intensity-time profiles of the 2–4 GV particles differ greatly from those of higher-energy particles. After the high-energy particle intensity attains a maximum, there is an abrupt decrease in intensity below the background level, followed by the phase of slow recovery to background values. The angular distribution of higher-energy particles across the celestial sphere shows, along with an increased intensity, also regions with decreased relative background intensity. Results of comparison of observed increase amplitudes with that calculated by different techniques, similar to previous case (see Fig. 9.5), also demonstrates a significant difference between the results of Dvornikov and Sdobnov (1998) and that of Cramp et al. (1993b, 1997). As noted by the authors (Dvornikov and Sdobnov 1998), these results are contradictory both to existing models for the propagation of particles in external electromagnetic fields of the heliosphere and to results of analysis of the NM network data obtained by using traditional propagation models and analytic technique (*e.g.*, Shea and Smart 1982).

Specifically, by analyzing the event of 22 October 1989, Cramp et al. (1993b, 1997) concluded that in the initial flare stage an increase in intensity occurs in a very narrow angular range, while in the other parts of the celestial sphere the intensity of all-energy particles remained at a background level. A similar conclusion was drawn by Stoker et al. (1995) by analyzing the GLE of October 19, 1989. As emphasized by Dvornikov and Sdobnov (1998), it is impossible to obtain a different conclusion in terms of the techniques used, because the assumptions made with reference to the form of the distribution function dictate that the solution can be only such that the variation amplitude of the directed differential intensity of primary cosmic rays in those regions of the celestial sphere from which the integral intensity close to background values is recorded, must be about zero. Thus, the reported results of analysis may lead one to put in doubt the validity of conventional approaches to the problem of SCR propagation in the heliosphere. These authors believe that to explain the entire set of phenomena in a unified context, it is necessary to consider the propagation process of accelerated particles in self-consistent, rather than external, electromagnetic field (Dvornikov and Sdobnov 1997).

It is timely to mention that this methodical debate seems to be continued. In development of their preliminary study (Cramp et al. 1993a, b), the same research group (Lovell et al. 1998) undertook more sophisticated analysis of the GLE of 29 September 1989 (Fig. 9.7) based on the data from 42 neutron monitors (NM) and six surface muon telescopes (MT).

Their technique for modeling NM responses to relativistic solar protons has been described by Cramp et al. (1997). To determine the asymptotic viewing directions of ground-based instruments (Flückiger and Kobel 1990), this group employed the geomagnetic field model of Tsyganenko (1989). Then, a least-squares technique was used to determine the apparent arrival directions of incoming solar protons; their pitch-angle distributions (PADs) and rigidity spectrum which best reproduce the observed responses. The model should minimize the difference between observed and calculated responses. In other words, it should produce null responses for those stations which did not record an intensity increase, as well as accurately present the observed increases. The method has been further refined to allow simultaneous use of NM and MT data.

It should be recognized that the spectra derived by this technique in the paper by Lovell et al. (1998) at 12:15, 13:25, and 16:00 UT of 29 September 1989 are considerably different from that of Cramp et al. (1993a, b), especially at rigidities below 1 GV. Thus, a difference between the 1 GV proton intensities at 12:15 and 16:00 UT obtained by Lovell et al. (1998) is about one order of magnitude less than that from Cramp et al. (1993a, b). Furthermore, the spectra by Lovell et al. (1998) at high rigidity have been extrapolated up to $\geq$25 GV. Notably, however, that in both papers the spectrum at 12:15 UT was found to be significantly harder than two other ones. The authors also argued in favour of that all those spectra are consistent with satellite data. Figure 9.7 shows proton and alpha particle data from *IMP* 8, *GOES* 6 and *GOES* 7 satellites (hourly averages), as well as a spectrum derived from hourly surface data (*shaded area*). The error range of the NM spectrum was estimated from contour plots of the variance of calculated and observed increases against the two parameters of the modified power-law spectrum.

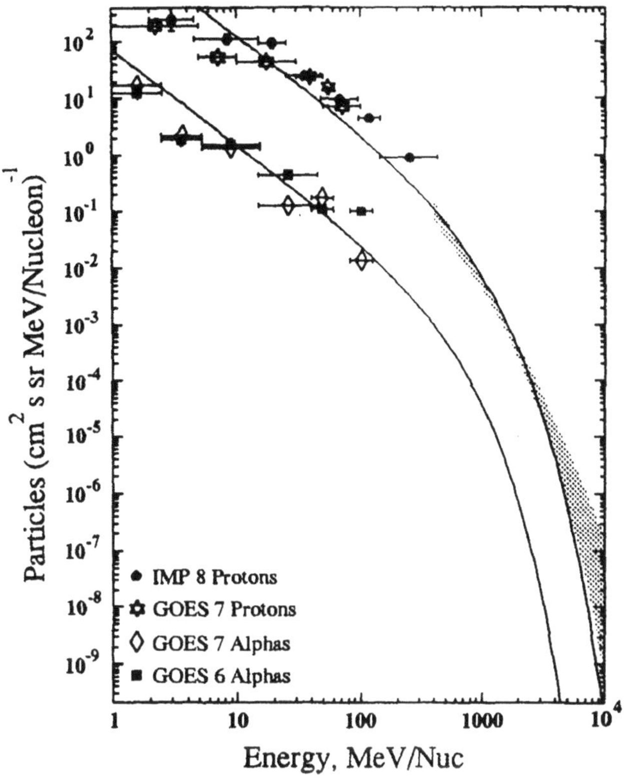

Fig. 9.7 Hourly proton and alpha particle data from *IMP* 8, *GOES* 6, and *GOES* 7 (as indicated) and neutron monitor spectrum (*shaded area*) for the GLE of 29 September 1989 from the paper by Cramp et al. (1998). Fitted curves correspond to spectral form resulted from shock acceleration model by Ellison and Ramaty (1985)

From Fig. 9.7 Lovell et al. (1998) concluded that the spectrum between a few MeV and a few tens of GeV is consistent with the form obtained from shock acceleration model by Ellison and Ramaty (1985) which is valid both for protons and alpha particles. The conclusions by Lovell et al. (1998) are in agreement with the findings of Cliver et al. (1993b) on the origin of gamma-ray emission (Fig. 7.22), but in variance with the results of Kahler et al. (1997) on temporal variation of shock-accelerated proton/alpha ratios at $E \geq 100$ MeV/nucleon in the same event (Fig. 2.9).

In the model by Cliver et al. (1993b), particle acceleration is assumed to proceed at the bow shock of a coronal mass ejection (CME-driven shock) that provides access for particles to magnetic field lines at some distance from the flare site. Meanwhile, this idea has been challenged recently by Klein et al. (1999, 2000). Comparing the time histories of coronal electron acceleration (mainly by the radio emission observations) and interplanetary energetic protons (>20 MeV), they found certain evidence of their co-evolution during the event. This means that the protons seen at 1 AU are accelerated behind the CME and its presumed bow shock. Those discrepancies between the results of different research groups (see also Sects. 7.4, 7.5, and 7.6) call for additional efforts in the modeling the complex proton events of the 29 September 1989 type, in particular, for a new mutual comparison of the results and efficiencies of the two techniques (Cramp et al. 1997; Lovell et al. 1998; Dvornikov and Sdobnov 1997, 1998).

9.7 New Modeling of Spectrum Dynamics

Rather promising developments in the methods of GLE analysis have come quite recently from a new computational modeling of incoming flux and anisotropy dynamics of solar cosmic rays during the GLEs of 7–8 December 1982 (Pchelkin et al. 2000) and 29 September 1989 (Vashenyuk and Pchelkin 1998; Vashenyuk et al. 2001). In two latter papers, the neutron monitor (NM) records obtained at 42 cosmic ray stations of world-wide network were compared with the NM calculated responses at the given values of the SCR flux. As a result, the spectra, pitch-angle distributions (PADs) and anisotropy of relativistic solar protons (RSP) in the interplanetary space have been determined at different stages of this outstanding event. Modeling was carried out with the purpose to derive the RSP intensity-time profiles in different energy ranges and to trace the dynamics of their spectra and pitch-angle distributions more exactly than in previous studies.

As shown in previous Chapters, large solar event of 29 September 1989 has been extensively studied during past 10 years, and at present more than 200 appropriate publications are available (as a review see Miroshnichenko et al. 2000). In particular, it was shown (Vashenyuk et al. 1997; Vashenyuk and Miroshnichenko 1998) that a number of the peculiarities observed in the event can be explained by two-fold ejection of relativistic protons from the Sun. Just at the early stage of this event very hard particles have been ejected with a strong anisotropy outward the Sun. At the second ejection that occurred about 1 h later the spectrum of RSP has become softer, and a bi-directional anisotropy was found (Vashenyuk et al. 1997; Vashenyuk and Miroshnichenko 1998; Vashenyuk and Pchelkin 1998) to exist in this latter case. By the methods of a computational modeling the event of 29 September 1989 has been analyzed by several research groups (Dvornikov and Sdobnov 1997; Vashenyuk and Pchelkin 1998; Lovell et al. 1998). Those researchers have estimated the parameters of primary flux of solar protons for three (Vashenyuk and Pchelkin 1998; Lovell et al. 1998) and four (Dvornikov and Sdobnov 1997) moments of time. In the paper by Vashenyuk et al. (2001) the modeling has been accomplished for 17 moments of time, that allowed to trace the flux dynamics of RSP in more detail.

Modeling procedure of the event of 29 September 1989 by the Earth's surface data included determination of asymptotic directions of approach for the RSP based on the calculation of trajectories for the particles of opposite sign launching from the Earth's surface in the model of geomagnetic field by Tsyganenko (1989). By the method of Runge–Kutta-Feldberg of 4–5th orders an equation of particle motion in geomagnetic field has been taken in the following form:

$$\gamma m\left(d^2 r/dt^2\right) = Ze(dr/dt) \times B \tag{9.25}$$

where γ, *r*, *Ze*, and *B* are the Lorenz-factor, radius-vector, charge, and vector of magnetic induction, respectively. A trajectory was traced up to the boundary of the magnetosphere. The direction opposite to the particle velocity vector at the launching point corresponds to the asymptotic direction of particle approach at given rigidity. Response function of a neutron monitor, $\Delta N/N$ (R_c), to anisotropic flux of solar protons at the point with a geomagnetic cut-off rigidity, $R_{c,}$ is given by the relation (Shea and Smart 1982):

$$(\Delta N/N\,)_j = K\int_{R_{Cj}}^{\infty} J_{||}(R)\, F\left(\left(\theta_j(R)\right)\, S(R)\right)\, dR \tag{9.26}$$

where $J_{||}(R) = J_O R^{-\gamma}$ is a rigidity spectrum of RSP flux in the direction of anisotropy axis, *S(R)* is a normalized specific yield function for the nucleon flux incident at the boundary of the atmosphere, as a function of rigidity *R*, and *F(θ)* is a pitch-angle distribution (PAD) of primary protons in the IMF admitted in the form of the function $F(u(R)) \sim exp(-u^2/C)$ (Shea and Smart 1982), where *u(R)* defines an angle between the direction of maximum intensity of particles and asymptotic direction of approach at a given rigidity.

Unknown parameters of solar proton flux are six quantities: normalization constant of the spectrum J_O; direction of the anisotropy axis (a pair of coordinates, *Φ* and *Λ*, in the GSE system); two constants in the relations for the rigidity spectrum form, γ *u*Δγ, and a constant of pitch-angle distribution $C = 2\sigma^2$. These parameters are determined with the optimization methods (Dennis and Schnabel 1988) by resolving a system of conditional equations. Then, we solved, in essence, a nonlinear least-square problem that was reduced to search a minimum of the function *SN* at the optimal set of parameters forming the relation (9.26):

$$SN = \sum\left((\Delta N/N)_{calc} - (\Delta N/N)_{jobserv}\right)^2 \rightarrow min \tag{9.27}$$

Inscriptions in the indexes in the relation (9.27) correspond to calculated and observed amplitudes of GLE at the *j*-th cosmic ray station. A quality of the results of optimization was estimated by a quantity of residual error determined with the formula (Dennis and Schnabel 1988):

$$\varepsilon = SN/\sum(\Delta N/N)^2{}_{observ} \tag{9.28}$$

Modeling procedure was applied to the parameters of September 29, 1989 event under two working hypotheses of (1) one-directional and (2) bi-directional

Table 9.3 Modeling parameters for relativistic solar protons on 29 September 1989

UT	1217	1237	1307	1327	1337	1347	1407	1417	1432	1447	1507
J_0	2.2	5.8	31.5	62.2	74.9	102.5	124.1	129.6	133.2	122.2	132.0
γ	1.08	1.64	2.97	3.50	3.70	4.06	4.31	4.36	4.44	4.44	4.53
$\Delta\gamma$	0.14	0.16	0.13	0.12	0.11	0.08	0.09	0.11	0.12	0.12	0.14
C	3.64	5.09	7.55	8.75	9.47	9.47	11.02	11.43	11.69	11.78	10.38
Lat.°	68	75	84	81	75	72	84	87	84	86	97
Long.°	258	255	249	261	258	257	258	260	265	267	273
ε, %	2.97	2.94	2.64	1.68	1.61	1.19	0.78	0.88	1.00	0.91	1.23
Δ_{max} %	55	93	91	74	79	67	42	43	41	32	39

anisotropy. In the latter case, a quality of optimization turned out to be better for the moment of second intensity peak (Vashenyuk and Pchelkin 1998). The combinations of mentioned parameters of RSP have been obtained for the 17 moments of time between 12:00 and 16:00 UT. Table 9.3 contains a summary of the determined parameters at the 11 selected moments of time.

In Fig. 9.8 (left panel) the energy spectra of RSP are shown being obtained at several subsequent moments of time. Thin lines mark the spectra observed after the first ejection from the Sun; solid line at 13:47 UT corresponds to the spectrum related to the second peak. For the four moments of time after the first ejection also the rigidity spectra of RSP have been constructed. Those latter spectra display an exponential dependence on rigidity, this feature being characteristic for particle acceleration by DC electric fields (*e.g.*, Miroshnichenko 1987). At the same time, the spectrum slope is seen to increase in time between 12:07 and 12:52 UT.

Rigidity spectra derived at the second peak for the four moments of time are shown in Fig. 9.8 (right panel). It should be noted that the spectrum of second ejection did not undergo considerable changes within the time interval between 13:17 and 14:07 UT. It also was shown that the flux of relativistic protons at the first peak was one-directional, while at the second peak (13:25 UT) anisotropy has displayed its obvious bi-directional behaviour. These findings, together with the difference in spectrum features in Fig. 9.8 (left and right panels) again clearly demonstrates two components of RSP, prompt and delayed ones (see Sect. 7.4), that, supposedly, have been accelerated on 29 September 1989 at coronal heights. Subsequently, they have been released from the Sun with a shift in time.

The questions of computational modeling of the RSP spectra in the two proposed sources that have given rise to the event of 29 September 1989 are discussed in a separate paper (Vashenyuk et al. 2000). In the same paper, the estimates of source parameters but only for the prompt component are given. As a whole, however, after a decade of intensive study no generally accepted acceleration model(s) was worked out (see Sect. 7.6). Apparently, detailed modeling efforts will be still required to construct a comprehensive picture of this historic event.

In conclusion of this discussion, it is worth noting that, in spite of such refined techniques applied to the analyses of the SCR dynamics in the Earth's

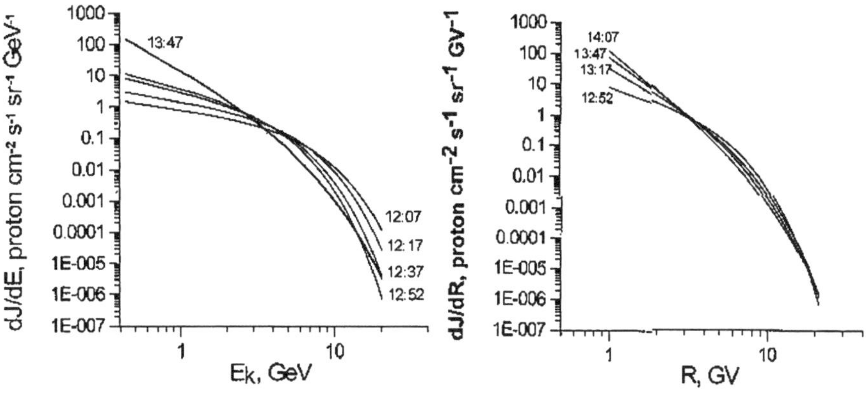

Fig. 9.8 *Left panel*: Inferred energy spectra of relativistic solar protons outside the Earth's magnetosphere (Vashenyuk et al. 2001) at different moments of time of 29 September 1989 for the first and second intensity peaks (*thin* and *thick lines*, respectively). *Right panel*: Dynamics of rigidity spectra during the second intensity increase (at 13:17, 13:47, and 14:07 UT) in the GLE of 29 September 1989 (Vashenyuk et al. 2001). Note the rather small changes of spectra during this period

environments, there are still challenging problems in this field. As an example, we mention latest results concerning the GLE of 2 May 1998 (Danilova et al. 1999). In this case, the two component structure of the RSP flux was recorded, too, but unexpected differences in the intensity-time profiles at several NM stations occurred, that compelled the authors to involve into their analysis a set of possible magnetospheric effects on the incoming relativistic particles (see Sect. 10.1).

9.8 Modern Basic Procedure

The NM network distributed over the globe can be considered as an integrated omnidirectional spectrometer for measuring relativistic SCR flux characteristics outside the Earth's magnetosphere. Modeling the NM response to an anisotropic SCR flux and solving the inverse problem, we can obtain the characteristics of relativistic solar protons outside the Earth's magnetosphere (Shea and Smart 1982; Humble et al. 1991a, b; Cramp et al. 1997; Vashenyuk et al. 2009, 2011). Data from 25 NM stations and a sufficient ground-level increase ($\geq$10 %) should be used in this basic procedure. Therefore, this procedure is as a rule used to study only rather large events. In this case, the main SCR characteristics (the energy spectrum, anisotropy, and pitch angle distribution) are determined by optimization methods when model NM responses are compared with observed responses. The SCR flux parameters determined at successive instants make it possible to trace the flux dynamics. The analysis methods include the determination of the SCR arrival asymptotic directions by calculating the trajectories of these particles in present-day geomagnetic field models. Following the newest narrative of the method (Vashenyuk et al. 2009, 2011), we cite below certain important details that are used to determine the SCR spectrum and anisotropy and describe other GLE properties.

A neutron monitor has a specific directional pattern. When the zenith angle increases, the particle flux weakening due to absorption is accompanied by an increase in the device spatial angle of reception. This results in the appearance of a pattern maximum, which is reached at zenith angles of $\theta = 20°$ and 18° for GCRs and SCRs, respectively. The CR arrival asymptotic directions are determined by integrating the motion equation for a negative test particle (see Fig. 1.9) with the proton mass emitted upward from an altitude of 20 km above a given station (this is the average altitude of production of secondary neutrons contributing to the NM counting). Optimization methods can be used to obtain the SCR parameters based on the NM network data. The expression for the function of the NM response to an anisotropic flux of solar protons has the form (Vashenyuk et al. 2009, 2011)

$$\left(\frac{\Delta N}{N_g}\right)_j = \frac{1}{8}\frac{\sum\limits_{(\phi,\vartheta)=1}^{8}\sum\limits_{R_{\min}}^{R_{\max}} J_{||}(R)\ F(\theta(R))\ S(R)\ A_{(\phi,\vartheta)}(R)\ \Delta R}{N_g} \tag{9.29}$$

where $(\Delta N/Ng)j$ is the relative increase in the NM count rate at station j; Ng is the GCR background before an increase; $J_{||}(R) = J_0 R^{-\wedge}(\gamma^*)$ is the rigidity differential spectrum from the direction of a source with a variable inclination; $\gamma^* = \gamma + \Delta\gamma(R-1)$ (where γ is the power spectrum index at $R = 1$ GV); $\Delta\gamma$ is the γ increment rate at 1 GV; and J_0 $(\mathrm{m}^2\ \mathrm{s\ sr\ GV})^{-1}$ is the normalization constant. Such a representation form makes it possible to describe the spectrum with an arbitrary shape using three parameters: γ, $\Delta\gamma$, and J_0 (Cramp et al. 1997). Other parameters in expression (1) are as follows: $S(R)$ is the specific collection function (Debrunner et al. 1984); $\theta(R)$ is the pitch angle at a given rigidity (more exactly, the angle between the asymptotic direction for a given rigidity and the anisotropy calculation axis, specified by coordinates Φ and Λ, in the solar–ecliptic coordinate system GSE); $A(R) = 1$ and 0 for allowed and forbidden trajectories, respectively; and $F(\theta(R)) \sim \exp(-\theta^2/C)$ is the SCR pitch angle distribution with characteristic parameter C.

The first sum in formula (1) takes into account the contribution of all eight sectors, into which the device spatial angle of reception is divided, to the NM response (Vashenyuk et al. 2009, 2011). The sum also takes into account the contribution of particles obliquely incident on NMs. The second sum results from the summation of all NM response parts with respect to all rigidities varying from 1 to 20 GV at the interval $\Delta R = 0.001$ GV. In expression (1), the NM responses are calculated successively at different values of the solar proton anisotropic flux parameters outside the Earth's magnetosphere (Φ, Λ, J_0, γ, $\Delta\gamma$, C).

The optimization method is subsequently used to determine the values of these parameters at each given instant by comparing the calculated responses of ground detectors with the observed ones. The following system of conditional equations for searching the function minimum is solved for this purpose:

$$F = \sum_j \left[\left(\frac{\Delta N}{N} \right)_j^{\text{вычисл}} - \left(\frac{\Delta N}{N} \right)_j^{\text{набл}} \right]^2 \Rightarrow \min \tag{9.30}$$

where superscripts correspond to the j-th NM responses, calculated using formula (1) and obtained from observations. The observed pitch angle distribution cannot always be described by a function close to the Gaussian function or by a combination of two oppositely directed fluxes, which is observed in the cases of the so-called bidirectional anisotropy. Vashenyuk et al. (2009, 2011) used the expression for the complex pitch angle distribution, which makes it possible to obtain good convergence of the optimization process:

$$F(\theta(R)) \sim \exp\left(-\theta^2/C\right)\left(1 - a\exp\left(-(\theta - \pi/2)^2\right)/b\right. \tag{9.31}$$

Such a function has a peculiarity when the pitch angle is close to $\pi/2$ and can theoretically take into account the pitch angle distribution peculiarities predicted by the theory of particle propagation in the IMF (Toptygin 1985; Bazilevskaya and Golynskaya 1989). According to its properties, expression (9.31) is close to the function that was used to describe complex pitch angle distribution cases (Cramp et al. 1997). When function (3) is used, two more parameters (a and b) are added to the six SCR flux parameters listed above. When a and b are zero, expression (9.31) is transformed into an ordinary Gaussian function.

We should note that many researchers have tried to modernize the basic GLE analysis procedure described above over the last decades (Lovell et al. 1998; Belov et al. 2005a, b; Bombardieri et al. 2006; Krymsky et al. 2008; Firoz et al. 2010; Andriopoulou et al. 2011). However, in contrast to the complex analysis of SCR time profiles during different GLE stages (Vashenyuk et al. 2009, 2011), most of the indicated works were mainly aimed at analyzing the properties of only one isotropic GLE stage. An alternative method of spectrographic global survey was proposed by Irkutsk researchers (Dvornikov and Sdobnov 1997). The method is based on the solution of a system of nonlinear algebraic equations. These equations take into account the global amplitude distribution of variations in the integral fluxes of different secondary components, coupling function between primary and secondary variations, changes in the geomagnetic cutoff rigidity planetary system during each observation hour, and other factors. The authors used data from more than 30 CR stations, including the data of the Sayan spectrographic complex and Irkutsk MT, in order to analyze GLE42 (September 29, 1989) (Fig. 9.6). A comparison with the results achieved by Cramp et al. (1993a, b) indicates that the two methods are substantially different, especially in the region of large amplitudes during the early GLE42 stages.

One of the serious common basic flaws in all above procedures consists in that the response functions of different (standard) ground detectors are known insufficiently. In particular, this is true for NMs in the region of comparatively low ($\leq$2 GeV) SCR energies (Struminsky and Belov 1997). The latter circumstance was mentioned again by the authors of the PAMELA direct space experiment (Adriani et al. 2011), when they tried to coordinate the spectral data of different detectors at energies varying from 80 MeV/nuc to 3 GeV/nuc based on the measurements performed during GLE70 (December 13, 2006). Taking into account the accuracy in estimating the absolute intensities of accelerated solar particle fluxes based on the NM data, Adriani et al. (2011) managed to reach a reasonable agreement between the fluxes measured during the PAMELA experiment and those estimated using the NM data. However, the PAMELA spectra were always harder than the spectra obtained from the NM data at low energies. This can indicate that the response functions for NMs are understated at energies of $\leq$700 MeV. During the second satellite pass over the polar cap, the indicated difference between the PAMELA and NM fluxes became larger, whereas the PAMELA data remained in very good agreement with the data of the IceTop ground-based experiment (Antarctica). Direct measurements of the SCR fluxes in the stratosphere also confirmed that the PAMELA data are correct.

The process of SCR penetration into the near-Earth space, in particular, at high geomagnetic latitudes and during geomagnetically active time periods, is still not be modeled accurately enough for certain analyses of ground-based cosmic ray measurements (*e.g.*, Bieber et al. 2013b). Thus, recent and further progress in the art of mapping the Earth's magnetosphere is highly appreciated by the cosmic ray community (*e.g.*, Flückiger and Kobel 1990; Cramp et al. 1997; Lovell et al. 1998; Pchelkin et al. 2000; Plainaki et al. 2010; Papaioannou et al. 2014).

9.9 Spectra of Prompt and Delayed Components

Using the procedure described above, Vashenyuk et al. (2009, 2011) analyzed 35 large GLE events that occurred from 1956 to 2006. Two components are present in each event with rare exceptions: the prompt component (PC) with an exponential energy spectrum and the delayed component (DC) with a power-law spectrum. We should note that the spectrum shape was not explicitly specified when the spectral parameters were determined, especially on the rigidity scale. The shape of the spectra, which were obtained when the inverse problem was solved, was subsequently determined based on better agreement with one of two representations: exponential or power ones. Table 9.4 presents the spectral parameters for each of 35 events (Vashenyuk et al. 2011). These are the J_0 and E_0 parameters of the exponential spectrum for the PC

$$J(E) = J_0 \, exp(-E/E_0) \tag{9.32}$$

and the J_1 and γ parameters of the power spectrum for DC

Table 9.4 Derived spectra for 35 GLEs of 1956–2006 (Vashenyuk et al. 2011)

GLE	Radio II type UT	Flare Imp.	Flare position	PC, J_0	PC, E_0	DC, J_1	DC, γ
05	03:36	3	N23 W80	7.4×10^5	1.37	5.5×10^5	4.6
08	10:17	3+	N13 W90	2.7×10^5	0.65	1.6×10^3	4.2
10	13:26	3+	N27 W04	–	–	7.5×10^3	4.1
11	02:22	3	N25 W35	–	–	1.0×10^5	5.3
13	09:47	3+	S07 W59	5.2×10^3	0.52	3.6×10^3	6.0
16	07:55	–	N22 W154	1.4×10^4	0.58	6.7×10^3	4.7
19	10:26	1B	N21 W87	1.2×10^4	0.58	2.6×10^3	5.5
20	09:04	2B	N13 W37	7.7×10^4	0.38	4.7×10^3	5.0
22	23:16	3B	N19 W49	3.4×10^4	0.45	8.7×10^3	5.8
23	19:34	–	S11 W120	–	–	4.7×10^3	5.4
25	15:19	3B	N14 W37	6.6×10^2	1.23	4.3×10^2	5.0
29	05:55	–	N10 W120	6.5×10^2	1.14	9.3×10^2	3.2
30	09:59	2B	N24 W40	1.5×10^4	0.77	1.1×10^4	4.7
31	03:27	1B/×2	N23 W82	3.5×10^4	1.11	1.3×10^4	4.0
32	09:58	3B/×1	N35 W50	–	–	7.0×10^2	4.7
36	06:24	2B/×3	S18 E31	1.7×10^3	1.21	–	–
38	23:44	1B/×2.8	S19 W86	5.7×10^3	0.65	7.2×10^3	4.5
39	09:00	–	– W132	–	–	5.2×10^4	5.9
41	01:03	2 N/×12.5	S15 W85	6.8×10^3	0.56	3.8×10^3	5.1
42	11:33	–/×9.8	– W105	1.5×10^4	1.74	2.5×10^4	4.1
43	12:49	3B/×13	S25 E09	4.0×10^4	0.53	3.0×10^4	4.8
44	17:44	2B/×2.9	S27 W31	7.5×10^4	0.91	1.5×10^4	6.1
45	18:00	2B/×5.7	S20 W57	2.4×10^4	0.72	1.1×10^5	4.9
47	22:12	2B/×5.5	N35 W36	6.3×10^3	1.13	2.7×10^3	4.3
48	21:00	1B/×9.3	N36 W76	2.8×10^4	0.60	9.1×10^3	4.3
51	02:05	2B/×12.5	N32 W15	2.6×10^3	0.83	3.3×10^3	4.8
52	08:14	3B/×12.5	N36 W70	–	–	5.8×10^3	4.6
55	11:53	2B/×9.4	S18 W63	8.3×10^3	0.92	2.8×10^4	4.6
59	10:19	3B/×5.7	N22 W07	3.3×10^5	0.50	5.0×10^4	5.4
60	13:48	2B/×14.4	S20 W85	1.3×10^5	0.62	3.5×10^4	5.3
61	02:17	–	– W120	2.5×10^4	0.52	1.2×10^3	3.6
65	11:02	4B/×17.2	S16 E08	1.2×10^4	0.60	1.5×10^4	4.4
67	17:14	2B/×8.3	S14 W56	4.6×10^4	0.51	9.7×10^3	6.3
69	06:44	2B/×7.1	N14 W61	2.5×10^6	0.49	7.2×10^4	5.6
70	02:51	2B/×3.4	S06 W24	3.5×10^4	0.59	4.3×10^4	5.7

$$J(E) = J_1 E^{-\gamma} \tag{9.33}$$

where J_0 and J_1 are given in $\mathrm{m}^{-2}\,\mathrm{s}^{-1}\,\mathrm{sr}^{-1}\,\mathrm{GeV}^{-1}$, whereas E and E_0 are measured in GeV. Extended analysis of all 70 GLEs (Table 2.1) have been carried out quite recently (Vashenyuk et al. 2011), and summary data on their energy spectra for

35 GLEs from 71 ones (Table 2.1) are presented in Table 9.4. The values of characteristic energy E_0 of exponential spectra are given in GeV; normalization constants of exponential (J_0) and power-law (J_1) spectra are measured in units proton/(m^2 s sr GeV).

The average values obtained from these data are as follows: $<E_0> = 0.52 \pm 0.15$ GeV and $<\gamma> = 4.85 \pm 0.25$. Note that it became possible to divide SCR fluxes into PC and DC and to study other fine details in an individual ground-level enhancement only because SNM-64 (super neutron monitors) register GLEs very accurately (Carmichael 1968). Notice also that from 59 GLEs of 1942–2000 (see Table 2.1), eleven events supposedly were due to the flares occurred behind the W-limb of the Sun, some of them (*e.g.*, February 16, 1984 and September 29, 1989) having been certainly correlated to the CMEs.

Chapter 10
Solar Cosmic Rays in the Geosphere

As well-known, during solar flares the very complex events and processes occur in terrestrial environments. They are due to the entire spectrum of solar ionizing radiation, including ultra-violet emission, X-rays and fast particle fluxes. In particular, there is some observational evidence of radio burst generation at ionospheric levels in the ultra-short wave range, at the frequency of 550 ± 40 MHz (*e.g.*, Musatenko 1980). Mechanism(s) of radio emission of near-Earth space at the frequencies much more high in comparison with the characteristic frequencies of ionospheric plasma are not yet well developed. Meanwhile, one of the contributors of this effect seems to be fast electrons from solar flares.

Over several decades an effect of polar cap absorption (PCA) of the short radio waves (~30 MHz) after solar flares has been extensively studied (*e.g.*, Bailey 1959, 1964). This phenomenon is due to additional ionization of the atmosphere (at altitudes about 30–110 km) by solar protons in the energy range of 10–30 MeV. The nature and main features of the effect were described at length earlier (*e.g.*, Dorman and Miroshnichenko 1968; Sakurai 1974; Bruzek and Durrant 1977; Dorman 1978; Miroshnichenko 1983c; Smart and Shea 1989b), therefore, we do not consider it here.

There are also certain indications of an active role of galactic and solar cosmic rays in the tropospheric processes (*e.g.*, Vitinsky et al. 1976; Loginov and Sazonov 1978; Tinsley et al. 1989; Tinsley and Deen 1991; Pudovkin and Raspopov 1992; Avdyushin and Danilov 2000). Therefore, it is necessary to consider available qualitative (theoretical) and quantitative (observational) premises of SCR influence on the geosphere. Below we concentrate mainly on some expected effects and/or poor-studied phenomena discovered within two to three last decades (for details and references see, *e.g.*, Miroshnichenko 2008).

L. Miroshnichenko, *Solar Cosmic Rays*, Astrophysics and Space Science Library 405, DOI 10.1007/978-3-319-09429-8_10

10.1 Geomagnetic Effects

In Chap. 8 one very peculiar geomagnetic effect was mentioned, namely, possible collective influence of energetic solar particles on the geomagnetosphere during intensive SPEs; this requires the solution of self-consistent problem of interaction between SCR and geomagnetic field (GMF) provided their energy density are comparable between them. Further, quasi-drift effects of high-energy solar cosmic rays in the magnetosphere were suggested (Shumilov et al. 1993) to explain some peculiarities of proton precipitation into the auroral zones and polar caps during the anisotropic phases of certain SPEs (for example, on February 16, 1984). These authors studied the structure of high-energy solar proton penetration zones in the polar and subpolar regions combining the data from riometers, transpolar satellite *DMSP-F*6, and neutron monitors. An extremely anisotropic GLE of 16 February 1984 displayed one rather unusual feature: a maximum in the latitude increase profile fell at Turku station which has a greater geomagnetic cutoff than Apatity and Oulu stations. To make all observational data consistent with this finding, Shumilov et al. (1993) postulated a precipitation of quasi-trapped particles drifting along the latitudinal direction from the entry site of the anisotropic flux at the dawnside of the magnetosphere. It is argued that moderate nonadiabaticity is necessary to be a cause of proton intensity maximum located at the auroral and nearby latitudes, i.e., to make the drifting particles precipitate to the Earth's surface.

In addition, a joint consideration of solar proton data obtained in the stratosphere and at the heights of 300–400 km most likely shows the presence of trapped protons with energy of E_p >200 MeV (Shurshakov et al. 1993). These authors have examined the energy spectra of solar particles derived from the measurements by the Lyulin dosimeter on board *Mir* station, by the sensors on board *GOES*-7 satellite and using radiosondes in the stratosphere during the March 23 and October 31, 1991 events. It was shown that fluxes of geomagnetically trapped particles at ~400 km heights may contribute to the Lyulin dosimeter counting rate. This is illustrated by Fig. 10.1 where energy spectra of solar protons on March 24, 1991 are represented. These phenomena should be taken into account, in particular, in a case of solar proton spectrum evaluation from latitude effect at the magnetospheric spacecraft orbits.

On the other hand, a significant progress took place during three last decades in the computer simulation of particle trajectories using a mathematical model of the Earth's magnetic field (GMF). It is well known that during large geomagnetic storms the propagation of cosmic ray particles in near-Earth space is affected considerably and that significant changes of the cutoff rigidities and asymptotic directions occur (see, *e.g.*, Flückiger et al. 1983, 1986; Kudo et al. 1987; Flückiger and Kobel 1990). In the majority of previous study (*e.g.*, Shea and Smart 1975; Smart et al. 1987a, b), the International Geomagnetic Reference Field (IGRF) model appropriate for a certain epoch was used to represent the GMF. Apart from few exceptions (*e.g.*, Gall et al. 1982), effects of the quiescent magnetosphere were generally not taken into account. Cosmic ray cutoff rigidities and asymptotic

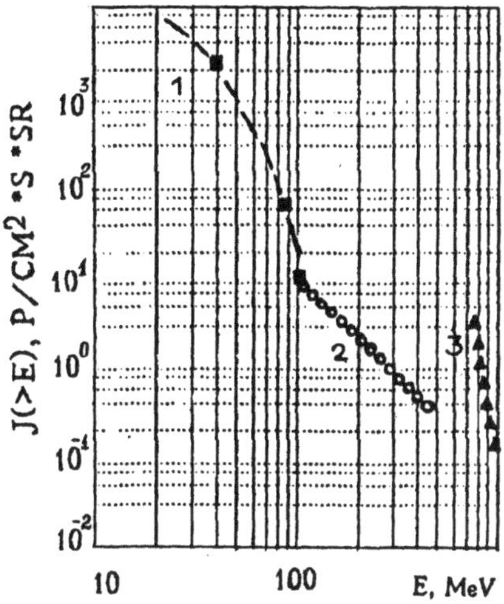

Fig. 10.1 Energy spectra of solar protons at 0,300–0,500 UT on March 24, 1991 inside the magnetosphere (Shurshakov et al. 1993): *1* spectrum estimated from *GOES* 7 measurements; *2* omnidirectional counter data from stratospheric observations; *3* spectrum deduced from Lyulin dosimeter measurements on board *Mir* station in the southern hemisphere

directions, however, are very sensitive to the secular changes in the internal geomagnetic field. Meanwhile, only a few studies have been performed on this subject utilizing the trajectory-tracing technique and combined models of the internal and external GMF (*e.g.*, Flückiger et al. 1983; Flückiger and Kobel 1990, and references therein).

As it was proved, this is a serious deficiency for the analysis of GLEs by the data of neutron monitors at high geomagnetic latitudes and/or during time periods of enhanced geomagnetic activity (Flückiger 1984). A very learning example for such an event is a GLE on December 7, 1982 (Smart et al. 1987a, b; Flückiger and Kobel 1990). This GLE occurred while a geomagnetic storm was in progress (its amplitude was $D_{st} \approx -80$ nT). At the time of the event the cosmic ray cutoff rigidities at middle latitudes lowered by about 0.3 GV. As shown by Flückiger and Kobel (1990), it turned out to be impossible to explain the neutron monitor data on this GLE without taking into account the effect of the external GMF. Therefore, a special study has been performed to calculate cosmic ray trajectories utilizing a model of the perturbed magnetospheric field (for details see Flückiger and Kobel 1990).

For the analysis of the GLE of December 7, 1982 the model worked out by Tsyganenko and Usmanov (1982) was used (see also Tsyganenko 1989). On the basis of their analysis, Flückiger and Kobel (1990) specified the limitations of any mathematical model of the main GMF in its application for cosmic ray studies. For such studies models of the Earth's internal and external magnetic field have to be combined. As an example, it was discussed a combination of the IGRF 1980.0 field model with the magnetospheric model by Tsyganenko and Usmanov (1982). Also, Flückiger and Kobel (1990) outlined the general procedure and problems of combining models of the Earth's internal magnetic field with models of the external GMF. Combined model was emphasized to be badly needed as a necessary prerequisite for many today's cosmic ray analyses.

As it follows from Chap. 9, the technique usually employed to determine the particle arrival directions, spectrum, and anisotropy in GLEs have been developed over many years (*e.g.*, Shea and Smart 1982; Humble et al. 1991a, b). Subsequent improvements have involved more accurate calculations of the asymptotic directions (*e.g.*, Flückiger and Kobel 1990) using better and more complex representations of the magnetospheric magnetic field. For example, the model by Tsyganenko (1989) includes parameters for six different levels of the geomagnetic index K_p. In a series of studies, Cramp et al. (1993a, b, 1995a, b, c, 1997) and Lovell et al. (1998) have made significant changes in the previous technique by two ways: (1) including the incorporation of the modern magnetospheric models (*e.g.*, Tsyganenko 1989) into the asymptotic direction calculations (Flückiger and Kobel 1990), and (2) using of a least squares method to determine the fitted parameters when deriving spectra, pitch angle distributions and particle arrival directions. This improved approach has been applied in studies of several highly anisotropic GLEs, namely, October 12, 1981; December 7, 1982; February 16, 1984; September 29, October 22, and October 24, 1989. Pchelkin et al. (2000) suggested to use another magnetospheric model which presents a magnetic disturbance in terms of three characteristic parameters, as follows: amplitude of D_{st} – variation, value of K_p – index, and dynamic pressure of solar wind. This allows to describe spatial behaviour of disturbed magnetosphere up to the distances of $10R_E$ (terrestrial radii). The advantages and deficiencies of such an approach have been demonstrated at the analysis of the GLE observed on 7 December 1982.

Those improvements turned out, however, to be insufficient to treat adequately some of recent GLEs. As an example, we mention here the results of analysis of the GLE of 2 May 1998 (Danilova et al. 1999). This event occurred against a background of disturbed interplanetary and magnetospheric conditions. The very complicated picture of the increase effects at 11 neutron monitor stations has been analyzed using the results of asymptotic direction calculations based on the geomagnetic field model by Tsyganenko (1989). It is remarkable that three stations (Goose Bay, Oulu and Newark) showed single intensity maxima during initial phase of the event (first peak), and the eight stations displayed the second peak only (second peak). Strikingly, the very short (pulse-like) increase observed by the NM at Oulu (10.4 %) contrasted with nearly absence of any increase at the neighbouring station at Apatity (~1.4 %). It means that the real asymptotic cone of acceptance for the Oulu station could have differed from calculated one (Danilova et al. 1999). In other words, the striking difference in responses of the Apatity and Oulu neutron monitors to relativistic solar protons, in this case, seems to be purely magnetospheric effect. The authors note, however, that the contribution of quasi-trapped particles drifting azimuthally more than 100° inside the magnetosphere (Shumilov et al. 1993) could not be sufficient. The only assumption to explain the discrepancy between observed increases at Apatity and Oulu is based on that those stations could found themselves at different sides of the border between the open and closed GMF lines, respectively, due to major geomagnetic disturbance, great negative B_Z – component of IMF, and strong compression of the magnetosphere (for further details see Danilova et al. 1999). Under such conditions, the anisotropic solar proton flux could directly reach the Oulu station, meanwhile the Apatity station could not accept the upward directed flux.

10.2 Atmospheric Impact of Energetic Solar Particles

Among the particle populations in the near-Earth space, of particular interest are those which possibly may be a link in the coupling of solar variability to the Earth's lower atmosphere. The solar activity control of weather and climate is a very controversial topic (*e.g.*, Avdyushin and Danilov 2000). However, convincing evidence of such control has accumulated recently, and its possible mechanisms are widely discussed (Reid 1991; Friis-Christensen and Lassen 1991; Labitzke and van Loon 1990; Pudovkin and Raspopov 1992; Avdyushin and Danilov 2000). According to Avdyushin and Danilov (2000), the most promising prospects in this field are bound up with the development of three possible versions of a trigger mechanism: dynamical (see references in Avdyushin and Danilov 2000), electrical (*e.g.*, Tinsley and Deen 1991), and optical (*e.g.*, Pudovkin and Raspopov 1992). Particle populations which affect the atmosphere in the most direct manner must be those which can penetrate deeply into the atmosphere. Thus, relativistic electrons appear to be among the most important candidates (Baker 1992). Other agents of this type, also heavily dependent on solar activity, are solar protons of MeV range energies.

Relativistic electrons are one of those factors affecting the atmosphere which are relatively faint in the sense of energetics, but on the other hand, vary greatly in association with solar activity (about an order of magnitude in the case of relativistic electrons), contrary to the "solar constant", S_0, characterizing the total energy flux from the Sun which varies only by ~0.1 % during a solar cycle. These factors directly affect only the dilute layers of the upper and middle atmosphere, down to the stratosphere. However, these influences may be very important. In particular, they do result in the ozone abundance variations at stratospheric heights, through intermediate abundances of odd nitrogen oxides.

Kropotkin (1996) tried to examine specific magnetospheric processes of transport and energization which might be involved in the forming of the fluxes of relativistic electrons precipitating into the middle and lower atmosphere. The highly variable character of those fluxes is due to the control of outer radiation belt by intense magnetospheric disturbances, *i.e.*, magnetospheric substorms.

It must be point out that the relative role of relativistic electrons in the stratosphere, in comparison with galactic cosmic rays and solar protons, remains a subject of considerable controversy. The most optimistic estimates (Callis et al. 1991), based on the measurements of relativistic electrons at the synchronous orbit, and their projection to the atmospheric heights with the use of a rather crude model, yields the ozone O_3 variations of tens of percent, being substantially higher than the effects due to other agents. In any case, both rather high fluxes of relativistic electrons in the magnetosphere, in the region of the outer radiation

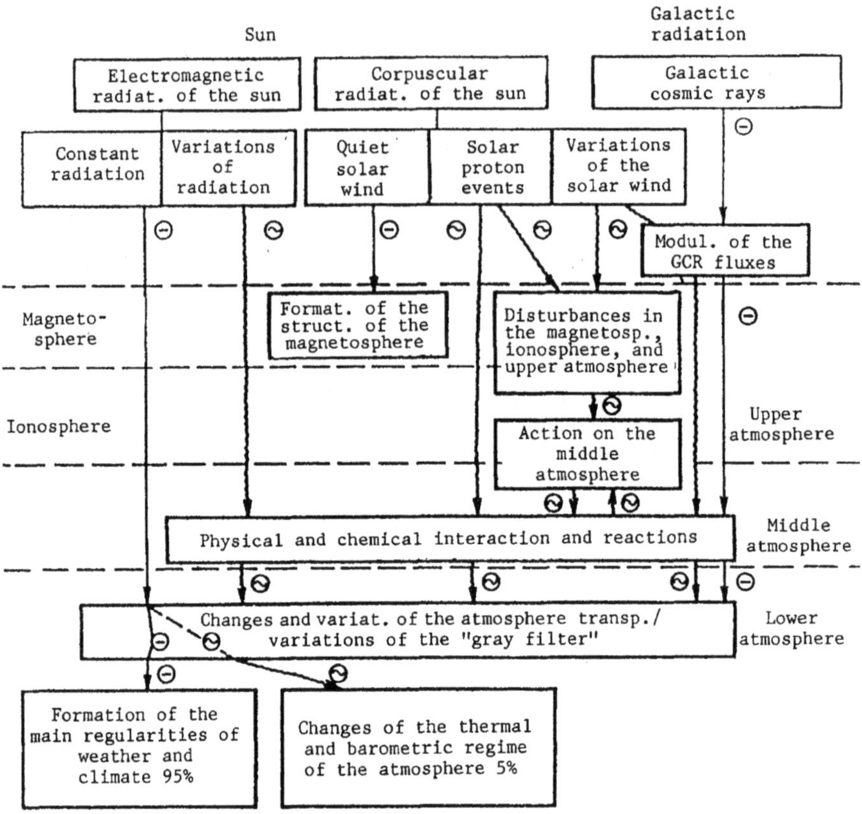

Fig. 10.2 General scheme of the action of solar activity and galactic cosmic radiation on the processes in the lower atmosphere, on the meteorological and climatic parameters (Pudovkin and Raspopov 1992)

belt, and intense and fast variations of those fluxes, are well established observational facts; up to now their nature lacks a reliable theoretical explanation (*e.g.*, Kropotkin 1996).

General scheme of the impact of electromagnetic radiation of the Sun, solar wind and solar cosmic rays, as well as galactic cosmic rays on processes in the lower atmosphere, on the meteorological and climatic parameters is depicted in Fig. 10.2 (Pudovkin and Raspopov 1992). Solar cosmic rays are apparently only one part of the complex mechanism(s) of outer space on the terrestrial environment. It is important to note that energetic solar particles may impact the Earth by two ways – directly (through a chain of physical interactions and chemical reactions in the upper and middle atmosphere) and indirectly (through the disturbances in the magnetosphere, ionosphere and upper atmosphere).

10.3 Depletion of Ozone Layer

Ozone variations are known to be important from two points of view: (1) they affect the UV flux penetrating to the Earth and strongly affecting living organisms; and (2) they have an influence upon the temperature variations at stratospheric heights. In turn, that affects propagation and damping of waves penetrating from the troposphere, *i.e.*, Rossby waves and internal gravity waves, and these are intimately associated with tropospheric circulation structures.

10.3.1 Observational Evidence

A certain decrease of ozone content related to the solar particle event (SPE) was recorded, for the first time, in the rocket measurements on November 2, 1969 at the height of 52 km (Weeks et al. 1972). Since then, during the two last solar cycles, the decreases of the ozone content have been observed in 13 cases of significant SPEs (Weeks et al. 1972; Heath et al. 1977; McPeters and Jackman 1985; Stephenson and Scourfield 1992; Zadorozhny et al. 1992).

The energy delivered by the fast solar particles to the Earth's atmosphere during individual SPE cannot be compared with the total radiation energy of the Sun and is insufficient to maintain, for example, a typical stratospheric perturbation. This disproportion, however, becomes less sharp if one bears in mind that the geomagnetic field forces the bulk of the SCR to concentrate at high latitudes where the effects of solar electromagnetic radiation are minimal, especially in conditions of the local winter. Therefore, it is natural to expect that the possible meteorological effects of the SCR will manifest themselves more distinctly at high latitudes and most clearly in the winter period. Let us illustrate this by concrete examples.

During the very powerful proton events of 4 and 7 August 1972 into the polar stratosphere in the form of SCR came an energy ~6×10^{-2} J cm^{-2} (Crutzen and Reid 1976). The events were accompanied by a heavy (>20 %) fall in the ozone content at the high latitudes in the stratosphere, above the level with the residual air pressure 4 mbar. Such a result was obtained in the observations of back-scattering of solar ultra-violet radiation on board the *Nimbus* 4 satellite (Heath et al. 1977). The dynamics of this fall is illustrated by Fig. 10.3. It is seen that the decrease in the ozone content continued for several weeks. The effect diminished in the zone from 55° to 65° N and was not noticed in the tropical zone. Similar effects, but with much greater changes, were noted in the southern hemisphere in the conditions of local winter.

A comparable effect in the total ozone content was recorded in observations of Shumilov et al. (1991, 1992a) in the northern hemisphere during a series of proton events in May 1990. These ground-based observations were carried out above several geophysical observatories – Barentzburg (Spitzbergen Island), Heiss Island and Murmansk. A general picture of the phenomena is represented in Fig. 10.4

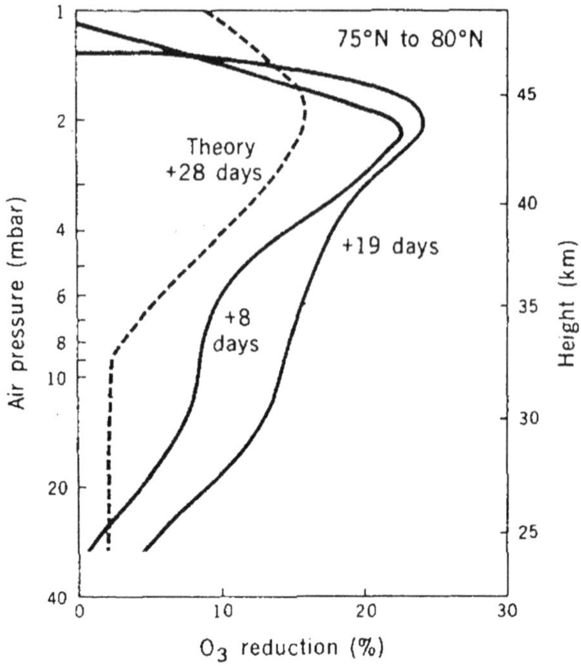

Fig. 10.3 Decrease in the ozone content in the Earth's stratosphere by the satellite measurements in the latitude range of 75–80°N (Heath et al. 1977) after the flare of 4 August 1972 within +8 and +19 days. The *dashed curve* represents calculated ozone content after +28 days

(Shumilov et al. 1995). The decrease in the ozone content for 2.5 days reached ~18 %. These data, in part, may answer the question whether an "ozone hole" can form above the Arctic (Zadorozhny et al. 1992; Kasatkina et al. 1992; Shumilov et al. 1992b, 1995, 1996a) similar to the widely discussed "hole" above the Antarctic.

Short-term depletions in the ozone total content were observed also during three other large proton events, namely, in March, September and October 1989 (Reid et al. 1991; Shumilov et al. 1995). To complete the picture of geophysical effects in these events, it is worth noting the response of the middle atmosphere to the SPEs of August-December, 1989. As reported by Reid et al. (1991), several major SEP events recorded during this period (August 12–18; September 29 – October 2; October 19–30; December 30, 1989 – January 2, 1990) were accompanied by greatly enhanced ionization rates and NO_x production in the polar regions of both hemisphere, with a subsequent impact on stratospheric ozone. According to estimates by Reid et al. (1991), the largest enhancements in stratospheric reactive nitrogen were expected to occur in the southern hemisphere, with corresponding peak ozone depletion of about 20 % near an altitude of 40 km in late October 1989. Reid et al. (1991) calculated substantial (10–70 %) increases in NO_2 near 30–40 km in the southern polar cap, which were observable with satellite measurements. They also estimated that maximum temperature decreases were about 3.0–3.5 °K in this region during October and November. In their opinion, the total influence of these SPEs on the middle atmosphere rivaled that of

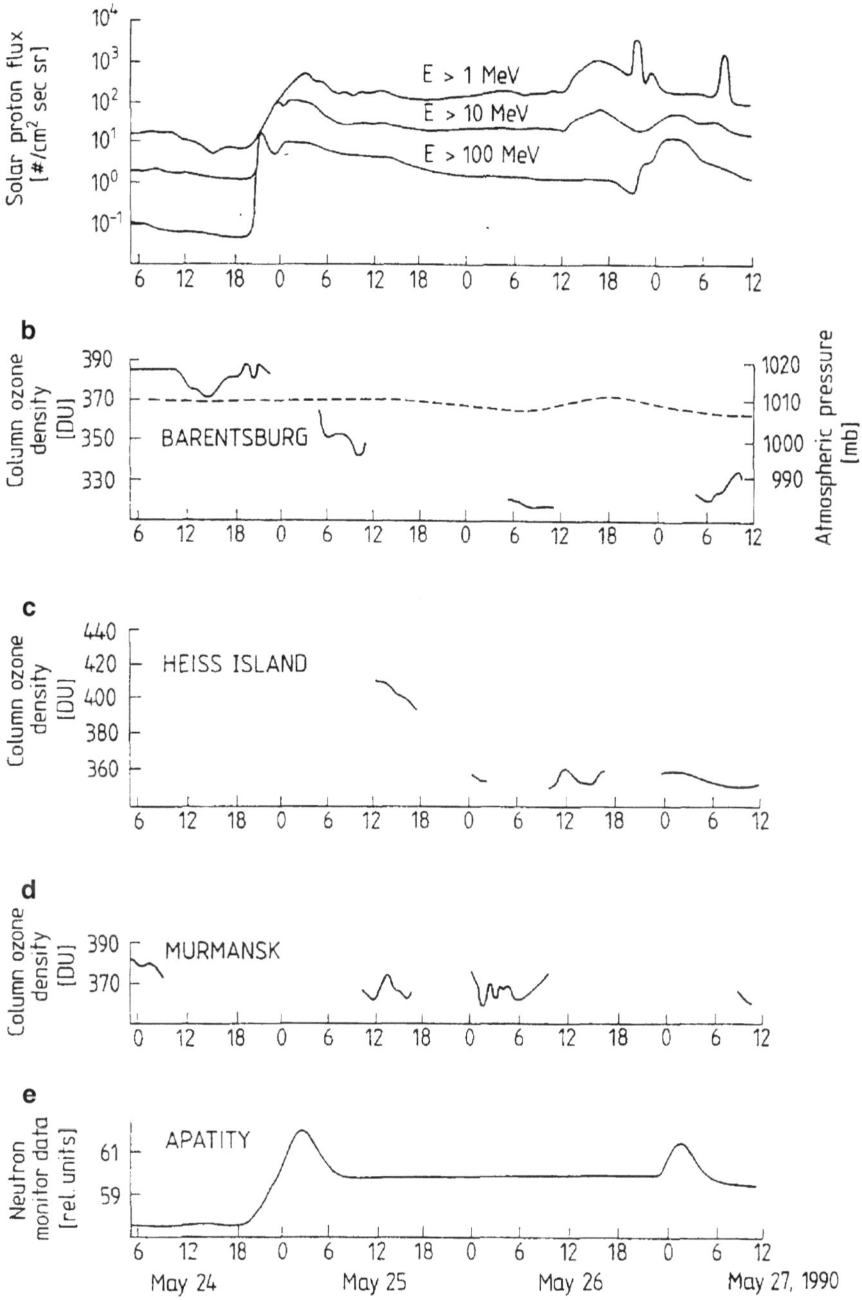

Fig. 10.4 Solar proton events in May 1990 and observations of total ozone content: (**a**) integral fluxes of solar protons from the data of *GOES*-7 satellite for energies >1 MeV, >10 MeV, and >100 MeV; (**b**) ozone record (*solid curve*) and atmospheric pressure (*dashed curve*) at Barentzburg station (Spitzbergen Island); (**c**) ozone record from Heiss Island; (**d**) ozone record from Murmansk; (**e**) neutron monitor data from Apatity (Adapted from Shumilov et al. 1995)

the August 7, 1972 event when an effect of stratospheric ozone depletion was observed for the first time (Heath et al. 1977). However, the share of the September 29, 1989 event in the total effect was rather small, no more than 11 %, if one compares the ratios of the proton fluences at >10 MeV in the four observational periods mentioned (*e.g.*, Shea and Smart 1993a, b). According to Reid et al. (1991), the main contribution (about 60 %) was due to proton fluxes observed between 19 and 30 October.

As a result of these investigations it was made clear that after SPE remarkable changes of ozone layer density are observed. Results obtained were confirmed independently by the observations above the Antarctic (Kodama et al. 1992; Stephenson and Scourfield 1992). From the other hand, recently it has been shown (Shumilov et al. 1991, 1995) that the intrusion of relativistic protons into the Earth's atmosphere during the proton events of GLE type results in the formation of ozone "mini-holes" (short-term decrease of total ozone content of >15 %) in the polar caps. In the case of a high SCR anisotropy, the fluxes of precipitating particles are different in the different hemispheres, *i.e.* an effect of N-S asymmetry exists (Rao 1976). Besides, based on the results of satellite measurements during some proton events it has been shown (Maeda et al. 1984; Stephenson and Scourfield 1992; Kasatkina et al. 1998) that a response of the ozone layer to the intrusion of solar protons is different in the northern and southern hemispheres.

10.3.2 Mechanism of Depletion

High-energy solar protons can produce in the stratosphere HO_x and NO_x constituents, and they can be of great importance for a temporal depletion of the ozone layer. One of the possible mechanisms for explaining the effect discussed was proposed by Crutzen et al. (1975) using homogeneous chemistry. Ionization produced in the stratosphere by SCR particles leads, in particular, to the production of a large amount of oxide of nitrogen (NO) through dissociation and dissociative ionization of the molecules of nitrogen. A few powerful flares during the year may produce NO in an amount comparable with that given by the main known source (oxidation of N_2O) and far greater than that given by galactic cosmic rays (GCR).

Ozone in the Earth's atmosphere below 45 km is degraded by nitrogen oxides due to the pair of catalytic reactions:

$$NO + O_3 \Rightarrow NO_2 + O_2 \tag{10.1}$$

$$NO_2 + O \Rightarrow NO + O_2 \tag{10.2}$$

The lifetime of HO_x in the middle atmosphere is only several hours and could not be compared with the lifetime of NO_x which varies from days to several months

(Shimazaki 1984). Therefore, for periods from days to months, only NO_x can have a significant influence on the ozone distribution. Although an individual increase in the flux of energetic solar protons lasts only a few days, the lifetime of NO_x in the stratosphere is rather large, and its impact on the ozone layer probably continues for several years. In addition, we would like to note that nitrogen dioxide NO_2 effectively absorbing solar radiation over the range 4,000–5,000 Å takes an active part in the redistribution of energy in the Earth's atmosphere (in particular, in the creation of so-called "green-house effect").

More detailed analysis of the depletion effect during the GLEs of May 1990 performed by Shumilov et al. (1995, 1996a, b) revealed, however, the limits of photochemical approach (Crutzen et al. 1975; Jackman et al. 1980) to the problem. The homogeneous photochemical model of ozone depletion during GLEs gives estimates that are one order of magnitude less than those observed. For the two SPE episodes, in August 1972 and July 1982, the model and experimental results turned out to be in a good agreement for altitudes above 40–50 km, meanwhile no depletion in total ozone content was noted (Jackman et al. 1980). It is remarkable, however, that these SPEs were not "pure" ones, because they occurred during noticeable Forbush decreases. The latter is characterized by a decrease of the GCR intensity and by a reduced production of odd HO_x and NO_x, and hence, by an increase of ozone content.

Therefore, Shumilov et al. (1995, 1996a, b) suggest that models which include only gas-phase chemistry can not be used to estimate decreases in *total* ozone content during GLEs. Most likely, "mini-holes" could be created by some stratospheric aerosol clouds forming due to the invasion of energetic solar protons. The evidence of aerosol content increase was obtained, for example, after the GLE of February 16, 1984 (see Shumilov et al. 1995, 1996a, b, and references therein). In stratospheric aerosol clouds, a complex set of heterogeneous processes takes place which are able to destroy stratospheric ozone efficiently.

Based on these considerations, Shumilov et al. (1995, 1996a, b) postulate that the GLEs seem to trigger the following sequence of events leading to significant temporal ozone decreases, or ozone "miniholes": an increase of incident flux of high-energy protons ⇒ increase of ionization in the stratosphere ⇒ increase of NO and HO content ⇒ preliminary decrease of stratospheric ozone content ⇒ decrease of stratospheric temperature (or increase of space electric charge in the stratosphere) ⇒ formation of stratospheric aerosol clouds (sometimes so-called Polar Stratospheric Clouds) ⇒ creation of stratospheric ozone "miniholes" (through a series of heterogeneous chemical reactions). Some evidence of this sequence may be seen in Fig. 10.5.

There are shown the temperature and ozone height profiles from balloon measurements above Antarctic station Mirny during the GLE42 of September 29, 1989. In this case, the small decrease of temperature at 18–20 km altitude coincides with a considerable decrease of ozone concentration (up to 50 %). The situation was probably similar to that observed during the GLEs of May 1990 (see Fig. 10.4), but, unfortunately, during this GLE episode balloon data were not available.

Fig. 10.5 Altitude profiles of ozone partial pressure (**a**) and temperature (**b**) above Antarctic station Mirny by stratospheric balloon measurement in September 1989 (Shumilov et al. 1995): *1* - September 26 (quiet day); *2* - September 29 (disturbed day)

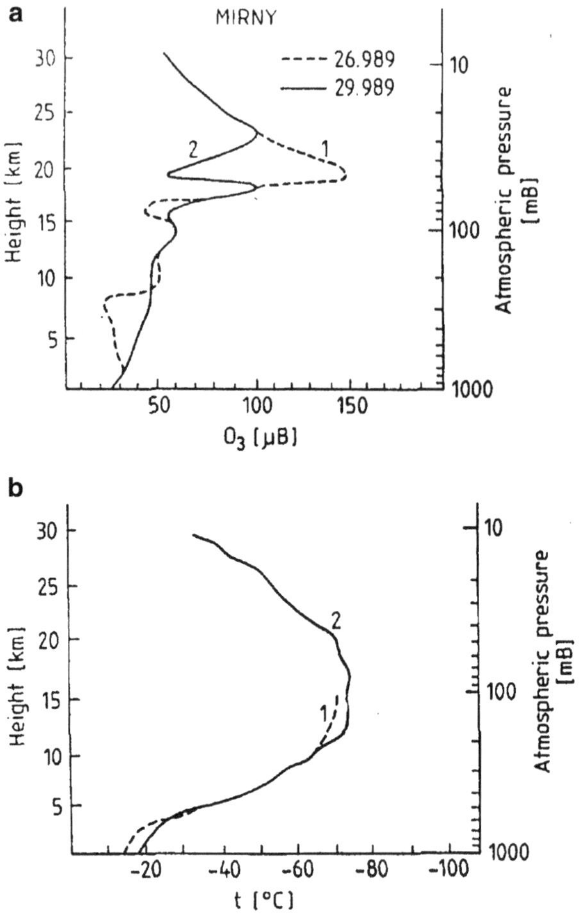

10.4 Perturbations in the Global Electrical Circuit

For about a 100 years the problem exists of a global atmospheric electricity, or a global electric current (GEC). The classical picture of atmospheric electricity assumes that the ionosphere is at a uniform potential with respect to the Earth's surface and that it effectively shields the underlying atmosphere from slowly varying fields outside of it because of its high electrical conductivity (for a review see, *e.g.*, Roble 1985). Such a picture does not account for the electrical perturbations induced by either the dynamical processes in the ionosphere (so-called ionospheric wind dynamo) or the solar wind -magnetosphere interaction (solar wind – magnetosphere dynamo). Meanwhile, both of these perturbations generate large-scale horizontal potential differences in the ionosphere that expand downward to lower altitudes.

In recent years, considerable progress has been made in improving our understanding of a number of solar-terrestrial interactions and their possible contributions to the global circuit of atmospheric electricity, especially with regard to the partial responses of the magnetosphere, ionosphere, and thermosphere. Now it is suggested that the problem under consideration includes various types of correlation between the lower-atmospheric electric fields and currents and auroras, sunspots, geomagnetic activity, solar flares, and solar magnetic sector boundary crossings. In many papers (*e.g.*, Roble 1985; Anisimov and Miroshnichenko 1989; Tinsley and Deen 1991) a set of external agents have been examined (for example, the known variations of galactic cosmic rays, energetic solar particles, auroral activity, ionospheric perturbations, *etc.*) with respect to their possible global or local influence on the electric properties of the lower atmosphere. It was found, in particular, that the global electric circuit (GEC) is perturbed for several days following certain solar flares and solar proton events.

As it was already mentioned, the August 4–9, 1972 solar flares produced one of the largest solar proton events ever recorded in terms of the total energy input into the middle atmosphere. Very large Forbush decrease in cosmic ray intensity also occurred during this event (see, *e.g.*, Duggal 1979). In addition to the vast quantity of particle and upper-atmosphere field data for this event (see, for example, the above discussed impact on the ozone layer), there were reported also middle- and lower-atmosphere electrical responses. In particular, Holzworth and Mozer (1979) observed order-of-magnitude variations of the vertical electric field at altitude of 30 km, in anticorrelation with the intensity of solar protons.

A unique set of balloon measurements of the air-Earth current density has been performed by at South Pole station (Cobb 1978) before and after the solar flare of November 22, 1977 occurred at 09:45 UT. Balloon-borne sensors were released at approximately 03:00 UT each day for 5 consecutive days during the period November 22–26, 1997. The measured air-Earth current density profiles are shown in Fig. 10.6. One can see, in particular, that on November 24 the entire air-Earth current density was enhanced by 70 % above preflare magnitude. Two days later the measured air-Earth current density profile returned to preflare level.

It is important that these measurements are consistent with those made from mountaintops at lower latitudes, though the causes of this correlation are not very clear (for details see Roble 1985). Some new experimental facts seem to confirm the increase of the vertical electric field in the stratosphere during SPEs (Holzworth et al. 1987; Zadorozhny et al. 1994). Such an unusual behaviour of the electric field during proton events may indicate significant electrification of the aerosols (Zadorozhny et al. 1994), thus supporting the aerosol hypothesis of Shumilov et al. (1995, 1996a, b) proposed to explain the creation of ozone "miniholes" (see Sect. 10.2).

As to the mechanism(s) responsible for the observed variations, they are not understood yet completely. Reagan et al. (1983) have shown that at times during the August 4–9, 1972 SPE episode the current carried by solar protons bombarding the middle atmosphere exceeded the normal air-Earth current flowing locally in the global circuit of atmospheric electricity This occurred when the vertical proton flux

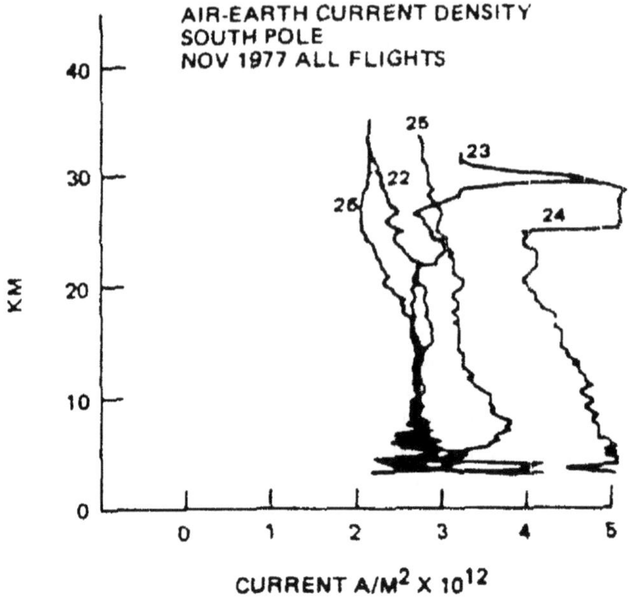

Fig. 10.6 Air-Earth current densities measured during balloon flights from the South Pole during November 22, 1977 (Cobb 1978). The days in November are indicated on each curve

was about $1.8 \times 10^3\,m^{-2}\,s^{-1}$ (it is equivalent to the current density of 3.0×10^{-12} A m^{-2}). The bombarding solar protons deposit a positive space charge within the middle atmosphere. Any accompanying electrons are mainly directed into auroral zone and usually stop above about 80 km. There is thus a charge separation during these particle precipitation events, and an equalizing conduction current should flow between the charge centers. If the bombarding protons and electrons are equal, the current closure takes place between the middle atmosphere and ionosphere. Any imbalance must be countered by a return current flow to space, since no net space charge can accumulate on the Earth over any length of time (see Roble 1985, and references therein).

As it follows from the above considerations, the ionosphere and upper-atmosphere responses to solar flare impact are various, complex and highly variable. There is usually enhanced magnetospheric convection, or dawn-to-dusk potential drops across the polar cap, those maps into the lower atmosphere. Further, there are also enhanced thermospheric winds and ionospheric disturbance dynamo effects and enhanced auroral activity and ionospheric current flows. At last, the upper atmospheric boundaries generally display rapid movements from their pre-flare position. Thus, possible electrical effects of all the known upper-atmosphere responses should be considered when attempting to evaluate the global electric response to solar flares.

10.5 Change of Atmospheric Transparency

As it was shown by many researchers (*e.g.*, Vitinsky et al. 1976; Tinsley et al. 1989; Pudovkin and Babushkina 1992a; Miroshnichenko 2008), magnetospheric disturbances associated with solar flares are followed by distinct changes in the state of the lower atmosphere. As a rule, atmospheric disturbances in their developments pass through two stages (*e.g.*, Schuurmans 1982): an initial, or "early", stage is followed by intensification of the zonal circulation and by decrease of the air temperature in the high-latitude stratosphere; a "late" one is notable for that the stratospheric temperature increases, and the zonal circulation diminishes. On the other hand, Veretenenko and Pudovkin (1993, 1994) have found some observational evidence in favour that the first stage of atmospheric disturbance is due to the bursts of solar cosmic rays (SCR) with the energy of E_p >90 MeV; whereas the second one – due to the Forbush decrease of galactic cosmic rays (GCR). In addition, Pudovkin and Babushkina (1992b) have shown that after the decrease of the GCR flux, an increase of the solar irradiation at the Earth's surface takes place, *i.e.*, the atmospheric transparency also enhances. It is reasonably to assume that an increasing of the cosmic ray intensity during solar proton events should result in the decreasing of the atmospheric transparency.

One of the first attempts to reveal this effect was undertaken by Roldugin and Vashenyuk (1994). Based on statistical data on occurrence rate of large SPEs and on variability of cloudiness in the northern hemisphere, these authors indicate that in some cases solar protons may serve as an active agent in the production of water aerosols above the polar caps and, consequently, in the enhancement of solar radiation absorption. However, it should be noted that Roldugin and Vashenyuk (1994) used the actinometric data from the Murmansk and Arkhangelsk stations where meteorological conditions are not favourable for this kind of research. Thus, their results, in fact, are rather uncertain, namely: from six proton events underwent to the study, in three cases there was found a decreasing of the solar irradiation (on April 30, 1976; April 3 and August 17, 1979); in two cases (April 8, 1978 and September 8, 1979) an increasing took place, and one case (May 11, 1978) showed no discernible effect.

Some later, a similar study was performed by Pudovkin et al. (1997) based on the actinometric data of the Olenek observatory located in the sub-auroral zone ($\varphi = 68.5°$). There were separated five intervals, with a duration 9 days each, when meteorological conditions allowed for everyday measurements of the solar irradiation, or solar constant S_0. These intervals were related to the corresponding proton events observed at the threshold energy E_p >90 MeV, namely: April 4, 1980; March 25, 1981; March 7, 1982; April 15, 1983, and March 14, 1984. Figure 10.7 shows the results obtained by using the superposed epoch method (C. Chree technique). As a 0 day was accepted the day of the event onset. It is seen that the event development displays, as it was expected, a decreasing of the S_0 value within 5–10 %. Unfortunately, a small number of the events and significant dispersion of the measured S_0 values did not allow to obtain a statistically

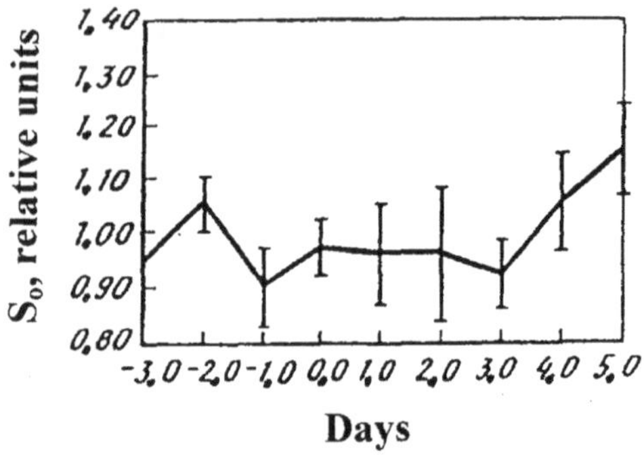

Fig. 10.7 Variations of the solar irradiation S_0 observed in the course of solar proton events in 1980–1984 (Pudovkin et al. 1997). A point $t = 0$ corresponds to the day of event onset; short vertical lines show the mean square deviations of the measured S_0 values

significant magnitude of dS. Nevertheless, a tendency of the S_0 value to diminish in the course of the event may be seen rather distinctively. Though the results of Roldugin and Vashenyuk (1994) and Pudovkin et al. (1997) are not very impressive, solar proton events are likely to cause the decreasing effect in the atmospheric transparency, probably in combination with the contributions from other cosmophysical agents.

Pudovkin and Raspopov (1992) developed a physical mechanism concept of solar activity influence on the lower atmosphere and climate based on the observational data. First of all, the data accumulated on the Russian net of actinometric stations as well as the data on global cloudiness distribution were summarized, and this allowed to show synchronous changes of cloudiness density after intensive solar flares. Hence, changes in opposite phase of the meteorological solar constant, S_0, *i.e.*, changes of atmosphere transparency, have been revealed. Second, they took into account that cosmic radiation (galactic and solar cosmic rays), in fact, is the only physical agent which is governed by solar activity and can influence on physical-chemical processes (reactions) in the lower atmosphere including cloudiness density changes. At last, it was suggested that the change of solar energy flux penetrated in lower atmosphere, *i.e.*, a variability of atmosphere transparency, is due to the variations of solar and galactic cosmic rays modulated by solar activity. Below we describe briefly their argumentation to prove advanced hypothesis.

Short-term and long-term changes of solar activity, geomagnetic field as well as the weather and climate have quasi-periodic variations nearly the same in period (*e.g.*, Vitinsky et al. 1976). Observational data demonstrate rather convincingly an existence of statistical connections between solar and geomagnetic variability, from one side, and climate changes, from the other one. However, the recognition of the physical nature of these connections meets serious difficulties. In particular, in some papers the very possibility of solar activity influence on weather and climate is rejected from energetic point of view (Avdyushin and Danilov 2000). Really, the characteristic power of the atmospheric processes which form the Earth's climate is of value 10^{26}–10^{27} erg/day. The average energy of solar wind injected into magnetosphere and, hence, into the Earth's atmosphere during geomagnetic disturbances after solar flares is equal to 10^{23} erg/day which turned to be of 3–4 orders less than the previous magnitude.

Table 10.1 Additional energy dissipating in the lower atmosphere during a geomagnetic disturbance

δ_0, degree	−20	−10	0	10	20
ΔW, 10^{26} erg/day	0.2	0.7	1.5	2.4	2.6

Thus, it is necessary to find a balance between additional energy injected into the atmosphere due to its transparency changes (during solar activity variations) and the energy required for the stimulation (excitation) of large-scale dynamic processes in the lower atmosphere, such as powerful zone circulation, changes of the flux of direct solar radiation, variations of air pressure near the Earth's surface, *etc.*

The amount of solar energy falling on the unit square of 1 cm^2 at the upper boundary of the atmosphere during 24 h is defined by the formula (see Pudovkin and Raspopov 1992, and references therein):

$$Q = 2\,S_0/\rho_0\,[\tau_0 \sin\varphi \sin\delta_0 + T/2\pi \cos\varphi \cos\delta_0 \,\sin(\omega\tau_0)] \qquad (10.3)$$

where S_0 is the solar constant, ρ_0 is the ratio of the Sun-Earth distance to its average value, δ_0 is the declination angle of the Sun, φ is the geographic latitude, T and ω are the period and angular velocity of the Earth's rotation, respectively, and τ_0 is the time of sunrise and sunset relative to the noon: $\cos(\omega\tau_0) = -\,\mathrm{tg}\varphi\,\mathrm{tg}\delta_0$.

The amount of additional energy ΔW entering and dissipating in the lower atmosphere during a geomagnetic storm can be estimated if instead of the S_0 value we substitute in (10.3) its variations, $\delta\langle S_\varphi \rangle$, during the disturbance and then integrate (10.3) over the part of terrestrial surface Σ where the variations of atmospheric transparency take place being stimulated by geomagnetic disturbances, i.e., in the latitude range $\varphi = 55–75°$. Let us suppose, further, that on the average, in the lower atmosphere only about 40 % of solar energy dissipates, and geomagnetic disturbances alter the amount of energy falling at the upper boundary of the atmosphere by about 6 %. In such a case one can write for ΔW a following equation (Shumilov et al. 1991)

$$\Delta W = 2.4 \times 10^{-2} \int Q \; d\Sigma \qquad (10.4)$$

The results of calculations of ΔW as a function of season, *i.e.*, the declination angle of the Sun δ_0, are given in Table 10.1.

It is evident that the amount of additional energy entering the lower atmosphere in 24 h during geomagnetic disturbances depends on the season and amounts to about 1.5×10^{26} erg on the average (at $\delta_0 = 0°$) for a geomagnetic storm lasting 3–4 days; the total amount of additional energy can reach approximately 10^{27} erg. This value is 3–4 orders times the amount of energy entering the magnetosphere from the solar wind ($\cong 10^{23}$ erg/day) and is commensurable with the energy required for the observed change in zonal circulation of the lower atmosphere: $\Delta Q = 5 \times 10^{26} - 2 \times 10^{27}$erg (Pudovkin and Babushkina 1992a).

Thus, the physical mechanism, suggested by Pudovkin and Raspopov (1992), provides self-consistent description of the influence of solar activity and other cosmophysical factors on the dynamic processes in the lower atmosphere, on the meteorological and climate parameters. Moreover, it seems to allow, for the first time, to explain an origin of the energy source of proposed mechanism which is always in operation and is controlled by solar and geomagnetic activities. The power of this source, from one hand, is much more than those of solar wind and subsequent processes provoked by solar wind in the magnetosphere and polar ionosphere; and from the other, it is quite enough to exert appreciable influence on the processes in the lower atmosphere.

It is also clear that solar energetic particles may contribute considerably into proposed global mechanism of solar-terrestrial relationships. Anyway, solar cosmic rays seem to play a certain role in the hierarchy of possible mechanisms operating in the Sun-weather-climate chain. Interdisciplinary studies of the influence of solar and geomagnetic variability on the terrestrial climate may allow to estimate relative contribution of SCR into the global mechanism(s).

10.6 Production of Nitrates

When penetrating into the upper atmosphere, solar cosmic rays are involved in the two main physical processes – ionization and nuclear interactions with nitrogen and oxygen, the two primary constituents of the Earth's atmosphere. The first process results in the production of negative NO_3 ions (nitrates), and the second one creates radionuclides, or cosmogenic isotopes (for example, well-known radiocarbon ^{14}C).

Because the majority of solar energetic particles are of low energy, they deposit most of their energy at high altitude within auroral zone. The formation of NO_x oxides and downward transport of reaction products formed in this zone tend to be spread over long period of time (up to 2 years) (see, *e.g.*, Dreschhoff and Zeller 1990, and references therein). A large portion of the nitrogen oxides produced are ultimately oxidized to nitric acid HNO_3 and incorporated in snow crystals, together with nitrates from tropospheric sources that also contribute to the general background. In particular, the Antarctic continent acts as a cold trap that effectively freezes out this signal and retains it in the stratigraphy of the ice shelves and the continental ice sheet.

During preliminary studies in 1985–1986 it was shown (Zeller et al. 1986; Laird et al. 1987) that the nitrate signal exhibits pulses that can be correlated with specific major solar flare events. This conclusion was tested more fully at two widely separated locations in Antarctic during the 1987–1988 field season, and these results led to the design of the high-resolution, on-site analysis project that was completed during the 1988–1989 Antarctic field season (Dreschhoff and Zeller 1990; Zeller and Dreschhoff 1995).

The measurements of nitrate concentration in samples of Antarctic ice and snow confirm the importance of studying nitrogen oxides NO, NO_2, NO_3, N_2O_5, and

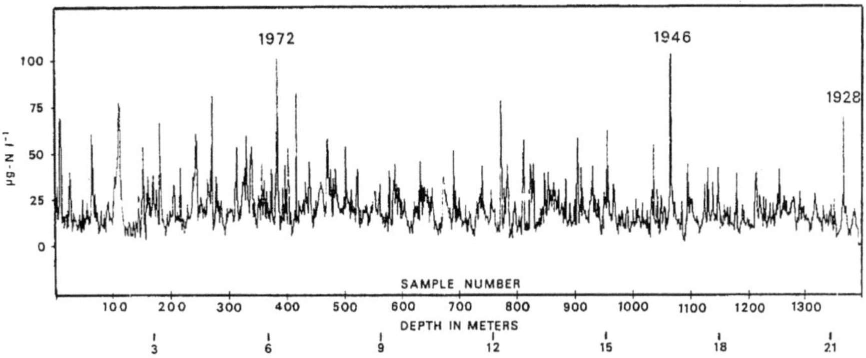

Fig. 10.8 Nitrate concentration profile from the Windless Bight core on the Ross Ice Shelf, Antarctic, by the data of Dreschhoff and Zeller (1990). The *x*-axis is proportional to true depth below the surface; the *y*-axis represents nitrate concentration in *mg* per unit of the entire length of the core. At least three major flares occurred in 1928, 1946, and 1972 are visible in the records as large concentration peaks

HNO_3 formed by the SCR and GCR (Dreschhoff and Zeller 1990). First of all, distinct 11- and 22-year variations in the content of NO during at least the last 200 years have been found. These variations bear a close similarity to those in the content of ^{14}C detected in the rings of trees in a period of ~1,200 years. In addition, with a resolution of ~3 months it has been shown that the effect of an individual solar flare may be detected in the corresponding snow layers, especially in the period of the Antarctic winter. Variations in concentration were at the level of 53 % corresponding to one standard deviation. Short concentration jumps were noted with increase within the limits of 11 standard deviations. On the average, over the whole series of data the accuracy of the measurements was not worse than 12 %. One of the ice samples (firn cores) was extracted by drilling from a depth of 21.7 m which correspond to its age ~62 years, i.e., exceeds the duration of the five solar cycles 16–21.

Analysis of the time series consisting of 1,393 individual measurement points indicates statistically significant modulation of the background signal clearly correlating with the variations in solar activity (Dreschhoff and Zeller 1990). These authors found several anomalously large jumps in the concentration of NO_x which could be confidently dated and tied to major proton events in August 1972 and in July 1946 and also to the "white" solar flare in July 1928. Three major peaks in the nitrate concentration profile are represented in Fig. 10.8. The increases above the series mean are 7, 11, and 4 standard deviations, respectively.

The conclusions drawn concerning the effects of the SCR are confirmed by calculations of the rates of NO_x transport in the atmosphere (Laird et al. 1987) on the basis of the real (measured) spectra of solar protons. There is no doubt that the significant share of NO_x in the samples studied is the final result of chemical reactions stimulated by additional ionization of the atmosphere by SCR and GCR. Later on, Gladysheva and Kocharov (1995) and Gladysheva (1996) examined a possibility of quantitative explanation of the first peak (August 1972) in the nitrate concentration profile in Fig. 10.8. They calculated the production rate of nitric oxides and estimated that about 1.5×10^{15} molecule/cm^2 in the atmospheric column between the Earth's surface and the altitude of 37 km are enough to explain the observed peak in the nitrate records of Dreschhoff and Zeller (1990).

Dreschhoff et al. (1997) and Dreschhoff and Zeller (1998) examined discrete nitrate events in the Antarctic and Greenland ice core records during the late nineteenth and early twentieth century. These results confirmed the major solar event association of enhanced peaks in the nitrate concentration in 1859 and 1909, as well as in a period of enhanced SPE activity toward the end of the last century. Thus, such a "nitrate method" is a very useful research tool, and the formation of nitrogen oxides is apparently a part of the physical mechanism linking the incursions of SCR into the stratosphere with the processes in the troposphere.

10.7 Periodicities in Solar Particle Fluxes

Although the data on the first SPEs (see Chap. 2) were rather fragmentary and overloaded with methodological uncertainties of various kind, they were of enormous interest to look for the mechanisms of acceleration and transport of SCR as well as of the SCR impact on the terrestrial ionosphere. However, until about 1955 the registration of energetic solar particles was episodic, due to a scarcity of the detectors (*e.g.*, Smart and Shea 1989b). Later, as data accumulated on the SPEs over long time intervals ($\geq$11 years), it became possible to identify the cyclic variations and to reveal some periodicities in the behaviour of SCR flux over 11-year scale. This allowed for working out the methods of long-term prediction of solar proton fluxes (Chap. 11).

10.7.1 Peculiarities in GLE Occurrence Rate

The first GLE events (before 1956) were registered at sparse stations equipped with ICs and MTs, which were mainly intended for measuring one hard (muon) component. Since the effective registration energy of NMs is lower than that of MTs and ICs, the latter detectors are less sensitive to SCRs. A special technique (Shea and Smart 1982; Humble et al. 1991a, b; Cramp et al. 1997; Vashenyuk et al. 2009), which takes into account the anisotropy of SCRs fluxes that approach the Earth, steep energy spectrum of SCRs, and the high NM sensitivity, is used to identify GLEs. At the same time, some weak GLEs (~1–10 %) were registered only at high-latitude or polar stations.

It is interesting that the last event in cycle 23 (GLE70; December 13, 2006) was registered not only at the worldwide NM network but also with substandard ground detectors, specifically, with the URAGAN muon hodoscope (Timashkov et al. 2008). Moreover, this GLE was also registered with the IceTop extensive air shower (EAS) detector, which is the component of the Ice Cube neutrino telescope in Antarctica (Abbasi et al. 2008). All GLEs registered from 1942 to 2012 are listed in Table 2.1.

Based on Table 2.1, we can assume that some weak GLEs were not registered in the early years of observations due to technical and methodical difficulties. If the average occurrence rate of these GLEs is η ~1.0 per year, the number of omitted events in 1942–1956 could be considerable (Miroshnichenko et al. 2012). A prolonged minimum of cycle 23 ended in December 2008; however, cycle 24 (started in January 2009) proceeds very slowly (flabbily), and sunspot formation and solar flare and proton activities are generally at a rather low level. Thus, only one GLE was registered after January 2009 up to now (the middle of 2014).

10.7.2 GLE Registration Frequency

The occurrence rate of so-called relativistic SPEs observed at the Earth's surface (or GLEs), *i.e.*, events with protons of relativistic energies (Table 2.1), may be seen in Fig. 10.9 (Miroshnichenko et al. 2012). Upper panel (*a*) shows monthly mean values of counting rates at the neutron monitors in Climax (1953–1963, *thin line*) and Apatity (1964–2000, thick line). A cyclic course of solar activity (measured in sunspot numbers, *W*) is shown at the bottom panel. The moments of GLEs are marked by triangles (*upper panel*) and parts of straight line (*bottom panel*). The most of relativistic events are apparently observed in periods of ascending and descending in sunspot numbers, less often at the maximum activity, and are virtually absent at the minimum. Such a quasi-regular behaviour of the occurrence rate of GLEs (on the average about 1 event per year) was found to be sharply broken in the past (22nd) solar cycle: in July 1989 – June 1991 as many as 13 GLEs were recorded. This means that against the background (quasi-periodic) variations of the GLE occurrence rate its individual manifestations may be subject to the strongest fluctuations.

Complete GLE statistics, accumulated during the 72 years of ground-based SCR observations (Table 2.1), makes it possible to study some problems related to the spatial and temporal variations in solar activity and the properties of the global solar magnetic field (GSMF). It is interesting to know, e.g., the distribution of GLEs over the heliolongitude of their sources (flares). It was established that the IMF is the "directing" factor when SCR fluxes are formed. Although relativistic particles are as a rule insignificantly scattered when moving toward the Earth (their path length can be comparable with 1.0 AU), the probability that they reach the Earth evidently strongly depends on the Parker spiral angle of the IMF. As a result we get a rather strong dependence of the registration frequency (η) on the source heliolongitude: most sources are related to the ~30–90° W interval of longitudes. However, it is striking that SCRs came to the Earth even from behind-the-limb sources in 12 cases. The source distribution for giant nonrelativistic SPEs has approximately the same

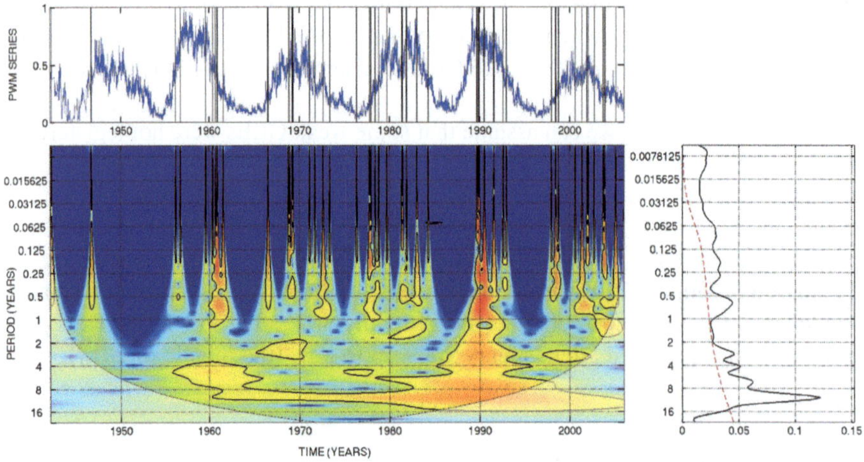

Fig. 10.9 Oscillations of occurrence rate of GLEs (Miroshnichenko et al. 2012). The PWM time series for the GLE registration rate was constructed using the Morlet method by the registration data of 70 events in 1942–2006 (*top panel*); a wavelet diagram for the oscillation spectrum is shown on the bottom panel (the periods in year fractions are presented along the ordinate axis). The oscillation power density spectrum in arbitrary units (the abscissa axis) is shown on the right-hand side depending on the period as a fraction of the year (the ordinate axis). *Dotted line* at the *right* hand corresponds to the power of a red noise

shape. The SPE sources that are supposedly related to the acceleration by shocks in the interplanetary space are distributed more uniformly and have a maximum at a ~30° W heliolongitude (Miroshnichenko et al. 2001).

Another interesting aspect, which characterizes the Sun as a star, was revealed as a result of a wavelet analysis of the GLE registration frequency (η), depending on the SA level (sunspot number) and solar cycle epoch (Miroshnichenko et al. 2012). Using the dates of the events from Table 2.1 and the Morlet method (Pulse Width Modulation, PWM), we constructed the PWM series for parameter η, which includes the statistically significant oscillation with a period of ~11 years. In this case, η oscillations to a certain degree cohere with the time series of the parameters of the photosphere (sunspot number S) and corona (coronal index CI) (Fig. 10.9). In spite of the limitations of the GLE statistics and the wavelet analysis method, these results can be interesting for understanding the periodic phenomena in the solar dynamo, solar atmosphere, interplanetary medium, and cosmic rays.

The tendency of GLEs to group mainly on the ascending and descending brunches of solar cycles is apparently caused by the specific features in the GSMF spatial-temporal structure. As is known, this field reverses its sign precisely near SA maximums. Therefore, we mention the results achieved by Nagashima et al. (1991). These authors used MT and NM data for 43 GLEs from 1942 to 1990 in order to analyze the above GLE tendency. They indicated that the flares that cause GLEs are basically forbidden during the cycle transient phase, when the

GSMF reverses its sign. The absence of GLEs at an SA maximum is explained by a decrease in the particle acceleration efficiency during the GSMF reconfiguration rather than by the suppression of SCR escape processes due to strong magnetic fields. Certain periodicities found in (Miroshnichenko et al. 2012) are coherent for parameters η, S, and CI; therefore, during this stage of the study, the conclusion can be made that oscillations are synchronized in different layers of the solar atmosphere: from the photosphere to the corona. This can indicate that the SCR generation (GLE) involves wide areas in the solar atmosphere rather than is a local (isolated) process.

If the threshold of registration by energy and intensity of the particles is lowered, then the occurrence rate of observed events increases sharply. For example, at the standard proton intensity $I_p(\geq 10$ MeV) ≥ 1 pfu in the descending branch (1980–1986) of the 21st cycle 104 events were recorded (Bazilevskaya et al. 1990a). For a comparison, we would note that 128 such events were observed (Sladkova et al. 1998) over all the 22nd solar cycle (1987–1996). Decreasing in the registration threshold and increasing in the sensitivity of the particle detectors, along with increasing of the duration of direct measurements in space using spacecraft, suggested that the Sun is, in essence, a permanent source of particles with an energy $E \geq 1$ MeV/nucleon (*e.g.*, Miroshnichenko and Petrov 1985).

The general association between the occurrence rates of SPEs with the 11-year cycle of solar activity is beyond doubt. Thus, according to the observation in cycles 18–20 (1940–1970), it was found (Hakura 1974) that the occurrence rate of the proton events has several peaks during one cycle. A similar distribution pattern of the SPEs was observed in the cycles 20–21, judging from the data of existing SPE Catalogues (Dodson et al. 1975; Akinyan et al. 1983; Bazilevskaya et al. 1990a; Sladkova et al. 1998).

The change in the proton activity of the Sun in 1965–1985 (cycles 20 and 21) may be traced, for example, from the observations of the fluxes of protons with $E_p \geq 10$ and ≥ 90 MeV on board the *Meteor* satellite (Avdyushin et al. 1987). These authors chose the events with the following intensity thresholds: $I(\geq 10$ MeV) ≥ 1 pfu; $I(\geq 90$ MeV) twice as large as GCR background. As it turned out, the number of events changes irregularly in different stages of these cycles. Rapid increase in the number of events is observed after the minimum of the 19th cycle. The largest number of SPE in the 20th cycle was recorded in 1967 (at the end of the ascending stage) and in 1968–1969 (epoch of the maximum). In the 21st cycle the maxima in the frequency distribution of the events were recorded in 1978 (ascending stage) and 1982 (descending stage), the number of events in the descending stage of this cycle being appreciably greater than in the corresponding period of the previous cycle.

To reveal any regular patterns in the behaviour of the SCR flux (long-term trends, quasi-periodic variations, *etc.*) it is common to average the flux $<F>$ over certain time interval or to sum (to integrate) the flux F_s over the whole event (in the last case the magnitude of F_s is called fluence). Specialists in particle physics usually have in mind the peak (maximum) flux observed in a separate energy channel of a solar particle detector. This may either the integral flux with energy above the given threshold in units of *pfu/MeV*. Individual events are, in the main, compared using identical energy channels. The peak flux, as a rule, is used to characterize solar proton events as a whole. Fluence F_s is an event-integrated number of particles with energy above the given threshold. The value of F_s may be determined either from the isolated direction in the units of *particle* $cm^{-2}\ sr^{-1}$, or in the form of the sum over all directions in the units of *particle* cm^{-2}. In general, fluence characterizes the total exposure radiation dose. If necessary, instead of fluence for one proton event fluence is used for a certain period of solar activity.

One of the first set of data on event-integrated fluences of solar protons was compiled by Reedy (1977) who took into consideration events with F_s (>10 MeV) $>10^7\ cm^{-2}$. In all 13 such events, observed in 19–20th cycles of solar activity, were identified. Later on, Goswami et al. (1988) identified 63 three similar events, of which six belong to the end of solar cycle 20 (1972–1975) and 57 to cycle 21 (1976–1984). Two years after, Shea and Smart (1990a, b) have published a summary of the 218 events between 1955 and 1986 with F_s (>10 MeV) $>10^7\ cm^{-2}$. Their list contains also the data (Feynman et al. 1990a) for the 135 events with F_s (>30 MeV) $>10^5\ cm^{-2}$.

In Fig. 10.10 (upper panel) we represent the proton fluences for individual events for the three solar cycles, from 19 through 21, based on the reported data by Reedy (1977) and Goswami et al. (1988). Event-integrated fluences, F_s(>10 MeV) and F_s(>30 MeV), are shown for each event. The solid curve in this figure represents the smoothed monthly-averaged sunspot numbers. Lower panel shows the proton fluences F_s(>30 MeV) for several last solar cycles, actually, from 21 through the middle of cycle 24 (Getselev et al. 2009, 2013).

An interesting aspect that can be noted from Fig. 10.10 is the fact that major proton events are relatively rare near sunspot maximum and occur mostly at the ascending and declining stages of solar cycle. From this figure and other findings (see Miroshnichenko and Petrov 1985; Miroshnichenko 2001) it follows that there is no close correlation between the values of cycle-averaged flux, $\langle F \rangle$, fluence F_s (>10 MeV), and *W*. The average $\langle F \rangle$ values are determined mostly by one (or several) of the largest events of the given cycle. For example, several consecutive events in August 1972 made a contribution of ~70 % to the magnitude $\langle F \rangle$ for the whole 20th cycle.

A similar situation was fixed in the cycle 22 when a series of large SPEs in September-October 1989 contributed about 50 % of the total fluence of F_s(>10 MeV) estimated for over the cycle (Shea and Smart 1993a). These authors also ranked the largest SPE series total fluences of F_s (>10 MeV) as follows: July 1946, November 1960, October 1989, and July 1959. The events of August 1972, with a particle fluence of $1.1 \times 10^{10}\ cm^{-2}$, has often been used as a fiducial mark for a "worst case" solar proton fluence of F_s(>10 MeV), primarily because it was the first extremely large event measured comprehensively in space (*e.g.*, Smart and Shea 1989b). However, since August 1972 there have been other events that had larger fluences of F_s(>10 MeV), such as the major episode of solar proton activity in October 1989, with a total fluence of F_s(>10 MeV) $= 1.9 \times 10^{10}\ cm^{-2}$ (Shea and Smart 1993b).

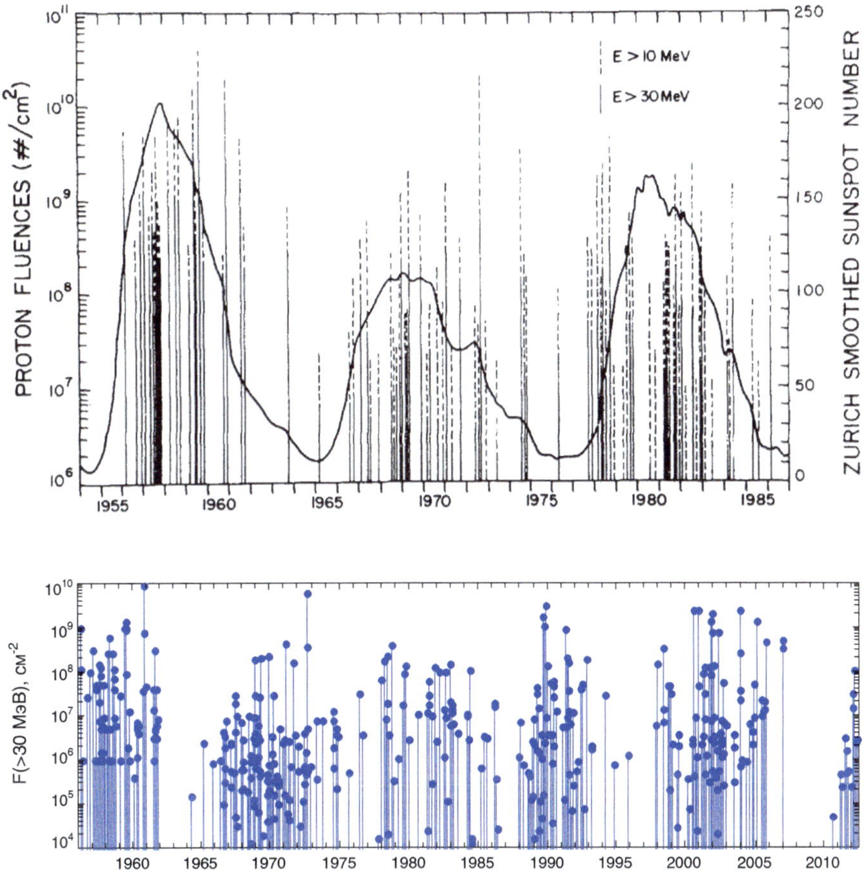

Fig. 10.10 *Upper panel*: The event-integrated fluences above 10 and 30 MeV and the smoothed monthly averaged sunspot numbers for the three solar cycle numbers 19–21 (Goswami et al. 1988). *Lower panel*: The proton fluences $F_s(>30\ \text{MeV})$ for several last solar cycles, actually, from 21 through the middle of cycle 24 (Courtesy by M.V. Podzolko, SINP MSU, 2014)

As to the occurrence rate of proton events treated in terms of fluences *versus* solar activity level, the results of recent studies seem to be controversial. Based on the data by Feynman et al. (1990a) for the events with $F_s(>30\ \text{MeV}) > 10^6\ \text{cm}^{-2}$, Nymmik (1999b) argued that the mean occurrence frequency of the >30 MeV events is proportional to W. Meanwhile, proceeding from the same data set, Dorman and Pustilnik (1999) found no similar regular feature. They stated that for great fluences ($>10^6\ \text{cm}^{-2}$) the occurrence frequency increases with the increasing of solar activity, indeed; the largest observed fluences, however, occurred in periods of intermediate solar activity ($W = 40$–80, 80–120, and 120–160).

There is no relation whatsoever would be the apparent link between the SPEs and a complex of solar activity parameters, their quasi-periodic regularities prompted Zil et al. (1987) to subject the data of several Catalogues (Dodson et al. 1975; Akinyan et al. 1983; Bazilevskaya et al. 1990a) to spectral analysis. The body of the data analyzed included peak fluxes of protons with an energy E_p >10, >30, and >60 MeV for 406 events for 1964–1984 and the fluences of the protons F_s(>30 MeV) for 297 events. From these data, calculations of power density spectra (PDS) were made. As it turned out, quasi-periodic variations are found in the distribution of the events by the fluence of the protons of F_s (>30 MeV) >10^5 cm^{-2}. Analysis indicated the presence of several distinct peaks in the PDS in the range of frequencies up to 100 nHz. Some results of the calculations (Zil et al. 1987) are given in Fig. 10.11.

It is seen that in the time series presenting different manifestations of solar activity there is, in particular, a peak at the frequency about 16.5 nHz corresponding to the QBO (quasi-biennial oscillation). A confidence level of 90 % was exceeded by the peaks corresponding to the periods of about 11, 3, 2 years, 14, 8, and 5 months. The most powerful peak in the low-frequency range roughly corresponds to the 11-year period of solar sunspot activity, and periods of ~5, ~8 months, ~2 and ~11 years exceeded a 95 % confidence level. These variations are in a good agreement with those of other parameters of solar activity both in the frequency and phase of the oscillation. Periodic variations in the occurrence rate of the SPEs and other characteristics of solar activity have been found in a great number of studies. Of special interest are oscillations with periods ~5 months (about 150–155 days) and ~2 years. Below we describe briefly some of the most important results.

Bai (1987) have analyzed the occurrence rate of major flares in solar cycle 19. It was found a periodicity of 51 days, which is one-third of the period found from the flares rates of solar cycles 20 and 21 by various authors. The statistical significance of the periodicity is estimated to be at 99.85 % confidence level. This periodicity is thought to be related to the 153 day periodicity. Rieger et al. (1984) discovered a periodicity of about 152 days by analyzing flares detected by the Gamma-Ray Spectrometer (GRS) on board SMM during the interval from February 1980 to August 1983. Kiplinger et al. (1983, 1984) also analyzed about 7,000 flares observed with Hard X-Ray Burst Spectrometer (HXRBS) on board the SMM during the period February 1980 to September 1984 and found periodicity of 158 days. Dröge et al. (1990) have examined the occurrence times of energetic (>10 MeV) solar flare electron events observed on board the *ISEE* 3 spacecraft during the years 1978–1982. They found strong evidence for a periodicity of 153 ± 2 days, confirming the discovery of a periodicity on the occurrence rates of solar flares producing hard X-rays (Kiplinger et al. 1983, 1984) and gamma rays (Rieger et al. 1984). The Rayleigh test for periodicity gave a probability of less than 10 that the times of the electron flares were drawn from a uniform distribution.

The best determined length of the period for the occurrence rate of proton events, 154.6 ± 0.6 days, was obtained by Bai and Cliver (1990) for solar cycles 19–21 (1955–1986). They selected 385 events with a peak flux of I(>10 MeV) >1.0 pfu from the Catalogues by Dodson et al. (1975) and Akinyan et al. (1983), as well as from the *IMP*-8 data. Two epochs have been identified that exhibit a 154 day

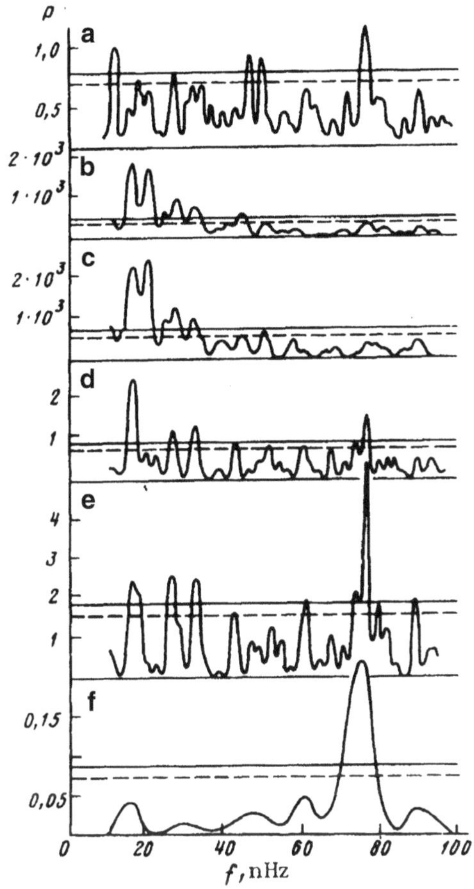

Fig. 10.11 Power density spectra of the variations of the proton fluences for the events with F_s (>30 MeV) $>10^5$ cm^{-2} (Zil et al. 1987) in the period 1964–1984 (**a**); intensity of galactic cosmic rays according to measurements in the stratosphere above the Mirny station (Antarctic) at two altitudes with residual pressure of 50 g cm^{-2} (**b**) and 20 g cm^{-2} (**c**); intensities of protons with E_p >0 MeV at two levels of discrimination, 10 pfu (**d**) and 1.0 pfu (**e**); and frequencies of solar gamma-ray bursts (**f**). Solid and dashed lines denote the 90 and 95 % confidence levels, respectively

periodicity. These epochs are a 14-year interval from January 1958 through December 1971 and a 5.5-year period from February 1978 to August 1983. The evidence was found that the phase of this periodicity changed between the above-mentioned two epochs by half a period ($\Delta\varphi = 0.5 \pm 0.16$).

Gabriel et al. (1990) have carried out power spectral analyses of the occurrence rate of SPEs with time-integrated flux of protons F_s (>30 MeV) $= 10^6$ cm^{-2} during the three solar cycles 19, 20, and 21 and also have revealed a periodicity around 154 days. This feature is prominent in all of the cycles combined, in cycles 19 and

21 individually, but is only weak in cycle 20. These results are consistent with the presence of similar periodicities between 152 and 155 days in the occurrence rate of major solar flares, the behaviour of so-called sunspot blocking function, P_s (it describes a depression of the solar irradiance due to sunspot passage through the solar disk), the 10.7 cm radio flux, $F_{10.7}$, and the sunspot number, W (see Gabriel et al. 1990, and references therein). This means that about 154-day period may be a fundamental characteristic of the Sun.

It should be noted, however, that the cause (mechanism) of the 154-day periodicity remains unknown (*e.g.*, Bai and Cliver 1990; Gabriel et al. 1990). Moreover, Kile and Cliver (1991) found no evidence for 154 day period in the occurrence rate of solar flares in 22nd solar cycle that began in September 1986. They have examined the Ottawa 2.8 GHz burst record from January 1955 to February 1990 and have found that the 154 day periodicity to be statistically significant only for the years from 1978 to 1983 corresponding to the activity maximum of solar cycle 21. From their revision and critical analysis of other relevant results, Kile and Cliver (1991) concluded that, other than for the period in cycle 21 in which it was first discovered, the evidence for the 154-day periodicity from flare-related data sets is contradictory and not compelling. They noted also that stronger evidence for the occurrence of the 154-day periodicity outside cycle 21 can be found in some recent studies (*e.g.*, Lean 1990) that examine parameters such as sunspot numbers and area that characterize solar active regions.

As it was noticed long ago (*e.g.*, Hakura 1974), the most powerful SPEs avoid the period close to the solar maximum. Actually, many solar, heliospheric and terrestrial parameters changing generally in phase with the solar activity are subjected to a temporary depression ("valley") close to the solar maxima (*e.g.*, Bazilevskaya et al. 1998, and references therein). This effect was called "Gnevyshev Gap" after the astronomer who initiated investigation of the double-peak structure of the solar activity cycle (Gnevyshev 1977, and references therein). The nature of such a structure is still obscure; nevertheless, the Gnevyshev Gap (GG) effect may be used to make some peculiarities of solar-terrestrial relations more clear.

Certain evidence of the GG effect on galactic cosmic rays (GCR) was obtained by Bazilevskaya et al. (1998) by the observations of 1954–1996 for such a transient phenomena as a Forbush decrease (the depletion of GCR intensity related to the passage of intense interplanetary disturbance). A more sophisticated analysis based on the power spectrum density (PSD) of 26–29 variations (1954–1996) allowed them to reveal the "valley" in the behaviour of GCR variability at the maximum phases of solar activity. It was also shown that the maximum flux of the >10 MeV protons observed in the Earth's environment in 1976–1996 displays the GG effect on yearly basis. In addition, it was demonstrated that the GLEs (1956–1996) usually avoid time intervals in which the GSMF reversal occurs (see Fig. 3.12), *i.e.*, an extended period after each sunspot maximum.

In more detail, the GG effect in solar cosmic rays was studied by Bazilevskaya et al. (1999) in attempting to trace the origin of the GG effect in GLEs and non-GLEs, or to separate solar energetic particles (SEPs) according to their flare or shock associations (see Sect. 2.7). They considered the >10 MeV proton events at the threshold intensity >1 pfu, without any preliminary selection of the events with respect to the flare association. It was shown that during the two last solar cycles (1975–1996) solar proton events with and without relativistic protons manifest the GG effect in different ways. While GLEs were absent, non-GLEs continued to occur, though with smaller particle intensity. Bearing in mind the role of interplanetary shocks in particle acceleration and geomagnetic storms, these authors used the data on SSC amplitude and D_{st}-variation in geomagnetic field as a proxy for interplanetary disturbances. As a result, Bazilevskaya et al. (1999) came to a conclusion that the GG effect in solar cosmic rays confirms a direct relation between GLEs and powerful solar flares, without regard to their impulsiveness. As to non-relativistic SEP events, they are more closely correlated with the amplitude of proxies for strong interplanetary disturbances.

Soon after, Miroshnichenko et al. (2001) have compiled more extended homogeneous data series for the period of 1955–1996, based on several Catalogues of solar proton events (Dodson et al. 1975; Akinyan et al. 1983; Bazilevskaya et al. 1990a; Sladkova et al. 1998). Above 1 *pfu* of the >10 MeV protons, we separated in all 320 events with a certain or probable flare association. Their size distributions are shown in Sect. 5.8 (Figs. 5.10 and 5.11). In the context of the issue under consideration, it is of interest to illustrate an occurrence rate of proton events on the time interval about four cycles of solar activity (42 years). The yearly numbers of proton events are plotted in Fig. 10.12 in comparison with the level of solar activity. The GG effect in solar cosmic rays is seen quite distinctly, in spite of a number of methodical difficulties concerning the selection of events at the "pre-spacecraft era" of the SPE observations (1955–1965) (for details see Miroshnichenko 2001, and references therein).

All above-mentioned findings are evidently of great importance in the studies of general proton emissivity of the Sun and long-term trends in the behaviour of solar magnetic fields. In addition, those data can be very helpful for elaboration methods for prediction the radiation conditions in space (Chap. 11). In particular, Zil et al. (1988) developed a technique of prediction the occurrence rate of the SPEs with the proton fluence of F_s (>30 MeV) $>10^5$ cm^{-2} as a function of calendar time for a period of up to 11 years. The technique was tested in the form of tentative forecast of the SPEs for a period from 1983 through 1986, and its preliminary results turned out to be rather promising.

10.8 Archaeology of Solar Cosmic Rays

As known, the understanding of climate oscillations or trends in the past and their prediction for the future require the long-term sets of various astrophysical and geophysical data. In this context, as one can conclude from the above considerations, there are of a certain interest also the data on long-term trends and cyclic variations in the SPE occurrence rate, spectra and SCR fluxes as a function of the

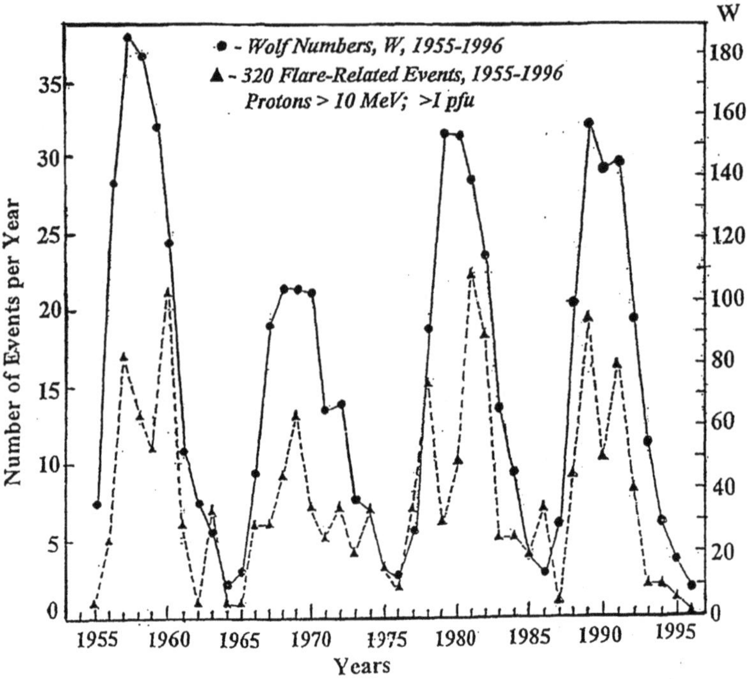

Fig. 10.12 Yearly numbers of the >10 MeV proton events at intensity threshold >1 pfu in comparison with the level of solar activity measured in Wolf numbers, *W*, for the period of 1955–1996 (Miroshnichenko 2001)

level of solar activity. Direct methods, however, do not permit one to establish the characteristics of the cosmophysical and ecological processes over a large time scale, excepting, probably, for the sunspot number variations observed since 1749. To solve these problems one has to use indirect methods of extracting the data from the physical "eyewitnesses" of the past which were capable not only of recording phenomena, but of retaining the relevant information in their memory in its original form. One of the "eyewitnesses" of such a kind are cosmogenic isotopes, produced in galactic and solar cosmic ray interactions with the material of the Moon and planets, meteorites, cosmic dust and the Earth's atmosphere (Dorman 1998). Cosmic ray sources and mechanisms of cosmogenic isotope production are represented schematically in Fig. 10.13.

The most detailed data available presently on the variations of the cosmic ray flux in the past were obtained by studying the ^{14}C and ^{10}Be isotopes produced in the Earth's atmosphere (see, *e.g.*, Kocharov 1991a, b). They are the radioisotopes with the highest atmospheric production rate (2.2 and 0.02 atoms $cm^{-2}\ s^{-1}$, respectively) of all long-lived isotopes (with a half-life $T_{1/2}$ ≥100 years). The major source for the radiocarbon is the reaction of the capture of thermal neutrons released in cosmic ray interactions in the atmosphere by nitrogen, $^{14}N(n,\ p)^{14}C$. Beryllium-10 is

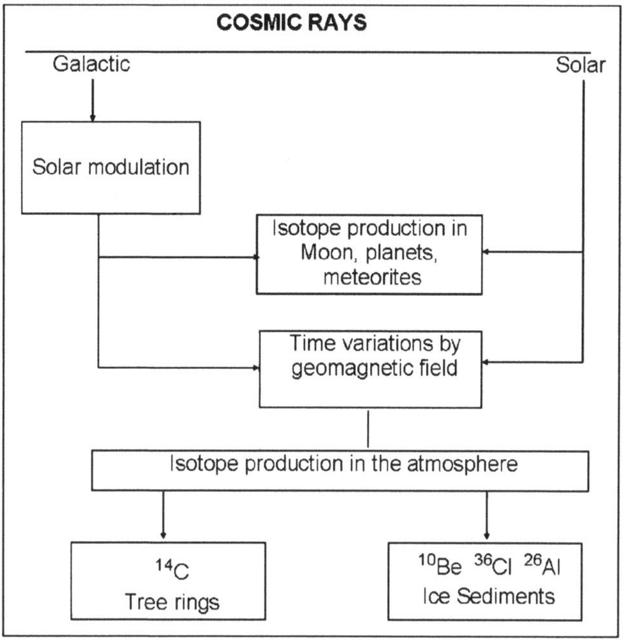

Fig. 10.13 Schematic representation of cosmic ray sources and cosmogenic isotope production mechanisms (After Kocharov (1991a, b))

produced in spallation reactions of nitrogen and oxygen by cosmic rays. These isotopes are rapidly oxidized to form CO_2 and BeO and enter subsequently into various geochemical and geophysical processes. Carbon dioxide distributes itself throughout the global carbon exchange system while beryllium oxide becomes attached to atmospheric aerosols and precipitate onto the Earth's surface with them. Thus, the samples used to study these isotopes are essentially different, namely, radiocarbon measurements which are primarily carried out on tree rings, whereas the Be content is determined in polar ices and oceanic sediments.

The high rate of ^{14}C production and comparatively short half-life time (~5,730 years) permit one to use well-developed radiometric techniques (see, *e. g.*, Dergachev 1994). The use of such techniques to measure ^{10}Be content in samples would involve formidable difficulties because of low ^{10}Be production rate and its long life ($T_{1/2} \sim 1.5 \times 10^6$ years). Therefore, in this case one employs a more sophisticated and expensive method of accelerator mass-spectroscopy. Besides, one can use another possibilities, for example, to choose different radionuclides in the analysis of long-term processes in the heliosphere, at the Sun, *etc.*

The most reliable series of data are compiled for the radiocarbon concentration over the last ~10,000 year (*e.g.*, Dergachev 1994). An essential difference in energy spectra makes it possible to discriminate between SCR and GCR effects in production of radioactive isotopes of ^{14}C, ^{81}Kr, ^{26}Al, and ^{53}Mn in meteorites and lunar rocks, as well as to estimate average SCR fluxes and to determine their spectral form as far as ~10^7 years back. Thus, Bhandari and Bhattacharaya (1975) used the data on the ^{26}Al content in lunar rocks to show that the spectrum of solar protons ~1.5×10^6 years ago can be represented in the form ~$\exp(-R/R_0)$, where $R_0 = 150$ MV is the characteristic spectrum rigidity. Then, the average proton flux with $E_p > 10$ MeV ($R > 143$ MV) is $<F> = 150$–180 cm^{-2} s^{-1}, with an accuracy within 25 %.

On the other hand, the values of $<F(>10\ \text{MeV})>$ in the same units obtained by Goswami et al. (1983) using a large amount of data on the content of cosmogenic isotopes in lunar rock are as follows: ~2×10^2 (for ~2×10^5 years, ^{81}Kr); ~125 (for ~10^6 years, ^{26}Al); ~ 70 (for ~10^7 years, ^{53}Mn). Similar estimates have been obtained for $<F(>10\ \text{MeV})>$ from solar flares in the three last solar cycles: 4×10^2 (cycle 19), ~90 (cycle 20), and ~65 (cycle 21). The SCR rigidity spectrum also varies significantly. If the spectrum is represented in the form ~$\exp(-R/R_0)$, then $R_0 = 100 \pm 20$ MV for the last ~10^6 years and $R_0 = 48 \pm 22$ MV for the years 1965–1982.

The data obtained suggest considerable long-term variations of SCR flux and its effective rigidity. Thus, the average SCR flux was steadily increasing for the last 10^7–10^4 years (approximately by a factor 3), whereas the value of $<F>$ for the last three solar cycles, on the contrary, became nearly 6.5 times smaller. The authors (Goswami et al. 1983) believe that these variations of SCR parameters may be caused by two factors: (1) long-term variations of solar activity with characteristic periods of ~10^5–10^6 years; (2) giant flares that produce SPE with the proton fluence of F_s ~10^{13} cm^{-2} (*e.g.*, flares of the 23 February 1956 type).

Both causes are plausible, though hypothetical (see Sect. 4.6). For example, the concentration of cosmogenic isotopes, ^{10}Be, ^{26}Al, and ^{53}Mn in ocean and lake sediments cores, in meteorites and lunar rocks shows that 2–4 million years ago their production rate was four times as high as nowadays (Kocharov 1978). On the one hand, this might be due to an increased GCR flux as a result of Supernova outburst (Kocharov 1982), to the geomagnetic field inversion (Dergachev 1994) and/or to an extremely low GCR modulation during the inversion of the global magnetic field of the Sun (Ustinova and Lavrukhina 1987). In turn, a weak GCR modulation may be indicative of a lowered solar activity. On the other hand, the effect noted by Kocharov (1978) might as well be the result of increased solar activity that is characterized by growing flare production and SPE intensity. In other words, the level of solar activity in the ancient times might have been quite different from the present-day situation.

In this context, it is of great interest to study the "ancient" acceleration processes which took place during the early evolution stage of the Sun when it was an active young star of the T-Tauri type, with a strong solar wind and a flare activity 10^3–10^5 times as high as at present. With this purpose, Caffee et al. (1987) used a high-sensitivity mass-spectrometer to measure the content of spallogenic noble gases (^{21}Ne and ^{38}Ar) in individual grains from the gas-rich meteorites. The grains containing the track of solar flare-generated heavy ions proved to be enriched with ^{21}Ne and ^{38}Ar compared with the grains without the tracks. The data on meteorites Murchison (the carbonaceous chondrite), Weston and Fayeteville (H-chondrites),

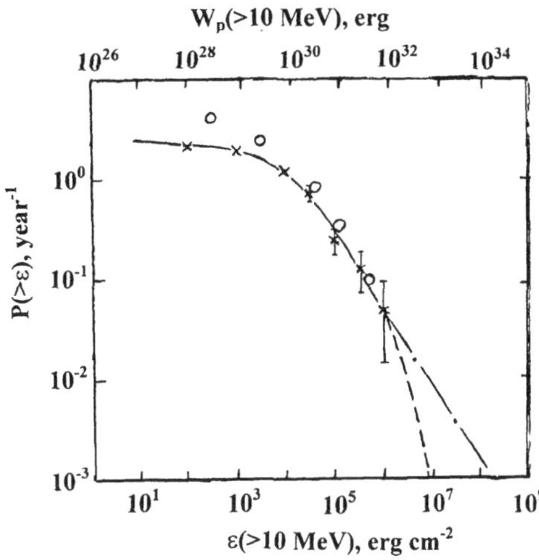

Fig. 10.14 Integral distribution of solar proton events in solar cycles 19–21 as a function of the >10 MeV proton energy fluence (Gladysheva et al. 1995). The circles represents the occurrence rate of proton events with total energy flux greater than ε; the crosses correspond to the seasonal distribution of energy fluence (averaged over the 3-month period). The dashed line only could be in agreement with available data on the nitrate content in the polar ices

and Kapoeta (achondrite) were analyzed. The contents found of the stable ^{21}Ne and ^{38}Ar imply a (100–200) × 10^6-year exposure to galactic cosmic rays near the surface. From other data it follows, however, that the exposure did not exceed 10^6 years. The authors have concluded that the >10 MeV proton fluence had to be some 10^{16}–10^{18} cm^{-2}, which correspond to a ~10^3 times as high irradiation by SCR flux for ~10^5 years during the T-Tauri stage as the irradiation during the last 10^6 years.

Thus, one can see that radiochemical methods, alongside a new "nitrate method" described in Sect. 10.6, are very effective tool for solar cosmic ray research. Although the analysis of tree rings, meteorites, returned lunar samples, oceanic sediments *etc*. is a more mature technology than the more recent analysis of polar ice cores (Dreschhoff and Zeller 1990) for the determination of historical proton events, each of these technique can be used to improve our knowledge of SPE occurrence prior to the middle of twentieth century and of some important features of solar cosmic rays. In the whole, this branch of space physics may be called "archaeology of solar cosmic rays" (see also Shea 1990; Shea and Smart 1990a).

As a recent example, it is worth to mention an interesting possibility to obtain upper limit of total energy induced by solar flare protons relying upon the data of nitrate abundance in the polar ice (Gladysheva et al. 1995). Crucial point of such an approach is a quantitative correlation between the abundance of the nitrate NO_X and total energy (fluence) ε delivered to solar protons in each solar flare in the past. In Fig. 10.14 the distribution of solar proton events for solar cycles 19–21 is represented (Gladysheva et al. 1995). The circles in Fig. 10.14 mark the occurrence rate of proton events with total energy fluence above the ε value, while crosses indicate the seasonal distribution of energy fluence (averaged over the 3-month period).

According to the data on cosmogenic isotope abundances in lunar samples (Reedy 1980), average flux of the >10 MeV solar protons for the past million is 100 ± 25 cm^{-2} s^{-1} which is close to the value obtained for the last several solar cycle (see above). Therefore, these authors conclude that the dotted-dashed line in Fig. 10.14 cannot not correspond to real SPE distribution in the past; on the contrary, the dashed line could be in agreement with available nitrate data. The abundance of the nitrate can be measured with the time resolution of about 3 months which is better than in the case of radiocarbon measurements. As it is seen from Fig. 10.14, averaging for 3 months practically does not change the distribution of the most powerful events. The measurements of the nitrate content for the last cycles of solar activity show that the threshold of sensitivity of this method is at the level of $\varepsilon = 3 \times 10^5$ erg cm^{-2}.

More recently, Peristykh and Damon (1999) presented evidence of intense solar proton events in the last decade of the nineteenth century (solar cycle 13) based on diverse solar and geophysical data. One of those events (15 July 1892) was observed as remarkable solar disturbance (white-light flare). Besides white-light flares, there were numerous storm sudden commencements (SSC) of high amplitude (>40 nT), noticeable enhanced annual sums of the *aa*-index, and more frequent observations of very bright auroras in North America. The event of 15 July 1892 is also revealed from data on nitrates in polar ice and cosmogenic isotopes in terrestrial archives.

10.9 Extreme Solar Proton Events

Long-term observations of solar proton events (SPEs), or solar energetic particle (SEP) events, give a number of evidence of that about once time during a given solar cycle, the event occurs whose fluence above a given energy (usually ≥10, ≥30, ≥60 and ≥100 MeV for protons) dominates that for the entire cycle (e.g., Shea and Smart 1990a, b). It may overlap the fluences from the other events and even determines, in fact, a total fluence for the cycle. Such rare phenomena are sometimes called by "rogue events" (Kallenrode and Cliver 2001a, b), in analogy to rogue ocean waves having unusually large amplitudes. Well-known examples of rogue SEP events at the Earth occurred on 14 July 1959, 4 August 1972, 19 October 1989, and 14 July 2000. Rogue events have been also observed in the inner heliosphere – with *Helios* 1 on 4 November 1980 at 0.5 AU and with *Ulysses* in March 1991 at 2.5 AU. Origin of the rogue events is thought to relate to the multiple coronal mass ejections (CMEs) and converging interplanetary shocks. If observed at the Earth's orbit, these rare extreme events become important geophysically and applicably (*e.g.*, radiation hazard for spacecraft).

In some cases extreme SEP events in non-relativistic energy range are accompanied by large fluxes of relativistic protons ($E_p \geq 500$ MeV), or solar cosmic rays

(SCRs). They are usually registered by neutron monitors (NMs) at the Earth's surface (GLE phenomena, or Ground Level Enhancement of SCR). Recently, it was suggested (Crosby 2009) to define rare Solar Extreme Events (SEEs) as those events in which the characteristics (field strength, speed, intensity of radiation, energies, etc.) of the associated phenomena (solar flares, CMEs, SEP events) are some orders of magnitude larger than in the most of other events (*e.g.*, event of 20 January 2005).

At present, the so-called "Carrington event" of 1–2 September 1859 (Townsend et al. 2003, 2006) seems to be added to the list of such "rare SEEs". As follows from (Smart et al. 2006, 2008), Carrington event (CE) had the largest integral fluence Φ of protons with the energy of $E \geq 30$ MeV (i.e., the energy-integrated fluence above a certain energy value) for about 450-year period starting from 1561. There is no doubt that the study of such rare events is of paramount importance. In particular, Townsend et al. (2003, 2006) suggest, henceforth, to consider CE with the integral fluence of $\Phi(\geq 30 \text{ MeV}) = 1.88 \times 10^{10}$ cm^{-2} as the best reference "worst-case" for the estimates of radiation hazard in space.

Indeed, the two nearest candidates for a role of "worst-case" – the events of 15 November 1960 and 4 August 1972 – were characterized by far less values of $\Phi(\geq 30 \text{ MeV})$, about 9×10^9 cm^{-2} and 5×10^9 cm^{-2}, respectively (Smart et al. 2006). Note, however, that those fluence values have been calculated by the data obtained in the epoch of historically fragmentary and indirect measurements of SEP fluxes. At present time it becomes clear that energy spectra applied earlier have the analytical forms quite different from the spectrum form established recently (Nymmik 2011). For this reason, fluence sizes for the events of 1960 and 1972 should be critically discussed in the light of new summary distribution function (see Sect. 12.5). Of great interest are also the estimates of occurrence probabilities of such rare events at present level of solar activity and possible extrapolation of obtained results for the remote past of the Earth (*e.g.*, Wdowczyk and Wolfendale 1977; Kiraly and Wolfendale 1999).

As we know by own long-term experience of studies of solar cosmic rays, rare large solar events do not form some specific "class" of solar phenomena. They seem to constitute a part of common ensemble of SEP events, since in reality there is no sharp boundary between this "class" and the rest of events. The SEP events are described by single distribution function, and SEEs naturally form its "tail" in the range of low probabilities. This point of view is confirmed, in particular, by recent results of Crosby (2009) analysis: SEEs are part of the global distribution of all events rather than "outliers" with their own special characteristics.

Our present study was greatly inspired by the publication of new data on proton fluences for a number of large events of 1561–1994 identified by so-called nitrate method (McCracken et al. 2001), as well as by the results of analysis and interpretation of those events (*e.g.*, Townsend et al. 2003, 2006; Smart et al. 2006, 2008). Distribution functions of SCR events on the proton fluences with the energy ≥ 30 MeV, or $\Phi(\geq 30 \text{ MeV})$, have been widely investigated (see, *e.g.*, Nymmik 2011, and references therein). These functions are constructed, as a rule, based on the data on SEP events whose sizes are determined by measurements on board the satellites *IMP* and *GOES* series. By present, the data sets available cover, depending of selection criteria, about 200 events with the fluence of $\Phi(\geq 30\ \text{MeV}) \geq 10^6\ \text{cm}^{-2}$ (Nymmik 2011). When describing the distributions, it is usually applied the power-law functions, sometimes with a break. Such approach, however, allows us to calculate the probabilities of event occurrence with certain fluence only down to a probability of ~0.5 % (~1/200) that is evidently not enough for extreme estimates.

Lately it has become obvious that the accumulation of new satellite data do not enable us to advance considerably in the problem of determination of the form of distribution function for SEP events on their fluences in the range of low probabilities. An attempt to involve the data on cosmogenic isotopes in the lunar soil (Reedy 1996), unfortunately, does not add a certainty in the resolution of this problem because of the isotope data are related to a total (summary) flux of SCR protons with the energy of $E \geq 10$ MeV for the past ~10 My, but not to individual SEP events.

Some progress in this problem has been outlined when the data on the fluences of large SEP events for the period of 1561–1950 have been obtained from the Greenland ice cores (McCracken et al. 2001). These authors have succeeded, in particular, in estimating the proton fluence for the largest event of that period, namely, Carrington event, occurred on 1–2 September 1859; its value was $\Phi(\geq 30\ \text{MeV}) = 1.88 \times 10^{10}\ \text{cm}^{-2}$. Nevertheless, even those data turned out to be not enough to determine a form of distribution function in a total diapason of changes of the fluence $\Phi(\geq 30\ \text{MeV})$. In fact, by those data it was impossible to determine an amount of small events that constitute initial part of distribution function in the range of fluences $\Phi(\geq 30\ \text{MeV}) = 10^6 \div 3 \times 10^9\ \text{cm}^{-2}$. Therefore, to calculate the probabilities of extra-large SEP events (and distribution function) from polar ice data it is needed to know how many single events of fluence $\Phi(\geq 30\ \text{MeV}) \geq 10^6\ \text{cm}^{-2}$ have occurred since 1561 up to now. To our opinion, the solution of this problem seems to exist (Miroshnichenko and Nymmik 2014).

With this goal in mind, we have again applied to the distribution function (3) interpolating it into the range of extremely low probabilities. When doing this, it was assumed that peak fluxes and fluences for large ensemble of SEP events are on average proportional one to another. Our estimates were based on the measurements onboard two spacecraft *IMP*-8 and *GOES* and on the data from Greenland ice core. We also have taken into account integral fluences of $\Phi(\geq 30\ \text{MeV})$ estimated by the data of Kiraly and Wolfendale (1999), with the extrapolation into the past for 1 and 100 My. Our final results are presented in Chap. 12.

In the course of this study, also deserve serious attention the estimates of the proton fluences at some other energies (besides 30 MeV), especially, for the understanding of flare (proton) activity of the Sun in the remote past. Many years ago, Wdowczyk and Wolfendale (1977) addressed the question on the long-term frequency of large solar energy releases and their possible effects, compared with other catastrophic events. The main body of their evidence appears still valid, although some details have changed. The very flat integral power-law fits (logarithmic slope around −0.5) suggest that several dramatic solar energy releases should be expected in geologically short times, if the trend continues.

Extrapolating their highest energies (>60 MeV) fit to long time scales, Kiraly and Wolfendale (1999) obtained some another estimates. It turns out that while the highest fluence measured up to 1999 (in about 30 years) was 3×10^9 cm^{-2}, one would expect in 1 My a few events above 10^{12} cm^{-2}, and in 100 My a few above 10^{13} cm^{-2}. This is far less than one would expect from flat slopes found by Wdowczyk and Wolfendale (1977), but still about two orders of magnitude higher than it follows from our estimates.

In fact, according to modern data on proton fluences at the energy $\geq$30 MeV, for the period from 1973 up to 2008 there were registered 205 events with the fluence $\geq 10^6$ cm^{-2} Nymmik (2011). If solar activity remains at modern (present) level, it means that for 1 and 100 My, respectively, we may expect for 6×10^6 and 6×10^8 of such events, and the probabilities of their realization would be ~1.7×10^{-7} and ~1.7×10^{-9}, respectively. According to our estimates (Fig. 12.10), for such long periods the events may appear with the fluences up to 6×10^{10} and 10^{11} cm^{-2}, respectively, that is for $1.5 \div 2$ orders of magnitude less that the estimates by Kiraly and Wolfendale (1999). Two triangles in Fig.12.10 depict our estimates of $\Phi(\geq 30$ MeV) based on the data by Kiraly and Wolfendale (1999) for the fluences of protons at the energy $\geq$60 MeV, with the extrapolation into the past for 1 and 100 My, respectively. Difference in the energies of protons (30 and 60 MeV) makes this discrepancy even much more.

The cause of this discrepancy is rather simple. As it was repeatedly noted (Nymmik 2006, 2007a, b, 2011), lognormal distribution function of SEP events (Feynman et al. 1993) that was applied by Kiraly and Wolfendale (1999), by no means does reflect a physical essence of SEP event distribution in the range of large fluences. Parameters of the model by Feynman et al. (1993) are determined mainly by subjective (random) magnitudes of the registration thresholds and selection of small SEP events; therefore, they can not serve for the extrapolation of the data into the range of extremely large events.

Quite recently, based on some indirect but totally independent data, we get a good possibility to verify our new methodical approaches and results. Miyake et al. (2012) have published the results of their ^{14}C measurements (so-called carbon-14 method) in annual rings of Japanese cedar trees from AD 750 to AD 850 with 1- and 2-year resolution. It was found a rapid increase of about 12 % in the ^{14}C content from AD 774–775, which was about 20 times larger than the change attributed to ordinary solar modulation. The authors, however, argue that neither a solar flare nor a local Supernova is likely to have been responsible for this increase. The reality of the AD775 event is confirmed by new measurements of ^{14}C in German oak (Usoskin et al. 2013). The authors, on the contrary to Miyake et al. (2012), argue in favour of that this event can be associated with a strong, but not inexplicably strong, solar energetic particle event (or a sequence of events).

However that may be, researchers from another group (Thomas et al. 2013) have decided to examine possible sources of a substantial increase of ^{14}C content in AD 774–775. First of all, the authors rejected a coronal mass ejection (CME) as a possible cause of the effect, because the required CME energy is not several orders of magnitude greater than known solar events. Further, they have modeled solar proton events (SPEs) with three different fluences and two different spectra. Finally, they concluded that the data may be explained by an event with fluence about one order of magnitude (about seven or more times greater) beyond the SPE of October 1989 (depending on the spectrum). Two hard spectrum cases considered by Thomas et al. (2013) may result in moderate ozone depletion, so no mass extinction is implied. At the same time, the authors do predict increases in erythema and damage to plants from enhanced solar UV. Also, they are able to rule out an event with a very soft spectrum that causes severe ozone depletion and subsequent biological impacts. As to nitrate enhancements expected in the period under consideration, they seem to be consistent with their apparent absence in ice core data.

Applying now to the data of October 1989 it should be emphasized that, in fact, at that time three separate SPEs have been registered (on 19, 22, and 24 October), with the fluences of $\Phi(\geq 30$ MeV) about 1.82×10^9, 7.44×10^8, and 3.95×10^8 cm^{-2}, respectively (*e.g.*, Nymmik 1999c). Their sum gives the fluence $\Phi(\geq 30\ \text{MeV}) = 2.96 \times 10^9$ cm^{-2}. This value, obviously, cannot compete with that for Carrington event (CE), 1.88×10^{10} cm^{-2}. However, when multiplying a total fluence for three events of October 1989 by a factor of $7 \div 10$ (Thomas et al. 2013), we get the values of $\Phi(\geq 30\ \text{MeV}) = (2.07 \div 2.96) \times 10^{10}$ cm^{-2} comparable with the CE fluence. As to integral flux of the protons $F(\geq 30$ MeV), its expected value for the event of AD775 just lies down the ULS curve when we multiply the corresponding flux value for the event of 19 October 1989 by the same factor of $7 \div 10$ (Miroshnichenko and Nymmik 2014).

Another approach to the interpretation of the event of AD775 has been undertaken by Cliver et al. (2014). The authors explored requirements for a solar particle event (SPE) and flare capable of producing the cosmogenic nuclide event of AD775, and reviewed solar circumstances at that time. It was found that a solar source for 775 would require a spectrum above 1 GV about 45 times stronger than that of the intense high-energy SPE of 1956 February 23 (GLE05). This implies a >30 MeV proton fluence (F_{30}) of ~8×10^{10} proton cm^{-2}, that is ~10 times larger than that of the strongest 3-month interval of SPE activity in the modern era (August-October 1989). This inferred F_{30} value for the 775 SPE is inconsistent with the occurrence probability distribution for >30 MeV solar proton events (Kovaltsov and Usoskin 2014). Figure 10.15 demonstrates this discrepancy. As can be seen in Fig. 10.15, the red point corresponding to the revised F_{30} value of ~8×10^{10} proton cm^{-2} for AD775 falls outside the 90 % confidence interval of occurrence probability distribution function recently constructed by Kovaltsov and Usoskin (2014). This function is based on direct satellite measurements of SPEs during the space era, measurements of sharp increases of ^{14}C and ^{10}Be concentration in tree rings and ice cores during the last 11,400 year, and cosmogenic nuclides in lunar rocks during the last half-million years.

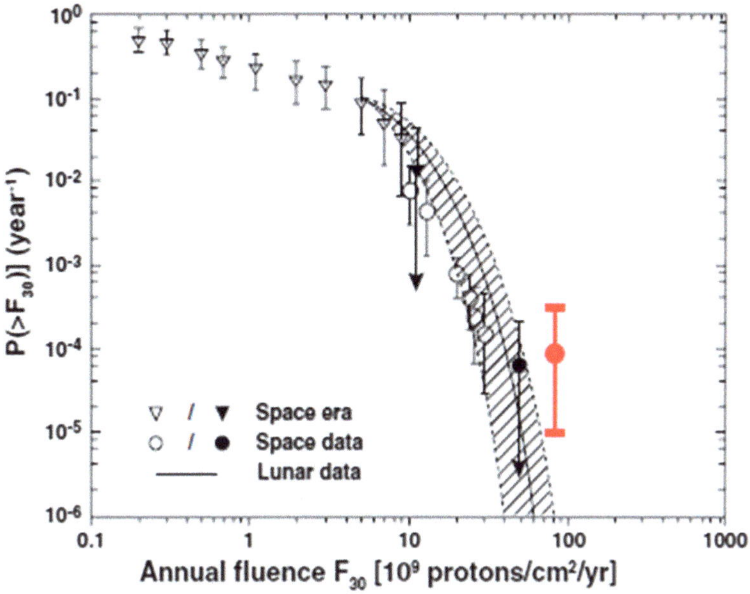

Fig. 10.15 Occurrence probability distribution function by Kovaltsov and Usoskin (2014) for >30 MeV SEP events (in units of 10^9 protons/cm^2/year). *Red point* corresponds to the revised estimate by Cliver et al. (2014)

Cliver et al. (2014) also estimated that the best guess value for the soft X-ray classification (total energy) of an associated flare is ~×230 (~9×10^{33} erg). For comparison, the flares on 4 November 2003 and 1 September 1859 had observed and inferred values of ~×35 (~10^{33} erg) and ~×45 (~2×10^{33} erg), respectively. The estimated size of the source active region for a ~10^{34} erg flare is ~2.5 times that of the largest region yet recorded. The AD775 event occurred during a period of relatively low solar activity, with a peak smoothed amplitude about half of that of the second half of the twentieth century. The ~1945–1995 interval, the most active of the last ~2,000 year, failed to witness a SPE comparable to that required for the proposed solar event in 775. These considerations challenge a recent suggestion that the AD775 event is likely of solar origin. Note that in the case of Carrington event we deal with real experimental data (see, however, Sect. 12.6), whereas for the event of AD775 all estimates are model dependent. On the other hand, the modern technological implications of such events may be extreme. Considering recent confirmation of super-flares on solar-type stars, this issue merits attention.

Chapter 11
Energetic Solar Particles and Radiation Hazard in Space

Since the middle of past century, when active human penetration into extraterrestrial space has begun, the researchers have clearly faced by a number of new problems associated with radiation hazard in space and biological action of cosmic rays: (1) radiation hazard in space flights; (2) radiation effects of cosmic rays in the search for traces of life on other planets, in the biological conquest of extraterrestrial territories; (3) indirect influence of cosmic rays on the biosphere through their action on the upper layers of the atmosphere (see Chap. 10). In more detail the problem of radiation hazard in space was reviewed in the monographs by Miroshnichenko and Petrov (1985) and Miroshnichenko (2003a). Our consideration below is mainly limited by principal physical concepts and mechanisms of the radiation impact of solar cosmic rays. In particular, we describe some models of radiation hazard taking into account a relative role (contribution) of solar cosmic rays.

11.1 Identification of Radiation Hazard

Although biological action of charged particles was known for some time before the beginning of manned space flights (see, *e.g.*, Grigoriev 1975), the radiation risk in space has been visually identified, for the first time, during the performance of the US lunar program Apollo (see, *e.g.*, Letaw et al. 1987, and references therein). The program did not include an extended study of the radiation conditions in space and biological effects of the radiation. Nevertheless, there were discovered two important phenomena.

First of all, a special radiochemical processing resulted in discovery of distinct tracks of heavy cosmic ray ions in the plastic material (lexan) of the helmets of the astronauts of Apollo 8 through *Apollo* 12. As an illustration, we show on the cover the sulfur-ion tracks from *Apollo* test helmet that has the properties similar to those of the human kidney cells. The helmets used for this experiment included one from

L. Miroshnichenko, *Solar Cosmic Rays*, Astrophysics and Space Science Library 405, DOI 10.1007/978-3-319-09429-8_11

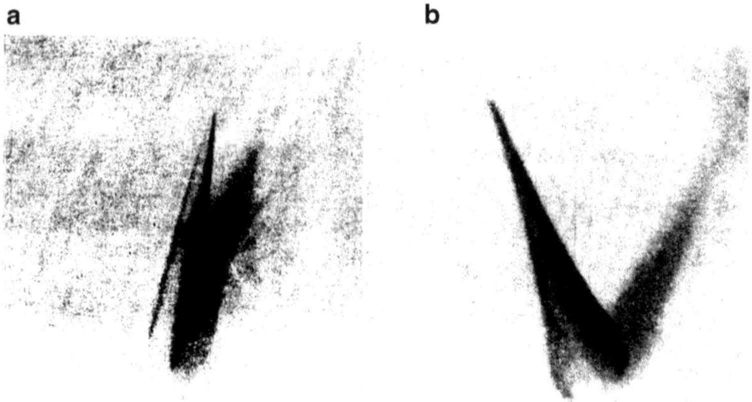

Fig. 11.1 Tracks produced by heavy cosmic ray nuclei on the inside of an Apollo space helmet (Comstock et al. 1971): (**a**) a track from a particle entering the helmet; (**b**) a ending track from a particle that crossed from the opposite side of the helmet an come to rest. The tracks are 500 and 700 μm in length, respectively

the *Apollo* 8, and all three from the *Apollo* 12 helmets. A control helmet was used to cosmic rays at a balloon altitude of 41 km at Fort Churchill (Canada). The helmets were stored in the dark to avoid ultraviolet enhancement effects.

The picture of observed radiation impact is shown in Fig. 11.1 (Comstock et al. 1971) for two cases: (A) a track from a particle entering the helmet; (B) an ending track, from a particle that has crossed from the opposite side of the helmet and come to rest (see also Sect. 2.6). The tracks are 500 and 700 μm in length, respectively.

As mentioned above, one of the most intense SEP events ever recorded occurred in August 1972, between two lunar missions, the *Apollo* 16 (since16 April) and *Apollo* 17 (since 7 December). If the event had coincided with the *Apollo* 17 flight, it would have made it necessary to abort operations on the lunar surface (Letaw et al. 1987).

Second, the crewmembers on the last seven Apollo flights (from *Apollo* 11 through *Apollo* 17) observed light flashes (*e.g.*, Pinsky et al. 1974). The flashes are tentatively attributed to cosmic ray nuclei (with the atomic numbers Z >6) penetrating the head and eyes of the observers and interacting in the retina, vitreous body (humor) and possibly the optic nerve (for some details see Chap. 7). Later on the light flash (LF) phenomena have been observed in detail by many cosmonauts (astronauts) on board the Space Stations *Apollo-Soyuz* and *Mir* (*e.g.*, Avdeev 2001). Visual pictures of light flashes observed by astronauts (cosmonauts) on board different spacecraft in 1969–1999 is shown in Fig. 11.2 (Pogorely 2001). Occurrence rate of LF depends on orbit parameters, especially on the high latitude, and grows in polar regions and in region of the South Atlantic Anomaly (SAA). The LFs are practically absent on the equator where the flux of charged particles is minimal.

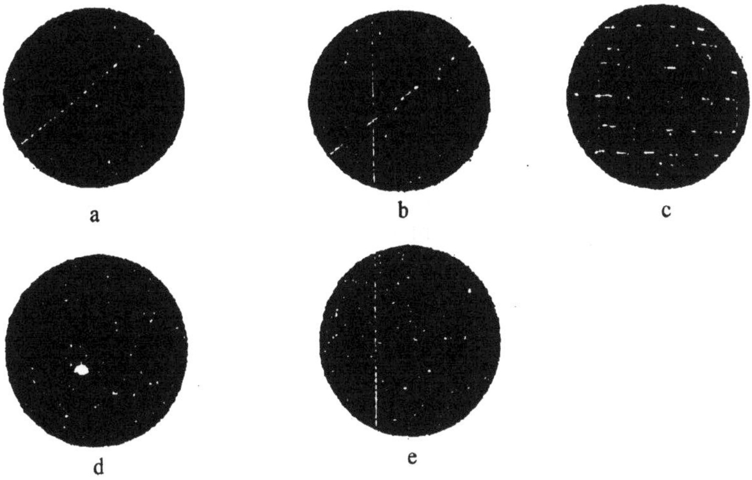

Fig. 11.2 Visual pictures of light flashes (LF) in various forms (**a**–**e**) observed by astronauts (cosmonauts) on board different spacecraft in 1969–1999 (Pogorely 2001)

The LF phenomenon consists of anomalous and unexpected visual sensation and appears to the astronauts as a faint spot or a streak of light (phosphene) in closed eyes after dark adaptation (Fig. 1.4). It is not known whether LFs are generated by particle interactions within retina or if they occur at different level in the visual system, from the optic nerve to the visual areas in the occipital regions of the cortex. As far as we know, LF reception is the only functional anomaly directly caused by a single particle. Therefore, it is important to determine simultaneously time, energy and trajectory of the particle passing through the cosmonaut's visual system, to recognize its kind and origin.

The LF phenomenon was studied in the past with dedicated observation programs on *Apollo*, *Skylab* and *Apollo-Soyuz* spacecraft. But new impulse has been given since 1995 with use, for the first time, of silicon detectors in SilEye-1 and SilEye-2 experiments on board the *MSS* (*Mir Space Station*) (e.g., Galper et al. 1996; Casolino et al. 2001; Avdeev 2001). The first prototype, SilEye-1, in 25 h of work with different astronauts, in 2 years, has tracked more than 50 LFs, correlating them with radiation background (Galper et al. 1996). The detector SilEye-2, since October 1997 up to August 1999, has collected more than seven million particles and hundreds of LFs (Galper et al. 2001) that partially clarify the problem but still do not answer many questions. In those experiments, a significant correlation between the particle fluxes and the number of perceived phosphenes have been demonstrated.

As mentioned above, one of the most intense SEP events ever recorded occurred in August 1972, between two lunar missions, the *Apollo* 16 (since 16 April) and *Apollo* 17 (since 7 December 1972). If the event had coincided with the *Apollo* 17 flight, it would have made it necessary to abort operations on the lunar surface (Letaw et al. 1987).

At present, practical cosmonautics has accumulated, in total, more than 50-year experience of manned space flights. In spite of this, up to now a humankind has a 2-week period only for *extra-magnetosphere* work of astronauts (Apollo program). Maximum radiation doses recorded at the near-Moon's orbit were 18 rem for the human skin and 8 rem for the blood (Silberberg et al. 1987).

As it well-known, together with geomagnetically trapped particles and galactic cosmic rays, solar-flare protons contribute to the terrestrial radiation environment (see Chaps. 4 and 10) and can present a threat to humans in space (*e.g.*, Grigoriev 1975; Fry and Nachtwey 1988) as well as to spacecraft sensors and sensitive components used in satellite subsystems and instrumentation (*e.g.*, Kreinin and Grigorieva 1979; Adams and Gelman 1984; Miroshnichenko and Petrov 1985; Gorney 1989, 1990; Tranquille and Daly 1992; Tylka et al. 1997b, and references therein).

We already know that the solar proton flux to locations in space for current and projected manned space missions is completely dependent on the characteristics of the spacecraft orbit (*e.g.*, Smart and Shea 1984). Depending on the mission requirements, it may be possible to use the Earth's magnetic field to shield partially (or even completely) against solar-flare particles. For example, the flux of solar protons to Earth-orbiting spacecraft will be limited to those particles that can penetrate through the Earth's magnetic shielding to the spacecraft position (*e.g.*, Zil et al. 1986). However, this situation changes dramatically for lunar and interplanetary missions where the spacecraft will be subjected to the full solar proton flux in the interplanetary medium.

As an introduction to the following discussion, we mention briefly that a number of significant solar flares associated with the recent solar cycle 22 may have been responsible for several spacecraft operational anomalies. In particular, such kind of anomalies was observed after powerful solar flares in March 1989 (Allen et al. 1989) and in March 1991 (Allen 1991). On the other hand, two soviet geostationary satellites experienced significant degradation of the solar-cell arrays (up to 25–30 %) during the period from February 1989 through February 1990 (Grigorieva GM 1990, private communications). Such an effect is equivalent to the annual impact of the particles from the radiation belts of the Earth.

Of special interest are the effects observed during the event of September 29, 1989. According to the data compiled by Shea (1990), the radiation monitors on the high altitude *Concorde* supersonic aircraft flying between Paris and Washington reached alert levels for the first time in the history of their operation. The alert level is set at an equivalent of a chest X-ray dose; this does not indicate a hazard to passengers and crew. Meanwhile, the *Magellan* spacecraft on its transit between the Earth and Venus experienced 5 % degradation in its solar cell power output. In addition, a number of spacecraft including *Magellan* observed interference in the guidance and command and control functions.

Those effects have caused concern among project engineers (see, *e.g.*, Tranquille and Daly 1992), who until recently have tended to understand the solar-proton problem. Furthermore, radiation protection will be a prime issue for extended missions to the planet Mars (*e.g.*, Pisarenko 1992) or for a return visit to the Moon. For these and other reasons, considerable interest has been shown in recent years in the prediction of solar proton fluences from data collected during past solar cycles (*e.g.*, Bengin et al. 1985; Miroshnichenko and Petrov 1985; Goswami et al. 1988; Smart and Shea 1989b; Shea and Smart 1990a; Feynman et al. 1990a, b, 1993; Getselev et al. 1988, 1992, 1996a, b; Feynman 1997). Dose fields produced by cosmic rays at high altitude in the stratosphere also have been extensively studied during several last decades (Kolomeets et al. 1997). In fact, these authors attempted, for the first time, to summarize numerous observational and theoretical data on the radiation hazard at the flight altitudes of supersonic aircraft and other airborne vehicles; the predicted levels of radiation dose for the crew and passengers on board those flying apparatus at the extreme flight conditions are also given.

Since solar proton events can adversely affect the terrestrial environment it is essential to learn as much as possible about the events in order to accurately predict their occurrence and severity. This objective assumes greater importance in planning for the safety of astronauts during long-term missions for space exploration.

11.2 Modern Concept of Solar Proton Event

As noted in Chap. 2, the Catalogues of Dodson et al. (1975), Akinyan et al. (1983), Bazilevskaya et al. (1986, 1990a), Sladkova et al. (1998), and Logachev et al. (2014) contain data of all recorded near the Earth increases of the flux of the $\geq$10 MeV protons with the intensity $I \geq 1$ pfu. The source of such protons not always can be identified with a certain solar flare (some share of increases is undoubtedly caused by shock waves in the interplanetary space). In those cases when the source of protons is a flare it is possible to speak of a solar proton event (SPE) as an increase of the flux of energetic particles (presumably protons) of solar origin at the Earth's orbit. In the course of compiling of the Catalogues mentioned above, it became clear that such a simple definition of SPE should be extended. A new concept of solar proton event was formulated in the middle of 80's (Miroshnichenko and Petrov 1985; Miroshnichenko 1986). If one includes into consideration also energetic solar electrons and nuclei heavier than the proton (*e.g.*, Smart and Shea 1989a; Feynman 1997) then we can treat SPE more extensively as a solar particle event, with corresponding modifications in their predictions.

At the modern level of our understanding, a solar proton (particle) event may be defined as follows: SPE is a set of interconnected processes preceding the moment of SCR generation, t_0 (with anticipation of T_0), occurring near this moment (synchronous effects), and accompanying SCR generation with a delay T with respect to the t_0 moment or to the interval $t_0 + \Delta t$, where Δt is the duration of acceleration. Such an approach makes it possible to formulate a problem of predicting SCR fluxes on a clear physical and methodical basis (*e.g.*, Bengin et al. 1985; Miroshnichenko and Petrov 1985). By this definition, it is implied to take into account source (flare) position, data on its electromagnetic wave emission, neutral radiation, specific features of the SPE time profile (for example, additional peaks due to interplanetary shock acceleration), *etc*. This concept of SPE by no means denies the importance of data on flare neutrons, gamma-quanta, electrons and nuclei heavier than the proton. It should, however, be stressed that within the energy interval of 10–200 MeV/nucleon (the most serious one from the point of view of radiation hazard) just SCR protons have the largest intensity (*e.g*., Miroshnichenko and Petrov 1985).

From the point of view of radiation conditions the main characteristics of SCR are (*e.g*., Miroshnichenko and Petrov 1985; Miroshnichenko 1990, 2003a): distributions of particles on their energies (spectrum), charges (composition) and directions (anisotropy) as well as temporal variations of a total fluxes (fluences) of SCR from one SPE to another, spatial-temporal evolution of the spectrum during certain SPE, and variations of SCR fluences within 11-year solar cycle. Radiation characteristics of SCR determining the magnitude and dose rate from individual event may be classified in a close association with the concept of SPE: (1) a time of propagation from the Sun to the Earth (the t_m value); (2) an index γ (or R_0) in the observed spectrum; (3) a maximum intensity of protons, I_m, with an energy exceeding the one assigned; (4) an extent of anisotropy, $A(E, t)$, as function of energy and time; (5) a decay rate of proton intensity after the maximum or characteristic time of the decay, $\tau_d(E)$; (6) a charge composition; (7) an intensity gradient of particles in the interplanetary space. All indicated parameters vary depending on the particle energy. It should be noted that in the energy interval of 10–200 MeV the proton propagation in the IMF may be described, in many case quite satisfactorily, by a diffusion model. In its turn, this helps to create procedures for predicting dynamics of SCR fluxes (see Sect. 11.5).

11.3 Dynamics of Radiation Dose

Let us consider in more detail dynamics of spectral characteristics of SCR as SPE develops. Energy spectra of solar particles form as a result of consecutive development of three main processes: acceleration of particles in a flare (source spectrum), transport in the solar atmosphere (ejection spectrum), and SCR propagation in the IMF (observed spectrum). But even if to confine oneself to the spectrum being observed, its shape and intensity of SCR vary significantly with time, especially fast during first hours after the flare. The evolution of the spectrum is due peculiarities of propagation of low- and high-energy particles - mainly due to energy dependence of their mean free path (see Chap. 8).

Before the study of Freier and Webber (1963) the spectrum was usually presented by a power-law energy function of the type (1.6). In a log-log scale the spectrum has a shape of a straight line with a slope -γ. In a narrow range of energies not exceeding one order of magnitude such a presentation is, within the limits of the errors, quite substantiated. But if the spectrum is measured within the range of two or more orders of energies, then deviations of the spectrum shape from the power-

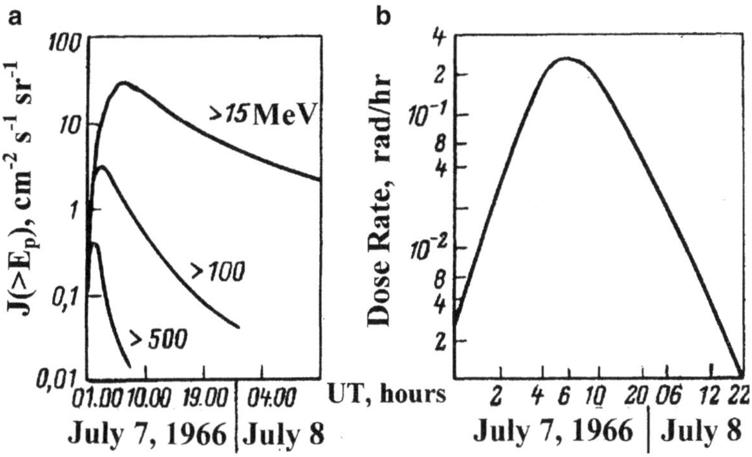

Fig. 11.3 Intensity-time profiles (**a**) of solar protons of different energies and dose rate (**b**) in open space with the shielding thickness of 1 g cm^{-2} Al during the event of July 7, 1966 (Baker et al. 1969)

law dependence become quite evident (see, *e.g.*, Fig. 4.4). Thus, the power-law index depends on both time and width of the interval of energies in question. This leads to considerable difficulties when comparing the spectra measured in different energy intervals. With this in mind, one cannot be sure of the accuracy of extrapolation of the spectrum obtained by different methods to the energy range overlapping a few orders of magnitude.

To overcome these difficulties, Freier and Webber (1963) have proposed a one-parameter exponential function of rigidity (1.8). As a result of analyzing spectra of solar particles for 53 time moments during 16 different proton events, they showed that the characteristic rigidity R_0 of the spectrum (1.8) varies from one SPE to another, but for the given event it depends on time only, $R_0(t)$. More later, investigating in detail the evolution of spectral characteristics of SCR during a particular event of September 28, 1961 in the rigidity range of 0.4 MV–1.0 GV (about 100–600 MeV in Fig. 1.1), Bengin et al. (1979) have found that the $R_0(t)$ value depends also on the rigidity interval under consideration. A similar behaviour of $R_0(t)$ was discovered (Kazaryan et al. 1979) for protons with E <24 MeV (R <210 MV) during the SPE of September 7, 1973 as well as for some other events. Such a pattern of evolution is inherent, mainly, of proton events with intensity-time profiles of diffusion type which are commonly observed from western flares.

As it was already mentioned, in the most of SPEs solar protons of high energy reach the Earth's orbit earlier than low-energy ones. Such a dispersion of their arrival velocities is seen, for example, in the event of July 7, 1966 (Fig. 11.3). Although this event was rather moderate (maximum intensity of the >15 MeV protons, for instance, did not exceed 30 pfu), its temporal and spectral

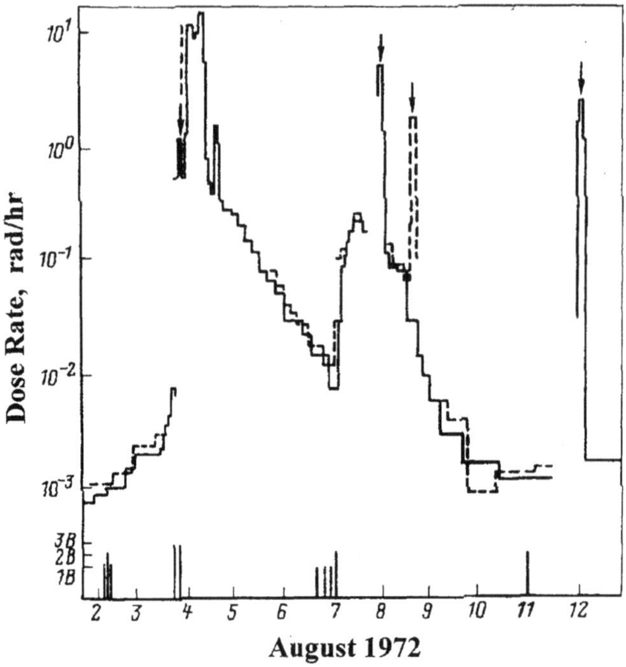

Fig. 11.4 Time profile of dose rate during solar proton events in August 1972 by the measurements on board the satellites *Prognoz* (*dashed line*) and *Prognoz*-2 (*solid line*). *Vertical lines* below denote observable solar flares (Savun and Sladkova 1976)

characteristics were typical for the proton events, so, it turns out to be very suitable to demonstrate the dose dynamics (Baker et al. 1969).

Two additional samples of temporal dynamics of the dose rate and dose value are given in Figs. 11.4 and 11.5, respectively. Figure 11.4 shows the results of measurements of the dose rate which were obtained by the ionization chamber behind the shielding of 6–7 g sm^{-2} Al on board the spacecraft *Prognoz* and *Prognoz* 2 during a series of the SPEs in August 1972 (Savun and Sladkova 1976).

In Fig. 11.4 we present the time variations of the dose calculated for a few major proton events of the 20th cycle of solar activity (Miroshnichenko and Petrov 1985). The calculations were performed for a spacecraft outside the geomagnetic field with a protection thickness of 1 g cm^{-2} Al. Arrows in Fig. 11.4 denote the moments of maxima of the intensity of the >30 MeV protons. It is seen that the dose increases with a greatest speed at the initial stage of SPE. At the same time, magnitudes of the dose increase smoothly enough. Thus, even in the case of the most intense among the events indicated in Fig. 11.5 (August 4, 1972) the dose increase approximately to 50 % of the maximum one for 18–20 h. Such pattern of the dynamics of the dose justifies the application of prediction methods based on the diagnostics of SPE parameters (see Sect. 11.5).

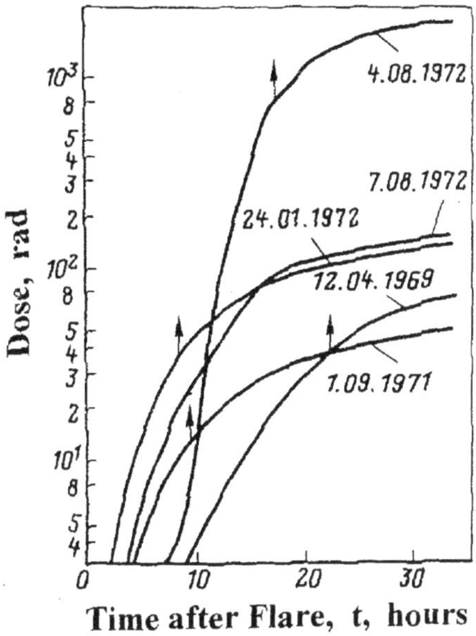

Fig. 11.5 Time profiles of radiation dose of solar cosmic rays from five major proton events of 20th cycle of solar activity (Miroshnichenko and Petrov 1985). *Arrows* mark the moments when the dose increase approximately to 50 % of the maximum one for each event

It is interesting to note that solar proton events during solar minimum are not significantly different from those recorded at other periods of the solar cycle (Gussenhoven et al. 1988). This follows from the dose measurements on board the *DMSP/F7* satellite at 840 km in 1983–1987 (solar minimum). In particular, the spectral features of protons observed between 1983 and 1987 at the energies <100 MeV seem to be similar to those found from earlier periods near solar maximum. Based on their findings, Gussenhoven et al. (1988) proposed a method characterizing the high energy particles (>1 GeV) using a power spectrum index and dose number from the <100 MeV protons. It can be useful in specifying polar radiation condition environments for the design of spacecraft.

Notice that proceeding from physical and/or practical reasons some researchers distinguish the most intense SPEs into a special group (*e.g.*, King 1974). If the main increase of SCR flux was observed in the non-relativistic range (for example, in July 1959, August 1972, October 1989) then such an event is most suitable for the modeling of "a worst case" from the point of view of radiation hazard (*e.g.*, Adams and Gelman 1984; Smart and Shea 1989b).

The gravity of such a hazard for spacecraft crews and equipment was re-affirmed on 20 October 1989 when the dose on board the orbital station *Mir* increased by 2 rad during 4 h (Tverskaya et al. 1991). These unique results are shown in Fig. 11.6 (Teltsov and Tverskaya 1992). On the other hand, according to Bengin et al. (1991), integral dose for the period of 17–27 October 1989 on board the Mir station was about 1.5 rem. This value did not exceed a limited dose for the flight of 0.5-year duration (37 rem), however, it was about 30 times higher than the background values of the dose at the altitude of the *Mir* station orbit. Such episodes emphasize the necessity to develop models of extreme radiation conditions in space. Last but

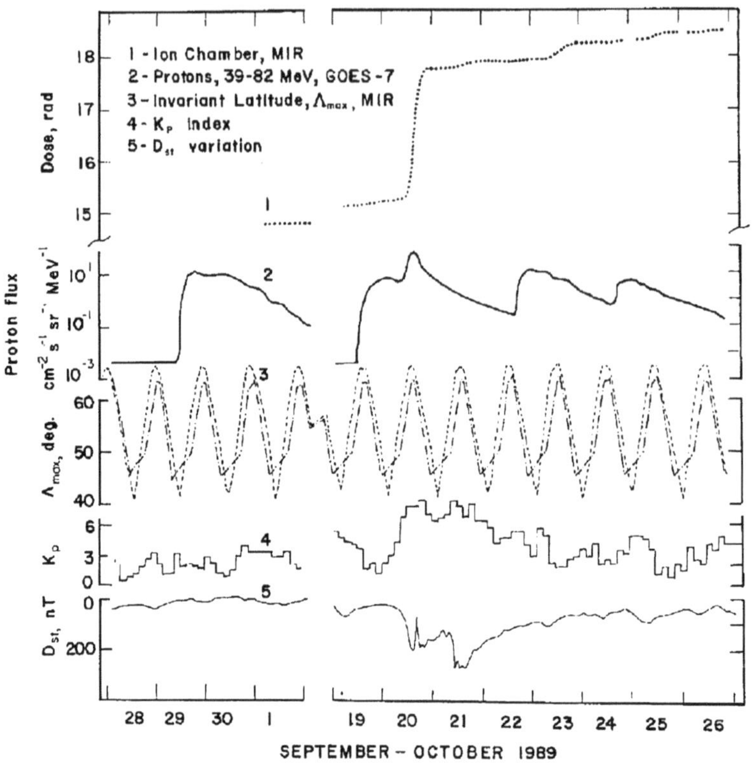

Fig. 11.6 Radiation conditions near the Earth in September-October 1989 (Teltsov and Tverskaya 1992): *1* radiation dose on board the station *Mir*; *2* solar protons with energies of 39–82 MeV by *GOES*-7 measurements; *3* maximum invariant latitude for the station *Mir* orbit; *4* K_p -index; *5* D_{st} -variation

not least, the potentially lethal effects of high-energy radiation on man also need to be considered (Fig. 11.6).

Future interplanetary manned missions will need to consider solar-flare activity very carefully due to the obvious detrimental effects of radiations on humans. Very high doses during the transit phase of a mission can result in radiation sickness or even death. This is equally true for extended visits to surfaces of other planets and Moons lacking a strong magnetic field capable of deflecting flare particle. The risk of developing cancer several years after the mission is somewhat more difficult to quantify, but must also be considered in mission planning.

Adequate radiation protection measures must be conceived for any lengthy interplanetary endeavours. Storm shelters will be necessary both on the transit spacecraft and on the planet surface (see, *e.g.*, Simonsen et al. 1991, and references therein). The latter can be provided to a certain extent by geological features of the body being visited. The design of radiation shielding for a spacecraft (Townsend et al. 1989) is much more difficult, given the inherent limitations associated with the construction of a manned interplanetary space vehicle.

11.4 Radiation Effects on Space Equipment

The two best-known elements of satellite systems most susceptible to damage by solar-flare protons are microelectronics (*e.g.*, Adams 1986; Ray and Mullen 1991) and solar cells (*e.g.*, Lanzerotti 1969; Kreinin and Grigorieva 1979). Both are affected by ionization and atomic displacements processes which in extreme cases can lead to the complete loss of a spacecraft. The Charge Coupled Devices (CCDs) used in modern spaceborne scientific instruments are also vulnerable to damage by SEPs (see, *e.g.*, Tranquille and Daly 1992). Below we give a short description of the mechanisms involved and illustrate some observed damage effects.

The physical parameter quantifying the ionization effects of radiation on both materials and man is dose, which is defined as the energy deposition per unit mass. Equivalent fluence is used to quantify displacement damage in solar cells and CCD detectors. The concept of equivalent fluence is used to convert integral proton and electron spectra into a monoenergetic 1 MeV equivalent electron flux, thereby providing a direct method of quantifying solar-cell degradation characteristics from easily simulated laboratory conditions.

In recent years the trend towards smaller and faster electronic components and more sensitive detectors has resulted in a need to understand and protect against solar-proton effects on spacecraft systems. In earlier satellites, larger components were used, which meant that single particles could only affect only a limited volume of the device, and thus only cumulative damage resulting from multiple particle interactions could lead to malfunction. With the size of modern devices minimized to improve processing speed and power consumption, a single particle can have a significant effect on, and can even cause irreversible damage to, an electronic device. Consequently, these devices are more susceptible to radiation effects than older components.

One mechanism that allows an energetic proton to deposit energy as it passes through matter is ionization. The energy given up by the incident particle results in the formation of electron-hole pairs, which in turn causes the device's performance to degrade.

Displacement damage occurs when the incoming energetic proton transfers momentum to atoms of the target material. If sufficient energy is transferred, the atom can be ejected from its location, leaving a vacancy or defect. The ensuing physical processes are varied and complex, but once again reduced device performance is the ultimate consequence. This is an important mechanism in solar-cell, CCD and material degradation.

The further mechanisms of importance are those of the Single-Event Upset (SEU) and the Single Event Latch-Up (SEL), which occur when an incident charged particle deposits a short but intense charge trail in the sensitive volume of a component. This charge trail is capable of reversing the logic state of a memory element (SEU) or causing destructive "latch-up" where a parasitic current path is created, allowing large currents to destroy the device (*e.g.*, Chenette and Dietrich

1984). This process is mainly limited to ions of higher atomic number than protons, since the Linear Energy Transfer (LET or *dE/dx*) of heavy ions is significantly greater than that of protons. However, energetic protons can undergo nuclear interactions with component materials and the short-range reaction products lead to increase local charge degradation, producing a SEU or even latch-up.

Variations in proton-induced upsets from large solar flares may be demonstrated by the data of systematic measurements of SEU on board the *TDRS*-1 (Telemetry and Data Retrieval Satellite) in geosynchronous orbit since April 1983 (*e.g.*, Normand and Stapor 1990; Wilkinson et al. 1991). The series of large solar flares in August, September, and October 1989 caused a substantial response in the *TDRS*-1 SEU rate. These flares were associated with particle events that produced five GLEs during this period (see Table 2.1), indicating an extremely high-energy component to the total flux. During these three episodes of high solar activity, August 12–17, September 29 – October 1, and October 19–25, on board the *TDRS*-1 there were recorded 23, 91, and 239 SEUs, respectively (Wilkinson et al. 1991).

Figure 11.7 shows the effect of the October 19, 1989 solar flare on the near-Earth space environment as recorded by the *GOES*-7 and the Deep River neutron monitor. The soft (1–8 Å) X-ray detector shows the flux going off-scale at 1,300 UT. An estimation of the peak flux resulted in an X13 classification. The proton panel in Fig. 11.6 shows the very steep onset of a solar particle event as measured by *GOES*-7. During energetic particle events, the *GOES* particle sensors overestimate the true flux by high-energy particles entering lower energy detectors (see Sect. 2.8). However, even after correction, the total fluence of this event is significantly more (see below, Sect. 11.5) in comparison to that of August 1972, the latter up to the 1989 events being considered the largest of the satellite era. Since then, the very large flares of September 29 and October 19, 1989 are now being considered to replace the flares of February 23, 1956 and August 4, 1972 as the "worst case" of radiation conditions in space (see Normand and Stapor 1990; Wilkinson et al. 1991, and references therein). The second enhancement of the proton flux on October 20, 1989 was due to arrival of an interplanetary shock.

The Deep River neutron monitor panel shows the signature of a GLE. This observation, together with others, indicates that the solar particle spectrum extended well beyond 1,000 MeV/nucleon. At the SEU panel one can see the *TDRS* upsets that resulted in operational anomalies in different devices of the satellite. On the other hand, based on proton integral flux measurements on board the *IMP*-8 satellite and the proton detector data of the *GOES*-7, Normand and Stapor (1990) have estimated expected upset rates for three Random Access Memory devices used on board the *TDRS*-1. It was shown that almost all of 239 SEUs were due to the solar protons because, during quiet times, *TDRS*-1 experiences about one upset every other day due to the galactic cosmic rays.

Figure 11.8 shows the results of calculations of the SEU rate for the six large SPEs of the 21st and 22nd solar cycles. Geomagnetic shielding was ignored and the devices were assumed to be shielded by about 2.5 mm of aluminum. The number of upsets/per day for one of the device were calculated also for the period 7 days after

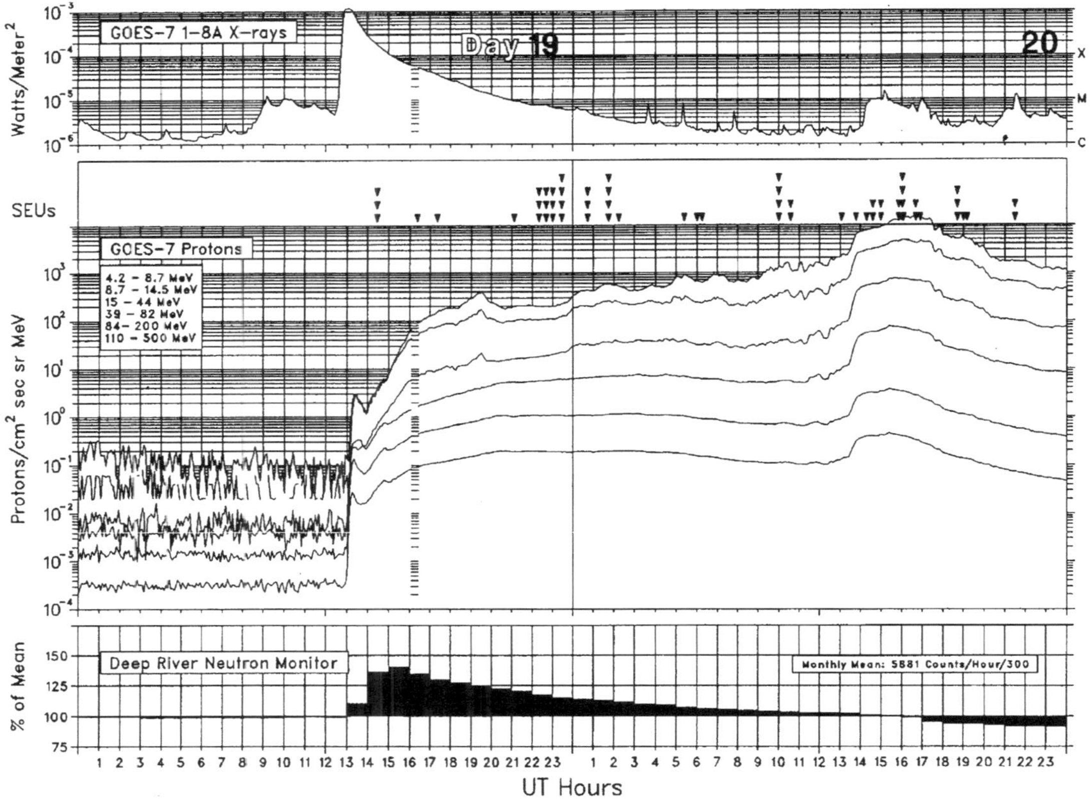

Fig. 11.7 Effects of the October 19, 1989 solar flare on the near-Earth space environment as recorded by *GOES*-7 and the Deep River neutron monitor. The SEU panel shows upsets from operational anomalies on board the *TDRS*-1 (Wilkinson et al. 1991). Each *arrowhead* represents one SEU

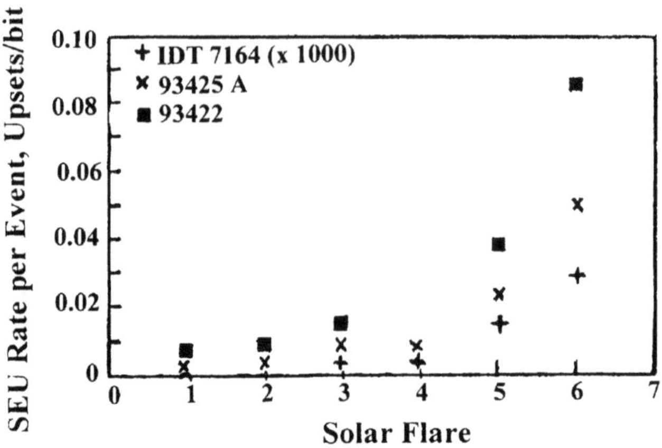

Fig. 11.8 Expected upset rates for three Random Access Memory devices on board the *TDRS*-1 calculated for the six large SEPs of the 21st and 22nd solar cycles (*1* February 13, 1978; *2* April 28, 1978; *3* September 23, 1978; *4* April 25, 1984; *5* September 29, 1989; *6* October 19, 1989). Almost all of 239 SEUs recorded during the October 19, 1989 event were due to the solar protons (Normand and Stapor 1990)

the onset of the October 19 event, at the shielding thickness of 2.5, 6.25, and 12.5 mm of Al. The SEU rate turned out to vary significantly when the integral proton flux changes; it depends also considerably on the shielding thickness.

Solar-cell performance is also adversely affected by the ionization and displacement mechanisms described above (Crabb 1981). Degradation results in the reduction of both the voltage and current output, which may have severe implications for the spacecraft lifetime. Usually, solar cells are made of silicon, although gallium-arsenide cells can provide enhanced efficiency at increased production cost. They are arranged in series and in parallel to provide the desire voltage and current levels, respectively, and collectively form the solar array. Thus, if a single cell fails in a string of cells an open circuit will develop, resulting in total power loss. Solar-cell strings can be arranged in such a way as to minimize power loss from a complete array, but degradation is inevitable.

Solar cells are protected at the front by coverglass, providing shielding against protons. Annealing processes can also offset performance degradation caused by the ambient radiation environment. This is especially true for silicon cells and to a lesser extent for gallium-arsenide cells, although the latter experience substantially less degradation.

As it was recently demonstrated, there are some negative effects of energetic particles on precise experiments on board the Earth-orbiting satellites (*e.g.*, Kahler 2001a). In turn, practical workers demand to improve the radiation standards in order to reduce the weight of space vehicle and total cost of the vehicle construction, as well as a total cost of space flight.

A serious practical problem now is related to various operation anomalies on board the numerous spacecraft of diverse purposes (*e.g.*, Vampola 1994; Wilkinson

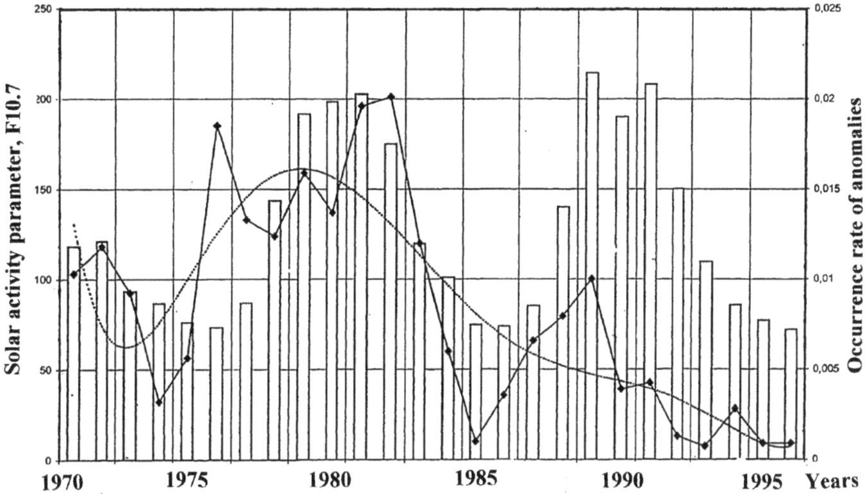

Fig. 11.9 Occurrence rate of various anomalies on board the standard low-orbital satellites of the Cosmos series (*broken line*), its polynomial approximation (*thin unbroken line*), and solar activity level (*histogram*) in the period since 1970–1997 (Chizhenkov 2002)

1994). Later on, Chizhenkov (2002) summarized the data on operation anomalies registered during the flights of standard Soviet (Russian) satellites of the *Cosmos* series since 1970 through 1997. All those satellites had circular orbits with the altitude of 800 km and inclination about 74°. In all, the data of 49 satellites have been selected and analyzed from the point of view of the anomalies' correlation with the solar-terrestrial disturbances ("space weather"). In Fig. 1.9 we demonstrate an occurrence rate of the anomalies, together with its polynomial approximation of 5th degree and solar activity parameter (radio flux F10.7 at the wavelength of 10.7 cm) in the period of 1970–1997.

One can see from Fig. 11.9 a distinct negative impact of solar activity (or space weather) on the rate of operation anomalies. In particular, total level of the anomalies in the periods of maximum solar activity is several times higher then that at the minimum of solar activity. The correlation coefficient between the annual values of occurrence rate and parameter F10.7 for the period of 1970–1989 was estimated to be about 0.6. Also, the correlation between the rate of operation anomalies and geomagnetic disturbances was investigated in detail, but possible contribution of SEPs into the rate of anomalies was not studied.

Noticeable that through 1970–1988 there was no active interference into the control process of *Cosmos* spacecraft functioning. Since the end of 1989, a general scheme of ground-based control has been changed and special technical and administrative measures were undertaken to diminish the negative effects of space weather on the equipment. Due to those improvements, average level of the operation anomalies apparently decreased, although time behaviour of the occurrence rate seems to remain the same, in general outline, as before 1989.

11.5 Diagnostics and Prediction of Solar Proton Events

One of the main problems associated with exploration and employment of the outer space is radiation safety of space crews and equipment. The solution of this problem requires, in particular, information on SCR fluxes obtained in time. It is important not only to estimate the value of expected peak proton flux, $F(>E)$, above a given energy E, but also to predict the SPE evolution, i.e., the space-time variations of proton intensity and energy spectrum (or dynamics of SCR fluxes).

11.5.1 Prediction of Proton Flux Dynamics

According to modern requirements of radiation safety in space flights, the probability (risk) of exceeding the standard dose level should not be more than a certain fixed value (for example, of 1 %). Thus, for predicting the radiation effect of each SPE, one has to estimate the accuracy and reliability of the proton flux values, $<F(>E)>$, with a given priority, T. The latter should be large enough to allow the necessary measures to be taken for the required reduction in the predicted radiation doses for astronauts. Permissible limits of change of T are usually from 1 to 24 h.

Most prediction methods are based either on the space-time distribution of SCR fluxes, or on empirical (heuristic) relationships between the SCR parameters and the characteristics of solar-flare electromagnetic emission (microwave radio bursts, X-ray bursts, *etc.*). The first kind of prediction is made with the method of recent measurements (*e.g.*, Baker et al. 1969; Miroshnichenko et al. 1974). As a typical example of the second method, it should be mentioned a quantitative diagnostics of SPEs (Akinyan et al. 1980) relying upon microwave radio burst data. Each of these methods has its own advantages and is efficient enough within the limits of its applicability. However, none of them satisfies all the above-mentioned requirements to the forecast and none adequately represents the physical nature of SCR generation and propagation, in particular, the probability character of SCR dynamics as a process subject to random fluctuations.

When applying the first group of methods, the prediction becomes possible merely after the beginning of increase of the SCR flux at the point of observations. The second group of methods yields the prediction 3 h in advance; however, the blunder of the prediction is rather large. Assessing the current methods from the point of view of the requirements to the forecast, other drawbacks may also be noted. Thus, basing on the data on radio bursts one can predict a number of SCR parameters: the onset and time of the intensity increase, maximum intensity, *etc.* However, in order to describe the dynamics of the event after the maximum, it is necessary to introduce an empirical parameter τ (characteristic time of intensity decay) which, in fact, is not connected with radio emission of flares, but is determined by the conditions of particle propagation in the interplanetary space.

11.5.2 Probability Prediction Technique

Prediction of the dynamics of SCR fluxes by the method of recent measurements was considered by many workers, however, without detailed discussion of the range of possible errors and reliability of the forecast. Some works (for the references see Miroshnichenko and Petrov 1985) determined the risk of exceeding of the assigned value of the flux or dose; however, such estimates do not give an idea of the dynamics of the event. Averaging of SCR fluxes and spectra over the total time of flight of the space vehicle cannot be used, either, for analyzing the dynamics of fluxes.

Further improvement in the quality of predicting radiation characteristics of SCR may be expected, to our opinion, if a probability description of the predicted values is used. In this case, the *a priori* information of SCR fluxes obtained from the solar-flare electromagnetic emission data should be used in combination with recent measurements taking into account the features of SPE dynamics. One of the possible versions of such a procedure based on Bayes theorem was proposed by Bengin et al. (1985). The main elements of the proposed scheme are the model of the event

$$F = f(t, \alpha) \tag{11.1}$$

describing the dynamics of SCR flux F (or intensity-time profile of the SPE), and the *a priori* probability density, $P_0(\alpha)$, of the expected α magnitudes

$$P_0(\alpha) = P_0(F_m, t_m, \sigma) = P_{01}(F_m)\ P_{02}(t_m)\ P_{03}(\sigma) \tag{11.2}$$

The multidimensional vector α corresponds to the set parameters of the model of SCR propagation (the number of accelerated particles, the mean free path in the interplanetary space, *etc*.). Here $P_{01}(F_m)$, $P_{02}(t_m)$, and $P_{03}(\sigma)$ are the distribution functions of the SPEs on the parameters of F_m, t_m, and σ, respectively, where F_m is a maximum proton flux near the Earth, t_m is a time of the peak intensity, and σ is an index of radial dependence of the mean free path in the interplanetary space. The derivation of these functions was made possible only due to accumulation of sufficient data bases on SPE parameters (SPE Catalogues).

The probability of the F flux being exceeded, *i.e*., the distribution function $\Phi(>F)$ for any time moment is determined by expression

$$\Phi(>F) = \int_G P_0(\alpha)\, \mathrm{d}\alpha \tag{11.3}$$

where G is the range of the P_0 values in which $F > f(t, \mathrm{a})$. If $\Phi(>F)$ is known, one can estimate the mathematical expectation of the magnitude of the predicted flux value

$$< F(t) >= \int_0^\infty F(t)\, \mathrm{d}\Phi(> F) \tag{11.4}$$

and other $F(t)$ characteristics (dispersion, the risk of exceeding the given flux value, *etc*.). The arrival of first solar protons at the detection site provides qualitatively new information, i.e., the measured SCR parameters. The data obtained from a certain SPE are a superposition of the determinate component $F = f(t, \alpha)$ and a random interference $\chi(t, F)$. The latter is due to the stochastic character of SCR propagation and detection.

Let us introduce the value $C(F, y)$ to take into account the random interference. This value is determined as the probability density of registration of y, the determinate basis of the process being expressed as F. It is implicitly suggested that the characteristics of the random function $\chi(t, F)$ depend only on F. The form of the function $C(F, y)$ is mainly determined by apparatus characteristics, as well as by the level of proton flux fluctuations in the interplanetary medium.

If $F(t)$, $P_0(\alpha)$, and $C(F, y)$ are known, one can calculate the probability density $P_1(\alpha)$, taking into account the additional information provided by SCR characteristics measured at a certain moment t_1 after the flare. If the measured flux value at this moment is y_1, then in accordance with Bayes formula (*e.g*., Korn and Korn 1968), we have:

$$P_1(\alpha) = P_0(\alpha) \frac{\mathrm{C}[f(t_1, \alpha), y_1)]}{\int_G P_0(\alpha) C[f(t_1, \alpha), y_1)] \mathrm{d}\alpha} \tag{11.5}$$

A number of measurements, k, allows the probability distribution function to be specified as

$$P_k(\alpha) = A_k P_0(\alpha) \prod_{i=1}^{\infty} C[f(t_i, \alpha), y_i] \tag{11.6}$$

where A_k is the normalization factor, y_i is the reading of the measuring device at a moment t_i. Substitution of Eq. (11.6) in Eq. (11.2) will give the relation

$$\Phi(> F) = \int_G P_k(\alpha)\, \mathrm{d}\alpha \tag{11.7}$$

which is the basis of the algorithm of the proposed procedure for prediction the SCR flux dynamics.

In accordance with the scheme developed by Bengin et al. (1985), in order to apply the procedure it is necessary to construct the $P_0(\alpha)$ function (11.2) for the *apriori* density distribution of SPEs on their characteristic parameters, *i.e*., to construct the functions $P_{01}(F_m)$, $P_{02}(t_m)$, and $P_{03}(\sigma)$. This scrupulous task has been worked out (for details see Miroshnichenko et al. 1986) relying upon all

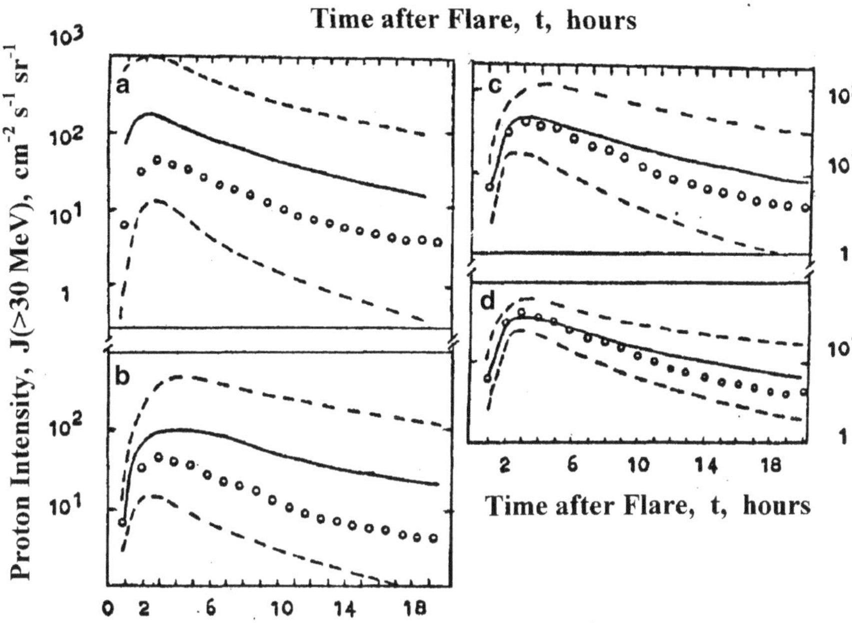

Fig. 11.10 Observed time profile of the proton flux with E_p >30 MeV on February 25, 1969 (*open circles*) in comparison with the prediction based on the *a priori* information P_0 on SCR characteristics (**a**) and taking into account one (P_1, **b**), two (P_2, **c**), and four (P_4, **d**) measurements of proton flux (*solid curves*). The *dotted line* is one percent limits of distribution (Bengin et al. 1985)

existing findings and developments of diagnostics and prediction methods for solar proton events existing before 1985 (*e.g.*, Belovsky and Ochelkov 1979; Smart and Shea 1979; Akinyan et al. 1980). In addition, the function $C(F, y)$ was chosen in the form of logarithmic-normal distribution with a mean-square deviation of about 25 %, which does not exceed the instrumental error during the >30 MeV proton measurements on *Explorer* satellite.

The results of evaluation of the method are illustrated in Figs. 11.10a–d using as an example the SPE of February 25, 1969. Solid lines show the mathematical expectation of the predicted proton flux value, circles correspond to real F (>30 MeV) values, and the dashed lines represent 1 % distribution limits, or, in other words, the reliability of prediction (the probability for the flux value to appear above the upper boundary and below the lower boundary is 1 %). The calculated curve in Fig. 11.10a is based only on *a priori* P_0 data. The spread of possible predicted flux value is seen to reach two orders of magnitude, which indicates to the ambiguity of initial data used for prediction. It can be noted that *a priori* enhancement estimates lie within the 1 % distribution limits, though the predicted fluxes turned out to be overestimated several times in comparison with observations.

Figures 11.10b–d have been plotted taking into account one (P_1), two (P_2) and four (P_4) first measurements (points), respectively. From Fig. 11.10b one can see that even the data from the first measurement reduce the ambiguity near the

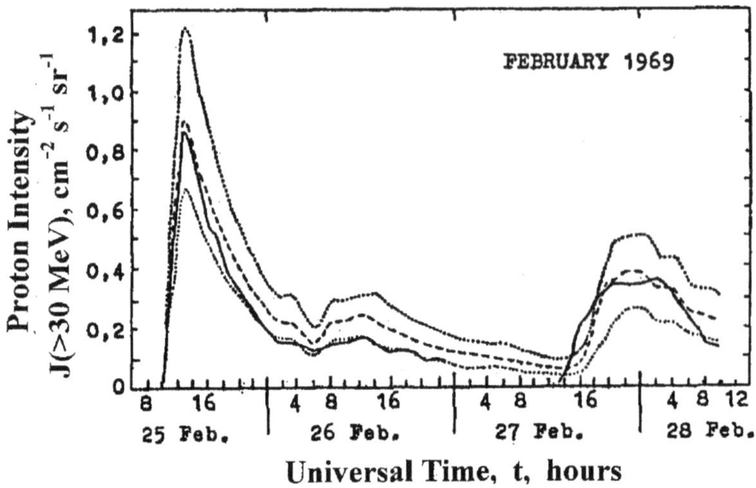

Fig. 11.11 The time profile of proton flux with E_p >30 MeV observed during the SPEs of February 25, 26 and 27, 1969 (*solid line*) and the forecast based on probability method (*dashed line*). One percent limits of the predicted flux distribution are shown with *dots* (Bengin et al. 1985)

enhancement maximum and allow it to be determined with a higher accuracy. When using two or more points, we practically apply an analog of the well-known method of prediction from the initial stage of the enhancement (method of recent measurements, see above) with additional estimation of the ambiguity zone of expected values. Figures 11.10c, d illustrate the process of improvement of the forecast while information on particle fluxes arrived at the observation site is accumulated.

Figure 11.11 shows the results of application of the method described above to the series of three SPEs - on 25, 26, and 27 February, 1969. The observed time variation of the >30 MeV proton flux is shown with a solid line, the mathematical expectation of predicted flux - with a dashed line, the predicted values, F_p, corresponding to 1 % quantiles of distribution functions with dots. The prediction was updated three times during the interval under consideration. The first prediction was based on the 3 GHz radio burst data and the observed proton fluxes at 10:00 UT and 11:00 UT on 25 February, the second one – on the suitable radio burst data and the observed total proton flux at 0,700 UT on 26 February, and the third one – on the radio burst data on 27 February only. Each of the two subsequent predictions contains the results of the previous one, Fig. 11.11 is collecting (but not summarizing) the results of three predictions.

As a whole, one can see that the predicted time variation is in a good agreement with the observed SCR flux dynamics. On the other hand, there exists a systematic overestimation of the predicted values with respect to the measured ones, and at some moments the observed proton flux appears beyond the calculated 1 % limits.

For the future developments of the proton prediction techniques it appears to be necessary, in particular, to specify and complement the concepts of SPE and radiation characteristics of SCR. Of special interest is a prediction of radiation effects from the rare, but very dangerous behind-the-limb flares of the September 29, 1989 type. The method of recent measurements can be also improved, for example, by optimization on parameter of σ in Eq. (11.2) and separation of diffusive and non-diffusive SPEs. Determination of SCR parameters may be more precisely by numerical solution of inverse problem with the improved Green functions, including a possibility of their numerical derivations from a general transport equation. It may result in the enlargement of the data (parameters) sets used and/or analyzed, sought for and/or expected (predicted).

Probability technique for the proton flux prediction (Bengin et al. 1985; Miroshnichenko and Petrov 1985; Miroshnichenko et al. 1986) may be further developed by construction of multi-dimensional distribution functions of SPEs on new invoked parameters of SCR in the source, corona and interplanetary space. Evidently, proposed method needs to include the data on electron component of SCR (*e*/*p*-ratio, efficiency of electron and proton leakage to interplanetary medium, correlations between electron and proton fluxes, *etc*.) with the specific features of coronal transport of electrons (*e.g*., Daibog et al. 1988; Wibberenz et al. 1989).

It remains actual to study temporal behaviour of the anisotropy vector for non-relativistic protons based on recent measurement data, on the analogy with the spacecraft observations by Petrov et al. (1987) during the event of July 9, 1985.

11.5.3 Prediction Based on Precursor Information

Protection from the radiation effects of SEPs for deep space missions requires a warning system to observe solar flares and predict subsequent charged particle fluxes. Such a system should relate precursor information observed in each flare to the intensity, delay, and duration of the subsequent SPE at other locations in the Solar system. A forecast and warning system of this type was established in the USA for the Apollo program and is now in operation at the NOAA Space Environment Services Center (SESC) in Boulder, Colorado. It has been used, in particular, to predict flare particle fluxes at the Earth in 1989 when solar conditions may resemble those during future space exploration missions (Heckman et al. 1992). In real time, the SESC collects observations of the Sun and variations in the Earth's environment resulting from solar activity. The data are used to warn of disturbances as they occur, to provide indices and summaries after the fact, and to make forecasts of activity levels in the future.

Typically, such forecasts use characteristics of solar flares that can be observed at electromagnetic wavelengths. From these observations, an estimate is made of the number of particles that might escape from the Sun in such a flare. After allowance for propagation of the particles away from the flare location, through the solar corona, and out into the interplanetary medium, a forecast can be made for the resulting particle flux at the Earth. These forecasts usually rely on a combination of some limited theory and considerable observations, combined in heuristic forecast model. A description of models of this type existing before 1980 was given in detail earlier (*e.g.*, Smart and Shea 1979; Miroshnichenko and Petrov 1985; Miroshnichenko et al. 1986; Miroshnichenko 2003a).

A new event-oriented prediction model has been developed by Smart and Shea (1989a); it was implemented at the US Air Force Space Environment forecast facility. The model allows to construct intensity-time profiles for a number of proton energy ranges; it is very important that this model is also capable of making predictions for the heavy ions (*e.g.*, the event of 2 January 1988). The SPE forecast algorithm in the SESC, called PROTONS (Heckman et al. 1992), is one of such heuristic (empirical) model.

The PROTONS model was first implemented in real time operation in 1972. It has two principal sections. The first provides an estimate of the number of energetic particles that are produced in a solar flare or other energetic solar phenomenon. For this purpose, the model uses a variation of a correlative relationship between the soft (1–8 Å) X-ray fluence from a flare and the resulting proton flux observed at the Earth. This kind of correlation was studied in detail by Huston and Kuck (1990) who compared the peak proton flux at geosynchronous orbit with the solar X-ray emissions using data collected by the *GOES* spacecraft.

The second major section of the model is an approximation for the propagation of energetic particles away from their solar source through the solar corona and the interplanetary medium to the Earth. The PROTONS is based on an anisotropic diffusion model described by Burlaga (1967). This model provides a time profile of a theoretical SPE as expected at the Earth, the maximum intensity of the particle flux being estimated as a function of solar flare location and interplanetary propagation conditions. The parameters required to apply this model were obtained by fitting the model to proton data from the IMP measurements in solar cycles 20 and 21.

Balch and Kunches (1986) provided a more detailed description of the current operational version of the PROTONS algorithm. A basic shortcoming of the algorithm is that it is not state-of-the-art physics, since much of the physics was not understood when it was first written and many aspects of the SPE problem remain descriptive. An advantage of the model, however, is that it involves a sequence of correlative relationships which, piece by piece, can be replaced by physical models as they are developed. Simultaneously, it offers a long-term evaluation of a model based on actual operational application over many SPEs. In Fig. 11.12 we demonstrate the results of using PROTONS to forecast the peak proton flux (PPF) for a set of SPEs observed in 1989. If the forecast were perfect, all symbols would lie on the diagonal line. The plots includes all events that had greater than 50 % probability of occurrence (see for details Heckman et al. 1992).

An example of known sources of error in the forecasts is shown in Fig. 11.13. A spike-like rise and fall of fluxes by a factor of 10 occurred in an event beginning on October 19, the first of three large October events. The spikes are typical of the passage of an interplanetary shock (*e.g.*, Cane et al. 1988), which is not included as a part of the PROTONS model. A geomagnetic storm sudden commencement (SSC), also known to be associated with the passage of shocks, was reported on the same day.

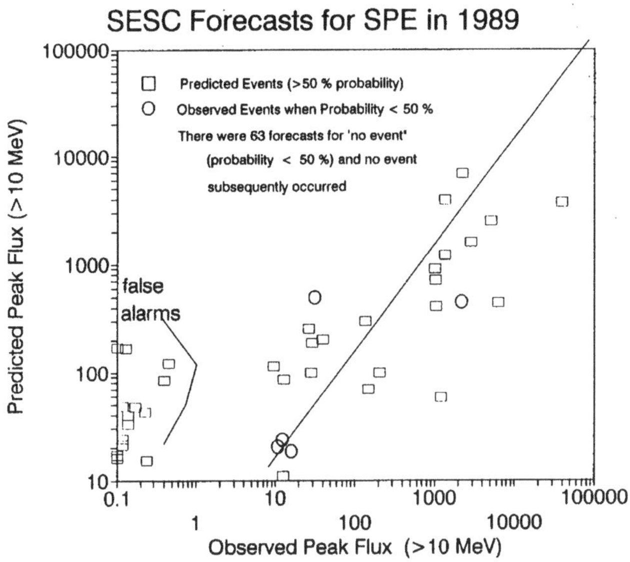

Fig. 11.12 Predicted peak particle fluxes (PPF) for SPEs in 1989 (Heckman et al. 1992). The diagonal line defines perfect forecasts

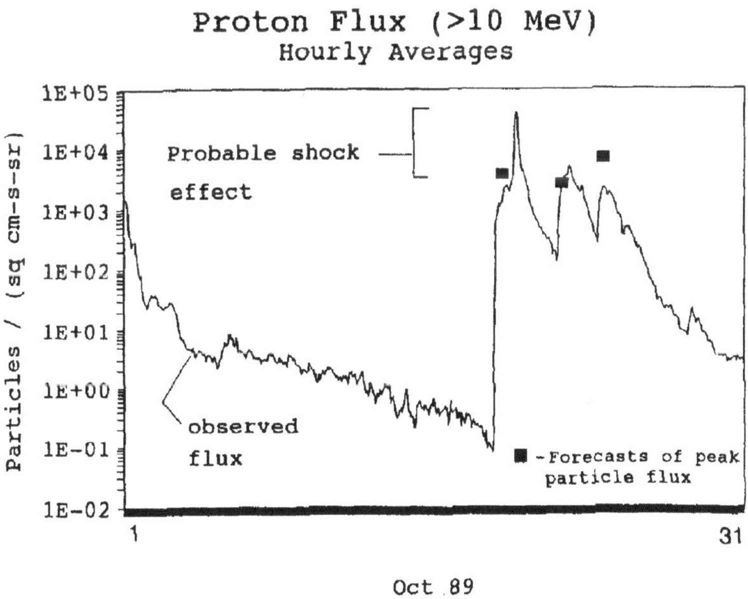

Fig. 11.13 Forecasts of the peak proton fluxes and observed intensity-time profiles of the >10 MeV protons for October 1989 (Heckman et al. 1992). The first large SPE shows a spike-like feature typical of the passage of an interplanetary shock

The effectiveness of the described warning system may be apparently improved, if one includes more complete observations of solar activity, especially data on the occurrences of solar mass ejections (CMEs), and up-to-date consideration of the effects of propagation conditions in the solar corona and interplanetary medium.

11.5.4 Coronal Mass Ejections and Prediction of Proton Fluxes

In the context of this consideration, it is timely to make clear a possible link between the problem of radiation hazard in space and so-called solar flare "myth" (see Sect. 3.2). Solar particle events are clearly associated with acceleration processes taking place at/near the Sun. There is, however, currently a major controversy as to whether the particle acceleration takes place in the flare itself or the particles are accelerated by associated coronal mass ejections (CMEs) (see Gosling 1993 and references therein). In any case, it has become evident that there are two different types of solar X-ray flares, gradual and impulsive. There is also increasing evidence that these different types of solar events are associated with different types of particle events in the energy range above 5 MeV (*e.g.*, Reames et al. 1994). The forecasters agree with the necessity of having CME information for predicting SPE (*e.g.*, Rust 1982; Heckman et al. 1992; Feynman 1997), but there are no reliable, continuous observations of CMEs available for real time use. Instead, Heckman et al. (1992), in their model PROTONS, uses Type II and Type IV solar radio bursts, at frequencies normally associated coronal emission, as a surrogate indicator of a mass ejection moving through the corona.

On the other hand, the largest solar proton fluence events often occur in association with a series of major gradual flares from a single active region as it is carried across the face of the Sun (*e.g.*, Malitson and Webber 1963; Feynman et al. 1993). For example, the major proton event of March 1989 was associated with a series of high velocity CMEs (see Feynman 1997, and references therein). It is widely believed these particles are accelerated by CME-driven shocks in the corona and lower solar wind (*cf.* Gosling 1993). Therefore, when predicting the particle fluence for a long period one should take into account the number of CMEs per year. The estimated number of shock producing CMEs that may produce proton enhancements at the Earth's orbit is 25–30 per year during the active period of the solar cycle (Feynman 1997).

Evidently, to predict and characterize short periods of high-flux and high-energy SPEs, new approaches are required. An empirical approach to predict the evolution of an SPE from observations of fast CMEs was suggested recently by Turner et al. (1998). As discussed above, the models employed by US Space Weather Forecast Centers (*e.g.*, Heckman et al. 1992) use soft (1–8 Å) X-ray fluence as a proxy for the process leading to an SPE. These models, however, have more than an order of magnitude scatter in predicted versus observed flux, a high false alarm rate (see, for example, Fig. 11.10), and a single template for the time to peak flux and time to decay.

Current understanding of SPEs suggests that a better model of the temporal variation of flux could be obtained by using observations of the shocks associated with fast CMEs. Turner et al. (1998) suggest a simple representation of the CME impact on the IMF structure to calculate the distance from the observer to the shock along a connecting field line. A history of the field lines contact with the shock and a representation of the propagation of high energy particles along the field line is used to predict the observed flux.

11.5.5 Prediction of Heavy Ion Fluxes and Anomalous Proton Events

The same principles involved for organizing and estimating the proton (ions with $Z = 1$) arrival and time-intensity profile seems to be also applicable to heavy ions. These data are conveniently organized by kinetic energy or momentum per unit charge (particle rigidity). It is reasonable to assume (Smart and Shea 1989a) that the same principles of coronal propagation and interplanetary propagation apply to all ions independently of the mass or atomic charge. There is a major problem in finding a simple common factor for the elemental abundance ratios.

There have been a number of papers reporting the variation of the elemental abundances in solar particle events (see, *e.g*., Reames 1996, 1999 for a review). A general summary may be that "small" events have the greatest variability in elemental composition, and the hydrogen to helium ratios are the most variable. The elemental abundance ratio seems to have a slight variation according to the energy of the measurement. This may be a reflection of the "size" of the particle event since small particle events would not have many heavy ions at high energies. For "large" events the heavier elemental abundance ratios seem to be in general agreement with the ratios expected from normal coronal material organized by first ionization potential (FIP). Unfortunately, most of the solar particle data currently available are for protons. As an expediency, Smart and Shea (1989a) proposed to utilize an assembly of the available solar flare heavy ion data (see, *e.g*., Cook et al. 1984; McGuire et al. 1986; Reames 1996, and references therein). If to normalize these data to hydrogen, it becomes possible to estimate the probable heavy ion fluence from the predicted proton fluence.

Such an approach was tested with the data on the first significant solar particle event of the 22nd solar cycle occurred on 2 January 1988. As a source of this moderate SPE, the 3B solar flare was identified at heliographic coordinates 38°S, 18°W, with an X-ray onset at 12:13 UT and X-ray classification X1.4. This flare produced a SPE with a peak flux of protons at energies > 10 MeV of 98 pfu. The initial prediction generated by PPS-87 (Smart and Shea 1989a) was "on time", but a factor of three too low in predicted peak flux at energies >10 MeV. The update

Table 11.1 Prediction summary for the SPE of 2 January 1988 at >10 MeV/nucleon (Smart and Shea 1989a)

Ion	Peak flux, pfu	Fluence, cm^{-2}
$Z=1$	98.0	6.0×10^6
$Z=2$	2.7	1.6×10^5
$Z=3$	1.2×10^{-1}	7.6×10^3

ability was utilized in order to normalize to the observed peak proton flux with energies >10 MeV, and from this the projected future particle time-intensity profile was generated. A summary of this prediction for this "average" solar particle event is given in Table 11.1.

As noted by Heckman (1997), improvements in current forecasts for manned space flight operations are at present based on observing activity at or near the Sun and predicting the occurrence of a solar particle event at the Earth's orbit. Apparently, these forecasts need to be distinguished between routine events and the possibility of a very large event. In this context, it is of great interest an empirical technique developed in IZMIRAN (Ishkov 1998, 1999) to predict large solar flares. It based on the observations of emerging magnetic fluxes (EMF) and their interactions with existing magnetic fields, as well as on the known features of building-up and evolution of the solar structures where important geoeffective phenomena take place. Among them, large solar flares are the most significant ones, thus their prediction is one of actual tasks of the solar-terrestrial physics.

It was found (*e.g.*, Ishkov 1998, 1999, and references therein) that a new magnetic flux starts to emerge about 2–3 days before the beginning of occurrence of the large solar flares. Moreover, there were identified the EMF group with a rapid evolution. Large and moderate flares occur if a new EMF is rather large (~10^{13} Wb), and its emerging velocity is $>10^9$ Wb s^{-1}. To predict a large flare in a real time forecaster needs to have a set of observational data on the increasing of the sunspot areas in the active region (AR) under consideration, changing of the sunspot configurations, behaviour of the sunspot polarity, etc. Active region usually has a period of energy release in the form of flares from 16 to 80 h, or on average about 55 ± 30 h which corresponds to 5–25 % of passage time of the active region through the solar disk. It is important to note that all large flares of certain active region occur in this time interval if the maximum AR area is S_m <1,000 millionth of the solar hemisphere (for example, AR 5533 in June 1989). If S_m >1,000 millionth of the solar hemisphere (for example, AR 5747 in October 1989), large flares occur in two stages (about 80 % and 20 %, respectively), with a time pause between them of 5 days. This technique, in its computer version, was applied successfully during the missions of Russian spacecraft *GRANAT*, *GAMMA* 1 and *CORONAS*-1.

11.6 Radiation Hazard at Different Heliospheric Distances

Extrapolation of Earth-based prediction methods to other location in space (in particular, to a Mars' mission) relies on some empirical data on the radial dependence of solar proton flux and fluence (e.g., Miroshnichenko 2005). For the simplicity, it is assumed that the maximum possible prompt solar proton flux would be at the position that is "well connected" to the solar flare source region. Using the intrinsic assumptions that the coronal particle intensity gradients control the particle flux observed around the Sun, it is possible to estimate the particle flux at any heliographic longitude (Smart and Shea 1993).

The arguments used for extrapolation of proton fluxes to other heliocentric distances rely on the assumption that the diffusion across magnetic field lines is negligible, and that the volume of the magnetic flux tube as the distance from the Sun increases expands in the manner expected from classical geometry. In this case, a power-law function of the form $\sim r^{-3}$ can be used to extrapolate to other distances (r is the radial distance from the Sun). Hamilton (1988) has analyzed the probable effects of diffusion, and his preliminary estimate was that the power-law function of $\sim r^{-3.3}$ would be an appropriate factor.

To extrapolate a prediction to other locations at 1 AU, it is necessary to use the Archimedean spiral and the coronal gradient concept. First, compute the longitude on the Sun from which the interplanetary magnetic field line passing through the spacecraft position would originate. Then determine the heliocentric angular distance between the location of the solar flare and the solar longitude of the "root" of the idealized spiral field line passing through the spacecraft. Next multiply the coronal gradient per radian by the heliocentric angular distance between the two positions in order to estimate the flux diminution. Finally, multiply the peak proton flux expected at the "favourable" propagation path by this flux reduction factor.

The above arguments rely on the volume of the magnetic flux tube behaving in a "classical" manner as the distance from the Sun increases. If "classical" behaviour is assumed, then a power-law function can be used to extrapolate to other distances. Any distortions of the magnetic flux tubes are unknown, so we have no way of accurately estimating. Because of this, there is no consensus view on the proper method for extrapolating solar particle fluxes and fluences from 1 AU to other distances in the heliosphere. The existing meager measurements are from comparison of Earth-orbiting satellite measured proton fluxes compared with space-probe measurements of the same event in the energy range of 10–70 MeV from 1 to 5 AU (*e.g.*, Beeck et al. 1988; Hamilton 1988).

For distances greater than 1 AU, Shea and Smart (1993a, b) recommend to extrapolate the expected proton flux at 1 AU using a simple form of $\sim r^{-3.3}$. This is the average solar proton radial gradient derived by Hamilton (1988) from a combination of *Voyager* and Earth-satellite data. The limited data available suggest that we should expect variations ranging from $\sim r^{-3}$ to $\sim r^{-4}$. For distances less than 1 AU one can extrapolate the expected proton flux using a functional form $\sim r^{-3}$. Again, the limited measurements available suggest that variations ranging from $\sim r^{-3}$ to $\sim r^{-2}$ should be expected. To extrapolate proton fluence from 1 AU to other distances in the heliosphere, it is recommended to use a functional form of $r^{-2.5}$, with expected variations ranging from $\sim r^{-3}$ to $\sim r^{-2}$.

Kahler (2001) has undertaken a new attempt to improve some existing models of SEP event prediction. Based on shock concept of SEP acceleration (see Fig. 3.5), he proposed to extend this basic scheme to predict SEP profiles at other regions of

Table 11.2 Solar connection parameters in the solar system (Kahler 2001)

Planet	Distance, AU	Φ, deg.	L, AU	T, min.	B/B_0
Venus	0.72	37	0.77	15	1.69
Earth	1.00	51	1.32	26	1.00
Mars	1.52	78	1.91	37	0.55
Asteroids	2.77	142	4.63	90	0.264
Jupiter	5.20	268	13.67	265	0.134

space that might be encountered by future space travelers. The high-energy solar particles at a point of a distance r AU from the Sun will be most intense when the source region is located at a west longitude of

$$\Phi(\text{deg}) = \Omega \times r/V = 51.4 \times r \tag{11.8}$$

where the average solar wind speed V is assumed to be about 450 km s^{-1} (Richardson et al. 1995), and the solar rotation rate is taken as $\Omega = 360/27$ deg/day. The SEPs must travel a distance L along the spiral field line to reach the observer where

$$L = \int \left[\sqrt{(1 + r^2\Omega^2/V^2} \right] dr \tag{11.9}$$

For above assumed values of V and Ω, the distance $L = 1.32$ AU at 1 AU. Travelers venturing to other solar system regions (Venus, Mars, Jupiter, *etc.*) would find the connection longitudes Φ (measured west of central meridian) and path lengths L indicated in Table 11.2. This table also shows T (minutes), the minimum time required for a 100 MeV proton to reach the distance r (AU) from the Sun, assuming a pitch angle cosine $\mu = 1$.

As astronauts venture further out into the Solar system from 1 AU, the magnetic connection longitude moves westward, but even for Mars-bound observers the connection region is still on the visible side of the west limb, allowing them to monitor solar activity that could result in a high-energy SEP event. However, for travelers beyond the asteroids the only signature of a solar eruptive event producing energetic particles would be a fast halo or large-angle west limb CME showing no associated front-side solar activity. Beyond Mars the path lengths and transit times of SEPs are considerably longer than at 1 AU due to the winding of the magnetic fields into a predominantly azimuthal direction.

The scattering of SEPs by magnetic irregularities (see Chap. 3) on their outbound trajectories results in significant decreases of peak SEP intensities with distance. An additional decrease of SEP intensity results from diverging magnetic field. For a population of high energy SEPs confined to a magnetic flux tube of flux $B \times A$ where B is the field intensity and A is the cross-sectional area, the decrease of the SEP intensity with distance should match the decrease of B. Using empirical equation of Burlaga (1995) for B, which varies roughly as r^{-2} near the Sun, but more nearly as r^{-1} beyond several AU, Kahler (2001) calculated the decrease of $B/$

B_0 with distance r (the last column of Table 11.2). At the distance of Jupiter the intensity decrease is less than a factor of 8 from the value at 1 AU.

Observational studies have determined how SEP intensities decrease with radial distance in the solar system. Hamilton et al. (1990) examined multiple spacecraft observations of five well-connected 10–20 MeV SEP events and derived power-law decreases for peak intensities as $r^{-3.3 \pm 0.4}$ and for fluences as $r^{-2.1 \pm 0.3}$. Ten years later, Lario et al. (2000) compared SEP events at the *WIND* spacecraft with those at *Ulysses* during 1997–1998 when *Ulysses* was near the ecliptic plane and at distance of 5.2–5.4 AU. It was found a rough correspondence between the major E ~10 MeV SEP events at the two spacecraft, despite the fact that the connection longitudes of each spacecraft to the source shocks varied significantly throughout the study period. Comparing the fourth largest event at each spacecraft, Kahler (2001) got a value of 2.7 for the ratio of the logarithms of the peak intensities, suggesting an $r^{-3.75}$ decrease of the peak intensity for the events. The event time scales at *Ulysses* clearly increase, however, so the decrease in the fluence will be less. These results appear consistent with earlier work of Hamilton et al. (1990).

Unfortunately, for space weather purposes, time profiles of the 10–30 MeV protons are often complicated by several factors. One is that the particle trapping region around the shock, which is not subject to the streaming limit, may or may not be the region with the highest SEP intensity, depending on whether shock acceleration continues or diminishes with distance (Reames, 1999a). At distances beyond 1 AU there can be interaction or merging of different transient shocks, and the corotating shocks begin to play a role, possibly by re-accelerating some of the SEPs from transient shocks (Lario et al. 2000).

In a mission to Mars, for example, the radial distance will vary according to the spacecraft trajectory chosen, and the flux radial dependence and SEP source locations are very important. As noted above, the flux of solar proton is expected to vary as a power law with radial distance from the Sun, and a power-law exponent of -3 would be expected from magnetic flux tube geometry. Since the radial distance to Mars is ~1.5 AU, then the flux at the orbit of Mars would be expected to be about 1/3 of the flux at 1.0 AU along the same spiral path. This variation should be contrasted with the average heliolongitudinal gradient of the order of magnitude per radian of heliocentric angular distance. A consideration of these expected variations suggests that the proton prediction problem for Mars is not dramatically different from the Earth (*e.g.*, Miroshnichenko 2005). Sensors on board the spacecraft viewing in the optical, radio and soft X-ray wavelengths should be able to provide useful prediction information.

The probability of a "surprise" (*i.e.*, a solar proton event being detected when there is no visible preceding solar activity), however, is significantly larger at the Mars orbit. At the Earth, about 20 % of the recorded SPEs are not associated with visually observed solar flares. It is presumed that the origin of "major" proton events not associated with visual solar flares have their source from solar activity from behind the western limb of the Sun as viewed from the Earth (Shea and Smart 1993a, b). Figure 11.14 shows assumed source locations on the Sun for relativistic SCR events (so-called GLEs) from 1956 through 1991 (*black circles*). Notice that 10 of 48 events

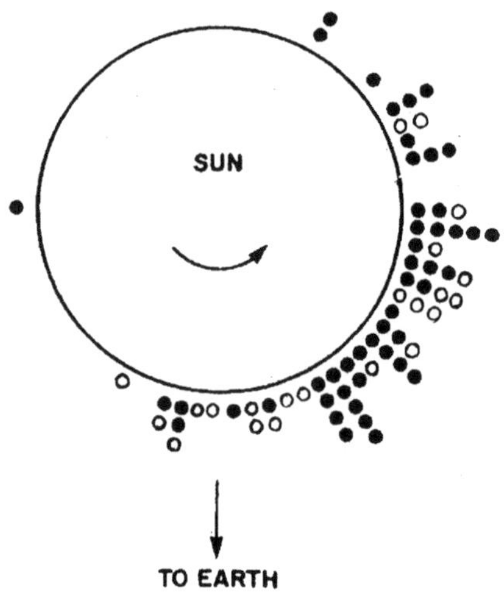

Fig. 11.14 The solar longitudes as viewed from the Earth of assumed sources for the GLEs occurred between 1956 and 1991 (Shea and Smart 1993a, b, *black circles*). *Open circles* are added by the author in 2011 based on the data from Table 2.1 for all 70 GLEs of 1942–2006

(*i.e.*, about 1/5) have been associated with solar activity (*i.e.*, an assumed flare) behind the west limb of the Sun. The same type of distribution is present, in general, for major non-relativistic SPEs (Shea and Smart 1993a, b). The shock-associated events, however, are distributed more uniformly, with a maximum at about 30°W (Miroshnichenko et al. 2001). Figure 11.14 shows assumed source locations on the Sun for relativistic SCR events (so-called GLEs) from 1956 through 1991.

Similarly, for the position of Mars, we would expect that about of 1/2 of the detected SPEs would have their source on the portion of the Sun that is not observable from Mars (Shea and Smart 1993a, b). It is of interest to estimate the probable "favourably connected" heliolongitude of Mars. At 1.5 AU distance the Sun-Mars transit time for a 400 km s^{-1} solar wind would be about 6 days. During that time the Sun would have rotated ~86°. This is essentially at the western limb of the solar disk visible from Mars. Assuming that the solar proton flare distribution is symmetrical in heliolongitude, then approximately 1/2 of the source solar proton flares cannot be observed from the Mars orbital distance (Shea and Smart 1993a, b).

This situation strongly argues for on-board particle and radiation sensors on a Martian mission. If we consider the intensity-time profile of a SPE, then the critical factor is the time from event onset to "hazardous" radiation levels. Depending on the propagation conditions even for "well-connected" events, this is likely to be of the order of an hour. The most "dangerous" particle radiation will be the ions that penetrate the shielding and stop in blood forming organs (thus depositing most of their energy in these organs). These will be the protons between 70 and 150 MeV assuming that there is ~5 g cm^{-2} of shielding provided by the body structure. The typical intensity-time profile observed in this energy range at 1 AU (and expected at 1.5 AU) provides for about an hour from particle onset until the maximum proton flux will be observed. We suggest that prudent mission planning would allow for movement of personnel to a more heavily shielded area or the re-distribution of mass on this time scale.

11.7 Relativistic Protons in Prediction Schemes

Several promising suggestions have been advanced to use relativistic protons ($R \geq 1$ GV) as a predictor of low-energy particle behaviour during the SPEs. In particular, Dorman et al. (1990) proposed a new approach to diagnostics of the interplanetary medium and prediction of the onset and development of solar proton events based on the method of numerical solution of the inverse problem (Miroshnichenko and Sorokin 1985, 1986). It is suggested, firstly, to reconstruct the transport parameters of solar cosmic rays (SCR) in the interplanetary magnetic field (IMF) by observations of Forbush-effect and anisotropy of galactic cosmic rays (GCR) after major solar flares. Then, with these data taken into account, the same technique (see Chap. 7) may be applied to reconstruct the ejection profiles of relativistic protons (after the GLE maximum is attained at the Earth) and to predict their flux variation at the decay stage of the event several hours beforehand. This seems to be reasonable, especially as a considerable share of SPEs occurs at the recovery phase of the Forbush-decrease (*e.g.*, Miroshnichenko and Petrov 1985).

Although methodical aspects of the proposed scheme appear to be rather grounded, it remains unclear how to check it by observational data. The approach suggested above requires GCR observations at several points in the interplanetary space and at ground-based stations at different longitudes. This also requires detailing of each particular event by taking into account the variety of the Green functions describing the cosmic ray flux transformation in the IMF. In addition, as seen in Figs. 4.2, 4.4, and 9.4, abnormal high magnitude of peak proton intensity at $R \geq 1$ GV does not mean yet that one should expect a similar catastrophic enhancement of SCR flux in non-relativistic energy range. It complicates the prediction procedure and diminishes the forecast accuracy.

Recently, Belov and Eroshenko (1996) developed a new empirical method to obtain proton spectra near the Earth within a wide energy range (from 10 MeV to 10 GeV) directly from observational data without any preliminary assumptions about the possible spectral shape (see Chap. 9). Their technique also allows to reconstruct a time profile of proton intensity for any selected energy. As well-known, the shape of the energy spectra is time-dependent. Its variations are well indicated by variations of the power-law spectral index, which means that proton intensity-time profiles differ for a different energy. For example, Fig. 11.15 shows the intensity-time profiles of proton fluxes reconstructed by Belov and Eroshenko (1996) for a number of energies for the GLE of October 19, 1989.

It is seen that, if the difference in energies becomes greater, then the shapes of the time profiles are more distinct. Any of these profiles has its own time of the flux maximum, which visibly varies with changes of energy. The high energy protons always do everything faster than the low-energy ones: a maximum of the effect is always close to the event onset for higher energies. A comparison of profiles for the 30 MeV and 3 GeV energies gives a good illustration of these dependencies: a difference between their time maxima is of about 13 h. In other words, the start and duration of the 3 GeV proton increasing is much earlier and shorter than those of the

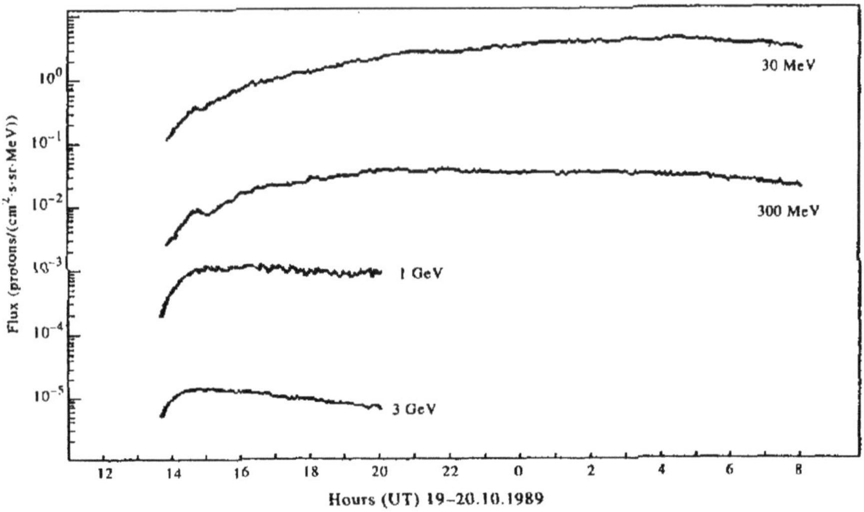

Fig. 11.15 The intensity-time dependencies of solar particle fluxes reconstructed at different energies by the data of various detectors for the event of October 19, 1989 (Belov and Eroshenko 1996). Curves for the 30 and 300 MeV protons are obtained by spacecraft data; two lower curves correspond to neutron monitor data

30 MeV protons. So, the 3 GeV protons might be a good predictor for the behaviour of the low-energy particles and, consequently, for the radiation dose estimates, at least, in large proton events. In their considerations Belov and Eroshenko (1996) did not take into account shock-accelerated or trapped particles that may arrive at the Earth a day or so after the initial particles.

11.8 Models of Proton Fluence at Large Time Scale

As follows from the above sections, short-term forecasts of SPEs are necessary for any tasks requiring extra-vehicular activity and the operation of radiation-sensitive scientific detectors. On the other hand, it is not possible to predict the exact occurrence, intensity or duration of solar proton events, and consequently mission planning on both a short-term and long-term basis can be rather problematic.

Long-term predictions of the radiation levels resulting from the SPEs are required if costly over-design or mission-threatening under-design are to be avoided. The dose accumulated over the mission lifetime is a function of the solar proton fluence (except for low near-Earth orbit, where geomagnetic shielding provides protection), and so a reliable estimate of this fluence is needed by a spacecraft engineer to optimize design parameters.

As with any form of long-term forecasting based on past observations, the statistical interpretation of data plays a central role in the final model definition.

The size of the data set used will always be limiting factor on the level of confidence associated with any solar proton model. As discussed in Chap. 2, the spacecraft measurements were not really systematic until about 1965 (see Fig. 2.1). Prior to that, solar proton fluences could only be inferred through ground-based or low-altitude measurements made by sounding rockets or balloons. Unfortunately, such techniques are prone to inaccuracy (*e.g.*, Smart and Shea 1989b, see also Sect. 2.8), so the data obtained for the last solar cycle before the advent of satellite technology can only be used with some precautions.

In Sect. 10.7 it have been mentioned a technique of prediction the occurrence rate of the SPEs with the proton fluence of F_s (>30 MeV) $>10^5$ cm^{-2} as a function of calendar time for a period of up to 11 years (Zil et al. 1988). Such a fluence model was based on quasi-periodic variations of the event occurrence rate. Below another approaches to the prediction problem at the long-term scale are considered. The three most known solar proton fluence models have been developed by King (1974), Getselev et al. (1988, 1992, 1996a, b), and Feynman and colleagues (Feynman and Gabriel 1990; Feynman et al. 1990a, b, 1993; Feynman 1997), referred to thereafter as the K-1974, G-1988, and JPL-1991, respectively. The first one was for a long time the standard model used by spacecraft engineers to predict mission-integrated solar-flare proton fluence. It has been coded and made available to the community by National Space Science Data Center (NSSDC). The latter has been recommended for use in NASA for future mission planning (see Tranquille and Daly 1992, and references therein).

The K-1974 model. It was constructed using data exclusively for the active years of solar cycle 20 (1966–1972). The purpose of the model was to provide solar-proton fluence levels for space missions that were in the planning stage for flight during the active years of the following cycle.

As a measure of solar activity (including flare activity and proton emissivity of the Sun) is usually considered an annual-mean sunspot number, W_a. Solar cycle 20 was different in nature from solar cycle 19 in two important respects. Firstly, the largest number W_a of cycle 19 was significantly higher than that of cycle 20 (and is indeed the highest on record up to date). Secondly, the SPE frequency and intensity of cycle 19 were much higher than those of the following cycle. Prediction for solar cycle 21 (*e.g.*, Lanzerotti et al. 1991) indicated that the sunspot number would most probably be less than that measured during cycle 20 (as noted by Tranquille and Daly 1992, in fact, this turned out to be false; the largest W_a value for solar cycle 21 was 155, compared with 107 for the previous cycle). Therefore, following the assumption by Webber (1967) that sunspot number W_a and annual-integrated solar-proton fluence F_a were linearly related, King (1974) chose to ignore the solar cycle 19 data set and took measurements only in cycle 20 as representative of the forthcoming cycle.

The data set was mainly obtained from proton measurements in the energy range 10–100 MeV made by instruments on the *IMP* series of satellites (4, 5 and 6). The data from any individual instrument or satellite were cross-calibrated with independent measurements (whenever possible), to check the mutual consistency of the complete database. As an example, it may be noted that the measurements made by four independent instruments on *IMP* 4 agreed to better than 25 %.

In total, King (1974) used 25 individual events, including the great proton event of August 1972, which accounted for about 70 % of the total >10 MeV fluence for the complete solar cycle 20. It should be noted that according to Akinyan et al. (1983), in August 1972, in fact, five SPEs were identified near the Earth, namely, on August 2, 4, 7, 11, and 16, the event of August 4 having been the largest one amongst them. Therefore, from the point of view of total proton fluence, King (1974) considered this five-event episode as an individual anomalously large SPE.

The statistical approach used by King (1974) was based on methods employed by Yucker (1972) and Burrel (1972) in their analyses of earlier solar-proton data. Yucker (1972) introduced the concept of compound probability to define the probability, P, of exceeding a specified fluence, f, of protons with energy greater than E during a mission lasting τ years, as

$$P(>F, E; \tau) = \sum_{n=1}^{\infty} p(n, \tau; N, T)\; Q(>F, E; n) \tag{11.10}$$

where $f = 10^F$, and N is the observed number of events occurring in T years. The probability, p, of observing exactly n events in t years is given by Burrel's extension of Poisson statistics

$$p(n, \tau; N, T) = [(n+N)!/n!N!]\;(\tau/T)^n/[1+(\tau/T)]^{1+n+N} \tag{11.11}$$

which valid for populations having a small number of samples. The probability, Q, that the logarithm of the combined fluence of n events will exceed F is given by

$$Q(>F, E; n) = \int_{-\infty}^{\infty} q(x, E)\, Q\left[>\log\left(10^F - 10^x\right), E; n-1\right] dx \tag{11.12}$$

where the recursive Q in the integrand is defined to be unity if the argument of the logarithm is less than or equal to zero, and to be zero if $x < F$ and $n = 1$, simultaneously. If the logarithmic fluences are assumed to be normally distributed, then

$$q(F) = \left[1/(2\pi)^{1/2}\sigma\right] \exp\{-(1/2)\,[(F - F^*)/\sigma]\} \tag{11.13}$$

where F^* is the mean logarithmic fluence and σ is the standard deviation. This distribution function is also implicitly a function of energy.

Since the large flare of August 1972 contributed such a large percentage of the total solar-cycle fluence, King (1974) decided to separate it from the remaining 24 events, and to class it as an anomalously large event (the other events being termed ordinary events). The energy spectrum of this single event was found to be best represented analytically by exponential in energy,

$$J(>E) = J_0 \exp[(30 - E)/E_0] \tag{11.14}$$

with $J_0 = 7.9 \times 10^9$ protons cm^{-2} and $E_0 = 26.5$ MeV. Mean logarithmic fluences and standard deviations for the 24 ordinary events were derived (King 1974) by assuming the normal distribution function described by $q(F)$ in Eq. (11.13). Unlike the August 1972 flare, the ordinary events were found to best approximated by an exponential in rigidity, R; the latter is related to kinetic energy by $R = (E^2 + 1876E)^{1/2}$ and is measured in units of MV.

The JPL-1991 model. With hindsight, several assumptions made by King (1974) were evidently questionable, and had to be addressed by Feynman and colleagues in their development of the new JPL-1991 model. Firstly, the omission of data from solar cycle 19 on account of the relationship between the cycle-integrated fluence and maximum annual sunspot number was clearly not justified given the eventual event history of cycle 21. Secondly, the separation of solar flares into anomalously large and ordinary classes seemed somewhat arbitrarily and unnecessary if the major events of cycle 19 and the current cycle are included in the analysis. Furthermore, the relatively low number of events recorded during cycle 20 could only provide a model with limited statistical accuracy.

The data set compiled for the three solar cycles 19–21 by Feynman et al. (1990a, b) includes observations reported by Yucker (1972) between 1956 and 1963, using detectors flown on rockets and balloons. These measurements were tabulated at three threshold energies (10, 30 and 100 MeV). Notice that an event in November 1960 provided higher solar-proton fluences than the August 1972 flare, and several other flares were responsible for significantly higher fluences than expected from ordinary flares, as classified by King (1974). After 1963, satellite monitoring of the near-Earth radiation environment became routine, and essentially continuous database has been constructed from measurements made by several spacecraft. A comparison between the data used by King (1974) and Feynman et al. (1990a, b) for the 24 ordinary flares identified by King (1974) displays some discrepancies. However, the differences are small enough not to be invalidating one data set in preference to the other.

The exact dates of solar maximum for the three solar cycles are 1957.9, 1968.9 and 1979.9. Using these dates as the zero reference year for each cycle, Feynman et al. (1990a, b) showed that the solar cycle can be divided into 7 years of high fluence, and 4 remaining years of low fluence. The years of high fluence begin 2.5 years prior to the zero reference date, and end 4.5 years after this date. Therefore, a asymmetry in the event frequency and intensity therefore exists with respect to the peak in solar activity. The JPL-1991 model only considers solar-proton fluences throughout the 7 hazardous years associated with a complete solar cycle. Fluences during the remaining 4 years are set to zero. Only events with an integrated proton fluence of greater than 10^7 and 10^6 cm^{-2} for threshold energies of 10 and 30 MeV, respectively, are considered in the model. Since the >100 MeV fluences are only available for solar cycle 19, they are omitted from the analysis.

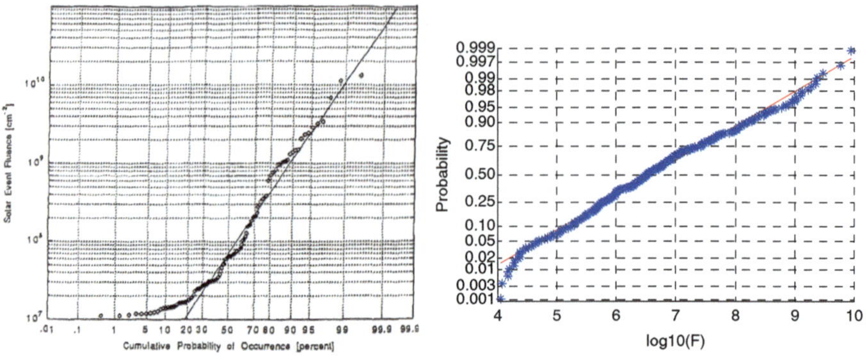

Fig. 11.16 Distribution of event-integrated fluences for the >10 MeV solar protons (*left panel*) during the active part of the solar cycle in 1963–1991 (Feynman 1997). The *straight line* is a Gaussian fit to the distribution of the large fluence events. *Right panel* shows a size distribution of SPE proton fluences above >30 MeV for >400 events from 1956 up to now. *Red direct line* corresponds to a log-normal distribution (Courtesy by M.V. Podzolko, SINP MSU, 2014)

The first step in the development of the JPL-1991 model was to assume that the event-integrated fluences follow a log-normal distribution. Let us define the cumulative frequency by $i/(n+1)$, where i is the rank of a particular event and n is the total number of events. Thus, by ordering the fluences in ascending order and plotting the cumulative frequency against the logarithm of the fluence, a straight-line relationship (a Gaussian distribution) should be obtained. For illustration, we represent in Fig. 11.16 (left panel) the resultant plot for proton energies >10 MeV during the active part of the solar cycle (Feynman 1997). For comparison, in the right panel we show recent findings by Getselev et al. (2009, 2013) on the proton fluences above 30 MeV. These authors have constructed extended database on >400 SPEs from 1956 up to the present time.

As expected, the data in Fig. 11.16 (left) lie approximately along a straight line, except for those events with smaller fluences where an evident upturn takes place. This can be explained by the fact that a log-normal distribution will under-estimate such minor events since, empirically, the number of smaller events increases as fluence decreases (Feynman et al. 1990a), whereas an inherent property of the distribution function requires that the opposite be true below the mean value. Thus, a second contributing factor is the fact that an upturn is an unavoidable consequence of truncating a data set.

Obviously, the distribution in Fig. 11.16 (left) is not Gaussian. However, the distribution is so steep at the high fluence end that the total fluence experienced in a mission will be determined by the number of very large events, if any occur. Because of this, it is only important to predict the large events accurately. To do this a Gaussian is used that fits the largest events well, as shown in Fig. 11.16 (left). This Gaussian is then used in Monte-Carlo calculations of fluences for various mission lengths (see Feynman et al. 1993). The distribution obtained by Getselev et al. (2009, 2013) for proton fluences >30 MeV can be well approximated by a log-normal law (red direct line at the right panel in Fig. 11.16).

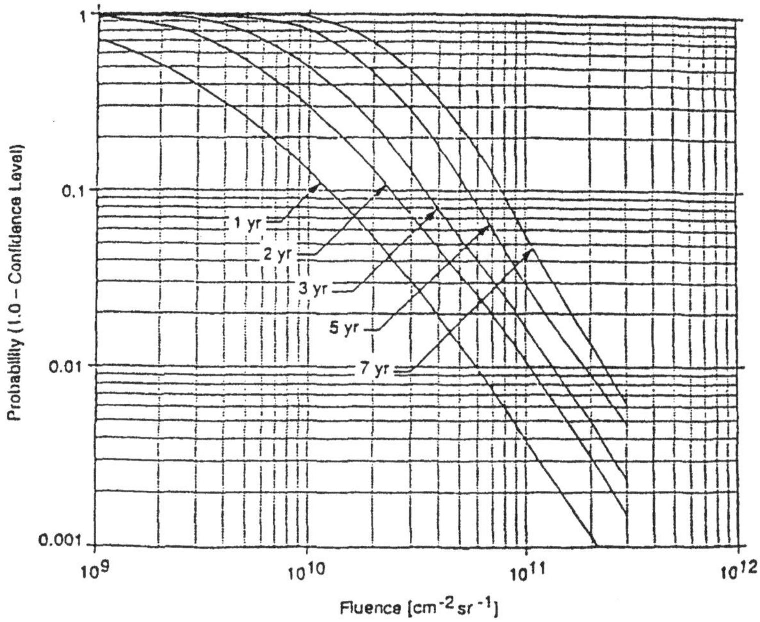

Fig. 11.17 The probability of exceeding a given fluence of the >10 MeV protons for several different mission lengths (Feynman 1997)

Figure 11.17 shows the probability of exceeding a given fluence for several different mission lengths for energy >10 MeV. This figure may be used directly for missions at 1 AU, no other modeling is needed. It is enough simply to count the number of years the mission will fly during the active part of the solar cycle, to choose the probability desired and to read the fluence from Fig. 11.14. For mission lengths longer than the 7 active years, Feynman (1997) suggests that no fluence will be collected during the quiet solar period, so one can simply add the fluence for the additional active years.

As noted by Tranquille and Daly (1992), the formulation of the JPL-1991 model is exactly the same as that used by King (1974) with the exception of the definition for the function $p(n, \tau; N, T)$. Due to the small sample of events available to King (1974), the Burrel extension to Poisson statistics was required (Eq. 11.11). However, the JPL-1991 data set has a significantly larger population of events, and so pure Poisson statistics are applicable, such that

$$p(n, \tau; N, T) = \exp(-\omega\tau)\ (\omega\tau)^{n}/n! \qquad (11.15)$$

where ω is the average number of events that occurred during the observation period (*i.e.*, *N*/*T*).

The G-1988 model. Similarly to Feynman and colleagues, when constructing their prediction model, Getselev et al. (1988, 1992, 1996a, b) divide a solar cycle into different phases and combine both data description in some way. They evaluate directly the flare proton fluence per year of the passive phase so that an annual fluence in the solar activity minimum is assumed to be at the level F_a (>10 MeV) ~10^8 cm^{-2} and F_a (>30 MeV) ~10^7 cm^{-2}. For the active phase of a cycle, lasting about 7 years, Getselev et al. (1996a, b) considered the available data from several Catalogues of SPEs (Akinyan et al. 1983; Bazilevskaya et al. 1986, 1990a). There were also used some unpublished data (Getselev et al. 1996b), including the data on all powerful flares of active phases of cycle 20, 21, and 22 as individual events.

As usually, it was assumed (Getselev et al. 1988) that the proton event fluences have a log-normal distribution

$$\varphi(F) = \left[1/(2\pi\sigma)^{1/2}\right] \exp\left\{-1/2[(F-\mu)/\sigma]^2\right\} \tag{11.16}$$

where $F = \log f$, σ is standard deviation, and μ is the mean log-fluence, while time intervals x between the neighbouring events obey a Weibull distribution

$$f(x) = 1 - \exp(-x^m/x_0) \tag{11.17}$$

where x and m are fitting parameters. Fitting parameters evaluated with the use of the data on three cycles are $\mu = 7.28$, $\sigma = 0.98$ for >10 MeV protons; $\mu = 6.15$, $\sigma = 1.23$ for >30 MeV protons; $x_0 = 11.38$, $m = 0.802$.

To illustrate how the G-1988 model does work, we describe one of its forecasts based on the following considerations. Since at least one powerful flare usually occurs during an active phase, it is appropriate to consider a forecast variant which is in some way includes a powerful flare. A "powerful" flare or "anomalously large" (AL) event is implied as that with fluence $\geq 10^{10}$ cm^{-2}. Getselev et al. (1996a) modeled such flare fluence by that of the October 19, 1989 event. If a powerful flare should occur during a space mission, radiation conditions could change radically for the worse, because its fluence not only exceeds substantially the integral annual fluence of ordinary flares but is also comparable with a total cycle fluence. Indeed, the simulated annual fluences of the >10 MeV protons turned out to be 4.5×10^9 cm^{-2} using the model distribution (11.16) for the entire data set of cycles 20–21, and 3.2×10^{10} cm^{-2} with influence of an AL-event taken into account.

Annual proton fluences have been also computed (Getselev et al. 1996a) by direct summing of fluences from individual flares occurring during one calendar year. In such a way the probability to observe an annual proton fluence exceeding a given level can be evaluated. This probability is defined as a ratio P of a number n of years in which a annual fluence exceeded a given level, to a total number N of years for which observational data exist, *i.e.*, $N = 24$. These "observed" probabilities are listed in Table 11.3.

It follows from these calculations that when a powerful flare occurs, the probability of the >10 and >30 MeV proton fluence to exceed 1.0×10^{10} cm^{-2} will be 6 and 1.0 %, respectively. From observational data these values are estimated as 12 and 8 %, while Feynman et al. (1990a) give 20 and 4 %. Getselev et al. (1996a)

Table 11.3 Probability to exceed a given fluence level (in %) during a 1-year space mission (Getselev et al. 1996a)

Probability	Calculated	Calculated	"Observed"	"Observed"
Fluence	>10 MeV	>30 MeV	>10 MeV	>30 MeV
5.0×10^7	99.8	92.2	100	71
5.0×10^8	90.7	37.7	75	42
1.0×10^9	74.8	20.4	62	20
1.0×10^{10}	5.9	1.0	12	8

believe that this discrepancy is likely to be caused by both deficiency of observational data and improper choice of the distribution function for proton event fluences.

JPL-1991 model assessment. Tranquille and Daly (1992) have reviewed and compared in detail two solar-proton event models. First of all, the JPL-1991 model was compared to the K-1974 model in order to identify similarities and differences in both the use of the model and in the computed fluences. Further, because the choice of a confidence level is somewhat arbitrary, this problem has been also investigated. Finally, the *GOES*-7 data obtained during the ascending phase of the 22nd solar cycle were compared to predictions made by the JPL-1991 model.

The K-1974 model is usually used by specifying the number of proton events (either ordinary or anomalously large) that will occur within a given number of years. If at least one anomalously large SPE is chosen, then any contributions from ordinary events will be small enough to be considered negligible. The K-1974 data set contains only one anomalously large event in a 7-year period, and so predicting the occurrence of such an event over a shorter interval can be regarded as pessimistic. Obviously, more pessimistic is to include more than one anomalously large event during a typical mission lifetime.

An alternative way to use the K-1974 model is to specify a confidence level, in much the same way as is required for the JPL-1991 model. Using the equation for p $(n, \tau; N, T)$, as given by Burrel extension (11.15) to Poisson statistics, it is possible to calculate the number of large events, n, that will occur over τ years, and hence determine the minimum number of events that need to be included depending on the confidence level used. For example, if a confidence level of 90 % is required for a space mission lasting 3 years, it would be necessary to include two anomalously large flares in the radiation analysis. As noted by Tranquille and Daly (1992), the use of Burrel statistics in this way for determining the number of events by the K-1974 model, is unrealistic and provides proton fluences which are conservative. On the other hand, allowing the model user to set the number of events can be very arbitrary, which is also not desirable.

The K-1974 model suffers from these problems by virtue of separating the anomalously large event of August 1972 from the rest of the events in the database. The statistical significance of observing a single event in a 7-year period is of limited value for predictive purposes. The JPL-1991 model does not suffer from this problem by virtue of the continuous distribution of event-integrated fluences. However, the choice of confidence level to use still needs to be addressed.

Tranquille and Daly (1992) noted, in addition, a difference in the spectral forms used to extrapolate solar proton fluences at energies other than 10 and 30 MeV. The anomalously large flare of the K-1974 model is represented by an exponential in energy (11.14), but the ordinary events from the same model follow an exponential in rigidity. The event-integrated fluences measured during the 22nd solar cycle also seem to show a preference for rigidity as the variable in the exponent (Lanzerotti et al. 1991), as will be discussed below.

Figure 11.18 constructed by Tranquille and Daly (1992) compares fluences derived from the K-1974 model with those obtained from the JPL-1991 model, using both spectral form, for intervals of 1, 3, 5 and 7 years. An arbitrary confidence level of 90 % is used for this comparison. Using Burrel statistics, the number of anomalously large events that need to be included at the specific confidence level, are 1 in a year, 2 over 3 years, 4 over 5 years and 5 in 7 years. As one can see, in general, the JPL-1991 model predicts higher fluences at energies below a few tens of MeV, but lower fluences at higher energies. Low-energy protons are most important for solar-cell degradation, and so the JPL-1991 model is more severe for predicting this effect. The discrepancy between the use of rigidity or energy to parameterize the spectral form used in the JPL-1991 model can result in differences of several orders of magnitude with increasing energy. This inconsistency is of paramount importance given that high-energy protons are most penetrating. This comparison highlights a few of the major differences between the two models, and illustrates how care must be taken when interpreting the predicting made using either model.

As mentioned above, the choice of confidence level in the JPL-1991 model is somewhat arbitrary. This allows the spacecraft engineers a wide range of predicted fluences which can accommodate specifications both under- and over-design. Furthermore, it is inherent that long missions will be less sensitive to the choice of confidence level when compared to shorter missions. This expectation is confirmed in Fig. 11.19 where the fluence is depicted as a function of confidence level estimated by Tranquille and Daly (1992) for intervals between 1 and 7 years for the two threshold energies. As expected, shorter missions require a more conservative choice of confidence level in order to account for the possibility of a large event occurring over a briefer period of time.

It is of great interest to compare predictions made by the JPL-1991 model to the *GOES*-7 data obtained during the rising phase of the 22nd solar cycle. The proton measurements were made in the following seven energy channels: 0.6–4.2 MeV; 4.2–8.7 MeV; 8.7–14.5 MeV; 15–44 MeV; 39–82 MeV; 84–200 MeV; and 110–500 MeV (Wilkinson 1992). These data have been used by Tranquille and Daly (1992) to compute event-integrated omni-directional energetic proton fluences at threshold energies of 1, 5, 10, 30, 50, 60, 100, and 800 MeV for the major events of 1989 and 1990. As known, several very large events have taken place during the period of interest, especially since March 1989. Seven of the events took place in 1989, including a major flare in October, which was close in intensity to the

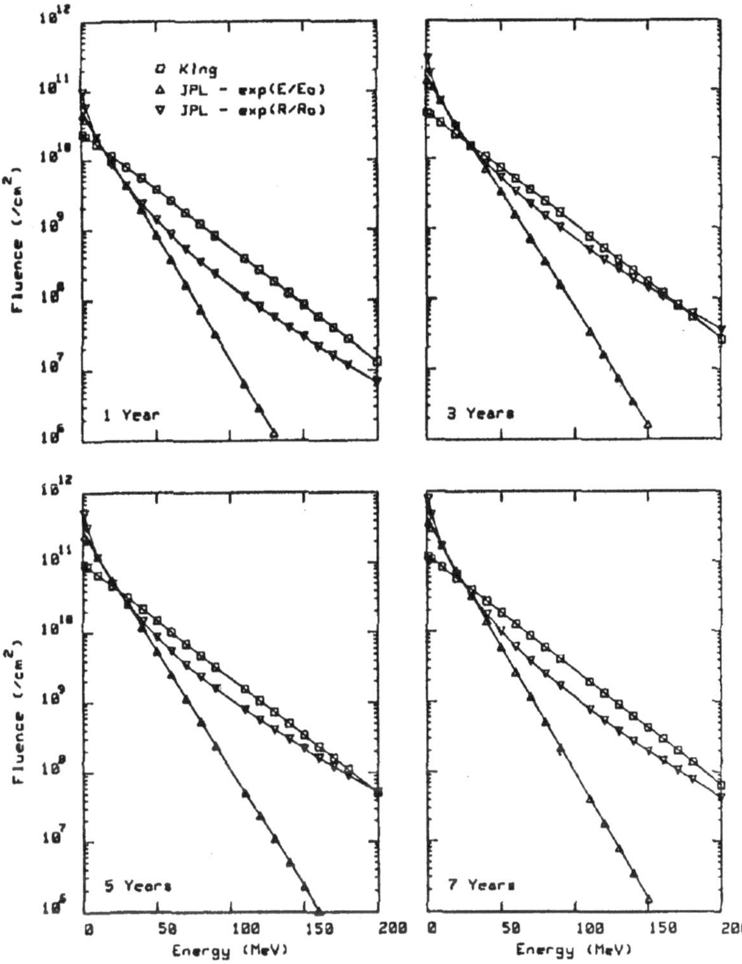

Fig. 11.18 Comparison of proton fluence spectra predicted with the K-1974 and JPL-1991 models (Tranquille and Daly 1992)

November 1960 eruption. According to findings by Tranquille and Daly (1992), all of these events are best represented by an exponential in rigidity. This observation further validates the choice of rigidity, and not energy, as the exponential parameter best characterizing solar-flare proton spectra. It should be noted, however, that the fits were made excluding data for proton fluences above 350 MeV.

In Fig. 11.20 the total fluence spectrum for the 2 years, 1989 and 1990, derived by Tranquille and Daly (1992) with the *GOES*-7 data, is compared directly with spectra obtained from the K-1974 and JPL-1991 models for a 2-year period. Two spectral forms are used (exponential in energy and in rigidity) to illustrate the

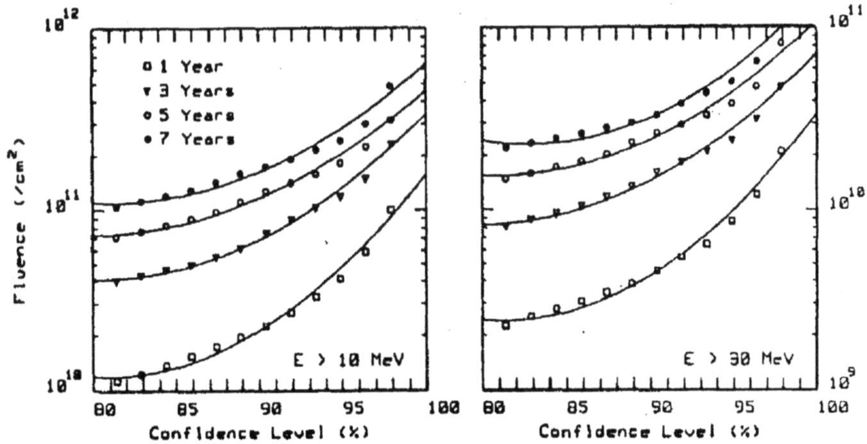

Fig. 11.19 Solar proton fluence as a function of confidence level to use with the JPL-1991 model for missions of different duration (Tranquille and Daly 1992)

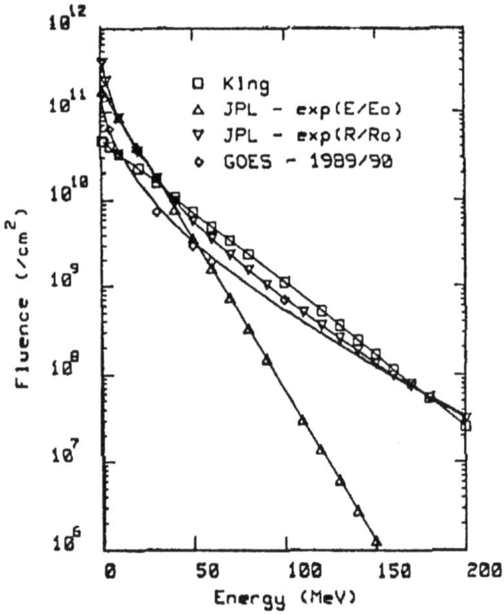

Fig. 11.20 Comparison of solar proton spectra modeled with the K-1974 and JPL-1991 prediction fluence models and derived from the *GOES*-7 data for the 1989–1990 proton events (Tranquille and Daly 1992)

significant differences that can arise at higher energies, depending on the choice made. The K-1974 model is run with two anomalously large events within 2-year period. This corresponds to a confidence level of about 96 %, which is very conservative. The JPL-1991 model is run with a confidence level 95 %. It follows from Fig. 11.20 that the modeled spectrum which best matches the *GOES*-7 data is that obtained from the JPL-1991 model with rigidity as the parameter used to define the spectral form. If energy is used instead of rigidity, there exists a discrepancy between modeled and measured fluence of several orders of magnitude at energies above 100 MeV. This will inevitably lead to misleading depth-dose or penetrating-fluence calculations. Tranquille and Daly (1992) conclude that due to the harder fluence spectrum derived from the *GOES*-7 measurements (*i.e.*, greater fluence at high energies) compared to the model-predicted spectrum, it is necessary to use higher confidence levels with increasing energy.

Finally, measurements made by *GOES*-7 during 1989–1990 indicate that predicted fluences obtained with the JPL-1991 model are not unrealistic. Meanwhile the K-1974 is deficient in low-energy solar protons, in particular, because the exponential in energy used to describe the spectral form of the August 1972 flare is inadequate for many other events. Nevertheless, it is clear that the JPL-1991 model must be used with care, notably in the choice of confidence level. No definitive guidelines exist, and the ultimate choice is essentially a compromise between reducing the risk factor to a minimum level and unrealistic over-design.

In conclusion of this discussion, it is necessary to note some issues which still remain rather disputable. One of them is an uncertainty in the distribution functions applied to predict proton fluence. Unlike a log-normal distribution used by Feynman et al. (1993) for the >10 MeV proton events (Fig. 11.13), Nymmik (1999a) proposed for the >30 MeV protons a power-law function with exponential steepening at large fluences. His model predicts the >30 MeV proton fluence range from 10^6 to 10^{11} cm^{-2}, meanwhile, according to Lingenfelter and Hudson (1980) and Gabriel and Feynman (1996), the corresponding distributions steepen considerably starting from fluences about 10^9 cm^{-2}. Also, the threshold effects of detection and separation of proton events may be important when low-fluence data are used in prediction work. Certain improvements of long-term prediction techniques are expected to come from the probability models for fluences and peak fluxes of solar energetic particles (*e.g.*, Xapsos et al. 1998; Nymmik 2011), as well as from a partial revision of some existing models for event occurrence rate, proton energy spectrum, heavy-ion-to-proton flux ratio, *etc.* (see, for example, Tylka et al. 1997b).

In some cases extreme SEP events in non-relativistic energy range are accompanied by large fluxes of relativistic protons ($E_p \geq 500$ MeV), or solar cosmic rays (SCRs). They are usually registered by neutron monitors (NMs) at the Earth's surface (GLE phenomena, or Ground Level Enhancement of SCR). Recently, it was suggested (Crosby 2009) to define rare Solar Extreme Events (SEEs) as those events in which the characteristics (field strength, speed, intensity of radiation, energies, etc.) of the associated phenomena (solar flares, CMEs, SEP events) are some orders of magnitude larger than in the most of other events (*e.g.*, event of 20 January 2005).

At present, the so-called "Carrington event" of 1–2 September 1859 (Townsend et al. 2003, 2006) seems to be added to the list of such "rare SEEs". As follows from (Smart et al. ,2006 2008), Carrington event (CE) had the largest integral fluence Φ of protons with the energy of $E \geq 30$ MeV (i.e., the energy-integrated fluence above a certain energy value) for about 450-year period starting from 1561 (McCracken et al. 2001). There is no doubt that the study of such rare events is of paramount importance. In particular, Townsend et al. (2003, 2006) suggest, henceforth, to consider CE with the integral fluence of $\Phi(\geq 30\ \text{MeV}) = 1.88 \times 10^{10}\ \text{cm}^{-2}$ as the best reference "worst-case" for the estimates of radiation hazard in space.

Indeed, the two nearest candidates for a role of "worst-case" – the events of 15 November 1960 and 4 August 1972 – were characterized by far less values of $\Phi(\geq 30\ \text{MeV})$, about $9 \times 10^9\ \text{cm}^{-2}$ and $5 \times 10^9\ \text{cm}^{-2}$, respectively (Smart et al. 2006). Note, however, that those fluence values have been calculated by the data obtained in the epoch of historically fragmentary and indirect measurements of SEP fluxes. At present time it becomes clear that energy spectra applied earlier have the analytical forms quite different from the spectrum form established recently (Nymmik 2011). For this reason, fluence sizes for the events of 1960 and 1972 should be critically discussed in the light of new summary distribution function (see Sect. 12.5). Of great interest are also the estimates of occurrence probabilities of such rare events at present level of solar activity and possible extrapolation of obtained results for the remote past of the Earth (*e.g.*, Wdowczyk and Wolfendale 1977; Kiraly and Wolfendale 1999). In more detail these arising problems are discussed in Chap. 12.

Chapter 12
Summary and Conclusion: Problems and Prospects

In the above Chapters the author attempted to summarize copious data on solar cosmic rays (SCR) and relevant solar, interplanetary and geophysical observations, to demonstrate the importance of solar energetic particles (SEPs) for a number of fundamental astrophysical and geophysical problems. Their applications to the tasks of practical astronautics also have been briefly discussed. One of the main goal of this book was to describe different (sometimes contradictory) approaches to the interpretation of those data and to the simulation of space plasma processes involved, at the contemporary (up-to-dated) level of our understanding of the particle acceleration at the Sun and SEP propagation in the interplanetary magnetic field (IMF). The author apologizes if some results of numerous space researchers were given rather briefly, or were not mentioned at all. Below we try to concentrate on actual problems in the study of SCRs, their sources, mechanisms of acceleration, and related theoretical, observational, and methodological issues. Especially, we pay attention to some unresolved, forgotten, and/or neglected problems.

12.1 Long-Standing Problems of Particle Acceleration at the Sun

Let us summarize, first of all, unresolved problems of particle acceleration at/near the Sun. The wealth of data on particle and electromagnetic emissions in solar flares which has become available within the last years has certainly led to significant progress in the understanding of particle acceleration in solar flares and of the transport of solar particles from the Sun to the Earth. No uniform picture, however, has emerged up to now which is consistent with all the details and interpretations of the observations (*e.g.*, Klecker et al. 1990b; Flückiger 1991, 2009; Reames 1996; Cliver 1996; Miller et al. 1997; Klecker 2009).

L. Miroshnichenko, *Solar Cosmic Rays*, Astrophysics and Space Science Library 405, DOI 10.1007/978-3-319-09429-8_12

The mechanisms proposed for *impulsive phase* particle acceleration include acceleration in DC electric fields, stochastic acceleration in turbulent plasmas, and maybe shock acceleration (Miller et al. 1997). Shock acceleration is assumed to be the preferred mechanism (*e.g.*, Berezhko et al. 2001) for the energetic ions observed in space following *gradual flares.* Two-step processes combining stochastic and shock acceleration of flare particles to high energies have been widely discussed (*e.g.*, Klecker et al. 1990b). A number of unanswered questions, however, still remain (see Chaps. 5 and 7). At the same time, very promising possibilities exist to interpret the data on solar energetic phenomena based on the concepts of periodic or random acceleration (*e.g.*, Aschwanden and Benz 1994), multiple (multi-step or multi-source) acceleration (*e.g.*, Mandzhavidze et al. 1996; Miroshnichenko et al. 2000; Miroshnichenko 2003b), and taking into account extended and fragmented magnetic structure of the SEP sources (*e.g.*, Kocharov et al. 1999).

12.1.1 List of Problems

In this Section we pay special attention to a number of poor-studied, forgotten and/or somewhat neglected questions and/or problems of solar gamma astronomy. In our opinion, their list may contain, at least, following issues to be investigated at modern observational and conceptual base:

1. Large-extended (*e.g.*, Vestrand and Forrest 1993) and long-lasting (*e.g.*, Kanbach et al. 1993; Kocharov et al. 1993, 1994; Akimov et al. 1996; Dunphy et al. 1999) gamma-events from the point of view of multiple acceleration processes at/near the Sun (see above Sects. 12.6 and 12.8). Do they provide new insight into the physics of particle acceleration and understanding of SEP/GLE relations? (See, *e.g.*, Pérez-Enriquez et al. 2000; Ryan et al. 2000).
2. Asymmetric geometry of magnetic field of the near-limb flares (*e.g.*, the flare of 23 July 2002), mechanism of ion acceleration and attenuation of gamma-ray flux (in particular, at the energy of 2.223 MeV) (*e.g.*, Ramaty et al. 1997; Kocharov et al. 2000; Miroshnichenko et al. 2000; Kichigin et al. 2010, 2014).
3. It seems to be timely to re-visit the event of 29 September 1989 (GLE42) that occurred due to behind-the-limb flare. Real position and properties of secondary neutron source in this case remain poor-known and in essential unexplained (*e.g.*, Miroshnichenko et al. 2000).
4. Abundance and expected preferential acceleration of ^{3}He ions and some other light elements at the Sun (*e.g.*, Roth and Temerin 1997). General physical and/or astrophysical aspects (acceleration mechanisms) and some cosmological aspects (post Big-Bang chemical evolution and baryon density of the Universe).
5. It deserves serious attention and thorough study a concept of multiple (and/or episodic) process of acceleration for considerable time during a certain solar

flare (*e.g.*, Hua and Lingenfelter 1987b; Struminsky et al. 1994; Mandzhavidze et al. 1996; Miroshnichenko 2003b).

6. Do gamma-ray studies make clearer the flare-CME physical links and eliminate the confrontation between flare acceleration (*e.g.*, Somov 1996) and CME-driven shock acceleration (*e.g.*, Reames 1999; Cliver 2009) models?
7. How to separate the contributions of penetrating and escaping particles?

Below we illustrate only some of mentioned problems and try to suggest modern ways for their solution.

12.1.2 Protons in Solar Flares

As it was noted earlier, we do not pretend to give here a comprehensive description of existing theoretical models of a solar flare. There are a number of excellent reviews of this fundamental problem (*e.g.*, Sakai and Ohsawa 1987; de Jager 1990; Somov 1992, 2012; Priest and Forbes 2000; Somov and Oreshina 2011; Zharkova et al. 2011; Zharkova 2012). Also, a special book has been recently published "High-Energy Aspects of Solar Flares: A *RHESSI*-inspired monograph" (Dennis et al. 2011; http://arxiv.org/abs/1109.5831), with modern view of the problem. The standard scenarios for explaining solar flares include current interruption, interaction between two loops or between a loop and newly emerging flux, and the eruption of a filament. These scenarios all require a specific geometry or pre-flare condition, and thus are not generally applicable to all flares. For example, the large two-ribbon flare of October 24, 1991 cannot unequivocally be explained by any of the foregoing models (see De La Beaujardière et al. 1995). Therefore, it is important to consider in detail some of disputable aspects of the flare physics from the point of view of particle acceleration at the Sun. In this section we address one of them, namely, the possible role of non-thermal protons as a means of transporting energy in solar (stellar) atmosphere (see, *e.g.*, Perez-Enriquez 1985; Simnett 1985, 1986, 1991, 1995, and references therein).

As noted by Perez-Enriquez (1985), the importance of energetic particles in the generation of solar flares and related phenomena has been underestimated (if not completely neglected) up to the middle of 80s. He has proposed a solar flare model relying on the early hypothesis (Elliot 1964; Schatzman 1966, and others) about an active (if not decisive) role of energetic particles in solar flare generation. According to this hypothesis, the cause of the solar flare (its energy source) seems to be energetic particles trapped in magnetic loops above active regions where they are first heated and then accelerated by absorbing part of the wave energy flowing upwards continuously from the convective zone. The instability of the magnetic trap may bring to the catastrophic release of particles into the chromosphere and the "ignition" of the flare proper. In contrast to the assumption of Elliot (1964) that the protons are acquiring the necessary energy slowly, during many hours or days, Perez-Enriquez (1985) showed that this process may take a few hours, and most observational features related to flares and particle acceleration are implicitly explained within such a new approach. The problem of this model is still, no doubt, that to this day no direct evidence exists for the presence of particles being accelerated in magnetic loops prior to the flare.

On the other hand, it is generally believed (see, *e.g.*, Simnett 1991, 1995) that energetic particles take a fundamental part in flare development, but it is controversial as to whether protons or electrons play the dominant role. Despite a wealth of high quality, high resolution data, there still remain significant unanswered questions regarding the precise physical processes responsible for a solar (stellar) flare. According to Simnett (1995), the non-thermal protons might provide some solutions.

The protons are of interest in a broader context on account of their ability to carry energy and momentum, without radiating significantly, over large distances. Their energy losses per unit distance and per unit time are easily predictable. Protons are almost "invisible" and "silent". Only at high energies, certainly well beyond the part of spectrum where most of the energy resides, do protons become easily "visible", for example, due to production of neutral emissions (gamma rays and neutrons) in the ambient matter. On the contrary, these properties are not shared by electrons. In fact, they not only do radiate profusely, but they are easily scattered; also, because of their high velocity per unit energy, in the same environment as the protons they traverse much more matter per unit time at the same energy.

Electrons have attracted most attention simply because of their radiative properties: in fact, most of flare wave radiation comes from electrons. However, as noted by Simnett (1995), this picture may be too simplistic, and the question has been raised as to whether, after all, the majority of energetic electrons might be *secondary*. To substantiate such a hypothesis, Simnett and Haines (1990) proceed from the recognition of two main facts: (1) most acceleration processes relevant to solar flares accelerate protons (theoretically) very much better than electrons, and (2) during the impulsive phase of flares there is *apparently* more energy in non-thermal electrons than in other particles. It is suggested that a neutralized ion (proton) and electron beam, with no net current, is accelerated by the process of magnetic reconnection in the corona. The beam propagates along the local magnetic field towards the chromosphere where it encounters the density discontinuity at the top of the transition region. At this level the beam electrons, which have the same velocity as the ions, scatter and effectively stop. The protons, with their larger momentum, continue. The situation is illustrated schematically in Fig. 12.1.

Because the electrons stop and the proton continue, an electric double layer will be established, unless some way is found to neutralize it. For low beam fluxes it is achieved due to an ample supply of cold chromospheric electrons. However, if the beam flux is large enough, the resistivity of the chromosphere is too high to supply sufficient electrons fast enough. In this situation a potential, Φ, develops in the transition region which accelerates the highest energy electrons available, namely those in the beam which have higher than average velocities.

This concept, though was criticized, has a number of attractive features which are consistent with flare observations, in particular, with respect to energy balance.

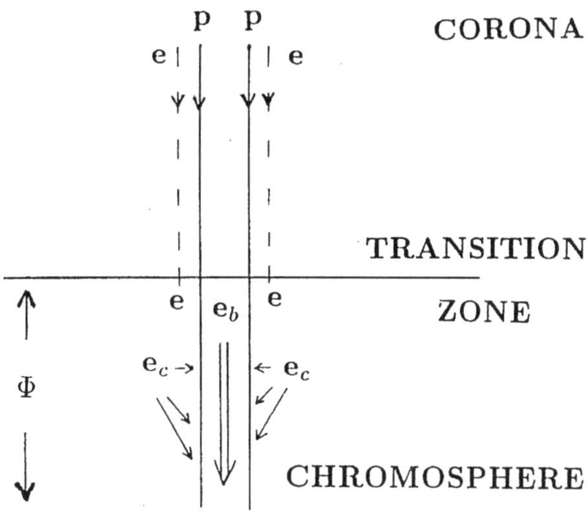

Fig. 12.1 The model of the development of an electric potential below the transition zone (After Simnett (1995)). A neutral beam of ions and electrons is incident on the transition zone from the corona. For low beam fluxes the potential is neutralized by cold chromospheric electrons, e_c. Above a certain flux threshold the resistivity of the chromosphere is too high for effective neutralization to occur, and beam electrons e_b experience runaway acceleration

If the electrons are totally secondary to the protons, the energy in the electron population *is not added* to the ion energy budget. In practice, filamentation of the beam may occur such that in the chromosphere *i* filaments may participate in the flare, each independently producing electron acceleration. In this case the energy in the non-thermal electrons is (Simnett 1995):

$$E_{nte} = \Sigma_i \int_{t_1}^{t_2} \eta_i(t) W(t) dt \tag{12.1}$$

where t_1 and t_2 are the start and stop times for the runaway conditions in the *i*-th filament, *W*(*t*) is the energy into protons from the primary reconnection, and $\eta_i(t)$ is the efficiency with which the ion energy is transferred to the electrons provided the runaway condition is satisfied. Simnett and Haines (1978) showed that $\eta_i(t)$ could easily be >0.9. This idea is virtually identical to the proposal by de Jager and de Jonge (1978) that flares are simply a collection of elementary flare bursts.

Simnett (1995) also notes that with a proton beam the typical energy per particle may be ~0.5 MeV. If the typical energy in a hypothetical electron beam is 25 keV, than a factor of 20 fewer particles are needed to transport the same amount of energy. As discussed by LaRosa and Moore (1993) and others (see Miller et al. 1997), the total number of particles required to provide flare energies of 10^{32} erg is a problem; any mechanism that has the potential of reducing this number by an order-of-magnitude is welcome.

In recognition of the need to consider the global implications of having a relatively invisible, but powerful, energy source in the non-thermal protons, Simnett (1995) discussed, in particular, their possible role in the onset of coronal mass ejections. Simnett and his co-authors (see references in Simnett 1995) discovered weak soft X-ray enhancements, often from the points separated by

$>10^5$ km, around the projected onset time of the CME. If a subsequent flare occurred it was from a point close to, but not identical with, one of the initial bright points. This implies that the chromosphere was sprinkled with small energy deposits. As the CME departed, more evidence of energy deposition was seen; sometimes it was a major flare, at other times two or more small events.

Based on some observations, it was suggested that protons in the 10^2–10^3 keV region could be responsible both for triggering the CME and for the chromospheric manifestations. These two phenomena may be explained in the following scenario. If protons are injected into a coronal loop of height 10^{10} cm, and a mean density $<n> = 4 \times 10^8$ cm^{-3}, a 130 keV proton will lose its energy through Coulomb collisions after around one traversal. A 500 keV proton would have a lifetime of around 9 min. The energy deposited will raise the temperature of the gas in the loop. Protons in the above energy range have the potential for transferring energy to the coronal gas which is unparalleled. An energy input of 10^{27} erg is sufficient to raise the temperature of 10^{29} cm^3 of the corona ($<n> = 4\ 10^8$ cm^{-3}) by 10^6 K. If this process does cause the onset of CMEs, this temperature increase in a coronal loop would approximately double the pressure.

In such a scenario the accelerated protons which are not trapped will immediately travel along the magnetic field to the footpoints of the loop where they would produce coincident soft X-ray or $H\alpha$ brightening. The departure of the CME will drive subsequent magnetic reconnection which could, if sufficient energy were involved, accelerate enough protons to power an associated flare.

The nuclear interactions of energetic protons give the most unambiguous evidence for their presence; they also give a physical insight into the atmospheric composition. However, because the steep energy spectrum above 30 MeV, which is inferred from direct measurement of the particles which escape into interplanetary medium, such protons are insignificant energetically to the total energy budget of a flare. It is difficult to extrapolate the spectrum deduced at high energies reliably into the sub-MeV region. Yet it is in the 0.1–1.0 MeV region that the bulk of the energy is believed to reside (Simnett 1986).

In flares, protons below the gamma-ray production threshold (about 10 MeV/nucleon) cannot be positively identified from observations of the intensity of emitted solar radiation. This does not mean that the low-energy component does not exist, merely that we must be ingenious in devising ways to detect it. The search until recently has been elusive, but there are now promising developments in diagnostic of low-energy proton beams. As it was shown (see Simnett 1995 and references therein), the interaction of protons of energies 10 keV–1 MeV with chromospheric hydrogen atoms results in two detectable effects: (a) asymmetry (towards the red-wing) in the $L\alpha$ profile, and (b) linear $H\alpha$ polarization in the flare to disk-center direction. The $L\alpha$ red-wing production has some constraints, so observations of $H\alpha$ polarization are the best and possibly the only practical way of consistently detecting the presence of protons <1 MeV in the solar atmosphere.

From interplanetary observations in the ecliptic plane it is difficult to derive the true proton energy spectrum below 250 keV, because such a spectrum may be associated with interplanetary shocks. Recently the *Ulysses* spacecraft has made

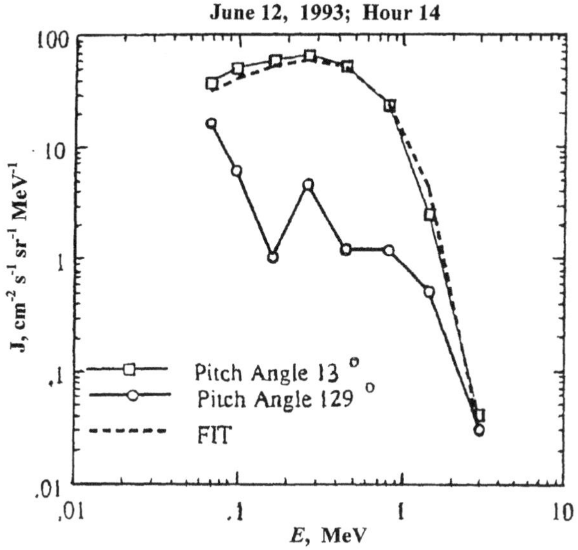

Fig. 12.2 Differential energy spectra of ions moving at pitch angles of 13° and 129° to the interplanetary magnetic field for the period 11:45–15:15 UT on June 12, 1993. The *dashed line* is a Maxwellian fit with a characteristic energy of 0.273 MeV. This is interpreted as a beam of particles injected from the corona into the interplanetary medium, which are propagating with negligible scattering (After Armstrong et al. 1994)

observations at high heliographic latitudes, well away from the heliospheric current sheet. Armstrong et al. (1994) have reported an event observed on June 12, 1993 at the latitude of 32°S. The source had a proton spectrum which peaked around 270 keV, and judging from the pitch angle distribution of the protons, they appeared to have been ejected from the corona. The ions spectra (predominantly protons) are shown in Fig. 12.2. As Simnett (1995) believes, if the outwardly-streaming spectrum is truly representative of that found at the Sun, then it shows that sub-MeV protons are energetically predominant.

In this context, the role of electrons in the flare processes should be studied more profoundly than before. Of course, the electrons are indispensable for our overall understanding of flares. In space, solar electrons do not exceed energies ~100 MeV. In this energy range electrons might be produced via interactions of relativistic protons when the latter are present. Hence, Simnett (1995) concludes that there is currently no evidence that an acceleration process produces electrons of this energy.

12.1.3 Splitting of Electron and Proton Spectra

A new approach to the problem under discussion was proposed by Miroshnichenko (1995) who suggested that the differences ("splitting") between proton (ions) and electron spectra are inevitable from the very beginning of the spectrum formation (if both these particle populations are accelerated in the same acceleration process).

As seen from Eq. (5.22), an evaluation of the real contribution of accelerated protons or electrons to the energetics of flares depends considerably on the form of SCR spectrum in the source. Due to the selectivity of the acceleration mechanisms

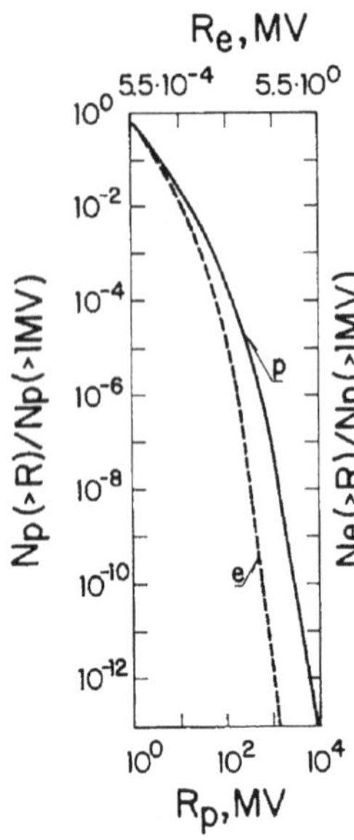

Fig. 12.3 Expected differentiation of the normalized rigidity spectra of accelerated particles owing to different efficiency of acceleration mechanisms and different pattern of energy losses of electrons and protons (*dashed* and *solid lines*, respectively). A spectrum curve for protons is in approximate accordance with the observational data; a similar one for electrons illustrates an assumed steepening of their spectrum (Miroshnichenko 1995)

and different nature of energy losses of electrons and protons, their source spectra should be subject to softening at different rates nearly from the very beginning of the acceleration process. This results in a differentiation of the two spectra initially closely coupled by a common acceleration mechanism (Miroshnichenko 1987, 1990). In this context, it is important to choose the most suitable parameter to compare the quantities of accelerated electrons and protons.

In the light of well-known proton hypothesis of Simnett (1986), a convenient parameter would be specific particle energy, *i.e.*, energy per unit of mass. Obviously, the equality of specific energies is possible, provided the condition of $V_p = V_e$ holds. It follows that the momenta (or rigidities) of two particles will obey the ratio of their masses, m_e/m_p. Therefore, it was suggested (Miroshnichenko 1995) to analyze the spectral differentiation on the condition of normalization of $R_p = 1{,}838\ R_e \cong 1.0$ MV, *i.e.* starting with $V_p \cong V_e$. A proton rigidity of 1 MV approximately corresponds to a thermal energy $E_p \cong 10^3$eV which would be the initial value for the acceleration of protons, for example, by a DC electric field.

It is natural to suppose that the amount of differentiation will increase with the increase of particle rigidity. Figure 12.3 shows the expected picture of the spectral differentiation, the curve for protons in general being compatible with observational data (Miroshnichenko 1990; Miroshnichenko et al. 1999). The normalized curve for electrons illustrates an assumed softening of their spectrum. To verify the effect under consideration it seems to be reasonable to combine the data of direct measurements of interplanetary electron flux and the observations of different types of wave radiation for the same flare with the results of calculations of electron energy losses. If the source spectrum of the electrons in the entire rigidity range of $R_e = 5.5 \times 10^{-4}$–5.5 MV can be reconstructed from this data set, then the comparison with the source spectrum of protons (Miroshnichenko 1990; Miroshnichenko et al. 1999) will yield information of major importance. As far as we know, such a problem has never been considered before.

Obviously, evaluation of the source spectrum and energy losses of the electrons from data of the wave radiation signature, even for a narrow region of the spectrum, will depend heavily upon model ideas on loss conditions in the source and on mechanisms of emission. Nevertheless, there are good reasons for a possibility of such evaluations. In particular, observations near the Earth show that electron energy spectra from 30 keV to 3.0 MeV have a non-monotonous decreasing form (see, *e.g.*, Kurt et al. 1981), the number of relativistic electrons being several orders less than the number of relativistic protons in the same flare. Such a difference seems to be explainable by electron losses to synchrotron radiation, which under flare conditions falls into the ultra-violet range (Bulanov et al. 1985). In connection with existing difficulties in interpreting of the electron component of solar cosmic rays, it appears to be reasonable to compile an Electron Source Spectrum Catalogue, similar to the Proton Source Spectrum Catalogue (Miroshnichenko et al. 1999).

12.1.4 Interacting and Escaping SEPs and Gamma-Rays

We illustrate here this point based on so-called "source spectra" of accelerated solar protons (Miroshnichenko et al. 1999; Miroshnichenko 2003b). These authors suggested a new method for the separation and localization of the sources of particles accelerated at/near the Sun. The method is relied upon the analysis of the data on source spectra of the interacting (precipitating) and escaping (propagating) protons (Fig. 12.4).

The source spectra are reconstructed by two ways. For the precipitating particles that interact with the solar atmosphere one can use the data on gamma-ray line emission. As to escaping particles (or SEPs), their source spectra may be reconstructed by the data of direct particle observations near the Earth's orbit with the subsequent reverse extrapolation of the intensity-time profiles to the moment of particle injection from the source.

By several examples of SEP events (see Fig. 12.4) it was found that total number of particles escaping from the solar atmosphere is methodically higher than that of precipitating (interacting) ones. Hence, when comparing the spectra for escaping and interacting particles, a possibility arises to separate their sources (impulsive flare or CME-driven shock). Such a separation becomes possible, but only in a few long-lasting solar events. On the other hand, in some events shock acceleration seems to be unsuitable for interpretation of the delayed component of gamma-emission (Akimov et al. 1996).

In fact, Miroshnichenko (2003b) attempted for the first time to separate reconstructed spectra depending on their assumed sources (impulsive or gradual flares, CME-driven shocks and other energetic solar/interplanetary phenomena).

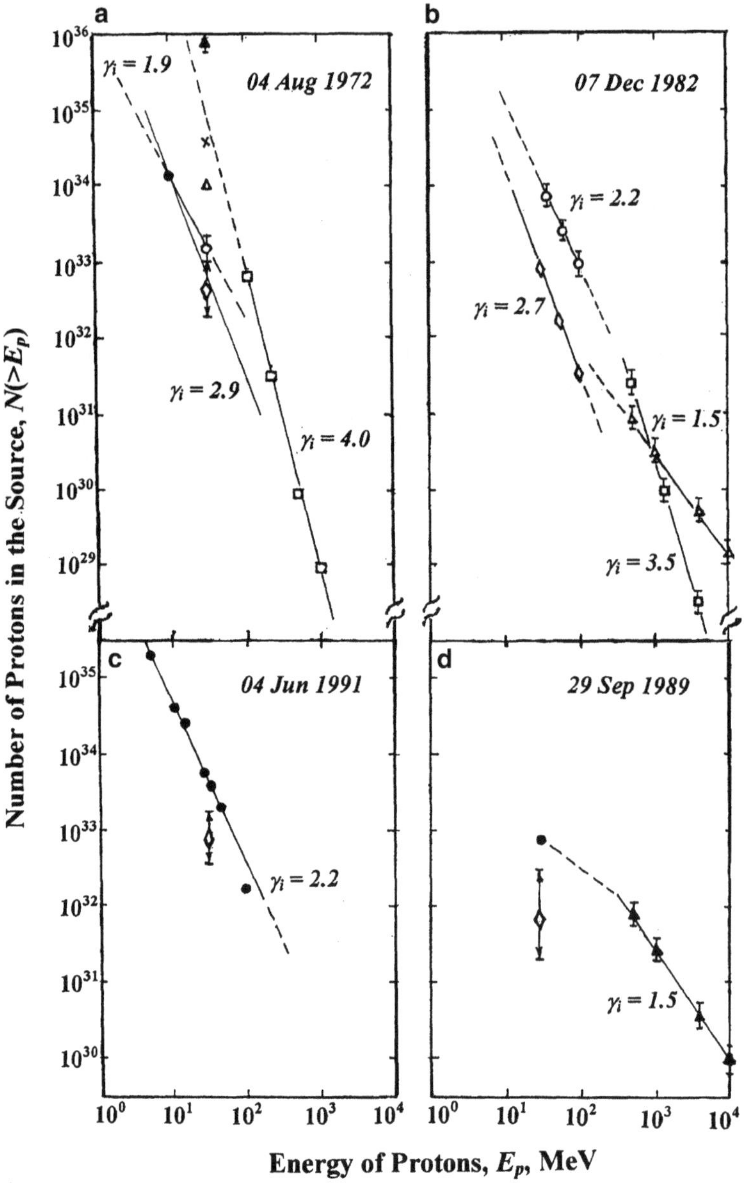

Fig. 12.4 Integral energy spectra of accelerated solar protons (with the exponent γ_i) in their sources (Miroshnichenko 2003b) by different estimates for the four SEP events: (**a**) 4 August 1972, interacting (*full and open circles, full diamond*) and escaping (*full and open triangles, cross and squares*) protons; (**b**) 7 December 1982, interacting protons (*triangles*); prompt (*diamonds*) and delayed (*squares*) relativistic components and non-relativistic escaping particles (*circles*); (**c**) 4 June 1991, interacting (*diamond*) and escaping (*circles*) particles; (**d**) 29 September 1989, interacting (*diamond*) and escaping (*circle*) protons; prompt relativistic component (*triangles*)

The question of principal interest we addressed here is whether the escaping and interacting particles are distinguishable in impulsive and gradual solar events, or do the SEPs observed near the Earth's orbit belong to the same population? As follows from the above analysis, the problem of separation of interacting and escaping SEPs can not be resolved accurately based on available fragmentary data on the source proton spectra only. Nevertheless, conceptually, the data on source proton spectra, at least, for large proton events, seem to be treated in terms of multiple acceleration processes in large-scale coronal structures. Those data must be also incorporated into modern models and scenarios of particle acceleration at/near the Sun.

12.2 Accelerated Particle and Atmospheric Density Models for the Sun

As it follows from our consideration (Chap. 6), in spite of limited number of available data, observed intensity of the line of 2.223 MeV may serve as a measure of the concentration of ^{3}He in the photosphere. Moreover, the time history of its fluence may be used to determine the altitude profile of plasma density deeply in the photosphere (*e.g.*, Kuzhevskij et al. 1998, 2005a). For the first time such a possibility was demonstrated by the data on the 22 March 1991 flare. Subsequent analysis of a set of other flares (6 November 1997, 16 December 1988, and 28 October 2003) confirmed an effect of plasma density enhancement (EDE) at the photospheric and sub-photospheric levels during those four flares mentioned (*e.g.*, Kuzhevskij et al. 2005a). Some later it was shown that the 2.223 MeV line data may be used to link the line fluences with a form of the spectrum of accelerated protons in the source (Kuzhevskij et al. 2005a; Troitskaia et al. 2007, 2009). Below we consider this possibility in more details.

12.2.1 Effect of Density Enhancement

Gamma-ray emission (de-excitation lines, neutron capture line of 2.223 MeV, annihilation line of 0.511 MeV and some others) provides information about the nature of accelerated particles and physical conditions in the medium where accelerated particles are produced or transported (*e.g.*, Ramaty and Murphy 1987; Murphy and Share 2005; Murphy et al. 2007). The neutron capture line of 2.223 MeV is of special interest.

Secondary flare neutrons are produced by accelerated particles in dense layers of the solar atmosphere, and then they are captured in the photosphere by hydrogen nuclei ^{1}H. As a result, there are produced a deuterium ^{2}H (stable isotope of hydrogen) and gamma-quantum of 2.223 MeV (sometimes this radiation is called as deuterium line): $^1\mathrm{H} + \mathrm{n} \rightarrow {}^2\mathrm{H} + \gamma(2.223\ \mathrm{MeV})$. By numerical simulation of the this process it was shown (Kuzhevskij et al. 1998) that the observed time profile of this line makes it possible to reconstruct the altitude profile of solar plasma density, simultaneously with the spectral index of accelerated primary protons. As known, the time constant for capture on ^{1}H is $\tau_H \approx 1/(n_H v_n \sigma_H)$, where n_H is the number density of hydrogen, v_n and σ_H are neutron velocity and cross sections for neutron capture, respectively. From this simple expression one may conclude that the observed intensity-time profile of gamma-emission of 2.223 MeV contain also the information on the neutron spectrum and plasma density in the region of neutron energy loss. Based on those findings, the authors have developed original approach and elaborated special method (SINP code) to determine the most probable profile of plasma density in the photosphere and adjoining levels during the period of a flare (for details see Kuzhevskij et al. 1998, 2005a; Troitskaia et al. 2007).

With several plausible models of the solar atmosphere density, this code was successfully applied to a number of large gamma-ray flares: 16 December 1988, 22 March 1991, 6 November 1997, and 28 October 2003 (Gan et al. 2003; Kuzhevskij et al. 2005a; Troitskaia et al. 2007, 2009). As a result, the authors have obtained some evidence of the effect of density enhancement (EDE) in the sub-flare region. This effect may be demonstrated, in particular, by the data on two large solar gamma-ray events, namely 28 October 2003 (Kiener et al. 2006) and 20 January 2005 (Arkhangelskaja et al. 2009a, b) (for details see Miroshnichenko and Gan 2012).

As a basic working model of the solar atmosphere it was used a standard astrophysical model – Harvard-Smithsonian Reference Atmosphere (HSRA) by Gingerich et al. (1971), in combination with a model of the solar convective zone (Spruit 1974). This basic density model (BDM) has a number 1 (No.1, or $m = 1$) in our numeration (Table 12.1). To investigate proposed declinations from the BDM by gamma-ray data, Kuzhevskij et al. (1998) have suggested four composed models ($m = 2$–5) representing smaller and larger densities at the photospheric and adjoining levels (Fig. 12.5, top panel). The only difference between the basic and modified models is in the fragments plotted by the dashed lines. A top of the photosphere corresponds to the level where the optical depth is $\tau = 0.005$ at a wavelength of 5,000 Å.

As noted by Share et al. (1996), an excellent measure of the atmospheric density is provided by the 3γ/2γ ratio from positron-electron annihilation. The 3γ continuum comes from annihilation of the triplet state of positronium while the 0.511 MeV line comes from either free positron-electron annihilation or from annihilation from the singlet state of positronium. It is important to note that the triplet is depleted at densities $>10^{14}$ cm^{-3}. Based on *SMM*/GRS measurements of the positronium continuum and annihilation line (Share and Murphy 1995) in 19 solar flares observed from 1980 to 1989, they showed that gamma-rays from flares with the lowest low-FIP to high-FIP ratios were produced deep in the chromosphere where the abundances should be close to photospheric. In the whole, Share et al. (1996) provide some evidence of that ions accelerated in different flares may interact at significantly different depths. It means that

Table 12.1 Selected density models of the solar atmosphere used in the SINP code

m	Main characteristics of the models	Height density profiles in more detail
1	Basic density model (BDM): a combination of the HSRA for the low chromosphere and photosphere (Gingerich et al. 1971) with the model of convection zone (Spruit 1974)	Smooth rise from 1.5×10^{16} cm^{-3} at the top of the photosphere to 2.0×10^{17} cm^{-3} at the level of 300 km below where $\tau = 1$ and rises sharply to $\tau = 10$ within 60 km deep down
2	Enhanced density inside and under the photosphere	Enhanced density up to 8.0×10^{17} cm^{-3} at the depths ~500 km under the top of the photosphere, i.e. in the deep sub-photospheric layers
3	Enhanced density inside and under the photosphere	Density under the photosphere rises more slowly and is of 6.0×10^{17} cm^{-3} at the same depths
4	Reduced density starting from the lower chromosphere and below	Reduced density starts above the photosphere; at its top the density reaches of 3.0×10^{15} cm^{-3} and then at the level of 300 km below it is about 2.0×10^{16} cm^{-3}
5	Enhanced density all over the depth of the photosphere	Enhanced density all over the depth of the photosphere is about 2.0×10^{17} cm^{-3}

m model number

gamma-rays in flares may be actually produced in regions ranging from the upper photosphere to the corona.

In this context, Gan et al. (2004) have studied the influence of the density of the annihilation region on the positronium continuum. A relation between the ratio $3\gamma/2\gamma$ and the density is explicitly given, with which one can derive directly from the observed $3\gamma/2\gamma$ the density where the annihilation occurs. A unique solution may be found from the observed width of the 0.511 MeV line. They applied the method to three major flares observed by *SMM*/GRS (3 June 1982, 24 April 1984, and 19 October 1989). It was shown that due to the measuring uncertainties in the 0.511 MeV line widths, one cannot distinguish a chromospheric source from a coronal source, though both are accurately localized. To improve the measuring accuracy of the 0.511 MeV line and the ratio $3\gamma/2\gamma$ will be an important step for a better understanding of the annihilation process in solar flares.

In above summary, we address several actual issues of solar gamma astronomy. First of all, based on our results for five flares, we suggest that EDE in deep layers of the photosphere may be rather a common feature of powerful solar flares on the whole. Implications and further prospects of the suggested method depend on involving new data on the flares observed during the last decade with high energy, time, and angular resolution. In particular, of great importance are the data on the events of 23 July 2002, October-November 2003, and 20 January 2005 from *RHESSI*, *INTEGRAL*, and *CORONAS-F* spacecraft. At the same time, it seems to be reasonable to incorporate into the SINP code new basic density models of the solar atmosphere (*e.g.*, Avrett 1981; Vernazza et al. 1981), or their combinations (see discussion below).

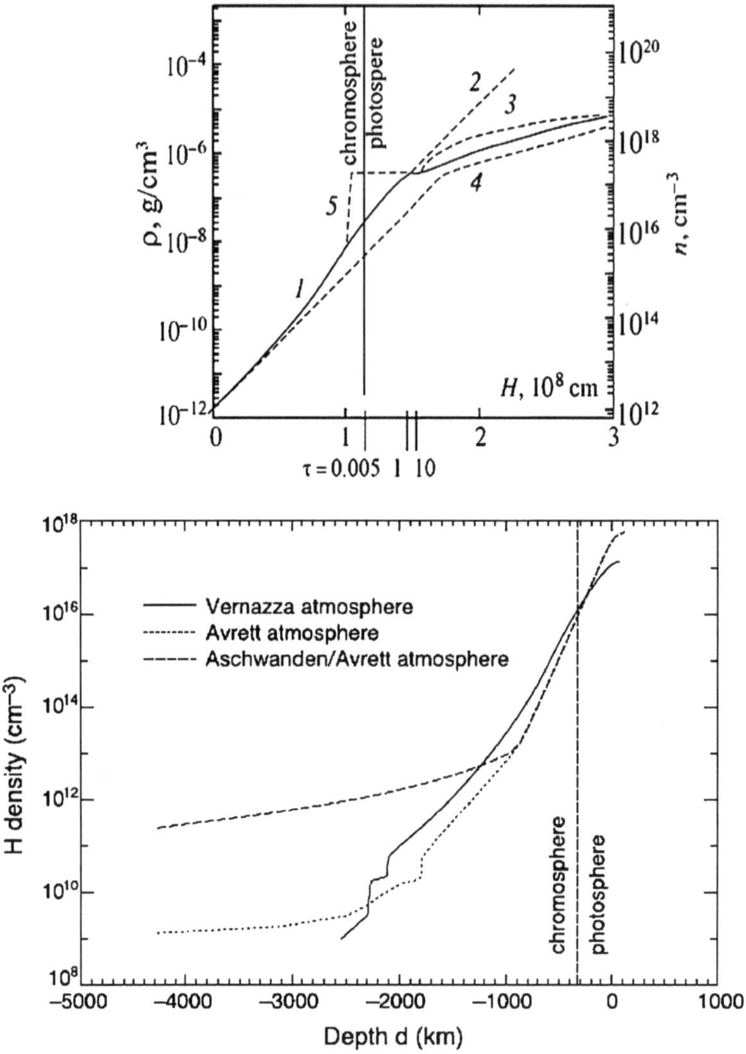

Fig. 12.5 Basic density model *1* of the solar atmosphere (*top panel*) and four distorted models *2–5* (only fragments differing from the curve *1* are shown). Parameter τ is the optical depth for a wavelength of 500 nm; the level $\tau = 0.005$ corresponds to the top of the photosphere (Kuzhevskij et al. 1998). Density *vs.* height in the solar atmosphere (*bottom panel*) for three atmospheric models considered by Murphy et al. (2007): Vernazza (Vernazza et al. 1981), Avrett (Avrett 1981), and *RHESSI*/Avrett (Aschwanden et al. 2002)

Some other possibilities arise from consideration of the pitch-angle scattering (PAS) of accelerated ions in the flare loop model and account for the depth dependence of neutron production in the solar atmosphere (Hua et al. 2002). The parameters directly associated with the ion transport are the level of PAS, the degree of loop magnetic convergence, and loop length. Based on the loop concept,

Murphy and Share (2005) present the gamma-ray line-production and loop transport models used in the calculations of high-energy emission. They discussed in detail the calculated interaction time history, the depth distribution, the interacting-particle angular distribution, and fluence ratios of the narrow gamma-ray line. It was shown that the pitch-angle distribution (PAD) of accelerated particles in the loop model is very important to estimate the fluences of gamma-ray lines. However, the effects of PAS (or PAD) on the depth dependence of neutron production could not be taken into account because of the SINP code limitations. Therefore, the present conclusions on EDE must be considered tentative until a new analysis, including PAS effects, is completed.

12.2.2 Density Profiles of Flaring Atmosphere

A number of interesting results concerning density-height profiles were obtained on measurements made with *RHESSI*/GeD (Share et al. 2004; Murphy et al. 2005; Aschwanden et al. 2002), as well as with *SMM*/GRS, *Yohkoh*/GRS, and *GRANAT*/PHEBUS (Kuzhevskij et al. 2005a). All of them have suggested that the flaring atmosphere within the loop may be considerably different from that of the quiet Sun. Therefore, Murphy et al. (2007) considered several density-height profiles for the solar atmosphere. In the sunspot model of Avrett (1981), the density at heights greater than several hundred kilometers is more tenuous than that of the quiet-Sun model of Vernazza et al. (1981), but below this height it is greater. Meanwhile, during a flare the atmosphere could be still considerably different (Share et al. 2004).

On the other hand, using *RHESSI* data on electron bremsstrahlung, Aschwanden et al. (2002) estimated the density-height profile in a solar flare. They found that density is enhanced relative to quiet-Sun models at heights greater than ~1,000 km, and the density scale height is considerably larger. Taking into account those results, Murphy et al. (2007) have constructed a hybrid solar-flare density model (the "*RHESSI*/Avrett model") that consists of the Avrett density profile below 1,000 km and the Aschwanden profile above (Fig. 12.5, bottom panel). They use the Avrett sunspot model at photospheric heights since flare loop structures are generally found above sunspots (*e.g.*, Hurford et al. 2003; Schrijver et al. 2006). As follows from *RHESSI* measurements, enhanced chromospheric density during flares could be due to filling of the loop by active processes prior to the main impulsive event. As it was clearly demonstrated by Murphy et al. (2007), the *RHESSI*/Avrett model provides larger chromospheric density and scale height. In fact, this is the first time that the dependence of solar-flare gamma-ray and neutron production on the atmospheric model has been systematically explored. Thus, a possibility arises to learn more about the structure of the solar atmosphere (at least, at the level of chromosphere) from gamma-ray measurements during flares.

SINP code calculations for the flare of 20 January 2005 lead the authors (Troitskaia et al. 2009) to the conclusion that the best fit to observational data may be reached with simulation parameters as follows: $\alpha T = 0.03$, $^3He/^1H = 1.4 \times 10^{-4}$, and $\alpha T = 0.1$, $^3He/^1H = 2.0 \times 10^{-4}$, with a photospheric density enhancement up to 2×10^{17} cm^{-3} (model $m = 5$). It is to note, that at the rise phase it was obtained the best approximation at the lesser value of ^{3}He content. Such a dynamics of ^{3}He abundance may be explained by gradual accumulation of the ^{3}He ions that were originally accelerated and then transported downwards (with some energy losses) into the lower chromosphere and photosphere.

When comparing existing atmospheric models applied by different research groups, it may be seen that the BDM ($m = 1$) and modified model $m = 5$ are rather close and seem to be complimentary to the models investigated by Murphy et al. (2007) (Fig. 12.5, bottom panel). A number of interesting results concerning density-height profiles were obtained on measurements made with *RHESSI*/GeD (Share et al. 2004; Murphy et al. 2005; Aschwanden et al. 2002), as well as with *SMM*/GRS, *Yohkoh*/GRS, and *GRANAT*/PHEBUS (Kuzhevskij et al. 2005a). All of them have suggested that the flaring atmosphere within the loop may be considerably different from that of the quiet Sun. Therefore, Murphy et al. (2007) considered several density-height profiles for the solar atmosphere. In the sunspot model of Avrett (1981), the density at heights greater than several hundred kilometers is more tenuous than that of the quiet-Sun model of Vernazza et al. (1981), but below this height it is greater. Meanwhile, during a flare the atmosphere could be still considerably different (Share et al. 2004).

Since 2002 the authors (Kuzhevskij et al. 2005a) apply improved version of the SINP code (*e.g.*, Gan et al. 2003), with energy spectra of solar flare neutrons calculated by Hua et al. (2002). Taking into account a number of recent improvements and modifications, it was found, in particular, that the density of disturbed solar atmosphere during the flare of 28 October 2003 changes more rapidly than in the model for the quiet Sun (Troitskaia et al. 2009). The density reaches enhanced values (EDE) in the photosphere or in sub-photospheric layers. Also, a hardening of accelerated particle spectrum was found at the decay phase of the gamma-ray burst.

Unlike this, in the case of 20 January 2005 flare, the modeling by the SINP code with necessary accuracy seems to become impossible, at least, within the standard SINP-code assumptions about the properties of solar atmosphere, the nature of nuclear reactions and initial parameters of neutron fluxes. Remind that this study was carried out by the *CORONAS-F*/SONG-D/AVS-F data with electronic AVS-F analyzing system. It was found that the results of model calculations are methodically higher than observed count rates. The EDE for this flare may be obtained only by involving an additional assumption (see Sect. 12.4) about enhanced abundance of ^{3}He nuclei in this extraordinary solar event (Arkhangelskaja et al. 2009c; Troitskaia et al. 2009, 2011). In spite of this controversial situation, these EDE findings (at least, for four major solar flares, see above) seem to be indirectly confirmed by recent optical results by Cessateur et al. (2010).

Solar flares presumably have an impact on the deepest layers of the solar atmosphere, but the observational evidence for such an impact is scarce yet. Using 10 years of measurements of the Na D1 and Na D2 Fraunhofer lines, measured by GOLF onboard *SOHO*, Cessateur et al. (2010) showed that this photospheric line is indeed affected by flares. The effect of individual flares is hidden by solar oscillations, but a statistical analysis based on conditional averaging reveals a clear signature. Although GOLF can only probe one single wavelength at a time, both wings of the Na line can nevertheless be compared. The varying line asymmetry can be interpreted as an upward plasma motion from the lower solar atmosphere during the peak of the flare, followed by a downward motion. Since the Na D line is mostly generated in the upper photosphere; this study thereby provides strong support for the influence of flares on the lower solar atmosphere.

At the modern level of our understanding, solar flare is predominantly a coronal phenomenon. It implies that the deeper layers of solar atmosphere are involved in the flare processes only in the cases of major flares (*e.g.*, in so-called white flares). Therefore, proposed effect of density enhancement in the photosphere may be expected, first of all, in the extreme solar events with large energy release ("syndrome of a big flare", Kahler 1982). In other words, EDE may be characteristic of the most powerful flares.

12.3 New Observation Techniques and Ideas

12.3.1 Concept of "Flagship" Stations

As known, cosmic ray stations with a cutoff rigidity of approximately 1 GV have asymptotic cones of acceptance that sweep around the globe near the equatorial plane (*e.g.*, Smart and Shea 1990a). Therefore, these stations are the most likely to record at least part of every solar cosmic ray increase. Neutron monitors with asymptotic cones of acceptance along the interplanetary field lines toward the Sun will probably record the highest increase during an anisotropic solar cosmic ray event. This impelled Smart and Shea (1990a) to suggest that the Deep River and Kerguelen Island neutron monitors, located in opposite hemisphere approximately 180° apart, are ideally situated for recording these unusual solar events. They proposed to designate them as "Flagship" monitors to provide a preliminary picture of any event for the cosmic ray community.

The concept of "Flagship" stations was taken from the geomagnetic observations where so-called *aa* index is determined from the data of two antipodal observatories to derive a daily index of activity that represents the magnetic conditions for the day. Smart and Shea (1990a) extended this concept to use the data from two strategically placed neutron monitors to typify transient cosmic ray phenomena, especially solar cosmic ray events. Inspection of the distribution of the world-wide network of cosmic ray neutron monitors shows that there exist two stations which are approximately antipodal, have approximately the same cutoff rigidities about 1.1 GV, where the maximum response of an isotropic solar cosmic ray event might be expected, and have wide asymptotic cones of acceptance which essentially sweep around the globe. These stations are Deep River in Canada (46.10°N,

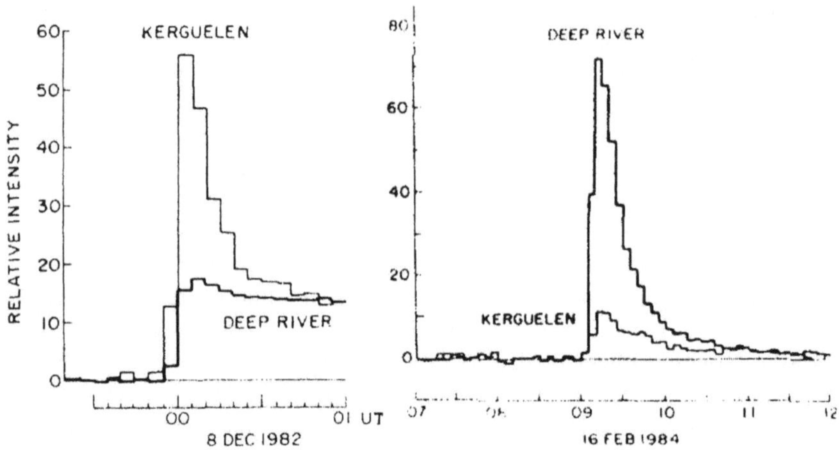

Fig. 12.6 Relative response of the Deep River (*heavy line*) and Kerguelen neutron monitors to the anisotropic solar cosmic ray flux during the GLEs of 7/8 December, 1982 (*to the left*) and of February 16, 1984 (*to the right*). These events occurred at the optimum time for a maximum response at Kerguelen and at Deep River, respectively (Smart and Shea 1990a)

282.50°E, a 48-NM-64 at geomagnetic cutoff of 1.15 GV) and Kerguelen Island located in the Indian ocean (49.35°S, 70.°E, an 18-NM-64 located at geomagnetic cutoff of 1.14 GV).

To illustrate the detection capabilities of this pair of "Flagship" stations, Smart and Shea (1990a) have calculated the response of these two neutron monitors to a high-energy ($E_p > 450$ MeV) anisotropic solar proton flux as a function of local time and anisotropy. Then, there were found two GLEs, observed on December 7/8, 1982 and February 16, 1984, which turned out to be relevant to demonstrate visually the proposed concept (see Fig. 12.6). From their results, Smart and Shea (1990a) conclude that there is a daily variation in the neutron monitor response to an anisotropic solar cosmic ray event. Nevertheless, a combination of "Flagship" stations such as the Deep River and Kerguelen Island neutron monitors would allow detection of even short-lived extremely anisotropic solar cosmic ray increases at any local time through the day (Fig. 12.6).

12.3.2 Potential of Neutron Monitor Data

In this context remember again an important remark of Flückiger (1994) about some open questions in the analysis and interpretation of neutron monitor data. In fact, some unusual intensity-time variations in the counting rate of neutron monitors still lack a satisfactory explanation, in particular, it can not be excluded that some small GLEs remain unidentified. Another aspect of the problem under consideration is that the technique developed for the analysis of cosmic ray ground level enhancements, may be applied to study other transient phenomena in interplanetary space. It was demonstrated, in particular, by Hofer and Flückiger (1998). Based on neutron monitor data they evaluated the spectral and directional variations in the cosmic ray intensity near the Earth during the March 24, 1991 Forbush decrease. It was found that the cosmic ray anisotropy vectors exhibit a rotational behaviour similar to the typical rotation of the magnetic field vector which is observed when a magnetic cloud is traversed. This study suggests that the potential of neutron monitor data to provide information about complex transient structures in the near-Earth interplanetary medium is significantly greater than generally assumed so far. Evidently, all these findings and propositions are very important for retrospective analysis of some very peculiar GLEs observed in the past (for example, in September-October 1989) and for the studies of future GLEs as well.

12.3.3 Optimized Network of Neutron Monitors

In spite of their large "working length of service" (see Fig. 2.1), neutron monitors remain today the state-of-the-art instrumentation for recording cosmic rays above 1 GV. The neutron monitor energy range is highly complementary to the upper range of energies measured by cosmic ray detectors flown in space (see, for example, Fig. 1.9). With their high count rate, neutron monitors excel at measuring the minor anisotropies associated with galactic cosmic rays. To this day, essentially all the information we have on steady state anisotropies is from measurements with ground-based detectors.

For measuring the larger anisotropies associated with transient cosmic ray events such as solar particle increases and Forbush decreases, a network of high-latitude neutron monitors is essential. High latitude sites uniquely offer uniform energy response and very well defined viewing directions. To exploit fully this capability, however, it is necessary to rationalize the word-wide distribution of monitors. One of possible scenarios of such a distribution has been presented by Bieber and Evenson (1995). They took into account some special features of high-latitude monitors.

As it is well-known, neutron monitors respond to primary cosmic ray particles above certain threshold rigidity. For mid-latitude and low-latitude stations, this threshold is the geomagnetic cutoff, R_c. At high latitudes, however, the geomagnetic cutoff becomes small, and the threshold is governed instead by atmospheric absorption. To illustrate this effect, Bieber and Evenson (1995) compare the response of Mawson, Antarctic with Newark, Delaware for a typical solar particle spectrum of $\sim R^{-5}$. The value of R_c for Newark is 2.08 GV, and the solar particle response rises steeply from this threshold. In contrast, the geomagnetic cutoff for Mawson is 0.19 GV, but because of atmospheric absorption the detector response remains negligible until the particle rigidity nears 1 GV.

As shown by Bieber and Evenson (1995), shifting the geomagnetic cutoff within the range from 0.0 to 0.6 GV has little effect on the solar particle response. As a

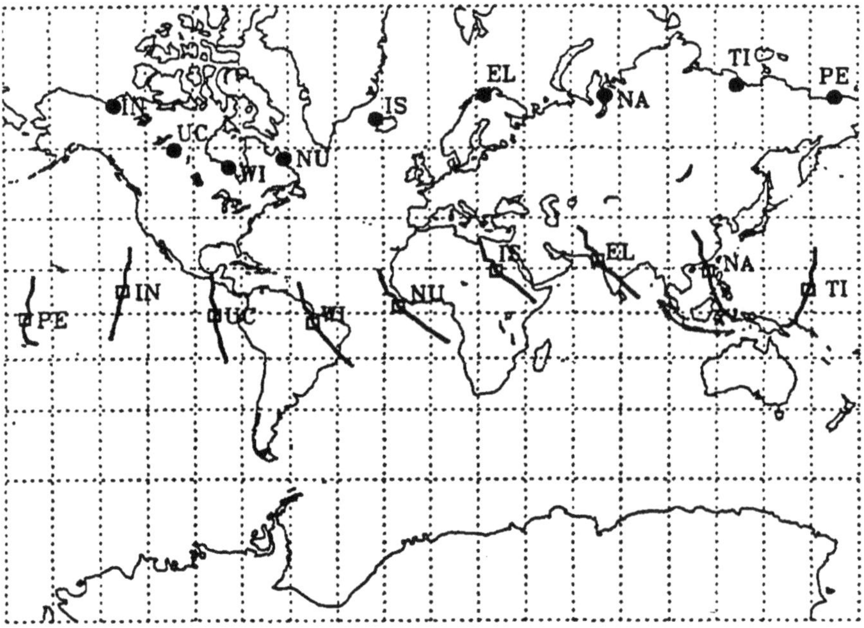

Fig. 12.7 Viewing directions for proposed network of nine Arctic/Subarctic high-latitude monitors (Bieber and Evenson 1995). For details see the text

result, all high-latitude neutron monitors located near sea level have nearly identical energy responses. Further, the atmosphere masks the near-cutoff region where cosmic ray asymptotic directions vary rapidly with energy. As a result, high-latitude monitors have much more confined viewing directions than mid- or low-latitude monitors. Due to this feature, at high latitudes, the effective angular resolution of a neutron monitor for solar particle events was estimated to be better than 20°.

In Fig. 12.7 one of the possible distributions of monitors is represented. Bieber and Evenson (1995) proposed a nine-station network comprising Inuvik, Canada (IN), Tixie Bay Russia (TI), and seven new stations as follows: Uranium City, Canada (UC), Winisk, Canada (WI), Nutak, Canada (NU), Isafjördur, Iceland IS), Elvebakken, Norway (EL), Napalkovo, Russia (NA), and Pevek, Russia (PE). These sites are used for illustrative purposes; their logistical suitability has not been investigated. Bieber and Evenson (1995) believe that the proposed network (as a combined "detector" with improved characteristics) would provide a high resolution measurements of the equatorial angular distribution during cosmic ray events. Together with data returned by Thule and the Antarctic stations, it would be possible then to have highly sophisticated "snapshot" observations of 3-dimensional cosmic ray fluxes with a degree of accuracy and completeness surpassing those provided by existing spacecraft detectors. One of the very successful applications of such "Spaceship Earth" observations has been recently demonstrated by Bieber et al. (2013a, b). The authors analyzed data from the

Spaceship Earth network, supplemented to comprise 13 polar NM stations registered giant GLE69 (20 January 2005) and obtained the time evolution of the relativistic proton density (two solar injection), energy spectrum (evolution in time), and three-dimensional directional distribution (very high initial anisotropy).

12.3.4 Muon Hodoscope for Studies in Solar-Terrestrial Physics

As well known, neutron monitors and standard muon telescopes have comparatively low angular resolution which does not allow to detect the direction of incoming cosmic ray particles with a sufficient accuracy. To overcome this deficiency, experimentalists from the Moscow Engineering Physics Institute (MEPhI) have developed the 512-channel large aperture muon hodoscope (Borog et al. 1995). The main goal of their experiment is to investigate some problems of solar-terrestrial physics by observing cosmic ray variations in the primary energy range above 10 GeV. These problems are: solar proton events at high energy; mechanisms of high-energy solar flares; processes in the interplanetary medium (shock waves, spatial non-uniformity of the IMF); cosmic ray interactions on the solar surface; geophysical effects (oscillations of the atmospheric density and gravity waves). The accuracy of measurements of cosmic ray muon directions is about 1–2°. The area of the hodoscope is 9 m^2 and its counting rate is about a thousand counts per second. The apparatus is on-line with a microcomputer. It is very important that the experimental setup is capable of being oriented in the Sun's direction. The operation time is 24 h/day (over all year), location depth is about 2 m. w. e., cutoff rigidity is 2.4 GV (Moscow). The hodoscope has been tested in cosmic ray muon flux, and the first results of its operation (April 1996 and October 1996 – May 1997) shows (Borog et al. 1997) that the setup performance is close to the expectations. The equipment operation is stable, and statistical accuracy about 0.5 % per 1-min time interval is provided. Registration of the time series in the form of angular matrix data improves the reliability of frequency response studies. It is interesting that the last event in cycle 23 (GLE70; December 13, 2006) was registered not only at the worldwide NM network but also with non-standard ground detectors, specifically, with another MEPhI installation, namely, the URAGAN muon hodoscope by (Timashkov et al. 2008).

12.3.5 Worldwide Network of Solar Neutron Telescopes

In January 1998 a new solar neutron detector was installed at Gornergrat, Switzerland (Flückiger et al. 1998), as the European cornerstone of a worldwide network for the study of high-energy neutrons produced in energetic processes at the Sun.

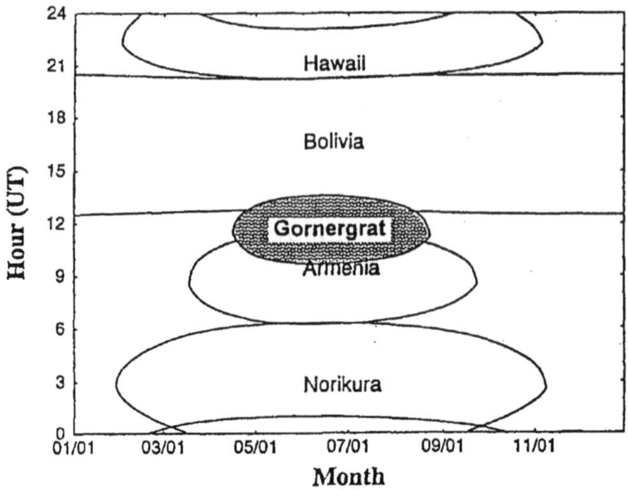

Fig. 12.8 World network of solar neutron telescopes (Flückiger et al. 1998): the UT hours and the time of the year for which a solar neutron event of the June 4, 1991 type can be detected with a statistical significance of $\geq 3\sigma$ by the individual detectors indicated

Observations of solar neutrons can provide unique information on the acceleration of particles in association with solar flares and coronal mass ejections. The first identification of a solar neutron event by ground-based detectors on June 3, 1982 (Debrunner et al. 1983; Chupp et al. 1987) initiated extensive theoretical and experimental work on the production of solar neutrons in high-energy processes at the Sun, their propagation to ~1 AU, and their detection near and at the Earth (see Chap. 6).

One part of the activities was setting up standardized neutron monitors at favourable observational locations, such as Haleakala, Hawaii (Pyle and Simpson 1991). Another part was the development of new ground-based detectors with enhanced sensitivity for solar neutrons (*e.g.*, Shibata et al. 1991; Muraki et al. 1993). Special "World Neutron Network" was initiated by the Solar-Terrestrial Laboratory of the Nagoya University. Solar neutron detectors were taken into operation at Mt. Norikura, Japan (138°E, 36°N, 2,770 m above sea level; Muraki et al. 1997); Mt. Chacaltaya, Bolivia (68°W, 16°S, 5,250 m above sea level, Matsubara et al. 1997a); Mauna Kea, Hawaii (155°W, 20°N, 5,200 m above sea level, Matsubara et al. 1997b); and Mt. Aragats, Armenia (45°E, 45°N, 3,500 m above sea level) (Fig. 12.8).

The neutron detector at Gornergrat (7.78°E, 45.98°N, and 3,135 m above sea level) is similar in type to the ones operating in Bolivia and Armenia (Matsubara et al. 1997a). The detector can discriminate neutrons from charged particles and is capable to determine the energy of incoming neutrons by measuring the energy deposition along the total track length of protons produced inside plastic scintillators by n-p reactions. Its effective are is of 4 m^2. It consists of a horizontal matrix of four 1 m × 1 m × 0.4 m plastic scintillators. Scintillation flashes in each scintillator are collected by a photomultiplier. The pulse height obtained by each photomultiplier is discriminated at four levels, which correspond to energy of a recoil proton of 40, 80, 120, and 160 MeV. The scintillators are surrounded at the top by 20 and at two opposite sides by 16 proportional counters which veto charged particles. Neutrons are identified by the scintillators in anticoincidence with the veto counters.

Detection efficiencies of the Bolivian-Armenian-Gornergrat type of neutron detectors have been determined in the energy range between 150 and 400 MeV using neutron beams at the Research Center for Nuclear Physics, Osaka University (Matsubara et al. 1997a). With the same experimental setup, expanded with a carbon slab to simulate the attenuation in the atmosphere, and theoretical studies the sensitivity of the detector to solar neutron incident at the top of the atmosphere was worked out (see references in Flückiger et al. 1998). Based on these results the significance of the Gornergrat neutron detector within the worldwide network of solar neutron telescopes can be evaluated. As an example, in Fig. 12.8 the UT hours and the time of the year are indicated for which a solar neutron event like the one on June 4, 1991 (Muraki et al. 1992) can be detected with a statistical significance of $\geq 3\sigma$. The figure demonstrate that during the Northern Hemisphere summer the Gornergrat detector indeed completes the network by filling the gap at European longitudes. The most impressive solar neutron event has been registered on 7 September 2005 (see Table 6.3 and Fig. 6.9).

12.3.6 New Technique for Analysis of Proton Spectra in GLEs

As well-known, although GLEs have been observed for more than 70 years, the physical processes responsible for accelerating protons to multi-GeV energies is still a matter of intense debate: compare, for example, Reames (1999) *vs.* McCracken et al. (2008) and Perez-Peraza et al. (2009). In the whole, proton acceleration to energies above ~500 MeV is a controversial and poorly understood aspect of SEP physics. Recently, Tylka and Dietrich (2009) have developed a new technique for analyzing data from the world-wide NM network. The authors have used the method to derive absolutely normalized event-integrated proton spectra for 53 of the 66 GLEs recorded since 1956. As a check on their results, they have compared the fluences from NM spectra to satellite measurements at ~300–700 MeV available from *IMP 8*, *SAMPEX*, and *GOES*. It was also found that the combined satellite and neutron-monitor measurements, ranging from ~10 MeV to ~10 GeV, can often be well-represented as a double power-law fit to the integral spectrum in rigidity, using the formulation given by Band et al. (1993). These comprehensive results are a useful starting point for investigations of the acceleration mechanism(s) in GLEs and for practical applications (*e.g.*, Usoskin et al. 2011; Kovaltsov et al. 2012; Usoskin and Kovaltsov 2012; Civer et al. 2014).

Unfortunately, compared to other techniques for analyzing GLEs, the method by Tylka and Dietrich (2009) has many shortcomings. Most importantly, the authors do not take careful account of anisotropies that are generally observed at the beginnings of events; instead, they rely on the world-wide network and wider time-binning to average out these effects, conceptually analogous to what happens with a spinning satellite. The authors also neglect other factors, such as rapid evolution of the spectrum in the initial stages of the event and careful numerical integration over each station's rigidity dependent asymptotic viewing cone. Finally, they rely on a semi-empirical model of GCR protons to set their normalization. Given these and other approximations in suggested method, it would be important to quantify the reliability of the results by Tylka and Dietrich (2009).

12.3.7 New Concept of Ground Level Enhancements

Analysis of unusual features in the intensity-time profiles, energy spectra and pitch angle distributions of relativistic solar protons (RSP) during the GLEs observed in the solar cycles 22–24, as well as retrospective overlook of more earlier RSP events, in fact, result in the formulation of a new concept of GLE as a separate (specific) class of solar proton events (see Chaps. 2 and 7). This may require significant modifications in today's models describing the occurrence of solar flares, particle acceleration at/near the Sun and the propagation of SCRs through the interplanetary medium and near-Earth space. In particular, new insight may be expected into the production of high-energy gamma rays and solar flare neutrons.

Another aspect of GLE problem arises in the context with its apparent evident weakness of the current solar cycle 24. As mentioned in Chap. 1, up to now (the middle of 2014) only one rather weak GLE71 was distinctly registered on 17 May 2012 (*e.g.*, Li et al. 2013; Papaioannou et al. 2014). At the same time, a number of large SEP events with the >10 MeV proton fluxes above 10^3 pfu have been registered during the current cycle. At least, the five most intensive events of them pretend to be the weak GLEs. So, it seems to be reasonably to search for observational evidence of the solar proton flux at the upper energy threshold of *GOES* detectors (700 MeV) and lower energy threshold of ground-based NMs (435–500 MeV). Preliminary analysis of such events by Li et al. (2014) gives some evidences of combined flare and shock acceleration of SEPs up to relativistic energies. In particular, combining multi-wavelength imaging data of the prominence eruption and CME, Li et al. (2013) obtained evidence that relativistic protons on 17 May 2012, with an estimated kinetic energy of ~1.12 GeV, were probably accelerated by the CME-driven shock when it travels to ~3.07 solar radii. The first results on the propagation of relativistic protons that have been recorded by NMs, as inferred by Papaioannou et al. (2014), also imply that they are most probably accelerated by the CME-driven shock.

Quite recently, International Neutron Monitor Data Base (MNDB) has published preliminary data on possible registration of GLE72 on 6 January 2014 (about 2.0 %

at NM Barentzburg and about 3 % at NM South Pole, by 10-min averaging of the data). As noted in STCE Newsletter (6 Jan 2014–12 Jan 2014), during 6–7 January, in fact, two proton events occurred. One started on January 6, passing the event threshold (10 pfu) at 09:15 UTC for >10 MeV energy levels and another one occurred on January 7. The first proton event was due to a strong flare erupting from NOAA AR 1936, which had already turned around the west limb. The second proton event was related to the X1.2-class flare emanated from an area closer to AR1943 (Credit: NASA/SDO). This event started during the declining phase of the proton flux of the first event. The proton flux reached a maximum value of 1,030 pfu during the second proton event.

At the modern level of our understanding (*e.g.*, Miroshnichenko and Perez-Peraza 2008; Perez-Peraza et al. 2009; Li et al. 2013), a flare and a CME front, both are able to accelerate particles to relativistic energies. So, in the case of so-called "weak GLEs" the problem is reduced to identify the SEP source by the properties of their intensity-time profile, anisotropy, energy spectrum etc. When the curve peaks suddenly, it is mainly the flare that gave its energy to the particles. When the curve increases more gradually, it is a CME front that accelerated the particles. Both mechanisms can be at work, of course, at the same time.

As it was shown long ago (*e.g.*, Dorman and Miroshnichenko 1968, and references therein), the source extended in time and space will deform the shape of intensity-time profile. Note also that the Sun's radius is <0.005 AU only (or about 0.266° in angular units), *i.e.*, we can consider a flare as a point-like source of accelerated particles. If one even relates SCR generation with the CME-driven shock, then, at a radial distance of the shock front from the Sun of ~3 solar radii (*e.g.*, Li et al. 2013), source dimensions would be <0.015 AU (~0.8°). Therefore, observed changes in the form of intensity-time profiles of SEP events seem to relate more closely with a prolonged acceleration (and/or release) of SEPs rather than with the finite source dimensions (i.e., with angular extension of a CME-driven shock). In fact, the shock mechanism requires certain time to accelerate seed particles to relativistic energies.

Role of prolonged acceleration (or exponential release) has been considered in diffusion approximation by Dorman and Miroshnichenko (1968). There were also cited some papers that took into account the finite dimensions of SEP sources, with a characteristic radius of 0.2–0.4 AU, where the SEPs are temporally trapped by coronal magnetic fields. For example, McCracken (1962c) has estimated possible spatial and angular dimensions of the source for the event of 4 May 1960 (GLE08), namely about 0.25 AU or ≈30° (for the observer at the Earth). Such a result seems to be applicable also to the event of 23 February 1956 (GLE05) (Dorman 1957). However, from the modern point of view, if we take into account observed dimensions of CMEs, the early estimates of the source dimensions looks far from to be realistic. More likely, we deal with some combination of particle acceleration in the reconnection region (prompt relativistic component, or PC) and at the shock front (delayed relativistic component, or DC).

Alternative version is the DC formation in the process of acceleration in magnetic bottle (Perez-Peraza et al. 1992). According to recent theoretical estimates (Oreshina and Somov 2011), in the region of magnetic reconnection particle acceleration both electrons and protons are accelerated to relativistic energies over very short time intervals, in thousandth and hundredth fractions of a second, respectively (see Fig. 5.8). Meanwhile, acceleration by CME-driven shock requires the time intervals of order minutes or tens of minutes, depending of the shock wave strength and CME speed. As a result, we again return to the scenario described in detail in Chap. 7 after the example of the event of 29 September 1989 (GLE42) (Miroshnichenko et al. 2000).

12.4 GLE Source: Flare and/or CME?

The GLE nature and the SCR acceleration sources and mechanisms have been discussed for several decades. The dilemma "flare or CME" is of special interest. The theoretical discussion includes the following question: what active process on the Sun – flare, CME, or their combination—is responsible for the SCR generation? It is apparently impossible to directly answer this question, and indirect arguments of adherents of any hypothesis do not yet lead to consensus. There is rather much evidence that the FC and SC of relativistic SCRs are related to a flare and CME, respectively. At the same time, some researchers consider that only CMEs accelerate high-energy solar particles. As an argument, they often use the characteristics of accelerated solar particles (SEPs) with energies lower than in the GLE case by 1–2 orders of magnitude (mainly protons with energies $\geq$10 MeV).

Data on the solar radioemission and X- and gamma rays, measurements of the SEP elemental composition and spectrum, etc., are also used. All these data are more or less thoroughly compared with the flare and/or CME characteristics. Meanwhile, it is well known that the SEP appearance in the Earth's orbit (SPE) is related to several previous (not always known) physical processes. In particular, observed SEPs are apparently affected by the multiple and/or prolonged acceleration processes in the source (Miroshnichenko 2003b) and by their propagation in the interplanetary medium. One of the last discussions regarding all these problems took place at two CDAW (Coordinated Data Analysis Workshop) working meetings in the United States (2009). The results of this discussion were used to prepare a special issue of the journal *Space Science Reviews* (2012, vol. 171). The journal's editors (Gopalswamy and Nitta 2012) note that GLE events account for only 15 % of the total number of giant SPEs during a solar cycle. Therefore, it is naturally interesting what special conditions should exist on the Sun for GLE generation. Most authors of the issue first of all relate GLE generation to CMEs. A detailed analysis of all papers in this issue is beyond the scope of this book; however, we will present below the most substantial results.

12.4.1 Problem of the First GLE Particles

An explosive energy release on the Sun generates a flare and a CME. It is considered that X and gamma rays are related to flares. Radioemission is a characteristic of disturbances propagating through the corona and interplanetary medium. Particles can acquire energy in flares and accompanying wave processes. Therefore, it is difficult to separate the characteristics of acceleration processes from particle observations. However, it is logical to consider that the GLE event early stage is the closest to the acceleration instant and the role of the interplanetary transfer is minimal for the first arriving particles.

Giant events make it possible to study the early phase best of all owing to a high signal-to-noise ratio, and relativistic solar protons are most applicable to the particle acceleration problem (Cliver et al. 1982) is a pioneer work on this problem, which was subsequently considered by Kahler et al. (2003), Bazilevskaya (2009), Firoz et al. (2010), Aschwanden (2012), and Gopalswamy et al. (2012). Bazilevskaya (2009) and Aschwanden (2012) indicated that the first relativistic particles (PC) leave the Sun at an instant close to the maximum of hard X-rays and high-energy gamma rays (Kuznetsov et al. 2011). These emission types are typical of the flare explosive phase. In contrast to the first particles, a DC appears 10–30 min after a PC, when a CME develops. No correlation between particle fluxes and CME characteristics was found in this case.

A shock (type II radio source) moves in front of an expanding CME. A shock can also accelerate particles, producing a power spectrum with $\gamma \sim 2.5$. The DC spectrum has $\gamma \sim 5$, which more corresponds to stochastic acceleration (Perez-Peraza et al. 2009). Particles trapped in loop-like magnetic structures within an expanding CME are accelerated while interacting with plasma turbulence. The adiabatic loss in such traps is small as compared to the acceleration effect (Perez-Peraza et al. 2009). Particles leave a trap when a CME appears in the high corona.

An absolutely unexpected aspect of the discussed problem of the first SCR particles was revealed by Struminsky and Zimovets (2009) when they analyzed the SCR effects in the ACS protection (the BGO scintillation detector weighing 512 kg), which shielded an SPI spectrometer onboard the *INTEGRAL* orbital astrophysical observatory. As is known, it is traditionally considered that the GLE onset registered by an NM at the worldwide network is the time of relativistic proton arrival. Inaccuracy and ambiguity in determining the time of solar proton arrival based on NM data is caused by the detector background (registration statistical accuracy) and variations in the geomagnetic cutoff rigidity threshold and the reception cone direction of arriving particles. The authors of this work paid attention to the fact that the ACS detector count rates increased evidently earlier than the ground NM count rates in some GLE events.

In two cases, an ACS SPI detector was more effectively used to observe the SPE–GLE onset in the Earth's orbit than the NM network: on January 17, 2005 (GLE68), and on December 13, 2006 (GLE70). According to the enhancement amplitude, these events were rather weak. In this case, the delay of the relativistic proton arrival to the Earth relative to the burst of hard X-rays was considered significant, which indicated that protons were accelerated later. Meanwhile, an increase in the ACS SPI count rate caused by the arrival of relativistic protons was observed earlier and corresponded to the SCR acceleration at the flare instant. This fact indicates that it is necessary to create space-based detectors of solar protons and electrons with a low natural background level. Such detectors should be used to measure low-intensity CR fluxes. Indeed, in contrast to the two weak GLEs mentioned above, solar protons arrived at the ACS SPI simultaneously with the beginning of anisotropic enhancement at the NM network; i.e., this arrival coincided with that of the prompt component (PC) of SCR.

12.4.2 GLE and Composition of Accelerated Particles

Kahler et al. (2012) studied several aspects of this problem. They compared the *e*/*p* and Fe/O ratios for several GLEs with the characteristics of the corresponding flares and CMEs. The authors proceeded from the fact that GLEs represent an extreme case of gradual SEP events (SPEs), which are related to shocks driven by wide and fast CMEs. The latter are in turn related to long-duration (>1 h) bursts of soft X rays (SXRs). However, it turned out that some large gradual SPEs, including GLEs, are related to short-duration flares (<1 h), the duration of which is comparable with that of impulsive low-energy SEP events enriched in heavy elements (*e.g.*, large Fe/O ratio), high particle (*e.g.*, Fe ion) ionization degree, and a large *e*/*p* ratio.

To determine how the *e*/*p* and Fe/O ratios, measured in two energy intervals, depend on the characteristics of the active regions (ARs) of the corresponding flares and CMEs, Kahler et al. (2012) statistically studied 40 GLE events registered from 1976. It turned out that abundance ratios tend to smaller and stable coronal values with increasing timescales (duration) of flares and peak fluxes of soft (thermal) and hard (bremsstrahlung) X rays, as well as with increasing AR dimensions. The authors assume that these results indicate that the flare effects are insignificant in these GLEs if the wide region of "heliolongitudes of coupling" between GLE sources with increased heavier elements is taken into account. The authors consider that SPEs accompanying GLEs are mostly accelerated at the fronts of CME-driven shocks and the relation of the flare power and time characteristics to the CME properties could explain the correlation of the SEP composition and flare properties.

Even if we assume that flares mainly contribute to GLEs, in this case, it is also unclear why Fe/O-type ratios weakly tend to decrease with increasing background SEP intensities. Therefore, the authors prefer an alternative interpretation (Tylka et al. 2005): a large Fe/O ratio characterizes the acceleration by a shock, which is quasi-perpendicular near the Sun; therefore, this shock mainly accelerates "a seed population" of flare particles (Fig. 12.9). Since higher injection energy is required in the case of quasi-perpendicular shocks, these shocks involve a generally smaller seed population in the acceleration process than quasi-parallel shocks. As a result, events with quasi-perpendicular shocks near the Sun will be generally characterized by

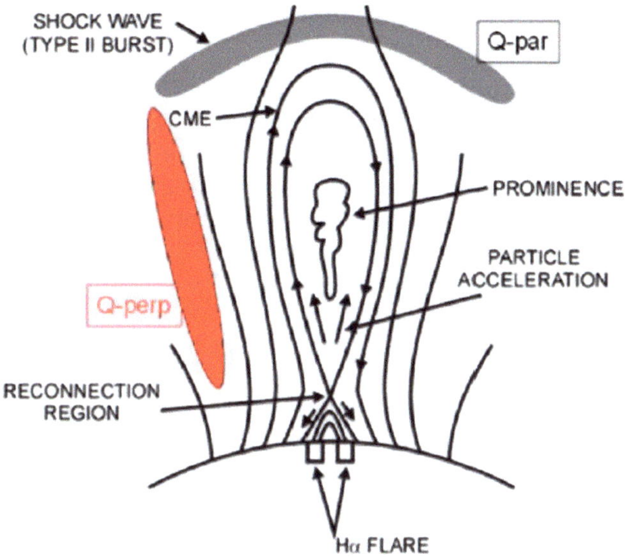

Fig. 12.9 Schematic showing where quasi-parallel and quasi-perpendicular shock acceleration might occur in a solar eruption (Cliver 2009). In the standard CSHKP picture for eruptive flares (Hudson and Cliver 2001), reconnection in the wake of the CME gives rise to a two-ribbon flare. If these flare particles escape the CME, they could become seed particles for the shock

smaller proton fluences at least at higher energies that were reached when a shock was closer to the Sun. In fact, we deal here with post-eruptive acceleration that was studied and discussed in some detail earlier (*e.g*., Chertok et al. 1996; Klein et al. 1999).

The data presented in (Miroshnichenko 2003b) can be added to this very elegant, but rather contradictory, pattern. We tried to divide the reconstructed solar proton emission spectra, depending on the proton sources (impulsive or gradual flares and CME-driven shocks). Using several SPEs as an example (including outstanding GLE42), we found out that the number of accelerated particles, which "precipitate" in the solar atmosphere and cause gamma-ray bursts in lines, is systematically smaller than that of runaway particles registered near the Earth as SEPs. This important fact is still insufficiently studied.

Based on the consideration of the problem as a whole, we tend to assume that a physical relationship between CME flares and GLEs undoubtedly exists. However, the regularities of this relation are not rigorously deterministic and most probably correspond to the "big flare syndrome," which was proposed and developed in several works (Kahler 1982; Kahler et al. 2012). Kahler et al. (2012) seem also partially to share this opinion: *"In this scenario, the tendency toward a decrease in the abundance ratio with increasing fluence of soft X-ray and radioemission fluxes at a 9 GHz frequency can be interpreted in terms of the "big flare syndrome" (Kahler* 1982*), which is reduced to the fact that all eruptive event emissions change their scale together: in this case, these events are SEP fluences and peak fluxes of flare electromagnetic emissions."*

12.5 New Distribution Function for SEP Events

Investigation of solar proton events over the past 72 years has been primarily devoted to their statistical studies or to research the characteristics of individual events. Although we do not yet understand how the Sun accelerates ions to relativistic energies, nor how to predict the fluence from individual flare, we have assembled enough data to be able to place preliminary, but nevertheless realistic, limits on the extent and severity of these events (*e.g.*, Shea and Smart 1993; Smart and Shea 1993). This book presents a summary of our knowledge of solar proton events as gained over the past incomplete eight cycles of solar activity (17–24). Hopefully this information may provide broad guidelines to space exploration planners (see Chap. 11).

Predicting solar proton activity can only be improved by extending the data set used to construct the models. Although we can make a statistical prediction of the mission integrated fluences of protons at 1 AU, the situation is less than satisfactory for other important parameters (Feynman 1997). For example, no models exist for the prediction of peak fluxes, although the peak flux distribution presented recently by Smart and Shea (1997) is in advance towards this goal. In this context, it is important also to continue statistical studies of proton events (see, *e.g.*, Miroshnichenko et al. 2001; Cliver et al. 2012) based on extended database – SPE Catalogues for 1955–2009 (see Chap. 2).

An overview of the Solar-Terrestrial Predictions Workshop-IV (Ottawa, Canada) contains extended summary of the problems existing in this field, in particular, the problem to identify which solar eruptions (solar flare and/or CME) will produce copious solar protons at the Earth (Hrushka et al. 1993). A number of research proposals are listed in the Radiation Group Report given in Hitachi (Japan) by Heckman (1997). One of special reviews of the problem was given in the monograph by Miroshnichenko (2003a).

Long-term observations of SPEs, or SEP events, give a number of evidence of that about once time during a given solar cycle, the event occurs whose fluence above a given energy (usually $\geq$10, $\geq$30, $\geq$60 and $\geq$100 MeV for protons) dominates that for the entire cycle (*e.g.,* Shea and Smart 1990a, b). It may overlap the fluences from the other events and even determines, in fact, a total fluence for the cycle. Such rare phenomena are sometimes called by "rogue events" (Kallenrode and Cliver 2001a, b), in analogy to rogue ocean waves having unusually large amplitudes. Well-known examples of rogue SEP events at the Earth occurred on 14 July 1959, 4 August 1972, 19 October 1989, and 14 July 2000. Rogue events have been also observed in the inner heliosphere – with *Helios* 1 on 4 November 1980 at 0.5 AU and with *Ulysses* in March 1991 at 2.5 AU. Origin of the rogue events is thought to relate to the multiple coronal mass ejections (CMEs) and converging interplanetary shocks. If observed at the Earth's orbit, these rare extreme events become important geophysically and applicably (*e.g.*, radiation hazard for spacecraft).

Distribution functions of SCR events on the proton fluences with the energy ≥30 MeV, or $\Phi(\geq 30$ MeV), have been widely investigated (see, *e.g.*, Nymmik 2011, and references therein). These functions are constructed, as a rule, based on the data on SEP events whose sizes are determined by measurements on board the satellites *IMP* and *GOES* series. By present, the data sets available cover, depending of selection criteria, about 200 events with the fluence of $\Phi(\geq 30\text{ MeV}) \geq 10^6\text{ cm}^{-2}$ (Nymmik 2011). When describing the distributions, it is usually applied the power-law functions, sometimes with a break. Such approach, however, allows us to calculate the probabilities of event occurrence with certain fluence only down to a probability of ~0.5 % (~1/200) that is evidently not enough for extreme estimates.

Lately it has become obvious that the accumulation of new satellite data do not enable us to advance considerably in the problem of determination of the form of distribution function for SEP events on their fluences in the range of low probabilities. An attempt to involve the data on cosmogenic isotopes in the lunar soil (Reedy 1996), unfortunately, does not add a certainty in the resolution of this problem because of the isotope data are related to a total (summary) flux of SCR protons with the energy of $E \geq 10$ MeV for the past ~10 My, but not to individual SEP events.

Some progress in this problem has been outlined when the data on the fluences of large SEP events for the period of 1561–1950 have been obtained from the Greenland ice cores (McCracken et al. 2001). These authors have succeeded, in particular, in estimating the proton fluence for the largest event of that period, namely, Carrington event, occurred on 1–2 September 1859; its value was $\Phi(\geq 30\text{ MeV}) = 1.88 \times 10^{10}\text{ cm}^{-2}$. Nevertheless, even those data turned out to be not enough to determine a form of distribution function in a total diapason of changes of the fluence $\Phi(\geq 30$ MeV). In fact, by those data it was impossible to determine an amount of small events that constitute initial part of distribution function in the range of fluences $\Phi(\geq 30\text{ MeV}) = 10^6 \div 3 \times 10^9\text{ cm}^{-2}$.

Therefore, to calculate the probabilities of extra-large SEP events (and distribution function) from polar ice data it is needed to know how many single events of fluence $\Phi(\geq 30\text{ MeV}) \geq 10^6\text{ cm}^{-2}$ have occurred since 1561 up to now. To our opinion, the solution of this problem seems to exist (Fig. 12.10). Our estimates (Miroshnichenko and Nymmik 2014) are based on the measurements onboard two spacecraft *IMP*-8 and *GOES* and on the data from Greenland ice core (McCracken et al. 2001).

In the context of this study, also deserve serious attention the estimates of the proton fluences at some other energies (besides 30 MeV), especially, for the understanding of flare (proton) activity of the Sun in the remote past. Many years ago, Wdowczyk and Wolfendale (1977) addressed the question on the long-term frequency of large solar energy releases and their possible effects, compared with other catastrophic events. The main body of their evidence appears still valid, although some details have changed. The very flat integral power-law fits (logarithmic slope around −0.5) suggest that several dramatic solar energy releases should be expected in geologically short times, if the trend continues.

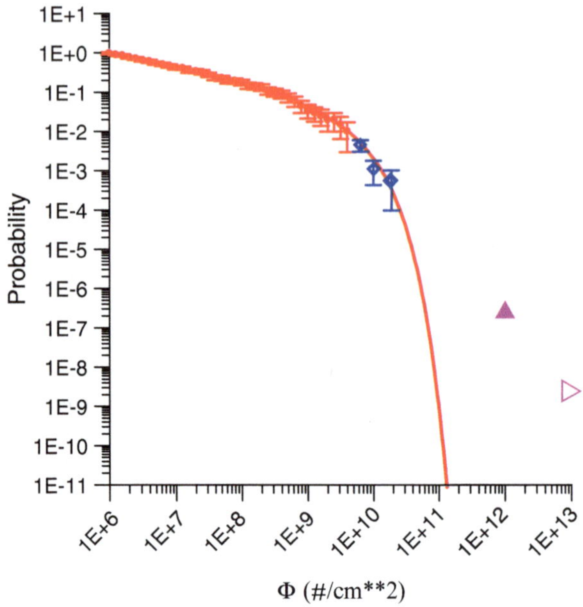

Fig. 12.10 Distribution function of SEP events on the integral fluences of $\Phi \equiv \Phi(\geq 30$ MeV), including the range of very small probabilities. Our estimates are based on the measurements onboard two spacecraft *IMP*-8 and *GOES* (*points*) and on the data from Greenland ice core (*blue diamonds*); *solid red line* – model distribution function (for details see Miroshnichenko and Nymmik 2014). *Full and open triangles* demonstrate the extrapolation of integral fluences of $\Phi(\geq 30$ MeV) estimated in the present work by the data of Kiraly and Wolfendale (1999) into the past for 1 and 100 My, respectively

Extrapolating their highest energies (>60 MeV) fit to long time scales, Kiraly and Wolfendale (1999) obtained some another estimates. It turns out that while the highest fluence measured up to 1999 (in about 30 years) was 3×10^9 cm^{-2}, one would expect in 1 My a few events above 10^{12} cm^{-2}, and in 100 My a few above 10^{13} cm^{-2}. This is far less than one would expect from flat slopes found by Wdowczyk and Wolfendale (1977), but still about two orders of magnitude higher than it follows from our estimates.

In fact, according to modern data on proton fluences at the energy $\geq$30 MeV, for the period from 1973 up to 2008 there were registered 205 events with the fluence $\geq 10^6$ cm^{-2} (Nymmik 2011). If solar activity remains at modern (present) level, it means that for 1 and 100 My, respectively, we may expect for 6×10^6 and 6×10^8 of such events, and the probabilities of their realization would be ~1.7×10^{-7} and ~1.7×10^{-9}, respectively. According to our estimates (Fig. 12.10), for such long periods the events may appear with the fluences up to 6×10^{10} and 10^{11} cm^{-2}, respectively, that is for $1.5 \div 2$ orders of magnitude less that the estimates by Kiraly and Wolfendale (1999). Two triangles in Fig. 12.10 depict our estimates of $\Phi(\geq 30$ MeV) based on the data by Kiraly and Wolfendale (1999) for the fluences of protons at the energy ≥60 MeV, with the extrapolation into the past for 1 and 100 My, respectively. Difference in the energies of protons (30 and 60 MeV) makes this discrepancy even much more.

The cause of this discrepancy is rather simple. As it was repeatedly noted (Nymmik 2006, 2007a, b, 2011), lognormal distribution function of SEP events (Feynman et al. 1993) that was applied by Kiraly and Wolfendale (1999), by no means does reflect a physical essence of SEP event distribution in the range of large fluences. Parameters of the model by Feynman et al. (1993) are determined mainly by subjective (random) magnitudes of the registration thresholds and selection of small SEP events; therefore, they can not serve for the extrapolation of the data into the range of extremely large events.

12.6 Geophysical Effects of SCR: Recent Development

Solar particles with energies about several tens and hundreds of MeV (SEPs) are substantial in many geophysical processes owing to their ionizing effect (Miroshnichenko 2008). The following geophysical processes are most known: the effects of ozone layer depletion and disturbance in the global circuit of atmospheric electricity, variations in the Earth's atmosphere transparency, generation of nitrates and cosmogenic isotopes, and many other less studied or still assumed (not proved reliably) phenomena. Below, we will briefly consider the contribution of relativistic solar protons to some of these effects. Note that the energy density and the total energy transferred by SCRs into the Earth's atmosphere are not comparable with any other energy that comes from the Sun to the near-Earth space. Therefore, SCRs are not the main cause of geophysical disturbances (as compared, *e.g.*, to CMEs and geomagnetic storms). However, the SEP arrival can be an important (trigger) component of the global solar–terrestrial coupling owing to its sporadic nature.

Penetration of SEPs into the polar atmosphere should inevitably modify the composition and physicochemical processes in the mesosphere and stratosphere (Quack et al. 2001; Kirillov et al. 2008). In particular, Quack et al. (2001) considered the SEP effect on the above processes for three GLEs, which were registered in October 1989, July 2000, and April 2001 in a wide range of energies and with regard to the time evolution of their spectra. They studied the generation of nitrogen (NO_x) and hydrogen (HO_x) oxides and variations in the ozone (O_3) content and compared the calculation results for different events. The analysis was based on a model, which took into account the penetration (precipitation) of particles into the atmosphere and the following modification of the atmospheric chemistry.

In October 1989, the ionization level in the lower stratosphere was first high, whereas the ionization in the mesosphere was lower by an order of magnitude. In due course, the ionization level in the lower stratosphere remained unchanged (since the high-energy particle intensity was almost constant); at the same time, the ionization in the mesosphere increased substantially due to the arrival of low-energy particles, the intensity of which continued increasing. The ionization variation pattern in July 2000 is more complex as compared to such a pattern in October 1989: the ionization was high in the middle mesosphere and was lower in the upper mesosphere. Since the intensity of high-energy particles starts decreasing rather early, the ionization rates in the stratosphere also decrease as an event evolves. Only in the upper mesosphere, the ionization rates increase in the course of time. In April 2001, the calculated ionization profiles are comparable in shape with the ionization profiles for the event in October 1989, although the ionization rate absolute values are smaller by a factor of ~2–3. Nevertheless, in due course, the ionization level decreases at all altitudes. The model gives identical results when the proton spectra are extrapolated to 500 or even 800 MeV.

Kirillov et al. (2008) performed studies in the same direction for GLE70 (December 13, 2006). They studied the effect of energetic solar protons on the chemistry of the middle atmosphere (20–80 km). The proton spectra were obtained based on the NM data and measurements in the stratosphere and on spacecraft. A one-dimensional model, which was previously developed by the authors with regard to its time dependence, was used to calculate the generation and loss of the content of minor atmospheric constituents during a GLE. The obtained ozone layer depletion rates were in good agreement with the measurements performed with a Microwave Limb Sounder (MLS) onboard the AURA spacecraft. The authors assumed that the ozone content in the middle atmosphere decreased when solar protons precipitated mainly due to the generation of odd HO_x components with the following recombination of ionization products.

Finally, let us consider well-known effect of cosmogenic isotope generation by cosmic rays in the Earth's atmosphere, using one of the latest works in this field (Webber et al. 2007) as an example. These authors performed new detailed calculations of the generation rate (power) of the ^{3}H, ^{7}Be, ^{10}Be, and ^{36}Cl cosmogenic isotopes, using the FLUKA (Monte Carlo code) software and taking into account recent data on the interaction cross-sections for vertically incident protons with energies varying from 10 MeV to 10 GeV. This made it possible to study the isotope generation due to SCRs and GCRs in the region of low energies, where the generation power is a very sensitive energy function. Based on the events in October–November 2003, it was indicated that the ^{10}Be generation rate reached a maximum during these events when the energy was ~100 MeV; at the same time, the ^{7}Be and ^{36}Cl isotopes were more intensely generated at an energy of ~25 MeV with a resonance cross-section of the process. If the SCR spectrum is steeper than such a spectrum in October–November 2003, the generation power maximum will shift toward lower energies. The generation peak will shift to higher energies if the spectrum is less steep (as was registered on January 20, 2005). For the events in 2003, the total integral generation values for the ^{7}Be and ^{36}Cl isotopes due to SCRs will be approximately three times as large as such values for the ^{10}Be isotope generation. This is explained by the resonance effect, which is formed when the proton energy is ~25 MeV. Precisely such protons generate the ^{7}Be and ^{36}Cl isotopes by splitting the nuclei of atmospheric nitrogen (^{14}N) and argon (^{40}Ar), respectively.

Only the extreme event of February 23, 1956, could substantially contribute to the annual generation of the ^{10}Be isotope. For the ^{36}Cl isotope, the annual generation values are ~2–5 times larger, depending on the SCR spectrum type. Webber et al. (2007) calculated the annual generation values for the generation of the ^{10}Be, ^{36}Cl, and other isotopes at >65° geomagnetic latitude for the 1940–2006 period, including six 11-year solar cycles. The average amplitude of the 11-year variation in the annual contents of these isotopes is ~1.77. If the latitudinal mixing is taken into account, this amplitude will decrease to 1.48 for the average global generation. Anyway, calculations of ^{10}Be production (*e.g.*, Webber et al. 2007; Usoskin and Kovaltsov 2012; Schrijver et al. 2012 and many others) provide an independent control of ancient SPE data obtained, for example, by so-called nitrate method (*e.g.*, McCracken et al. 2001).

In this context, it is timely to highlight one modern version of well-known "The Great Debate" related to "The Solar Flare Myth" (see Chap. 3). Advanced knowledge of the occurrence probability of extreme events related to solar energetic particles (SEPs) is very important and acute (Hudson 2010). This is important from different aspects: from purely astrophysical questions of the highest possible energy released in solar flares (*e.g.*, Schrijver et al. 2012) to the geo-environment (*e.g.*, Thomas et al. 2013), and even to the technological risk assessments (*e.g.*, Shea and Smart 2012). Direct observations of SEPs cover the past six decades with ground-based and space-borne instruments. Thus, an extension of the SEP data back in time is needed for a better estimate of the occurrence probability distribution function in the range of strong SEP events. Such an extension is possible only on the basis of indirect proxies.

One potential proxy was based on nitrate measured in polar ice (*e.g.*, McCracken et al. 2001; Shea et al. 2006), but recently Wolff et al. (2012) have found that nitrate data from Greenland, unfortunately, cannot be used as a quantitative proxy for SEP events. In particular, the nitrate event identified as 1859 in the GISP2 H core (McCracken et al. 2001) is most likely the same event that more recent Greenland cores identify at 1863. The parallel event in other cores, as well as all other significant nitrate spikes in those cores, has an unequivocal fingerprint of a biomass burning plume from North America. Although the authors cannot prove that this is true for the 1859 event in the GISP2 H-core, it seems to them overwhelmingly likely. In any case, the GISP2 H core is the only one of the eight Antarctic and six Greenland cores with high resolution discussed by Wolff et al. (2012) that claims a spike in 1859. Estimating the frequency of extremely energetic solar events, based on solar, stellar, lunar, and terrestrial records, Schrijver et al. (2012) have also had to conclude that nitrate concentrations in polar ice deposits cannot, at present, be used to extend the direct observational records of SEP events to a longer time base without at least significantly more study.

Another potential proxy is related to cosmogenic radionuclides ^{14}C and ^{10}Be in terrestrial independently dated archives, where peaks can be associated with strong SEP events, or in lunar rocks (Usoskin et al. 2006; Usoskin and Kovaltsov 2012; Kovaltsov and Usoskin 2014; Cliver et al. 2014). Accordingly to Schrijver et al. (2012), concentrations of select radionuclides measured in natural archives may prove useful in extending the time interval of direct observations up to ten millennia. But their calibration to solar flare fluences depends on multiple poorly known properties and processes. Therefore, these proxies cannot presently be used to help determine the flare energy frequency distribution. In order to empirically quantify the frequency of uncommonly large solar flares the authors suggest to base on extensive surveys of stars of near-solar age, because the flares on magnetically active, young Sun-like stars have energies and frequencies markedly in excess of strong solar flares (cf., however, Fig. 4.9).

To our opinion, this discussion cannot be completed without taking into account, at least, two important points. First of them is a great diversity (variety) of observed SCR spectra. Meanwhile, modern calculations of radionuclide concentrations are based mainly on two SPE scenarios – SPE56 (GLE05) and SPE72 (GLE24). The second one is related to independent nitrate data (Kepko et al. 2009) obtained with a high-resolution continuous flow analysis (CFA) of a Summit, Greenland ice core drilled in the summer of 2004. The authors identified the 1947 Hekla eruption in the conductivity data and were able to date the bottom 5 m of core with high confidence. They additionally included data from Windless Bight, Antarctica providing an inter-hemispheric comparison. As well-known, during the period of 1940–1950, four SCR relativistic events (from GLE01 through GLE04) were recorded by ground-based ionization chambers. These very large, very energetic solar proton events certainly generated NO_x at all levels of the polar atmosphere. All four of these events were time associated with significant impulsive nitrate enhancements in both the Greenland and Antarctic ice core data within 1–2 months after each GLE. This inter-hemispheric correlation is strong evidence for a global response. Additional impulsive nitrate enhancements were identified in the high-resolution 2004 Greenland core and appear temporally associated with significant large solar events (flares of importance 3 and 3+) and long-duration strong polar cap blackouts (PCA). As opposed to the understandable skepticism by Wolff et al. (2012) and restrained pessimism by Schrijver et al. (2012), the results by Kepko et al. (2009), in our opinion, provide independent experimental support of the McCracken et al. (2001) argument that there is a short time delay between the generation of massive amounts of NO_x throughout the polar atmosphere and their deposition in polar ice.

12.7 Matters Arising

As it was discussed in Chap. 2, solar proton events have been classified by different schemes based on different physical and/or methodical approaches. Meanwhile, recently several new research results appeared which are still not incorporated into any of the classification systems. Below we describe briefly two of them, namely, the so-called super-events and the oscillations of energetic particle fluxes in the heliosphere.

12.7.1 Super-Events in the Heliosphere

According to Müller-Mellin et al. (1986), super-events are ~10 MeV proton events characterized by long duration (~40 days) and weak intensity variations with heliolongitude. The most prominent super-events originate in extended episodes (0.5–2 months) of fast CMEs and solar energetic particle (SEP) events from single active regions or from narrow ranges of active longitudes. Super-events are initially observed in the inner heliosphere and propagate to the outer heliosphere. From the midpoints of super-event profiles at successive radial distances Dröge et al. (1992) determined transit speeds of ~700–1,000 km s^{-1} for five prominent super-events occurred in 1974–1985. If the onsets of the super-events at 1 AU are used instead of the mid-point in the speed determinations, the resultant lower limits speeds to *Pioneer* 10 for these five events range from 430 to 750 km s^{-1}. These speed values are comparable or greater than typical annual averages of the solar wind observed near solar maximum. In the outer heliosphere, super-events represent a mixture of SEPs and particles accelerated locally at interplanetary shocks (Dröge et al. 1992).

A series of association studies indicates (see Cliver et al. 1993a, and references therein) that super-events represent particularly strong transient disturbances of the heliosphere. In general, the presence of fast (>400 km s^{-1}) CMEs during super-events can be inferred from the nearly 100 % association between individual SEP events and such CMEs (Kahler et al. 1984). In their recent study, Cliver et al. (1993a) examined the role of the interplanetary disturbances associated with super-events in establishing the 11-year intensity variation of >1 GV cosmic rays at 1 AU for the period 1974–1985. They compared interplanetary particle flux profiles of intense super-events with the neutron monitor records at the times of long-term steps in the modulation of galactic cosmic rays (GCR). It was found that of six intense super-events, one occurred coincident with the onset of a step (June-July 1982), two occurred midway through step decreases, and three occurred at the end of step decreases.

In terms of convection/diffusive models of cosmic ray modulations, the poor association of the largest super-events with long-term GCR intensity decreases during the period 1974–1985 suggests (Cliver et al. 1993a) that the background level of more frequently occurring, and less energetic, CMEs is more important for establishing the 11-year cycle than are the sporadic, and relatively short-lived, outbreaks of major CME activity that give rise to super-events. In other words, super-events are not reliable signatures of so-called global merged interaction regions which are defined to be the large-scale solar wind structures that produce step decreases. At the same time, origin of the super-events and their solar-interplanetary associations are still of significant research interest.

12.7.2 Oscillations of Particle Fluxes

Due to long-term set of data obtained in charge particle measurements on board of the *Ulysses* and *Voyager* spacecraft, recently it was revealed a new interesting association of SEPs variations with the Sun's physics (Thompson et al. 1995, 1998). Traditionally, time variations in the particle fluxes have been ascribed solely to sources such as the ejection and propagation of particles from solar disturbances (for example, flares), the release of particles from planetary magnetospheres, the acceleration of particle at interplanetary shock waves and fluctuations in the magnetic field intensity and direction in interplanetary space. All these effects, as noted by Thompson et al. (1995) imply a continuous fluctuation spectrum, not discrete features (frequencies).

Contrary to this, Thompson et al. (1995) revealed in the *Ulysses* and *Voyager* data many periodic components similar to that of global oscillations of the Sun. Using the data from the *Voyager* 2 spacecraft (during 1985) and from the *Ulysses* spacecraft (1992–1994), they have analyzed the time series of fluxes of low-energy interplanetary particles (in the range of 80 keV–145 MeV/nucleon), whose ultimate origin is the Sun. From 1 to 140 μHz, the spectral components of the flux oscillations are consistent with those estimated (but not confirmed) for gravity-mode oscillations (*g*-modes) of the Sun; from 1,000 to 4,000 μHz, the spectral lines closely match the frequencies of known solar pressure *p*-modes. These concordances imply that the solar wind and the interplanetary magnetic field transmit solar oscillations and thus might be used to probe the interior structure of the Sun. In other words, in spite of the dynamic nature of interplanetary space, analysis of time series of charged particle fluxes and magnetic fields in the medium demonstrate clearly that the heliosphere is not completely fractal or chaotic environment, but is composed of many discrete frequencies (Thompson et al. 1998).

What physical mechanism(s) could explain this apparent modulation of the solar wind by solar oscillations? Thompson et al. (1995) suggest that such a solar control may be due to the super-granulation motions. It has been assumed that these motions are not completely random on the solar surface, but partially are due to the effect of many *g* modes. It has been predicted also (see Thompson et al. 1995 and references therein) that magnetic flux "frozen" in the supergranulation is responsible for the observed transverse magnetic fields in the polar regions of the Sun. Periodic components in the in the interplanetary magnetic field would cause periodic modulation of the charged-particle flux. The nature of the spectrum of the particle fluctuations is of evident importance in the context of interplanetary particle propagation. The results of Thompson et al. (1995) imply, in particular, that charged particles are being scattered in the interplanetary medium not only by a continuum spectrum of IMF fluctuations (see, *e.g.*, Eq. (8.1)), but also, and perhaps, more importantly, by discrete wave frequencies. Similarly, those results may impose severe constraints on plasma turbulence, and fractal or chaotic models of the IMF.

Despite some skepticism about this conclusion on theoretical grounds (*e.g.*, Kumar et al. 1996), continued analysis of interplanetary data provides new evidence (Thompson et al. 1998) for both the discrete line features in the data and the requirement for a driving source, of which the Sun is the most obvious. In particular, some of the very long periods found in the interplanetary data are consistent with frequencies reported from analysis of solar irradiance measured on *Nimbus* 7 and *Pioneer Venus Orbiter* and attributed to solar *r*-modes (toroidal oscillations with periods of the order of the Sun's rotation). As to the shorter periods, Thompson et al. (1998) draw attention to the data on emission lines from at least two comets that displayed oscillations on the time scale of minutes, those oscillations being not related to the Earth's atmosphere or to cometary activity, but rather to fluctuations of the solar wind.

12.8 Concluding Remarks (Instead of Epilogue)

Along with the formation of relevant databases, the researchers of different countries make great efforts in developing the models of terrestrial radiation environment during the individual SEP events, as well as in SEP flux forecasting for the periods from several hours to ~11 years (*e.g.*, Miroshnichenko 2003a, b; Perez-Peraza et al. 2011; Miroshnichenko et al. 2013). Of special interest is a latest issue of "Cosmos Model" (SINP MSU 2007) where the main ideas about electromagnetic and corpuscular emissions of the Sun are summarized, with taking into account their great importance for the development of practical cosmonautics. One of the recent publications on this topic (Vainio et al. 2009) gives a detailed review of the physical processes that determine the dynamics of radiation conditions in the Earth's vicinity. There are also expounded a number of scientific and engineering forecast models that are in operation and/or presently under development for this purpose. Three components of the radiation environment, *i.e.*, galactic cosmic rays (GCRs), solar energetic particles (SEPs) and terrestrial radiation belts, are considered separately. Models of the effects of particle radiation on the atmosphere are also reviewed. In the case of SEP models, the authors make a distinction between statistical flux/fluence models and those aimed at forecasting events. In spite of certain progress in developing the models of both types, we are still rather far from a reliable forecasting of the SEP events. To reach this goal, we need to identify the most important mechanisms of particle acceleration at the Sun and make clearer their relations to observed precursor signals at different wave lengths of solar electromagnetic emission. Therefore, for the further progress in our abilities for the real Space Weather forecasting there are still necessary additional profound investigations of SEP events.

In conclusion of this section, we list the most important sites that contain primary information on cosmic ray variations (including detailed data on the GLEs) through the long periods of time. Notice, in particular, the following NM databases: http://www.nmdb.eu/ (*e.g.*, Klein et al. 2009; Mavromichalaki et al. 2011); general GLE database http://data.aad.gov.au/aadc/gle/ (Australian Antarctic Data Center): partial NM data sets, *e.g.*, http://cosmicrays.oulu.fi/GLE.html (Usoskin et al. since 1966), http://neutronmonitorta3.sk (see also Kudela and Langer 2008) and http://neutronm.bartol.udel.edu/ (Bartol Research Institute, USA); World Data Center B, Russia; http://cr0.izmiran.rssi.ru/common/links.htm; DataBase at World Data Center C, Japan; ftp://cr0.izmiran.rssi.ru/Cosray!/FTP_NM/C/, IZMIRAN. Basic procedure for the calculation of SCR spectra are described in detail at the site of NM Apatity station http://www.pgia:81/cosmicray/gle. Note that a Forecast Center of IZMIRAN, in addition to its own methods and databases (*e.g.*, http://dec1.sinp.msu.ru/~osipenko/, uses also the measurements onboard the spacecraft *GOES*, *ACE*, *SDO*, *STEREO*, as well as the materials of several other centers of Space Weather: NOAA/SESC/SWPC, USA, http://www.swpc.noaa.gov; SIDC, Europe, http://sidc.oma.be/) and http://89.250.81.98/solar2010/index.htm (Institute of Ionosphere, Kazakhstan).

***

Life on the Earth becomes increasingly more dependent on technology. This ranges from communications to electrical power, from manufacturing to enjoying sporting events (homing pigeons). Over the past century and in particular since the space era started in 1957, our technological advances have been phenomenal. There is every expectation that the rapid increase in technological developments will be continuing into the foreseeable future, and this irrepressible progress will be take into account and rest on the undeniable achievements of science, including very impressive findings in the field of solar and solar-terrestrial physics.

*****

P.S. "Aquellos que observan Cielo, regresan a las cosas mundanas muchas mas sublimes. . .". *Marcus Tulius Cicerón (106–43 BC).*

Appendix 1. Acronyms

ACE	Advanced Composition Explorer
ALTEA	Anomalous Long Term Effects in Astronauts (project)
AR	Active Region
AU	Astronomical Unit
BNO	Baksan Neutrino Observatory (Northern Caucasus, Russia)
BUST	Baksan Underground Scintillation Telescope
CCD	Charge Coupled Device
CGRO	Compton Gamma Ray Observatory
CIR	Corotating Interacting Region
CME	Coronal Mass Ejection
COSPAR	Committee on Space Research
CR	Cosmic Rays
CREME	Cosmic Ray Effects on Microelectronics
CRRES	Combined Release and Radiation Effects Satellite
CSM	Central Solar Meridian
DC	Delayed Component of GLE
EC	European Communities
ERB	Earth's Radiation Belt
ESA	European Space Agency
ESTEC	European Space Research and Technological Center
EUV	Extreme Ultra Violet
EVA	Extra Vehicular Activity
FIP	First Ionization Potential
FSMB	First Significant Microwave Burst
FWHH	Full Width at Half Height
FWHM	Full Width at Half Maximum (Height)
GCR	Galactic Cosmic Rays
GEC	Global Electric Circuit
GLE	Ground Level Enhancement (Event)

L. Miroshnichenko, *Solar Cosmic Rays*, Astrophysics and Space Science Library 405, DOI 10.1007/978-3-319-09429-8

GMF	Geomagnetic Field
GOES	Geostationary Operational Environmental Satellite
GPS	Global Positioning System
GRE	Gamma Ray Event
GRO	Gamma Ray Observatory
GRL	Gamma Ray Line
GRS	Gamma Ray Spectrometer (SMM)
GSE	Geocentric Solar-Ecliptic coordinates
GSFC	Goddard Space Flight Center
GSM	Geocentric Solar-Magnetospheric coordinates
GSMF	Global Solar Magnetic Field
HCS	Heliospheric Current Sheet
HESSI	High Energy Solar Spectrometer Imager
HWHM	Half Width at Half Maximum
IC	Ionization Chamber
ICE	Interplanetary Cometary Explorer
ICRP	International Council on Radiological Protection
ICRU	International Commission on Radiation Units and Measurements
IGRF	International Geomagnetic Reference Field
IGY	International Geophysical Year
IMF	Interplanetary Magnetic Field
IMP	Interplanetary Monitoring Platform
INTEGRAL	International Gamma-Ray Astrophysics Laboratory
IQSY	International Quiet Sun Year
IRI	International Reference Ionosphere
IRPA	International Radiation Protection Association
ISEE	International Sun-Earth Explorer
ISM	Interstellar Medium
ISO	International Standard Organization
ISS	International Space Station
ISTP	International Solar-Terrestrial Program
IZMIRAN	Institute of Terrestrial Magnetism, Ionosphere, and Radio Wave Propagation of the Russian Academy of Sciences
JPL	Jet Propulsion Laboratory
LASCO	Large Angle and Spectroscopic Coronograph (on board SOHO)
LBL	Lawrence Berkeley National Laboratory
LDE	Long Duration Event
LDEF	Long Duration Exposure Facility
LET	Linear Energy Transfer
LISM	Local Interstellar Medium
LT	Local Time
MHD	Magnetohydrodynamics
MIR-MSS	Mir Space Station (Soviet-Russian Space Station)
MLT	Magnetic Local Time

MM	Maunder Minimum
MONSEE	Monitoring Sun-Earth Committee
MT	Muon Telescope
NASA	National Aeronautics and Space Administration
NCAR	National Center for Atmospheric Research
NCDC	National Climate Data Center
NCRP	National Council of Radiation Protection and Measurement
NGDC	National Geophysical Data Center
NM	Neutron Monitor
NMDB	Neutron Monitor Database
NOAA	National Oceanographic and Atmospheric Administration
NRC	National Research Council (USA)
NSSDC	National Space Science Data Center
OGO	Orbiting Geophysical Observatory
PAMELA	Payload for Antimatter Matter Exploration and Light-nuclei Astrophysics
PAD	Pitch Angle Distribution
PC	Prompt Component of GLE
PCA	Polar Cap Absorption
PDS	Power Density Spectrum
PFU	Proton Flux Unit (1 pfu = 1 p cm^{-2} cm^{-1} s^{-1})
PSD	Power Spectral Density
QBO	Quasi-Biennial Oscillation
QLT	Quasi-Linear Theory
RCL	Reconnecting Current Layer
RCS	Reconnecting Current Sheet
RHESSI	Reuven Ramaty High Energy Spectroscopic Imager
RMS	Root Mean Square
RSA	Russian Space Agency
SA	Solar Activity
SAMPEX	Solar Anomalous Magnetospheric Particle Explorer
SC	Spacecraft
SCOSTEP	Scientific Committee on Solar-Terrestrial Physics
SCR	Solar Cosmic Rays
SEC	Sun-Earth Connection
SEE	Single Event Effect
SEL	Single Event Latch-up
SEPs	Solar Energetic Particles
SESC	Space Environment Service Center
SETI	Search for Extraterrestrial Intelligence
SEU	Single Event Upset
SFU	Solar Flux Unit (1 sfu = 10^{-22} W m^{-2} Hz^{-1})
SGD	Solar-Geophysical Data
SI	Sisteme International (units)

SMM	Solar Maximum Mission
SMY	Solar Maximum Year
SOHO	Solar and Heliospheric Observatory
SPE	Solar Particle (Proton) Event
SPENVIS	Space Environment Information System
SSC	Sudden Storm Commencement
STP	Solar-Terrestrial Physics
STS	Space Transportation System (Space Shuttle Program)
SV	Space Vehicle
SWO	Space Weather Operation (in NOAA SESC)
TRACE	Transition Region and Coronal Explorer
UT	Universal Time
UTC	Coordinated Universal Time
UV	Ultraviolet light
WDC	World Data Center
WIND	WIND spacecraft
WMO	World Meteorological Organization

Bibliography

Abbasi R, Ackermann M, Adams J et al; IceTop Colaboration (2008) Solar energetic particle spectrum on 2006 December 13 determined by IceTop. Astrophys J Lett 689:L65–L68

Achard P, Adriani O, Aguilar-Benitez M et al (L3 collaboration) (2006) The solar flare of the 14th of July 2000 (L3+C detector results). Astron Astrophys 456:351–357

Achatz U, Dröge W, Schlickeiser R, Wibberenz G (1993) Interplanetary transport of solar electrons and protons: effect of dissipative processes in the magnetic field power spectrum. J Geophys Res 98(A8):13261–13280

Adams JH Jr (1986) Cosmic-ray effects on microelectronics, part IV, Naval Res. Lab., Washington, D.C., NRL Memo Rept. 5901, 31 Dec 1986

Adams JH Jr (1987) Cosmic-ray effects on microelectronics, Part IV, Naval Res. Lab., Washington, D.C., NRL Memo Rept. 5901, 31 Dec 1987

Adams JH Jr, Gelman A (1984) The effects of solar flares on single event upset rates. IEEE Trans Nucl Sci NS-31(6):1212–1216

Adams JH Jr, Hulbert EO (1983) The variability of single event upset rates in the natural environment. IEEE Trans Nucl Sci NS-30(6):4475–4480

Adams N (1950) A temporary increase in the neutron component of cosmic rays. Philos Mag Ser 41(316):503–505

Adriani O, Barbarino GC, Bazilevskaya GA et al; Collaboration PAMELA (2011) Observations of the 2006 December 13 and 14 solar particle events in the 80 MeV/n – 3 GeV/n range from space with the PAMELA detector. Astrophys J 742:102 (11 pp). doi:10.1088/0004-637X/742/2/102

Aglietta M, Badino G, Bologna G et al (in all 21 authors) (1991) Search for neutrinos from solar flares with the Mont Blanc Detector. In: Proceedings of 22nd international cosmic ray conference, vol 3. Dublin, pp 728–731

Ahluwalia HS, Xue SS (1993) Mean attenuation length for solar proton of 29 September 1989. In: Proceedings of 23rd international cosmic ray conference, vol 3. Calgary, pp 757–760

Ahluwalia HS, Xue SS, Kavlakov SP (1991) The ground level enhancement of September 29, 1989. In: Proceedings of 22nd international cosmic ray conference, vol 3. Dublin, pp 93–96

Akimov VV, Afanassyev VG, Belousov AS et al (in all 34 authors) (1991) Observation of high energy gamma-rays from the Sun with the GAMMA-1 telescope ($E > 30$ MeV). In: Proceedings of 22nd international cosmic ray conference, vol 3. Dublin, pp 73–76

Akimov VV, Belov AV, Chertok IM et al (in all 7 authors) (1993) High-energy gamma- rays at the late stage of the large solar flare of June 15, 1991 and accompanying phenomena. In: Proceedings of 23rd international cosmic ray conference, vol 3. Calgary, pp 111–114

Akimov VV, Ambroz P, Belov AV et al (in all 13 authors) (1996) Evidence for prolonged acceleration based on a detailed analysis of the long-duration solar gamma-ray flare on June 15, 1991. Solar Phys 166:107–134

L. Miroshnichenko, *Solar Cosmic Rays*, Astrophysics and Space Science Library 405, DOI 10.1007/978-3-319-09429-8

Akinyan ST, Chertok IM, Fomichev VV (1980) Quantitative forecasts of solar protons based on solar flare radio data. In: Donnelly RF (ed) Solar-terrestrial prediction proceedings, vol 3. - Washington, D.C., pp D14–D26

Akinyan ST, Bazilevskaya GA, Ishkov VN et al (in all 10 authors) (1983) In: Logachev YuI (ed) Catalogue of solar proton events 1970–1979. IZMIRAN, Moscow, Nauka, p 184

Alexander P, Valdes-Galicia JF (1998) A further search on waves generated by solar energetic protons. Solar Phys 183:407–418

Alexeenko VV, Chernyaev AB, Chudakov AE, Khaerdinov NS, Semenov AM, Szabelski J, Voevodsky AV (1993) GLE (ground level enhancement) at Baksan Air Shower Array (BASA). Proceedings of the 23rd international cosmic ray conference, vol. 3. Calgary, 163–166

Alexeyev EN, Karpov SN (1993) Short-term increases of muon intensity at Baksan Underground Scintillation Telescope correlated with solar flares? In: Proceedings of 23rd international cosmic ray conference, vol 3. Calgary, pp 167–170

Alexeyev EN, Karpov SN (1994) Short-term bursts of muon intensity at the Baksan Underground Scintillation Telescope. Geomagn Aeron 34(2):143–147

Alexeyev EN, Zakidyshev VN, Karpov SN (1992) Analysis of data of Baksan Underground Scintillation Telescope recorded during the September 29, 1989 solar flare. Geomagn Aeron 32(5):189–191

Alfvén H, Carlqvist P (1967) Currents in the solar atmosphere and a theory of solar flares. Solar Phys 1(2):220–228

Allen JH (1991) Technological effects at Earth and in space of solar activity in March 1991. Eos Trans AGU 72:377

Allen JH, Sauer H, Frank L, Reef P (1989) Effects of the March 1989 solar activity. Eos Trans AGU 70:1486

Aller LH (1961) Abundance of the elements. Interscience Publ, New York

Andreev AI, Efimov LN, Samoznaev LN, Chashei IV, Bird MK (1997) Characteristics of coronal Alfvén waves deduced from Helios Faraday rotation measurements. Solar Physics 176:387–402

Andriopoulou M, Mavromichalaki H, Plainaki C, Belov A, Eroshenko E (2011) Intense Ground-Level Enhancements of solar cosmic rays during the last solar cycles. Solar Phys 269:155–168

Anglin JD, Venkatesan D, Ananth AG (1995) Ulysses, Voyager 1 and IMP-8 low energy proton fluxes: Interplanetary particle transport. In: Proceedings of 24th international cosmic ray conference, vol 4. Rome, pp 281–284

Anisimov SV, Miroshnichenko LI (1989) Electricity in the atmosphere. Zemlya y Vselennaya (The Earth and the Universe) (4):14–22

Arkhangelskaja IV, Arkhangelsky AI, Kotov YuD, Kuznetsov SN, Glyanenko AS (2006) The flare catalogue in low-energy gamma-ray range based on the AVS-F instrument data onboard the Coronas-F satellite in 2001–2005. Astronomicheskii Vestnik (Astronomy Herald), Moscow, Nauka, 40(2):150–159 (in Russian)

Arkhangelskaja IV, Kotov YD, Arkhangelsky AI, Glyanenko AS (2009a) Thin structure of temporal profiles of solar flares January 15, 17 and 20, 2005 by data of AVS-F apparatus onboard CORONAS-F satellite. Adv Space Res 43(4):542–546

Arkhangelskaja IV, Arkhangelsky AI, Kotov YuD et al (in all 5 authors) (2009b) AVS-F observations of γ-ray emission during January 20, 2005 solar flare up to 140 MeV. Adv Space Res 43(4):589–593

Arkhangelskaja IV, Arkhangelsky AI, Troitskaya EV, Miroshnichenko LI (2009c) The investigation of powerful solar flares characteristics by analysis of excited states of 12C and various neutrons capture lines. Adv Space Res 43(4):594–599

Armstrong T (1993) National Space Science Data Centre (NASA/GSFC), Data Set 73-078A-08G-IMP8 H and J

Armstrong TP, Haggerty D, Lanzerotti LJ et al (in all 14 authors) (1994) Observation by Ulysses of hot (~270 keV) coronal particles at 32° south heliolatitude and 4.6 AU. Geophys Res Lett 21 (17):1747–1750

Aschwanden MJ (2012) GeV particle acceleration in solar flares and Ground Level Enhancement (GLE) events. Space Sci Rev 171:3–21

Aschwanden MJ, Benz AO (1994) Periodic or random acceleration in solar flares? Space Sci Rev 68:193–198

Aschwanden MJ, Dennis BR, Benz AO (1998a) Logistic avalanche processes, time structures, and frequency distributions in solar flares. Astrophys J 497:972–993

Aschwanden MJ, Dennis BR, Schwartz RA, Benz AO, Kliem B, Schwarz U, Kurths J (1998b) Elementary time scales in solar flares explored with wavelet analysis and logistic models. Eos Trans AGU Spring Meet Suppl 79(17):S280 (abstract)

Aschwanden MJ, Brown JC, Kontar EP (2002) Chromospheric height and density measurements in a solar flare observed with RHESSI – II. Data Anal Solar Phys 210:383–405

Aushev V, Antonova V, Belov A, Eroshenko E, Kryakunova O, Struminsky A (1999) Search for solar neutron events in Alma-Ata NM data. In: Proceedings of 26th international cosmic ray conference, vol 6. Salt Lake City, pp 50–53

Avdeev SV (2001) Experimental studies of particles inducing a light flash phenomenon in the cosmonaut's eyes by position-sensible semiconductor strip detectors. Candidate dissertation (Ph.D. thesis), Moscow Engineering and Physical Institute (MEPhI), Moscow, p 120

Avdeev SV, Galper AM, Vavilov NR et al (in all 26 authors) (2002) Spectrometer of charged particles based on silicon strip detectorsfor radiation measurements on board the International Space Station. Izvestiya RAN, Phys Series 66(11):1670–1672

Avdyushin SI, Danilov AD (2000) The sun, weather and climate: modern look at the problem (review). Geomagn Aeron 40(5):3–14

Avdyushin SI, Pereyaslova NK, Nazarova MN, Petrenko IE (1987) On the relationship between cosmic ray variations and solar activity. In: Proceedings of 20th international cosmic ray conference, vol 3. Moscow, pp 105–108

Avdyushin SI, Kulagin YM, Nazarova MN, Pereyslova NK, Petrenko IE, Svidsky PM (1984) Registration of solar cosmic rays by Cherenkov detectors onboard the system of Meteor satellites. Izv Acad Sci USSR Ser Phys 48(11):2254–2258

Avrett CW (1981) Models of the Solar Atmosphere. In: Cram LE, Thomas JH (eds) The physics of sunspots. AURA, Sacramento Peak Observatory, pp 235–254

Axford WI (1965) Anisotropic diffusion of solar cosmic rays. Planet Space Sci 13:1301–1309

Axford WI, Leer E, Scadron G (1977) The acceleration of cosmic rays by shock waves. In: Proceedings of 15th international cosmic ray conference, vol 11. Plovdiv, pp 132–137

Badhwar GD, O'Neil PM (1994) Long-term modulation of galactic cosmic radiation and its model for space exploration. Adv Space Res 14(10):749–757

Bahcall J (1990) Neutrino astrophysics. Chicago University Press, Chicago, p 567

Bai T (1986) Two classes of gamma-ray/proton flares: impulsive and gradual. Astrophys J 308:912–928

Bai T (1987) Periodicities of the flare occurrence rate in solar cycle 19. Astrophys J Lett 318:L85–L91

Bai T, Cliver EW (1990) A 154-day periodicity in the occurrence rate of proton flares. Astrophys J 363:299–309

Bai T, Sturrock PA (1989) Classification of solar flares. Ann Rev Astron Astrophys 27:421–467

Bailey DK (1959) Abnormal ionization in the lower ionosphere associated with cosmic-ray flux enhancements. Proc IRE 47:255–266

Bailey DK (1964) Polar cap absorption. Planet Space Sci 12:495–541

Baisultanova LM, Belov AV, Dorman LI, Yanke VG (1987) Magnetospheric effects in cosmic rays during Forbush-Decreases. In: Proceedings of 20th international cosmic ray conference, vol 4. Moscow, pp 231–234

Baker DN (1992) Solar wind coupling with the magnetosphere/atmosphere system. In: Fischer S, Vandas M (eds) Proc. 1st SOLTIP symposium, Liblice, Czechoslovakia, 30 September–5 October 1991, vol 2. Astronomical Institute CSAV, Prague, pp 54–66

Baker MB, Santina RE, Masley AJ (1969) Modeling of solar cosmic ray events based on recent observations. AIAA J 7(11):2105–2110

Balch C, Kunches J (1986) SESC methods for flare forecasts. In: Simon P, Shea MA, Heckman GR (eds) Solar-terrestrial predictions: proceedings of a workshop at Meudon, France, June 18–22, 1984, 1986th edn. NOAA, Boulder, pp 353–356

Band D, Matteson J, Ford L, Schaefer B, Palmer D, Teegarden B, Cline T, Briggs M, Paciesas W, Pendleton G, Fishman G, Kouveliotou C, Meegan C, Wilson R, Lestrade P (1993) BATSE observations of gamma-ray burst spectra. I – Spectral diversity. Astrophys J 413:281–292. doi:10.1086/172995

Baranov DG, Gagarin YF, Dergachev VA, Lukin VA, Yakubovsky EA (1997) Energy spectra of heavy nuclei from solar flares by observations at the orbital station "MIR" in October 1989. Geomagn Aeron 37(2):96–99

Bazilevskaya GA (1984) Solar protons with the energy above 100 MeV by the data of measurements in the stratosphere. Doctoral dissertation, Physical Lebedev Institute (FIAN), Moscow

Bazilevskaya GA (2009) On the early phase of relativistic solar particle events: are there signatures of acceleration mechanism? Adv Space Res 43:530–536

Bazilevskaya GA, Golynskaya RM (1989) The propagation of solar cosmic rays in the interplanetary medium with adiabatic focusing taken into account. Geomagn Aeron 29 (2):204–209

Bazilevskaya GA, Golynskaya RM (1990) Features of focused diffusion of solar cosmic rays. Geomagn Aeron 30(5):725–727

Bazilevskaya GA, Makhmutov VS (1983) Determination of absolute fluxes of solar protons with E >100 MeV based on the data of measurements in the stratosphere and by neutron monitors. Geomagn Aeron 23(3):373–377

Bazilevskaya GA, Makhmutov VS (1988) An upper limit to the energy of particles accelerated in a solar flare. Geomagn Aeron 28(2):169–172

Bazilevskaya GA, Makhmutov VS, Stozhkov YI, Svirzhevskaya AK, Svirzhevsky NS (2010) Solar proton events recorded in the stratosphere during cosmic ray balloon observations in 1957–2008. Adv Space Res 45:603–613

Bazilevskaya GA, Sladkova AI (1986) Azimuthal distribution and release of accelerated particles from the solar corona. Geomagn Aeron 26(2):187–190

Bazilevskaya GA, Sladkova AI (1997) Two groups of large solar proton events. In: Proceedings of 25th international cosmic ray conference, vol 7. Durban, pp 333–336

Bazilevskaya GA, Vashenyuk EV (1979) Some features of coronal and interplanetary propagation of solar cosmic rays of high energy. In: Proceedings of 16th international cosmic ray conference, vol 5. Kyoto, pp 156–161

Bazilevskaya GA, Vashenyuk EV, Ishkov VN et al (in all 10 authors) (1986) In: Logachev YuI (ed) Catalogue of energetic spectra of solar proton events 1970–1979. IZMIRAN, Moscow, Nauka, p 235

Bazilevskaya GA, Vashenyuk EV, Ishkov VN et al (in all 10 authors) (1990a) Catalogue of solar proton events 1980–1986, (ed) Logachev YuI, Moscow, World Data Center B-2, p 160 (part I), p 204 (part II).

Bazilevskaya GA, Debrunner H, Makhmutov VS, Flückiger E (1990b) Disturbances of the intensity of cosmic rays after small proton flares on the Sun. Geomagn Aeron 30(2):253–255

Bazilevskaya GA, Stozhkov YI, Struminsky AB (1994) The influence of interplanetary shocks on solar proton measured in the stratosphere. Adv Space Res 14(10):717–720

Bazilevskaya GA, Flückiger E, Makhmutov VS, Mizin SV (1996) Omnidirectional and vertical fluxes of charged particles in the Earth's atmosphere during solar proton events. Radiat Meas 26(3):443–446

Bazilevskaya G, Krainev M, Makhmutov V, Sladkova A, Storini M, Flückiger E (1998) The Gnevyshev gap in cosmic ray physics. In: Medina J (ed) Rayos cosmicos-98 (proc. 16th European cosmic ray symposium, Alcala University Press, Spain). Departamento de Física. Universidad de Alcalá, Alcalá de Henares, pp 83–86

Bazilevskaya GA, Flückiger EO, Krainev MB, Makhmutov VS, Sladkova AI, Storini M (1999) The Gnevyshev gap effect in solar cosmic rays. In: Proceedings of 26th international cosmic ray conference, vol 6. Salt Lake City, pp 240–243

Bednazhevsky VM, Miroshnichenko LI (1982) Integral multiplicities of generation for the neutron component and accuracy of estimation the spectrum of solar cosmic rays. Geomagn Aeron 22 (1):125–126

Beeck J, Mason GM, Hamilton DC et al (1988) A multi-spacecraft study of the injection and transport of solar energetic particles. Astrophys J 322:1052–1072

Beeck J, Mason GM, Marsden RG, Hamilton DC, Sanderson TR (1990) Injection and diffusive transport of suprathermal through energetic solar flare protons (35 keV to 20 MeV). J Geophys Res 95(A7):10279–10290

Beland P, Russel DA (1976) Biotic extinctions by solar flares. Nature 263(5574):259

Bell AR (1978) The acceleration of cosmic rays in shock fronts. Mon Not Roy Astron Soc 182:147–156

Belov AV, Eroshenko EA (1996) Proton spectra of the four remarkable GLEs in the 22nd Solar Cycle. Radiat Meas 26(3):461–466

Belov A, Eroshenko E, Mavromichalaki H, Plainaki C, Yanke V (2005b) A study of the ground level enhancement of 23 February 1956. Adv Space Res 35:697–701

Belov A, Kurt V, Mavromichalaki H, Garcia H, Gerontidou M (2005a) Proton enhancements and their relations to the X-ray flares during the three last solar cycles. Solar Phys 229:135–159. doi:10.1007/s11207-005-4721-3

Belov AV, Livshits MA (1995) Neutron burst on May 24, 1990. Astron Lett Russia 21:37–40

Belovsky MN, Ochelkov YP (1979) On some peculiarities of generation of electromagnetic and corpuscular radiation in solar flares. Izv AN SSSR Phys Ser 43(4):749–752

Bengin VV, Miroshnichenko LI, Petrov VM (1979) On applicability of diffusion model for describing spectral characteristics of solar cosmic rays. Geomagn Aeron 19(2):193–201

Bengin VV, Miroshnichenko LI, Petrov VM (1985) Dynamics and prediction of radiation characteristics of solar cosmic rays. Kosm Issled (Space Research) 23(1):123–133

Bengin VV, Petrov VM, Chernykh IV (1991) Operative dosimetric control of the crew at the Mir station during solar proton events in October 1989. Kosmich Biologia y Aviakosmich Meditsina (3):62–63

Berezhko EG, Taneev SN (2003) Shock acceleration of solar cosmic rays. Astron Lett 29(8):530–542

Berezhko EG, Petukhov SI, Taneev SN (2001) Acceleration of solar cosmic rays by shock waves in the solar corona. Izv RAN Phys Ser 64(3):339–342

Berezinsky VS, Bulanov SV, Dogel VA, Ginzburg VL, Ptuskin VS (1990) Astrophysics of cosmic rays. Elsevier Science Publishers, Amsterdam, p 534

Bespalov PA, Trakhtengerz VY (1974) On the turbulent dispersion of a cloud of fast particles in a plasma. Zh Eksp Teor Fiz 67(9):969–978

Bhandari N, Bhattacharaya S (1975) Shape of the solar flare proton spectrum in 10–100 MeV region in the past. In: Proceedings of 14th international cosmic ray conference, vol 12. München, pp 4245–4246

Bhatnagar A, Jain RM, Burkepile JT, Chertok IM, Magun A, Urbarz H, Zlobec P (1996) Transient phenomena in the energetic behind-the-limb solar flare of September 29, 1989. In: Proceedings of the IAU Coll. No.154, January 1995, Pune, India, vol 243(1). Astrophysics and space science, pp 209–213

Bhattacharya R, Barman P, Roy M (2013) Response of neutron monitors to cosmic ray counts: a statistical approach. Int J Eng Sci Techn (IJEST) 5(09):1713–1721

Bieber JW, Burger RA (1990) Cosmic-ray streaming in the Born approximation. Astrophys J 348:597–607

Bieber JW, Evenson PA (1991) Determination of energy spectra for the large solar particle events of 1989. In: Proceedings of 22nd international cosmic ray conference, vol 3. Dublin, pp 129–132

Bieber JW, Evenson PA (1995) Spaceship Earth – an optimized network of neutron monitors. In: Proceedings of 24th internation cosmic ray conference, vol 4. Rome, pp 1316–1319

Bieber JW, Matthaeus WH (1991) Cosmic ray pitch angle scattering in dynamical magnetic turbulence. In: Proceedings of 22nd international cosmic ray conference, vol 3. Dublin, pp 248–251

Bieber JW, Matthaeus WH (1992) In: Zank GP, Gaisser TK (eds) Particle acceleration in cosmic plasmas. AIP conference proceedings 264. AIP, New York, pp 86

Bieber JW, Earl JA, Green G, Kunow H, Müller-Mellin R, Wibberenz G (1980) Interplanetary pitch angle scattering and coronal transport of solar energetic particles: new information from Helios. J Geophys Res 85(A5):2313–2323

Bieber JW, Evenson PA, Pomerantz MA (1986) Focusing anisotropy of solar cosmic rays. J Geophys Res 91(A8):8713–8724

Bieber JW, Evenson PA, Matthaeus WH (1987) Magnetic helicity of the Parker field. Astrophys J 315:700–705

Bieber JW, Evenson PA, Pomerantz MA (1990) Unusual cosmic ray spike. Antarctic J 25(5):277–278

Bieber JW, Matthaeus WH, Smith CW, Wanner W, Kallenrode M-B, Wibberenz G (1994) Proton and electron mean free paths: the Palmer consensus revisited. Astrophys J 420:294–306

Bieber JW, Wanner W, Matthaeus WH (1995) 2D turbulence and the quasilinear theory "discrepancy". In: Proceedings of 24th international cosmic ray conference, vol 4. Rome, pp 269–272

Bieber JW, Clem J, Evenson P, Pyle R, Sáiz A, Ruffolo D (2013a) Giant ground level enhancement of relativistic solar protons on 2005 January 20. I. Spaceship Earth observations. Astrophys J 771:92 (13pp). doi:10.1088/0004-637X/771/2/922,3

Bieber J, Clem J, Evenson P, Oh S, Pyle R (2013b) Continued decline of South Pole neutron monitor counting rate. J Geophys Res 118:6847–6851. doi:10.1002/2013JA018915

Black DJ (1967) Cosmic ray effects and faunal extinctions at geomagnetic reversals. Earth Planet Sci Lett 3(3):225–236

Blandford RD, Ostriker JP (1978) Particle acceleration by astrophysical shocks. Astrophys J Lett 221:L29–L32. doi:10.1086/182658

Boberg RP, Tylka AJ, Adams JH (1995) The source plasma of solar energetic particles in gradual events. In: Proceedings 24th international cosmic ray conference, vol 4. Rome, pp 466–469

Bombardieri DJ, Duldig ML, Michael KJ, Humble JE (2006) Relativistic proton production during the 2000 July 14 solar event: the case for multiple source mechanisms. Astrophys J 644:565–574

Bombardieri DJ, Duldig ML, Humble JE, Michael KJ (2008) An improved model for relativistic solar particle acceleration applied to the 2005 January 20 and earlier events. Astrophys J 682:1315–1327

Bondarenko VA, Zil MV, Kolomensky AV, Kolosov DE (1986) Spectra of solar proton events in solar activity cycles 20 and 22. World Data Center B-2, Moscow, pp 46

Borog VV, Burinsky AYu, Dronov VV, Gvozdev AV, Dronov VV, Petrukhin AA (1995) Large aperture muon hodoscope for studies in solar-terrestrial physics. In: Proceedings of 24th international cosmic ray conference, vol 4. Rome, pp 1291–1294

Borog VV, Burinsky AYu, Gvozdev AV (1997) Angular and temporary cosmic ray muon flux characteristics with large aperture scintillation hodoscope. In: Proceedings of 25th international cosmic ray conference, vol 2. Durban, pp 449–452

Borovkov LP, Lazutin LL, Shumilov OI, Vashenyuk EV (1987) Injection characteristics of energetic particles on the Sun during GLE. In: Proceedings of 20th international cosmic ray conference, vol 3. Moscow, pp 124–127

Bostanjyan NKh, Chilingarian AA, Eganov VS, Karapetyan GG (2007) On the production of highest energy solar protons on 20 January 2005. Adv Space Res 39:1454–1457

Bruzek A, Durrant CJ (eds) (1977) Illustrated glossary for solar and solar-terrestrial physics. D. Reidel Publ. Co, Dordrecht, p 207

Bulanov SV, Sasorov PV (1975) Energy spectrum of particles accelerated in the vicinity of the null line of magnetic field. Astrono Zh 52(4):763–771 (in Russian)
Bulanov SV, Kurnosova LV, Ogulchansky YY, Razorenov LA, Fradkin MI (1985) Acceleration of ultrarelativistic electrons in solar flares. Astron Lett Russia 11(5):383–389
Burkepile JT, St. Cyr OC (1993) A revised and expanded catalogue of mass ejections observed by the Solar Maximum Mission coronograph. NCAR technical note, NCAR/TN- 369+STR. Boulder, Co.
Burlaga L (1967) Anisotropic diffusion of solar cosmic rays. J Geophys Res 72(17):4449–4466
Burlaga LF (1995) Interplanetary magnetohydrodynamics. Oxford University Press, New York
Burlaga LF, Lepping RP, Jones JA (1990) Global configuration of a magnetic cloud. In: Russel CT, Priest ER, Lee LC (eds) Physics of magnetic flux ropes, vol 58, Geophysical monograph series. AGU, Washington, D.C, p 373
Burrel VO (1972) The risk of solar proton events to space missions. NASA report TMX- 2440
Caffee M, Goswami JN, Hohenberg CM, Swindle TW (1987) Solar flare irradiation in meteorites provides evidence for an early active Sun. In: Proceedings of 20th international cosmic ray conference, vol 4. USSR, Moscow, pp 299–302
Callis LB, Baker DN, Blake JB et al (1991) Precipitating relativistic electrons: their long-term effect on stratospheric odd nitrogen levels. J Geophys Res 95(D2):2939–2976
Cameron AGW (1973) Abundances of the elements in the Solar system. Space Sci Rev 15 (1):121–146
Cane HV (1998) Energetic particles in the solar wind: propagation, acceleration and modulation. In: Potgieter MS, Raubenheimer BS, van der Walt DJ (eds) Proceedings 25th international cosmic ray conference: invited, rapporteur, and highlight papers, vol 3. World Scientific, Singapore/New Jersy/London/Hong Kong, pp 135–150
Cane HV, Reames DV (1990) The relationship between energetic particles and flare properties for impulsive solar flares. Astrophys J Suppl 73:253–258
Cane HV, McGuire RE, von Rosenvinge TT (1986) Two classes of solar energetic particle events associated with impulsive and long-duration soft X-ray flares. Astrophys J 301:448–459
Cane HV, Reames DV, von Rosenvinge TT (1988) The role of interplanetary shocks in the longitude distribution of solar energetic particles. J Geophys Res 93(A9):9555–9567
Cane HV, Reames DV, von Rosenvinge TT (1991) Solar particle abundances at energies of greater than 1 MeV per nucleon and the role of interplanetary shocks. Astrophys J 373:675–682
Cane HV, Sheeley NR, Howard RA (1987) Energetic interplanetary shocks, radio emission, and coronal mass ejections. J Geophys Res 92:9869–9874
Cargill PJ, Goodrich CC, Vlahos L (1988) Collisionless shock formation and the prompt acceleration of solar flare ions. Astron Astrophys 189:254–262
Carmichael H (1962) High-energy solar-particle events. Space Sci Rev 1:28–61
Carmichael H (1968) Cosmic rays (instruments). In: Annals of the IQSY, vol 1. MIT Press, Cambridge, MA, pp 178–197
Casolino M, Bidoli V, De Grandis E, et al (in all 28 authors) (2001) Cosmic ray measurements on board space station Mir with SILEYE-2 experiment. In: Proceedings of 27th international cosmic ray conference, vol 9. Hamburg, pp 4011–4014
Cessateur G, Kretzschmar M, Dudok de Wit T, Boumier P (2010) The influence of solar flares on the lower solar atmosphere: evidence from the Na D absorption line measured by GOLF/ SOHO. Solar Phys 263:153–162
Chambon G, Hurley K, Niel M, Talon R, Vedrenne G, Estuline IV, Likine OB (1981) The November 22, 1977 solar flare: evidence for 2.3 and 4.33 MeV line emission from the Signe 2MP experiment. Solar Phys 69(1):147–159
Charakhchyan AN, Bazilevskaya GA, Svirzhevskaya AK, Stozhkov YI, Charakhchyan TN (1973) 11-year cycle of intensity of cosmic rays in the stratosphere and its dependence upon solar activity. Izv AN SSSR Phys Ser 37(6):1258–1263

Chebakova EA, Kolomeets EV, Sevostyanov VN (1985) Investigation of cosmic ray propagation in interplanetary space. In: Proceedings of 19th international cosmic ray conference, vol 4. La Jolla, pp 356–359

Chen W, Gan WQ (2011) The calculation of solar gamma-rays by TALYS. Acta Astrono Sin 52:219–232

Chenette DL, Dietrich WF (1984) The solar flare heavy ion environment for single-event upsets: a summary of observations over the last solar cycle, 1973–1983. IEEE Trans Nucl Sci 31 (6):1217–1222

Chertok IM (1993) Solar coronal transients. Astron Zh 70(1):165–187 (in Russian)

Chertok IM (1995) Post-eruption particle acceleration in the corona: a possible contribution to solar cosmic rays. In: Proceedings of 24th international cosmic ray conference, vol 4. Rome, pp 78–81

Chertok IM (1996) Yohkoh data on CME-flare relationships and post-eruption magnetic reconnection in the corona. In: Bentley RD, Mariska JT (eds) Proceedings of Yohkoh conference on observations of magnetic reconnection in the solar atmosphere. Astronomical Society Pacific conference series, vol 111. San Francisco, pp 369–374

Chertok IM (1997a) The role of coronal mass ejections and post-eruption energy release in solar high-energy phenomena. J Moscow Phys Soc 7:31–40

Chertok IM (1997b) Some features of the post-energy release in the solar corona. In: The corona and solar wind near minimum activity, proceedings of the fifth SOHO workshop, Oslo, Norway, 17–20 June 1997, ESA SP-404, September 1997, pp 269–273

Chiba N, Dion GM, Hayashida N, Honda K, Inoue N, Kadota K, Kakimoto F, Kamata K, Kawaguchi S, Kawasumi N, Matsubara Y, Nagano M, Ohoka H, Teshima M, Tsushima I, Yoshida S, Yoshii H, Yoshikoshi T (1992) Possible evidence for >10 GeV neutrons associated with the solar flare of 4 June 1991. Astropart Phys 1(1):27–32

Chih PP, Lee MA (1986) A perturbation approach to cosmic ray transients in interplanetary space. J Geophys Res 91(A3):2903–2913

Chirkov NP, Filippov AT (1977) Acceleration of energetic particles to relativistic energies in the interplanetary medium. Izvestiya (bulletin) of AN SSSR. Phys Serv 41(9):1776–1781

Chizhenkov VA (2002) Analysis of the relation between operation anomalies of the equipment on board the standard satellites of the Cosmoc series and solar-terrestrial activity. Candidate dissertation (Ph.D. thesis), Institute of the Physics of the Earth, Russian Academy of Sciences, Moscow, p 112

Chuchkov EA, Ermakov SI, Kontor NN, Lyubimov GP, Pavlov NN, Tulupov VI, Shcherbovskiy BY (1996) Interplanetary medium diagnostics during the large event in March 1991 based on GRANAT records. Radiat Meas 26(3):437–442

Chupp EL (1984) High-energy neutral radiation from the Sun. Ann Rev Astron Astrophys 22:359–387

Chupp EL (1987) High-energy particle production in solar flares (SEP, gamma-ray and neutron emissions). Phys Scr T18:5–19

Chupp EL (1990) Emission characteristics of three intense solar flares observed in cycle 21. Astrophys J Suppl 73:213–226

Chupp EL (1996) Evolution of our understanding of solar flare particle acceleration: (1942–1995). In: Ramaty R, Mandzhavidze N, Hua X-M (eds) High energy solar physics, vol 374, AIP conference proceedings. AIP, New York, pp 3–31

Chupp EL, Forrest DJ, Ryan JM, Heslin J, Reppin C, Pinkau K, Kanbach G, Rieger E, Share GH (1982) A direct observation of solar neutrons following the 0118 UT flare on 1980 June 21. Astrophys J Lett 263:L95–L98

Chupp EL, Debrunner H, Flückiger E, Forrest DJ, Golliez F, Kanbach G, Vestrand WT, Cooper J, Share GH (1987) Solar neutrons emissivity during the large solar flare on 1982 June 3. Astrophys J 318:913–925

Clem JM, Dorman LI (2000) Neutron monitor response functions. Space Sci Rev 93:335–359. doi:10.1023/A:1026508915269

Cliver EW (1995) Solar flare nomenclature. Solar Phys 157(2):285–293
Cliver EW (1996) Solar flare gamma-ray emission and energetic particles in space. In: Ramaty R, Mandzhavidze N, Hua X-M (eds) High energy solar physics, vol 374, AIP conference proceedings. AIP, New York, pp 45–60
Cliver EW (2009) History of research on solar energetic particle (SEP) events: the evolving paradigm. In: Gopalswamy N, Webb DF (eds) Proceedings of IAU symposium no. 257, 2008, Universal Heliophysical Processes. pp 401–412
Cliver EW, Ling AG (2007) Electrons and protons in solar energetic particle events. Astrophys J 658:1349–1356. doi:10.1086/511737
Cliver EW, Webb DF (1997) LASCO data dazzle Chapman conference. Eos Trans AGU 78 (10):107, 112
Cliver EW, Kahler SW, Shea MA, Smart DF (1982) Injection onsets of ~2 GeV protons, ~1 MeV electrons, and ~100 keV electrons in solar cosmic ray flares. Astrophys J 260:362–370
Cliver EW, Kahler SW, Cane HV, Koomen MJ, Michels DJ, Howard RA, Sheeley NR Jr (1983) The GLE-associated flare of 21 August, 1979. Solar Phys 89(1):181–193
Cliver EW, Forrest DJ, Cane HV, Reames DV, McGuire RE, von Rosenvinge TT, Kane SR, MacDowall RJ (1989) Solar flare nuclear gamma-rays and interplanetary proton events. Astrophys J 343:953–970
Cliver E, Reames D, Kahler S, Cane H (1991) Size distribution of solar energetic particle events. In: Proceedings of 22nd international cosmic ray conference, vol 3. Dublin, pp 25–28
Cliver EW, Dröge W, Müller-Mellin R (1993a) Superevents and cosmic ray modulation, 1974–1985. J Geophys Res 98(A9):15231–15240
Cliver EW, Kahler SW, Vestrand WT (1993b) On the origin of gamma-ray emission from the behind-the-limb flare on 29 September 1989. In: Proceedings of 23rd international cosmic ray conference, vol 3. Calgary, pp 91–94
Cliver EW, Kahler SW, Neidig DF, Cane HV, Richardson IG, Kallenrode M-B, Wibberenz G (1995) Extreme "propagation" of solar energetic particles. In: Proceedings of 24th international cosmic ray conference, vol 4. Rome, pp 257–260
Cliver EW, Ling AG, Belov A, Yashiro S (2012) Size distributions of solar flares and solar energetic particles events. Astrophys J Lett 756:L29 (4 pp)
Cliver EW, Tylka AJ, Dietrich WF, Ling AG (2014) On a solar origin for the cosmogenic nuclide event of 775AD. Astrophys J 781:32. doi:10.1088/0004-637X/781/1/32
Cobb WE (1978) Balloon measurements of the air-Earth current density at the South Pole before and after a solar flare. In: Conference on cloud physics and atmosphere electricity. American Meteorological Society, Issaquah, Washington. 31 July – 4 August 1978
Compton AH, Wollan EO, Bennet RD (1934) A precision recording cosmic ray meter. Rev Scientif Instrum 5(12):415–422
Comstock GM, Fleischer RL, Giard WR, Hart HR Jr, Nichols GE, Price PB (1971) Cosmic-ray tracks in plastics: the apollo helmet dosimetry experiment. Science 171:154–156
Cook WR, Stone EC, Vogt RE (1984) Elemental composition of solar energetic particles. Astrophys J 279:827–838
Cooke DJ, Humble JE, Shea MA, Smart DF, Lund N, Rasmussen IL, Byrnak B, Goret P, Petrou N (1991) On cosmic-ray cut-off terminology. Nuovo Cimento 14C(3):213–234
Crabb RL (1981) Solar array radiation damage. ESA Bulletin (56):72
Cramp JL, Duldig ML, Humble JE (1993a) The GLE of 29 September 1989. In: Proceedings of 23rd international cosmic ray conference, vol 3. Calgary, pp 47–50
Cramp JL, Duldig ML, Humble JE (1993b) The GLE of 22 October 1989. In: Proceedings of 23rd international cosmic ray conference, vol 3. Calgary, pp 51–54
Cramp JL, Duldig ML, Humble JE (1995a) Neutron monitor responses during highly anisotropic Ground Level Enhancements. In: Proceedings of 24th international cosmic ray conference, vol 4. Rome, pp 248–251

Cramp JL, Duldig ML, Humble JE (1995b) Rigidity dependent pitch angle distribution in modeling ground level enhancements. In: Proceedings of 24th international cosmic ray conference, vol 4. Rome, pp 285–288

Cramp JL, Duldig ML, Humble JE (1995c) The effect of near-Earth IMF structure on the modeling of ground level enhancements. In: Proceedings of 24th international cosmic ray conference, vol 4. Rome, pp 289–292

Cramp JL, Duldig ML, Flückiger EO, Humble JE, Shea MA, Smart DF (1997) The October 22, 1989, solar cosmic ray enhancement: an analysis of the anisotropy and spectral characteristics. J Geophys Res 102(A11):24237–24248

Crosby NB (2009) Solar extreme events 2005–2006: effects on near-earth space systems and interplanetary systems. Adv Space Res 43:559–564

Crosby NB, Aschwanden MJ, Dennis BR (1993) Frequency distributions and correlations of solar X-ray flare parameters. Solar Phys 143(2):275–299

Crutzen PJ, Reid GC (1976) Reply. Nature 263(5574):259

Crutzen PJ, Isaksen ISA, Reid GC (1975) Solar proton events: stratospheric sources of nitric oxide. Science 189(4201):457–458

Daibog EI, Kurt VG, Stolpovsky VG (1981) Spectrum of flare protons at small energies. Kosm Issled 19(5):704–711

Daibog EI, Kurt VG, Stolpovsky VG (1984) On the problem of energy dependence of mean free path for flare particles. Kosm Issled 22(5):763–773

Daibog EI, Kurt VG, Logachev YuI, Stolpovsky VG (1985) Two types of electron events in solar flares. In: Proceedings of 19th international cosmic ray conference, vol 4. La Jolla, pp 62–65

Daibog EI, Logachev YI, Stolpovsky VG, Kolomeets EV, Sevostyanov VN, Chebakova EA (1986) Analysis of five solar cosmic ray flares occurring in 1977. In: Variations of cosmic rays and investigations of space. IZMIRAN, Moscow, pp 7–13

Daibog EI, Kurt VG, Logachev YI, Stolpovsky VG, Melnikov VF, Podstrigach TS (1988) Multiple correlation and regression analysis of the relation between amplitude and spectral characteristics of proton events and parameters of microwave bursts. Izv AN SSSR Phys Ser 52:2403–2406

Daibog EI, Logachev YuI, Stolpovsky VG, Melnikov VF, Podstrigach TS (1990) Effectiveness of energetic particle escape into interplanetary medium. In: Proceedings of 21st international cosmic ray conference, vol 5. Adelaide, pp 96–99

Danilova OA, Tyasto MI, Kananen H, Tanskanen P (1997) The cosmic ray asymptotic directions for station Oulu in the magnetic field of the Tsyganenko 1989 model. In: Proceedings of 25th international cosmic ray conference, vol 2. Durban, pp 369–372

Danilova OA, Tyasto MI, Vashenyuk EV, Gvozdevsky BB, Kananen H, Tanskanen P (1999) The GLE of May 2, 1998: an effect of disturbed magnetosphere on solar cosmic rays. In: Proceedings of 26th international cosmic ray conference, vol 6. Salt Lake City, pp 399–402

De Jager C (1963) The Sun as a source of interplanetary gas. Space Sci Rev 1:487–521

De Jager C (1969) Solar flares: properties and problems. In: Proceedings COSPAR symposium on solar flares and space research. North-Holland Publ. Co, Amsterdam, pp 1–15

De Jager C (1986) Solar flares and particle acceleration. Space Sci Rev 44(1/2):43–90

De Jager C (1987) Energetic phenomena in impulsive solar flares. In: Proceedings of 20th international cosmic ray conference: invited talks, vol 7. Moscow, pp 66–76

de Jager C (1990) An explanation of the 'granulation boundary' in the HR diagram. Solar Phys 126:201–205

De Jager C, de Jonge G (1978) Properties of elementary flare bursts. Solar Phys 58(1):127–137

De Koning CA (1994) Analysis of the ground level enhancement of September 29, 1989. Master's thesis, University of Calgary, Calgary

De Koning CA, Bland CJ (1995) The GLE of September 29, 1989: analysis of the pitch angle distribution I. In: Proceedings of 24th international cosmic ray conference, vol 4. Rome, pp 212–215

De Koning CA, Mathews T (1995) The GLE of September 29, 1989: analysis of the pitch angle distribution II. In: Proceedings of 24th international cosmic ray conference, vol 4. Rome, pp 216–219

De Koning CA, Mathews T (1996) Diffusion analysis of the GLE (Ground-Level Enhancement) of September 29, 1989. Canadian J Phys 74:290–294

De La Beaujardière J-F, Canfield RC, Hudson HS, Wülser J-P, Acton L, Kosugi T, Masuda S (1995) The 1991 October 24 flare: a challenge for standard models. Astrophys J 440:386–393

Debrunner H, Flückiger EO, Lockwood JA (1982) Response of neutron monitors to solar cosmic rays events. In: 8th European cosmic ray symposium, book of abstracts, Rome

Debrunner H, Flückiger E, Chupp EL, Forrest DJ (1983) The solar cosmic ray event on 1982 June 3. In: Proceedings of 18th international cosmic ray conference, vol 4. Bangalore, pp 75–78

Debrunner H, Flückiger EO, Lockwood JA, McGuire RE (1984) Comparison of the solar cosmic ray events on May 7, 1978 and November 22, 1977. J Geophys Res 89(A2):769–774

Debrunner H, Flückiger E, Grädel H, Lockwood JA, McGuire RE (1988) Observations related to the acceleration, injection, and interplanetary propagation of energetic protons during the solar cosmic ray event on February 16, 1984. J Geophys Res 93(A7):7206–7216

Debrunner H, Flückiger EO, Stein P (1989) On the sensitivity of neutron monitors to solar neutrons. Nucl Instrum Method Phys Res A278:573–578

Debrunner H, Lockwood JA, Ryan JM (1992) The solar flare event on May 24, 1990: evidence for two separate particle accelerations. Astrophys J Lett 387:L51–L54

Debrunner H, Lockwood JA, Ryan JM (1993) Solar neutron and proton production during the 1990 May 24 cosmic-ray flare increases. Astrophys J 409:822–829

Debrunner H, Lockwood JA, Ryan JM (1995) Relationship between the timing of solar flare gamma-ray emission and high-energy neutron production. In: Proceedings of 24th international cosmic ray conference, vol 4. Rome, pp 167–170

Decker RB (1988) Computer modeling of test particle acceleration at oblique shocks. Space Sci Rev 48(3–4):195–262

Decker RB, Vlahos L (1986) Numerical studies of particle acceleration at turbulent, oblique shocks with an application to prompt ion acceleration during solar flares. Astrophys J 306:710–729

Delcroix JL (1965) Plasma physics. Wiley, New York, p 61

Dennis BR (1996) X-ray aspects of the high energy solar physics workshop: rapporteur paper I. In: Ramaty R, Mandzhavidze N, Hua X-M (eds) High energy solar physics, vol 374, AIP conference proceedings. AIP, New York, pp 519–532

Dennis J, Schnabel R (1988) Numerical methods of absolute optimization and resolution of non-linear equations. Moscow, Mir, 440 pp

Dennis BR, Emslie AG, Hudson HS (eds) (2011) High-energy aspects of solar flares (A RHESSI-inspired monograph). http://arxiv.org/abs/1109.5831

Dergachev VA (1994) Radiocarbon chronometry of geomagnetic field. Geomagn Aeron 34 (4):433–449 (in English)

Dietrich W, Lopate C (1999) Measurements of iron reach SEP events using the University of Chicago IMP-8 instrument. In: Proceedings of 26th international cosmic ray conference, vol 6. Salt Lake City, pp 71–74

Ding L, on behalf of the L3 Collaboration (2001) Search for possible enhancement in the flux of high energy muons due to the solar flare of 14 July 2000 with the L3+Cosmic Muon Spectrometer. In: Proceedings of the 27th international cosmic ray conference, vol 8. Hamburg, pp 3372–3374

Djantemirov HM, Galper AM, Khodarovich AM, Luchkov BI, Ozerov YuV, Zemskov VM, Zverev VG (1995) The temporal and spectral characteristics of solar gamma-flares on the data of GAMMA-1 experiment. In: Proceedings of 24 international cosmic ray conference, vol 4. Rome, pp 94–97

Dodson HW, Hedeman ER, Kreplin RW et al (in all 8 authors) (1975) In: Svestka Z, Simon P (ed) Catalogue of solar proton events 1955–1969. D. Reidel Publ. Co., Dordrecht, 430 pp

Dorman LI (1957) Cosmic ray variations. Moscow, Gostekhteorizdat, 429 pp (in Russian). English version: Transl Techn. Doc. Liaison Office, Wright-Patterson Airforce Base, USA, 1958
Dorman LI (1958) The generation and propagation of solar cosmic rays. Nuovo Cimento 8(Suppl 2):391–402
Dorman LI (1963a) Geophysical and astrophysical aspects of cosmic radiation. In: Wilson JG, Wouthuysen SA (eds) Progress in elementary particles and cosmic ray physics, vol 7. North-Holland Publ. Co, Amsterdam, p 358
Dorman LI (1963b) Cosmic ray variations and space exploration. Nauka, Fizmatgiz, Moscow, p 1073
Dorman LI (1972) Acceleration processes in space, vol 7. VINITI, Moscow, p 205
Dorman LI (1978) Cosmic rays of solar origin, vol 12. VINITI, Moscow, p 186
Dorman LI (1998) Cosmic rays and cosmogenic nuclides: 1. In space, inside bodies, in atmospheres; 2. Radiocarbon method and elements global mixing and exchange on the Earth. In: Shapiro MM, Silberberg R, Wefel JP (eds) Towards the millennium in astrophysics, problems and prospects. World Scientific, Singapore/New Jersy/London/Hong Kong, pp 303–350
Dorman LI, Katz ME (1973) In: GE Kocharov (ed) Leningrad international seminar on cosmophysics. Leningrad, Nauka, pp 311–316
Dorman LI, Katz ME, Shogenov VK (1973) Kinetic description of cosmic ray interactions with shock waves. Izv Acad Sci USSR Ser Phys 37(6):1254–1258
Dorman LI, Kolomeets EV (1961) Small cosmic-ray bursts on quiet and disturbed days associated with type II and III solar radio bursts. Geomagn Aeron 1(6):1015–1016
Dorman LI, Miroshnichenko LI (1966) On a procedure for determining the solar cosmic ray spectrum in the range of high energies. Geomagn Aeron 6(2):215–222
Dorman LI, Miroshnichenko LI (1968) Solar cosmic rays. Moscow, Nauka, Fizmatgiz, pp 468 (in - Russian). English Edition for NASA by Indian National Scientific Documentation Center, Delhi, 1976
Dorman LI, Pakhomov NI (1979) The dependence of the integral generation multiplicity of neutron component at various depth in the atmosphere on zenith angle of primary particle incidence. In: Proceedings of 16th international cosmic ray conference, vol 4. Kyoto, pp 416–420
Dorman LI, Pustilnik LA (1999) Statistical characteristics of FEP events and their connection with acceleration, escaping and propagation mechanisms. In: Proceedings of 26th international cosmic ray conference, vol 6. Salt Lake City, pp 407–410
Dorman LI, Valdes-Galicia JF (1999) Numerical simulation and analytical description of solar neutron transport in the Earth's atmosphere. J Geophys Res 104(A10):22417–22426
Dorman LI, Venkatesan D (1993) Solar cosmic rays. Space Sci Rev 64(3/4):183–362
Dorman LI, Yanke VG (1981) The coupling functions of NM-64 neutron supermonitor. In: Proceedings of 17th international cosmic ray conference, vol 4. Paris, pp 326–329
Dorman LI, Katz ME, Fedorov YI, Shakhov BA (1979) On the character of the energy change of galactic and solar cosmic rays in the interplanetary space. Izv AN SSSR Phys Ser 43 (4):778–792
Dorman LI, Katz ME, Nosov SF, Fedorov YI, Shakhov BA (1988) Acceleration of cosmic rays in randomly gyrotropic electromagnetic field. Izv AN SSSR Phys Ser 52(12):2358–2360
Dorman LI, Miroshnichenko LI, Sorokin MO (1990) Cosmic ray data as a basis for predicting the onset and development of solar proton events. In: Thompson RJ, Cole DG, Wilkinson PJ, Shea MA, Smart DF, Heckman GH (eds) Solar-terrestrial predictions: proceedings of a workshop at Leura, Australia, October 16–20, 1989, vol 1. NOAA ERL, Boulder, pp 386–390
Dreicer H (1959) Electron and ion runaway in a fully ionized gas, I. Phys Rev 115(2):238–249
Dreicer H (1960) Electron and ion runaway in a fully ionized gas, II. Phys Rev 117(2):329–342
Dreschhoff GAM, Zeller EJ (1990) Evidence of individual solar proton events in Antarctic snow. Solar Phys 127:333–346
Dreschhoff GAM, Zeller EJ (1998) Ultra-high resolution nitrate in polar ice as indicator of past solar activity. Solar Phys 177(2):365–374

Dreschhoff GAM, Shea MA, Smart DF, McCracken KG (1997) Evidence for historical solar proton events from NO(X) precipitation in polar ice cores. In: Proceedings of 25th international cosmic ray conference, vol 1. Durban, pp 89–92
Dröge W (1994a) Transport of solar energetic particles. Astrophys J Suppl 90:567–576
Dröge W (1994b) Heliospheric propagation of cosmic rays. In: Leahy DA, Hicks RB, Venkatesan D (eds) Proc. 23rd international cosmic ray conference: invited, rapporteur and highlight papers. World Scientific Pub. Co. Pte. Ltd, Singapore, pp 185–212
Dröge W (1995) Solar energetic electrons: I. Multiple spacecraft observations; II. Comparison with gamma-ray emission. In: Proceedings of 24th international cosmic ray conference, vol 4. Rome, pp 187–194
Dröge W (1996) Energetic solar electrons spectra and gamma-ray observations. In: Ramaty R, Mandzhavidze N, Hua X-M (eds) High energy solar physics, vol 374. AIP Conference Proceedings, AIP, New York, pp 78–85
Dröge W, Schlickeiser R (1986) Particle acceleration in solar flares. Astrophys J 305:909–919
Dröge W, Meyer P, Evenson P, Moses D (1989) Electron acceleration in solar flares. Solar Phys 121(1):95–103
Dröge W, Gibbs K, Grunsfeld JM, Meyer P, Newport BJ, Evenson P, Moses D (1990) A 153 day periodicity in the occurrence of solar flares producing energetic interplanetary electrons. Astrophys J Suppl 73:279–283
Dröge W, Müller-Mellin B, Cliver EW (1992) Superevents: their origin and propagation through the heliosphere from 0.3-35 AU. Astrophys J Lett 387:L97–L100
Dröge W, Ruffolo D, Klecker B (1995) Probable detection of electrons from the decay of solar flare neutrons. In: Proceedings of 24th international cosmic ray conference, vol 4. Rome, pp 183–186
Dröge W, Ruffolo D, Klecker B (1996) Observation of electrons from the decay of solar flare neutrons. Astrophys J Lett 464:L87–L90
Dröge W, Ruffolo D, Khumlumlert T (1997) The rigidity dependence of solar electron mean free paths in the range 0.3 to 20 MV. In: Proceedings of 25th international cosmic ray conference, vol 1. Durban, pp 137–140
Dryer M (1996) Comments on the origin of CMEs. Solar Phys 169(2):421–429
Duggal SP (1979) Relativistic solar cosmic rays. Rev Geophys Space Res 17(5):1021–1058
Duggal SP, Pomerantz MA (1971) The propagation of energetic solar particles during highly anisotropic ground level events. In: Proceedings of 12th international cosmic ray conference, vol 2. Adelaide, pp 533–537
Duggal SP, Guidi I, Pomerantz MA (1971) The unusual anisotropic solar particle event of November 18, 1968. Solar Phys 19:234–246
Dulk GA, Smerd SF, MacQueen RM, Gosling JT, Magun A, Stewart RT, Sheridan KV, Robinson RD, Jacques S (1976) White light and radio studies of the coronal transient of 14–15 September 1973: I. Material motions and magnetic field. Solar Phys 49(2):369–394
Dunphy PP, Chupp EL (1991) High-energy gamma rays and neutrons from the solar flare of 1989 March 6. In: Proceedings of 22nd international cosmic ray conference, vol 3. Dublin, pp 65–68
Dunphy PP, Chupp EL, Schneid EJ, Bertsch DL, Gottesman S (1995) High energy neutrons and gamma-rays from the solar flare of 1991 June 11. In: Proceedings of 24th international cosmic ray conference, vol 4. Rome, pp 203 (abstract)
Dunphy PP, Chupp EL, Bertsch DL, Schneid EJ, Gottesman SR, Kanbach G (1999) Gamma-rays and neutrons as a probe of flare proton spectra: the solar flare of 11 June 1991. Solar Phys 187 (1):45–57
Dvornikov VM, Sdobnov VE (1993) A spectrographic global survey technique for studying cosmic-ray distribution function variations and the planetary system of geomagnetic cutoff rigidities. In: Proceedings of 23rd international cosmic ray conference, vol 3. Calgary, pp 797–800

Dvornikov VM, Sdobnov VE (1995a) The GLE of 29 September 1989: time variations of the cosmic-ray rigidity spectrum. In: Proceedings of 24th international cosmic ray conference, vol 4. Rome, pp 232–235
Dvornikov VM, Sdobnov VE (1995b) The GLE of 29 September 1989: time variations in angular distribution of different-energy particles. In: Proceedings of 24th international cosmic ray conference, vol 4. Rome, pp 2362–2239
Dvornikov VM, Sdobnov VE (1997) Time variations of the cosmic ray distribution function during a solar proton event of September 29, 1989. J Geophys Res 102(A11):24209–24219
Dvornikov VM, Sdobnov VE (1998) Analyzing the solar proton event of 22 October 1989, using the method of spectrographic global survey. Solar Phys 178:405–422
Dvornikov VM, Sdobnov VE, Sergeev AV (1983) Analysis of cosmic ray pitch angle anisotropy during the Forbush-effect in June 1972 by the method of spectrographic global survey. In: Proceedings of 18th international cosmic ray conference, vol 3. Bangalore, pp 249–252
Dvornikov VM, Sdobnov VE, Sergeev AV (1984) Recurrent double bursts of pitch angle anisotropy and their relationship to solar proton events. Izv AN SSSR Phys Ser 48(11):2140–2142
Earl JA (1976a) The effect of adiabatic focusing upon charged-particle propagation in random magnetic fields. Astrophys J 205:900–919
Earl JA (1976b) Non-diffusive propagation of cosmic rays in the solar system and in extragalactic radio sources. Astrophys J 206:301–311
Earl JA (1992) Effect of magnetic helicity upon rectilinear propagation of charged particles in random magnetic fields. Astrophys J 395(1):185–191
Earl JA (1995) Propagation of charged particle pulses immediately after injection. In: Proceedings of 24th international cosmic ray conference, vol 4. Rome, pp 293–296
Efimov YE, Terekhov MM (1988) Solar neutron propagation in the atmosphere. Geomagn Aeron 28:832–834
Efimov YE, Kovaltsov GA, Kocharov GE, Kocharov LG, Usoskin IG (1993a) Detection of solar neutrons by the neutron monitor network. Izv RAN Phys Ser 57(7):142–145
Efimov YuE, Kocharov GE, Kudela K (1983) On the solar neutrons observation on high mountain neutron monitor. In: Proceedings of the 18th international cosmic ray conference, vol. 10. Paris, pp 276–289
Efimov YuE, Kocharov LG, Kovaltsov GA, Usoskin IG (1993) Detection of solar neutrons by the neutron monitor network. In: Proceedings of 23rd international cosmic ray conference, vol 3. Calgary, pp 159–162
Elliot H (1952) The variations of cosmic ray intensity. In: Wilson JG, Wouthuysen SA (eds) Progress in cosmic ray physics, vol 1. North-Holland Publ. Co, Amsterdam, pp 453–514
Elliot H (1964) The nature of solar flare. Planet Space Sci 12(6):657–660
Ellison MA (1963) Energy release in solar flares. J Roy Astron Soc 4(1):62–73
Ellison DC, Ramaty R (1985) Shock acceleration of electrons and ions in solar flares. Astrophys J 298:400–408
Ermakov SI, Lyubimov GP, Tulupov VI (1996) A model of the solar proton event of 23–31 March 1991. Radiat Meas 26(3):433–436
Erofeeva IN, Kolomeets EV, Murzin VS, Sevost'yanov VN (1983) Neutrino generation during the flares at visible and invisible sides of the solar disk. In: Kolomeets EV, Gonchar GA, Lyakhova AK, Petrukhin AA, Chebakova EA (eds) Study of muons and neutrino in large water volumes. Kazakh State University, Alma- Ata, pp 24–33 (in Russian)
Erukhimova TL, Suvorov EV, Trakhtenherz VY (1990) High-frequency electromagnetic emission of auroral ionosphere. Geomagn Aeron 30(1):74–82
Evenson P, Meyer P, Yanagita S (1982) Solar flare shocks in interplanetary space and solar flare particle events. J Geophys Res 87:625–631
Evenson P, Meyer P, Pyle KR (1983) Protons from the decay of solar flare neutrons. Astrophys J 274:875–882
Evenson P, Meyer P, Yanagita S, Forrest DJ (1984) Electron-rich particle events and the production of gamma-rays by solar flares. Astrophys J 283:439–449

Evenson P, Kroeger R, Meyer P, Reames D (1990) Solar neutron decay proton observations in cycle 21. Astrophys J Suppl 73:273–277
Falcone A, Ryan J (1999) MILAGRO as a solar observatory. Astropart Phys 11:283–285
Fedorov YuI (1995) Propagation of solar cosmic rays at anisotropic initial angle distribution. Kinemat Phys Celest Bodies, Kiev, 11(5):34–45
Fedorov YI (1997) Anisotropy of solar cosmic rays at different ejection regimes of energetic particles. Geomagn Aeron 37(6):1–13
Fedorov YuI, Shakhov BA (1993) Solar cosmic rays in homogeneous regular magnetic field. In: Proceedings of 23rd international cosmic ray conference, vol 3. Calgary, pp 215–218
Fedorov YuI, Kyzhyurov YuV, Nosov SF, Shakhov BA (1995) Solar cosmic ray transport under anisotropic injection. In: Proceedings of 24th international cosmic ray conference, vol 4. Rome, pp 297–300
Fedorov YuI, StehlikM, Kudela K, Kassovicova J (1997) The GLE of May 24, 1990: kinetic approach to anisotropic event. In: Proceedings of 25th international cosmic ray conference, vol 1. Durban, pp 193–196
Fermi E (1949) On the origin of the cosmic radiation. Phys Rev 75(8):1169–1174
Feynman J (1997) Proton fluence prediction models. In: Heckman G, Marubashi K, Shea MA, Smart DF, Thompson R (eds) Solar-terrestrial prediction-V, proc. of a workshop at Hitachi, Japan, January 23–27, 1996. RWC Tokyo, Hiraiso Solar Terrestrial Research Center, Communications Research Laboratory, Hitachinaka, pp 457–469
Feynman J, Gabriel S (1990) A new model for calculation and prediction of solar proton fluences. In: 28th aerospace sciences meeting. AIAA papers, No.0292. Reno. Jan 8–11, 1990, pp 1–7
Feynman J, Armstrong TP, Dao-Gibner L, Silverman S (1990a) New interplanetary proton fluence model. J SpacecrRocket 27(4):403–410
Feynman J, Armstrong TP, Dao-Gibner L, Silverman S (1990b) Solar proton events during solar cycles 19, 20 and 21. Solar Phys 126(2):385–401
Feynman J, Spitale G, Wang J, Gabriel S (1993) Interplanetary proton fluence model: JPL 1991. J Geophys Res 98(A8):13281–13294
Fichtel CE, Guss DE, Ogilvie KW (1963) Details of individual solar particle events. In: McDonald FB (ed) Solar proton manual. NASA Rept. TR- 169. NASA, Washington, D.C. Sept 1963, pp 29–55
Filippov AT, Chirkov NP (1977) Spectrum of relativistic particles accelerated in the interplanetary medium. In: Proceedings of 15th international cosmic ray conference, vol 5. Plovdiv, pp 208–213
Filippov AT, Chirkov NP (1978) Acceleration of particles in the interplanetary medium up to relativistic energies. Izv (Bull) AN SSSR Phys Ser 42(5):1078–1081
Filippov AT, Krivoshapkin PA, Transky IA, Kuzmin AI, Krymsky GF, Niskovskikh AS, Berezhko EG (1991) Solar cosmic ray flare on September 29, 1989 by data of the Yakutsk Array Complex. In: Proceedings of 22nd international cosmic ray conference, vol 3. Dublin, pp 113–116
Firoz KA, Cho K-S, Hwang J et al (2010) Characteristics of ground-level enhancement- associated solar flares, coronal mass ejections, and solar energetic particles. J Geophys Res 115:A09105
Fisk LA (1978) ^{3}He-rich flares: a possible explanation. Astrophys J 224:1048–1055
Fisk LA (1979) The interactions of energetic particles with the solar wind. In: Parker EN, Kennel CF, Lanzerotti LJ (eds) Solar system plasma physics, vol 1. North Holland Pub Co., Amsterdam, p 177
Flückiger EO (1984) Effects of magnetospheric currents on cosmic rays. In: Kudela K, Pinter S (eds) Invited talks, 9th European cosmic ray symposium. Slovak Academy of Sciences, Institute of Experimental Physics, Kosice, Czechoslovakia, pp 105–119
Flückiger EO (1991) Solar cosmic rays. Nucl Phys B Proc Suppl 22:1–20
Flückiger EO (1994) Open questions in the analysis and interpretation of neutron monitor data. In: 14th European cosmic ray symposium, Balatonfured, Hungary, August 28–September 3, 1994. Symposium program and abstracts. Contributed paper 1-SH-13C. Balatonfured

Flückiger EO (2009) Ground level events and terrestrial effects (cutoffs, cosmic rays in the atmosphere, cosmogenic nuclides). In: Caballero R, D'Olivo JC, Medina-Tanco G, Valdés-Galicia JF (eds) Proceedings of the 30th international cosmic ray conference, vol 6. Universidad Nacional Autonóma de México, Mexico City, pp 239–253

Flückiger EO, Butikofer R, Chilingaryan A et al (in all 9 authors) (2001) Search for solar neutrons in association with the July 12 and July 14, 2000, solar flares. In: Proceedings of 27th international cosmic ray conference, vol 6. Hamburg, pp 3044–3047

Flückiger EO, Hofer MY (1998) Interplanetary shocks, magnetic clouds, and cosmic rays – Findings from an analysis of the 1991 March 24 Forbush decrease. Eos Trans AGU Spring Meet Suppl 79(17):S294 (abstract)

Flückiger EO, Kobel E (1990) Aspects of combining models of the Earth's internal and external magnetic field. J Geomagn Geolectr 42:1123–1136

Flückiger EO, Smart DF, Shea MA (1983) The effect of local perturbations of the geomagnetic field on cosmic ray cutoff rigidities at Jungfraujoch and Kiel. J Geophys Res 88 (A9):6961–6968

Flückiger EO, Smart DF, Shea MA (1986) A procedure for estimating the changes in cosmic ray cutoff rigidities and asymptotic directions at low and middle latitudes during periods of enhanced geomagnetic activity. J Geophys Res 91(A7):7925–7930

Flückiger EO, Butikofer R, Muraki Y, Matsubara Y, Koi T, Tsuchiya H, Hoshida T, Sako T, Sakai T (1998) A new solar neutron telescope at Gornergrat. In: Medina J (ed) Rayos cosmicos-98 (proc. 16th European Cosmic Ray Symposium Alcala University Press, Spain). Departamento de Física. Universidad de Alcalá, Alcalá de Henares, pp 219–222

Fomichev VV, Chertok IM (1985) Relation between γ-ray emission, radio bursts, and proton fluxes from solar flares. Sov Astron 29(5):554–559

Forbush SE (1946) Three unusual cosmic-ray intensity increases due to charged particles from the Sun. Phys Rev 70:771–772

Forbush SE, Stinchcomb TD, Schein M (1950) The extraordinary increase of cosmic- ray intensity on November 19, 1949. Phys Rev 79:501–504

Forman MA, Ramaty R, Zweibel E (1986) The acceleration and propagation of solar flare energetic particles: Ch.II. In: Sturrock PA (ed) Physics of the Sun. D. Reidel Publ. Co., Dordrecht, pp 249–289

Forrest DJ, Chupp EL (1983) Simultaneous acceleration of electrons and ions in solar flares. Nature 305(5932):291–292

Freier PS, Webber WR (1963) Exponential rigidity spectrum for solar-flare cosmic rays. J Geophys Res 68(6):1605–1629

Friedman M, Hamberger SM (1969) Plasma turbulence in solar flares as an explanation of some observed phenomena. Solar Phys 8(1):104–114

Friis-Christensen E, Lassen K (1991) Length of the solar cycle: an indicator of solar activity closely associated with climate. Science 254:698–700

Fry RJ, Nachtwey DS (1988) Radiation protection guidelines for space missions. Health Phys 55:159

Gabriel S, Feynman J (1996) Power-law distribution for solar energetic proton events. Solar Phys 165(2):337–346

Gabriel S, Evans R, Feynman J (1990) Periodicities in the occurrence rate of solar proton events. Solar Phys 128(2):415–422

Gall R, Orozco A, Marin C, Hurtado A, Vidargas G (1982) Tables of approach directions and points of entry of cosmic rays for higher latitude cosmic ray stations. Instituto de Geofisica, UNAM, Mexico, p 421

Gallegos-Cruz A, Perez-Peraza J (1995) Derivation of analytical particle spectra from the solution of the transport equation by the WKBJ method. Astrophys J 446(1):400–420

Gallegos-Cruz A, Perez-Peraza J, Miroshnichenko LI, Rodriguez-Frias MD, del Peral L, Vashenyuk EV (1995) Acceleration of solar articles by short wavelength turbulence. In: Proceeding of 24th international cosmic ray conference, vol 4. Rome, pp 14–17

Galper AM, Dmitrenko VV, Grachev YuV et al (in all 13 authors) (1996) Electrons with energy exceeding 10 MeV in the Earth's radiation belt. In: Lemaire JF, Heynderickx D, Baker DN (eds) Radiation belts: models and standards, AGU geophysical monograph 97, pp 169–180

Galper AM, Luchkov BI, Ozerov YuV, Khodarovich AM, Rinchinov SB, Kovaltsov GA, Usoskin IG (1997) Phases of charged particle acceleration during powerful solar gamma flares detected by the GAMMA-1. In: Proceeding of 25th international cosmic ray conference, vol 1. Durban, pp 169–172

Galper AM, Vavilov NR, Ozerov YuV et al (in all 18 authors) (2001) Direct measurements of characteristics of charged particles producing light flashes in the cosmonaut eyes with the SILEYE-2 apparatus on board the space station MIR. Izvestiya RAN Ser Phys. 65(3):397–399

Gan WQ (2004) On both the time histories and the 0.511 MeV line and 2.223 MeV line from the X4.8 flare of 23 July 2002 observed with RHESSI. Solar Phys 219:279–287

Gan W, Kuzhevskij BM, Miroshnichenko LI, Troitskaia EV (2003) Time profile of the 2.223 MeV line emission and some features of the 16 December 1988 solar event. In: Proceeding ISCS 2003 symposium, "Solar variability as an input to the earth's environment", Tatranska Lomnica, Slovakia, ESA SP-535, pp 655–657

Gan WQ, Chang J, Li YP, Su Y, Miroshnichenko LI (2004) On the positronium continuum and 0.511 MeV line in solar flares. Chin J Astron Astrophys 4(4):357–364

Gentile LC (1993a) Relativistic solar proton database for the ground level enhancements during solar cycle 22. J Geophys Res 98(A12):21107–21109

Gentile LC (1993b) Ground level enhancement database for solar cycle 22: an update. In: Proceeding of 23rd international cosmic ray conference, vol 3. Calgary, pp 63–66

Getselev IV, Gubar YI, Ermakov SI, Zosim LE, Kontor NN, Moszhukhina AR, Timofeev GA (1988) Long-term forecasting of solar flare proton fluences. Kosm Issled 26:949–953

Getselev IV, Ignatiev PP, Kabashova NA, Kontor NN, Moszhukhina AR, Timofeev GA, Khotilovskaya TG (1992) A model of radiation conditions during spacecraft flights in the interplanetary space and in the Earth's magnetosphere. Adv Space Res 12(2/3):441–444

Getselev IV, Morozova EI, Moszhukhina AR, Pisarenko NF, Polyanskaya MG, Timofeev GA (1996a) Long-term solar flare fluence forecasting. Radiat Meas 26(3):455–456

Getselev IV, Kontor NN, Pavlov NN (1996b) A database for solar cosmic ray measurements. Radiat Meas 26(3):457–459

Getselev IV, Podzolko MV, Veselovsky IS (2009) Optimization of the interplanetary energetic proton flux database and its application in modeling radiation conditions. Solar Syst Res 43 (2):136–142. doi:10.1134/S0038094609020063

Getselev IV, Podzolko MV, Okhlopkov VP (2013) Passive longitudes of solar cosmic ray sources. J Phys 409:012203. doi:10.1088/1742-6596/409/1/012203

Gingerich O, Noyes RW, Kalkofen W, Cuny Y (1971) The Harvard-Smithsonian reference atmosphere. Solar Phys 18:347–365

Giovanelli RG (1947) Magnetic and electric phenomena in the Sun's atmosphere associated with sunspots. Month Notices Royal Astron Soc 107(4):338–355

Gladysheva OG (1996) Production of nitrates during major solar flares. Geomagn Aeron 36 (5):147–150

Gladysheva OG, Kocharov GE (1995) Solar protons from August 1972 flare and nitrate abundance in Antarctic snow. In: Proceeding of 24th international cosmic ray conference, vol 4. Rome, pp 1126–1128

Gladysheva OG, Kocharov GE (1996) On a possibility to study the solar activity in the past. Izv RAN Ser Phys 60(8):121–124

Gladysheva OG, Iwasaka Y, Kocharov GE, Muraki Y (1995) Unique possibility to obtain upper limit of total energy induced by solar flare protons. In: Proceeding of 24th international cosmic ray conference, vol 4. Rome, pp 1129–1131

Gnevyshev MN (1977) Essential features of the 11-year solar cycle. Solar Phys 51(1):175–183

Goldstein ML, Matthaeus WH (1981) The role of magnetic helicity in cosmic ray transport theory. In: Proceedings of 17th international cosmic ray conference, vol 3. Paris, pp 294–297

Gombosi TI, Owens AJ (1981) Numerical study of solar flare particle propagation in the heliosphere. Adv Space Res 1(3):115–119

Gomez-Herrero R, del Peral L, Sequeiros J, Rodrigues-Frias MD, Medina J, Kunow H, Müller-Mellin R, Sierks H (1998) COSTEP observation of SPEs in November 1997. In: Medina J (ed) Rayos cosmicos-98 (Proceeding of 16th European cosmic ray symposium, Alcala University Press), pp 153–156

Gopalswamy N, Nitta NV (2012) Preface. Space Sci Rev 171:1–2

Gopalswamy N, Xie H, Yashiro S, Akiyama S, Mäkelä P, Usoskin IG (2012) Properties of ground level enhancement events and the associated solar eruptions during solar cycle 23. Space Sci Rev 171:23–60

Gorchakov EV, Timofeyev GA, Morozova TI (1975) Energy dependence of the scattering mean free path. Geomagn Aeron 15(6):758 (English version)

Goriely S, Hilaire S, Koning AJ (2008) Improved predictions of nuclear reaction rates with the TALYS reaction code for astrophysical applications. Astron Astrophys 487:767–774

Gorney DJ (1989) Solar cycle effects on near-Earth plasmas and space systems. J Spacecr Rocket 26(6):428–438

Gorney DJ (1990) Solar cycle effects on the near-Earth space environment. Rev Geophysics 28 (A10):315–335

Gosling JT (1993) The solar flare myth. J Geophys Res 98(A11):18937–18949

Gosling JT (1994a) Correction to "The solar flare myth". J Geophys Res 99(A3):4259–4260

Gosling JT (1994b) The solar flare myth in solar-terrestrial physics. In: Solar system in space and time, vol 84, Geophysical monograph. AGU, Washington, D.C., pp 65–69

Gosling JT (1995) Reply. J Geophys Res 100(A3):3479–3480

Gosling JT, Hundhausen AJ (1995) Reply. Solar Phys 160:57–60

Goswami JN, Lal D, Macdougall JD (1979) Composition and energy spectra of solar flare heavy nuclei during the early history of the solar system. In: Proceeding of 16th international cosmic ray conference, vol 5. Kyoto, pp 116–121

Goswami JN, Jha R, Lal D, Reedy RC, McGuire RE (1983) Secular variations in solar flare proton fluxes. In: Proceeding of 18th international cosmic ray conference, vol 2. Bangalore, pp 373–376

Goswami JN, McGuire RE, Reedy RC, Lal D, Jha J (1988) Solar flare protons and alpha particles during the last three solar cycles. J Geophys Res 93(A7):7195–7205

Grevesse N, Sauval AJ (1998) Standard solar composition. Space Sci Rev 85:161–174

Grigoriev YG (1975) Radiation safety of space flights (radiobiological aspects). Atomizdat, Moscow, p 256

Gubarev VS (2004) Belyi arkhipelag Stalina, (Stalin White Archipelago). Molodaya Gvardiya, Moscow

Guglenko VG, Kocharov GE, Kovaltsov GA, Kocharov LG, Mandzhavidze NZ (1990a) Generation of high-energy neutral radiation in flare loops, II. Monte Carlo simulation and comparison with observations. Solar Phys 125(1):91–123

Guglenko VG, Efimov YE, Kocharov GE, Kovaltsov GA, Mandzhavidze NZ, Terekhov MM, Kocharov LG (1990b) Neutron and gamma-ray emission on 1982 June 3: the possibility of fitting the data on the assumption of one population of accelerated particles. Astrophys J Suppl 73:209–211

Gussenhoven MS, Brautigam DH, Mullen EG (1988) Characterizing solar flare high energy particles in near-Earth orbits. IEEE Trans Nucl Sci 35(6):1412–1419

Guzik TG, Chen J, Wefel JP (1995) Temporal variations of helium energy spectra and 3He/^{4}He ratios for March and June 1991 SEP events. In: Proceeding of 24th international cosmic ray conference, vol 4. Rome, pp 123–126

Haisch BM (1989) An overview of solar and stellar flare research. Solar Phys 121:3–18

Haisch BM, Strong KT, Rodoño M (1991) Flares on the Sun and other stars. Ann Rev Astron Astrophys 29:275–324

Hakura Y (1974) Solar cycle variation in energetic particle emissivity of the Sun. Solar Phys 39(2):493–497

Hamilton DC (1988) The radial dependence of the solar energetic particle flux. In: Feynman J, Gabriel S (eds) Proceedings of the JPL workshop in the interplanetary charge particle environment, vol 1. NASA JPL Publication Jet Propulsion Laboratory, Pasadena, p 86

Hamilton RJ, Petrosian V (1992) Stochastic acceleration of electrons, I, Effects of collisions in solar flares. Astrophys J 398:350–358

Harrison RA (1996) Coronal magnetic storms: a new perspective on flares and the "solar flare myth" debate. Solar Phys 166(2):441–444

Hasselmann K, Wibberenz G (1968) Scattering of charge particles by random electromagnetic fields. Z Geophys 34:353–388

Hasselmann K, Wibberenz G (1970) A note on the parallel diffusion coefficient. Astrophys J 162:1049–1051

Hatton CJ (1971) Chapter 1: The neutron monitor. In: Wilson JG, Wouthuysen SA (eds) Progress in elementary particles and cosmic ray physics, vol X. North-Holland Publ. Co, Amsterdam, pp 3–100

Hatzky R, Wibberenz G (1995) Pitch angle distribution of solar energetic particles (SEPs) and the transport parameters in interplanetary space: II. Local scattering properties. In: Proceeding of 24th international cosmic ray conference, vol 4. Rome, pp 265–268

Hatzky R, Wibberenz G, Bieber JW (1995) Pitch angle distribution of solar energetic particles (SEPs) and the transport parameters in interplanetary space: I. Properties of steady-state distributions. In: Proceeding of 24th international cosmic ray conference, vol 4. Rome, pp 261–264

Hatzky R, Kallenrode MB, Schmidt JM (1997) The effect of adiabatic deceleration on angular distribution of energetic particles. In: Proceeding of 25th international cosmic ray conference, vol 1. Durban, pp 245–248

Heath DF, Krueger AJ, Crutzen PJ (1977) Solar proton event: influence on stratospheric ozone. Science 197:886–889

Heckman GR (1997) Radiation working group report. In: Heckman G, Marubashi K, Shea MA, Smart DF, Thompson R (eds) Solar-terrestrial prediction-V. Proceeding of a workshop at Hitachi, RWC Tokyo, Hiraiso Solar Terrestrial Research Center, Communications Research Laboratory, Hitachinaka, Ibaraki, pp 18–21, 23–27 Jan 1996

Heckman GR, Kunches JM, Allen H (1992) Prediction and evaluation of solar particle events based on precursor information. Adv Space Res 12(2/3):313–320

Hedgecock PG (1975) Measurements of the interplanetary magnetic field in relation to the modulation of cosmic rays. Solar Phys 42(2):497–527

Heristchi D, Trottet G, Perez-Peraza J (1976) Upper cutoff of high energy solar protons. Solar Phys 49(2):151–175

Hey JS (1946) Solar radiations in the 4–6 m radio wave-length band. Nature 157(3976):47–48

Heyvaerts J, Priest ER, Rust DM (1977) An emerging flux model for the solar flare phenomenon. Astrophys J 216:123–137

Hofer MY, Flückiger EO (1998) Cosmic ray spectral and directional variations near the Earth during the 24 March 1991 Forbush decrease. In: Medina J (ed) Rayos cosmicos-98 (Proceeding of 16th European cosmic ray symposium, Alcala University Press), pp 79–82

Holman GD (1996) Particle acceleration by DC electric fields in the impulsive phase of solar flares. In: Ramaty R, Mandzhavidze N, Hua X-M (eds) High energy solar physics, vol 374, AIP conference proceedings. AIP, New York, pp 479–492

Holman GD, Kundu MR, Kane SR (1989) Soft X-ray, microwave and hard X-ray emission from a solar flare: implications for electron heating and acceleration in current channels. Astrophys J 345(2):1050–1059

Holzworth RH, Mozer FS (1979) Direct evidence of solar flare modification of stratospheric electric fields. J Geophys Res 84:363–367

Holzworth RH, Norville KW, Williamson PR (1987) Solar flare perturbations in stratospheric current systems. Geophys Res Lett 14:852–855

Hrushka J, Shea MA, Smart DF, Heckman G, Hirman JW, Coles GL Jr (1993) Overview of the solar-terrestrial predictions workshop-IV. In: Hrushka J, Shea MA, Smart DF, Heckman G (eds) Solar-terrestrial prediction-IV, proceeding of a workshop at Ottawa, vol 1. NOAA ERL, Boulder, Co., pp 3–10, 18–22 May 1992

Hua X-M, Lingenfelter RE (1987a) Solar flare neutron and accelerated ion angular distributions. Astrophys J 323:779–794

Hua X-M, Lingenfelter RE (1987b) Solar flare neutron production and the angular dependence of the capture gamma-ray emission. Solar Phys 107:351–383

Hua X-M, Kozlovsky B, Lingenfelter RE, Ramaty R, Stupp A (2002) Angular and energy-dependent neutron emission from solar flare magnetic loops. Astrophys J Suppl 140:563–579

Huang YH, Wang XY (1995) Relativistic electron spectrum and plasma wake field acceleration in large solar flares. In: Proceedings of the 24th international cosmic ray conference, vol. 4. Rome, pp 5–8

Hudson HS (1978) Threshold effect in second-stage acceleration. Solar Phys 57(1):237–240

Hudson HS (1985) Ion energy storage for post-flare loops. In: Proceeding of 19th international cosmic ray conference, vol 4. La Jolla, pp 58–61

Hudson HS (1995) Solar flares: no "myth". EOS Trans AGU 76(41):405–406

Hudson HS (2010) Solar flares add up. Nat Phys 6:637–638.[DOI], [ADS]

Hudson HS, Cliver EW (2001) Observing coronal mass ejections without coronagraphs. J Geophys Res 106(A11):25,199–25,213 (2000JA004026)

Hudson HS, Haisch BM, Strong KT (1995) Comment on "The solar flare myth". J Geophys Res 100(A3):3473–3477

Humble JE, Duldig ML, Smart DF, Shea MA (1991a) Detection of 0.5-15 GeV solar protons on 29 September 1989 at Australian stations. Geophys Res Lett 18(4):737–740

Humble JE, Duldig ML, Smart DF, Shea MA (1991b) The 29 September 1989 event as observed by Australian stations. In: Proceedings of 22nd international cosmic ray conference, vol 3. Dublin, pp 109–112

Hundhausen J (1993) The sizes and locations of coronal mass ejections: SMM observations from 1980 and 1984–1989. J Geophys Res 98:13177–13200

Hurford GJ, Schwartz RA, Krucker S, Lin RP, Smith DM, Vilmer N (2003) First gamma-ray images of a solar flare. Astrophys J Lett 595:L77–L80

Hurford GJ, Krucker S, Lin RP, Schwartz RA, Share GH, Smith DM (2006) Gamma-ray imaging of the 2003 October/November solar flares. Astrophys J Lett 644:L93–L96

Huston SL, Kuck GA (1990) An early warning prediction scheme for solar proton events. In: 20th intersociety conference on environmental systems, SAE Technical Paper Series, No.901348, Williamsburg, pp 1–6, 9–12 July 1990

Ilencik J (1979) Analysis of solar cosmic ray increases during the 20th cycle of solar activity. Candidate dissertation, Ustav Experimentalnei Fyziky SAV, Kosice

Ilencik J, Dubinsky J, Miroshnichenko LI (1978) Absolute spectra of solar cosmic rays for the Ground Level Events of 1966–1976, Physica Solariterrestris, Potsdam, No.8, pp 11–20

Ishkov VN (1998) Emerging magnetic fluxes – a key to the forecast of major solar flares. Izv RAN Ser Phys 62(9):1853–1857

Ishkov VN (1999) Forecast of geoeffective solar flares: potentialities and limitations. Izv RAN Ser Phys 63(11):2148–2151

Ivanov KG, Harshiladze AF, Eroshenko EG, Styazhkin VA (1989) Configuration, structure, and dynamics of magnetic clouds from solar flares in light of measurements on board VEGA 1 and VEGA 2 in January- February 1986. Solar Phys 120(2):407–419

Jackman CH, Frederick JE, Stolarski RS (1980) Production of odd nitrogen in the stratosphere and mesosphere: an intercomparison of source strengths. J Geophys Res 85:7495–7505

Jokipii JR (1966) Cosmic ray propagation. I. Charged particles in a random magnetic field. Astrophys J 146(2):480–487

Jokipii JR (1971a) Deceleration and acceleration of cosmic rays in the solar wind. Phys Rev Lett 26(11):666–669
Jokipii JR (1971b) Propagation of cosmic rays in the solar wind. Rev Geophys Space Phys 9 (1):27–87
Jokipii JR, Kota J (1997) Galactic and anomalous cosmic rays in the heliosphere. In: Proceeding of 25th international cosmic ray conference, Invited, Rapporteur, and Highlight Papers, vol 8. Durban, pp 151–174
Jokipii JR, Levy EH, Hubbard WB (1977) Effects of particle drift on cosmic-ray transport. I. General properties, application to solar modulation. Astrophys J 213:861–868
Kahler SW (1982) The role of the big flare syndrome in correlation of solar energetic proton fluxes and associated microwave bursts parameters. J Geophys Res 87(A5):3439–3448
Kahler SW (1992) Solar flares and coronal mass ejections. Ann Rev Astron Astrophys 30:113–141
Kahler SW (1993) Coronal mass ejections and long risetimes of solar energetic particle events. J Geophys Res 98(A4):5607–5615
Kahler SW (1994) Injection profiles of solar energetic particles as functions of coronal mass ejection heights. Astrophys J 428:837–842
Kahler SW (1996) Coronal mass ejections and solar energetic particle events. In: Ramaty R, Mandzhavidze N, Hua X-M (eds) High energy solar physics, vol 374, AIP conference proceedings. AIP, New York, pp 61–77
Kahler SW (2001) Origin and properties of solar energetic particles in space. In: Song P, Singer HJ, Siscoe GL (eds) Space weather, vol 125, Geophysical monograph. AGU, Washington, D. C, pp 109–122
Kahler SW, Reames DV (1991) Probing the magnetic topologies of magnetic clouds by means of solar energetic particles. J Geophys Res 96(A6):9419–9424
Kahler SW, Reames DV, Sheeley NR Jr (2001) Coronal mass ejections associated with impulsive solar energetic particle events. Astrophys J 562:558–565
Kahler SW, Shea MA (1997) A search for magnetic helicity effects in rise times of Ground Level Events. Proceedings of 25th international cosmic ray conference, vol 1. Durban, pp 197–200
Kahler SW, Sheeley NR Jr, Howard RA, Koomen MJ, Michels DJ, McGuire RE, von Rosenvinge TT, Reames DV (1984) Associations between coronal mass ejections and solar energetic proton events. J Geophys Res 89(A11):9683–9693
Kahler SW, Cliver EW, Cane HV, McGuire RE, Stone RG, Sheeley NR Jr (1985) Energetic protons from a disappearing solar filament. In: Proceeding of 19th international cosmic ray conference, vol 4. La Jolla, pp 94–97
Kahler SW, Reames DV, Sheeley NR Jr (1990) Coronal mass ejections and injection profiles of solar energetic particle events. In: Proceeding of 21st international cosmic ray conference, vol 5. Adelaide, pp 183–186
Kahler SW, Shea MA, Smart DF, Cliver EW (1991) Ground level events from impulsive flares. In: Proceeding of 22nd international cosmic ray conference, vol 3. Dublin, pp 21–24
Kahler SW, Kunches J, Smith DF (1995) Coronal and interplanetary magnetic sector structure and the modulation of solar energetic particle events. In: Proceeding of 24th international cosmic ray conference, vol 4. Rome, pp 325–328
Kahler SW, Smart DF, Sauer H (1997) Temporal variation of shock-accelerated proton/alpha ratios at E $\geq$100 MeV/n in a large solar energetic particle event. In: Proceeding of 25th international cosmic ray conference, vol 1. Durban, pp 305–308
Kahler SW, Simnett GM, Reiner MJ (2003) Onsets of solar cycle 23 ground level events as probes of solar energetic particle injections at the Sun. In: Proceeding of 28th international cosmic ray conference, vol 6. Tsukuba, pp 3415–3418
Kahler SW, Cliver EW, Tylka AJ, Dietrich WF (2012) A comparison of ground level event e/p and Fe/O ratios with associated solar flare and CME characteristics. Space Sci Rev 171:121–139
Kai K et al (1983) Radio observations of small activity prior to main energy release in solar flares. Proc Astron Soc Japan 35:285–287

Kallenrode M-B (1993a) Shocks as mechanism for the acceleration and propagation of energetic solar particles. Adv Space Res 13(9):341–350
Kallenrode M-B (1993b) Neutral lines and azimuthal “transport” of solar energetic particles. J Geophys Res 98(A4):5573–5591
Kallenrode M-B (1993c) Particle propagation in the inner heliosphere. J Geophys Res 98 (A11):19037–19047
Kallenrode MB (2001) Charged particles, neutrals and neutrons, in Solar Encounter: the first solar orbiter workshop, Puerto de la Cruz, Tenerife, 14–18 May 2001, ESA SP- 493, Sep 2001, pp 23–34
Kallenrode M-B (2003) Current views on impulsive and gradual solar particle events. J Phys G Nucl Part Phys 29:1–17
Kallenrode M-B, Cliver EW (2001a) Rogue SEP events: observational aspects. In: Proceedings of 27th international cosmic ray conference, vol 8. Hamburg, pp 3314–3317
Kallenrode M-B, Cliver EW (2001b) Rogue SEP events: modeling. In: Proceedings 27th international cosmic ray conference, vol 8. Hamburg, pp 3318–3321
Kallenrode M-B, Cliver EW, Wibberenz G (1992a) Composition and azimuthal spread of solar energetic particles from impulsive and gradual flares. Astrophys J 391:370–379
Kallenrode M-B, Wibberenz G (1991) Particle injection following solar flares on 1980 May 28 and June 8: evidence for different injection time histories in impulsive and gradual events? Astrophys J 376:787–796
Kallenrode M-B, Wibberenz G (1993) Propagation and injection of particles accelerated at an interplanetary shock. In: Proceedings of 23rd international cosmic ray conference, vol 3. Calgary, pp 298–301
Kallenrode M-B, Wibberenz G (1997) Propagation of particles injected from interplanetary shocks: a black box model and its consequences for acceleration theory and data interpretation. J Geophys Res 102(A10):22311–22334
Kallenrode M-B, Wibberenz G, Hucke S (1992b) Propagation of relativistic electrons in the inner heliosphere. Astrophys J 394(1):351–356
Kaminer NS (1967) On taking into account the barometric effect of the neutron component during cosmic ray flares. Geomagn Aeron 7(5):806–809
Kananen H, Tanskanen PJ, Gentile LC, Shea MA, Smart DF (1991) A quarter of a century of relativistic solar cosmic ray events recorded by the Oulu neutron monitor. In: Proceeding of 22nd international cosmic ray conference, vol 3. Dublin, pp 125–128
Kananen H, Kocharov LG, Kovaltsov GA, Tanskanen P, Usoskin IG (1995) The June 3, 1982 solar flare as a test for neutron monitor sensitivity for solar neutrons. In: Proceeding of 24th international cosmic ray conference, vol 4. Rome, pp 147–150.
Kananen H, Kovaltsov GA, Tanskanen P, Usoskin IG (1997) Direct deduction of the number of neutrons injected from the Sun on the basis of neutron monitor response. In: Proceeding of 25th international cosmic ray conference, vol 1. Durban, pp 49–52
Kanbach G, Bertsch DL, Fichtel CE et al (in all 19 authors) (1993) Detection of a long- duration solar gamma-ray flare on June 11, 1991 with EGRET on COMPTON-GRO. Astron Astrophys Suppl Ser. 97:349–353
Kane SR, Chupp EL, Forrest DJ, Share GH, Rieger E (1986) Rapid acceleration of energetic particles in the 1982 February 8 solar flare. Astrophys J Lett 300:L95–L98
Kane SR, Hurley K, McTiernan JM, Sommer M, Boer M, Niel N (1995) Energy release and dissipation during giant solar flares. Astrophys J Lett 446:L47–L50
Karpov SN, Miroshnichenko LI (2008) Method of additional fluctuations for search of weak signals. In: Proceednig of 30th international cosmic ray conference, vol 1. Merida, pp 241–244
Karpov SN, Miroshnichenko LI, Vashenyuk EV (1995a) On the possibility of observation of extremely high-energy particles by the Baksan Underground Scintillation Telescope during solar flares. Bull Acad Sci Russia Phys Ser 59:52–57 (in Russian)

Karpov SN, Miroshnichenko LI, Vashenyuk EV (1995b) Extremely high-energy solar protons by the data of Baksan Underground Scintillation Telescope'. In: Proceeding of 24th international cosmic ray conference, vol 4. Rome, pp 50–53

Karpov SN, Miroshnichenko LI, Vashenyuk EV (1997a) On a search for the extremely high-energy particles from the Sun by the Baksan Underground Scintillation Telescope data. In: Proceeding of 25th international cosmic ray conference, vol 1. Durban, pp 157–160

Karpov SN, Miroshnichenko LI, Vashenyuk EV (1997b) Intensity of protons with energy above 500 GeV during GLEs. In: Proceeding of 25th international cosmic ray conference, vol 1. Durban, pp 205–208

Karpov SN, Miroshnichenko LI, Vashenyuk EV (1997c) Intensity of protons with the energy above 500 GeV during solar flares. Bull (Izv) RAN Phys Ser 61(7):1466–1470

Karpov SN, Miroshnichenko LI, Vashenyuk EV (1998) Muon bursts at the Baksan Underground Scintillation Telescope during energetic solar phenomena. Nuovo Cimento C 21(5):551–573

Karpov SN, Miroshnichenko LI, Vashenyuk EV (2005) The muon bursts with energy $\geq$200 GeV during GLE events of 21–23 solar activity cycles. In: Proceedings of the 29th international cosmic ray conference, vol. 1. Pune, pp 197–200

Kartavykh Yu.Yu, Ostryakov VM, Stepanov IYu, Yoshimori M (1997) Simulation of energy spectra of ions from impulsive solar flares. In: Proceedings of the 25th international cosmic ray conference, vol. 1. Durban, pp 69–72

Kasatkina EA, Shumilov OI, Raspopov OM, Henriksen K (1992) Ozone "mini- holes" and location of the polar cap. In: Proceeding of 19th Annual European meeting on atmospheric studies by optical methods, IRF science report 209, Kiruna, pp 11–16

Kasatkina EA, Shumilov OI, Raspopov OM, Henriksen K (1998) Ozone "mini- holes" initiated by solar protons in the northern and southern polar caps. Geomagn Aeron 38(2):30–36

Kawakami S, Aikawa Y, Ikeda N et al (1999) Observation of cosmic ray modulation and possible detection of the solar flares with GRAPES III muon telescopes at Ooty. In: Proceedings of the 26th international cosmic ray conference, vol. 7. Salt Lake City, pp 171–174

Kazaryan MS, Kuzhevsky BM, Pisarenko NF (1979) Spectrum of protons in the flare of September 7, 1973. Geomagn Aeron 19(5):922–925

Kepicova O, Miroshnichenko LI, Stehlik M (1982) Analysis of solar cosmic ray increases in September 1977 based on ground-level data, vol 19. Physica Solariterrestris, Potsdam, pp 40–52

Kepko L, Spence H, Shea MA, Smart DF, Dreschhoff GAM (2008) Observations of impulsive nitrate enhancements associated with ground-level cosmic ray events 1–4 (1942–1949). In: Caballero R, D'Olivo JC, Medina-Tanco G, Nellen L, Sánchez FA, Valdés-Galicia JF (eds) Proceeding of 30th international cosmic ray conference Universidad Nacional Autónoma de México, vol 1. Mexico City, (SH), pp 729–732

Kepko L, Spence H, Smart DF, Shea MA (2009) Interhemispheric observations of impulsive nitrate enhancements associated with the four large ground-level solar cosmic ray events (1940–1950). J Atm Solar-Terr Phys 71:1840–1845

Kichatinov LL (1983) New mechanism for acceleration of cosmic particles in the presence of a reflective non-invariant turbulence (RNIT). Zh Eksp Teor Fiz (Pis'ma) 37(1):43–45

Kichigin GN, Miroshnichenko LI, Sidorov VI, Yazev SA (2010) Peculiarities of the major solar event of 23 July 2002: source model for energetic particles. In: Stepanov AV (ed) Solar and solar- terrestrial physics-2010. Pulkovo, Sankt Petersburg, pp 201–204

Kichigin GN, Miroshnichenko LI, Sidorov VI, Yazev SA (2014) Dynamics of accelerated ions in coronal loops and model of a gamma-ray source. ISSN 1063-780X. Plasma Phys Rep 40 (3):178–193

Kiener J, Gros M, Tatischeff V, Weidenspointner G (2006) Properties of the energetic particle distributions during the October 28, 2003 solar flare from *INTEGRAL*/SPI observations. Astron Astrophys 445:725–733

Kile JN, Cliver EW (1991) A search for the 154 day periodicity in the occurrence rate of solar flares using Ottawa 2.8 GHz burst data, 1955–1990. Astrophys J 370:442–448

King JH (1974) Solar proton fluences for 1977–1983 space missions. J Spacecr Rocket 11 (6):401–408
Kiplinger AL, Dennis BR, Emslie AG, Orwig LE (1983) Millisecond time variations in hard X-ray solar flares. Astrophys J Lett 265(2):L99–L104
Kiplinger AL, Dennis BR, Frost KJ, Orwig LE (1984) Fast variation in high energy X-rays from solar flares and their constraints on nonthermal models. Astrophys J Lett 287(2):L105–L108
Kiraly P, Wolfendale AW (1999) Long-term particle fluence distributions and short- term observations. In: Proceeding of 26th international cosmic ray conference, vol 6. Salt Lake City, pp 163–166
Kirillov AS, Balabin YuV, Vashenyuk EV, Fadel Kh, Miroshnichenko LI (2008) Effect of solar protons on the middle atmosphere composition during the GLE of 13 December 2006. In: Proceeding of 30th international cosmic ray conference, vol 2. Merida, Yucatan, pp 129–132
Kirsch E, Wilken B, Mall U, Gloeckler G, Galvin AB, Cierpka K (1997) Corotating interacting regions observed by the particle experiment SMS on the WIND-S/C (E = 0.5-226 keV/e). In: Proceeding of 25th international cosmic ray conference, vol 1. Durban, pp 309–312
Klecker B (2009) Sun, Corona, and Transient Phenomena in the Heliosphere. In: Caballero R, D'Olivo JC, Medina-Tanco G, Valdés-Galicia JF (eds) Proceedings of the 30th international cosmic ray conference, vol 6. Universidad Nacional Autonóma de México, Mexico City, pp 225–237
Klecker B, Hovestadt D, Gloeckler G, Ipavich FM, Scholer M, Fan CY, Fisk LA (1984) Direct determination of the ionic charge distribution of helium and iron in Fe- rich solar energetic particle events. Astrophys J 281:458–462
Klecker B, Rieger E, Hovestadt D, Forrest DJ, Cliver EW, Dröge W (1990a) A correlative study of solar gamma-ray continuum bursts and interplanetary electron events. In: Proceeding of 21st international cosmic ray conference, vol 5. Adelaide, pp 80–83
Klecker B, Cliver E, Kahler S, Cane H (1990b) Particle acceleration in solar flares. EOS Trans AGU 71(39):1102–1103
Klein LW, Burlaga LF (1982) Magnetic clouds at 1 AU. J Geophys Res 87(A2):613–624
Klein K-L, Chupp EL, Trottet G, Magun A, Dunphy PP, Rieger E, Urpo S (1999) Flare-associated energetic particles in the corona and at 1 AU. Astron Astrophys 348:271–285
Klein KL, Trottet G, Benz A, Kane SR (1988) Extended coronal shock waves and electron acceleration during solar flares. In: Proceedings of the Joint Varenna-Abastumani International School and workshop on plasma astrophysics, vol. 1, ESA SP-285. Varenna, pp 157–160
Klein KL, Trottet G, Chupp EL, Dunphy PP, Rieger E (2000) Temporal evolution of a gradual SEP event and the possible role of coronal particle acceleration. In: Ramaty R, Mandzhavidze N (eds) High energy solar physics: anticipating HESSI, ASP conference series, vol 206. pp 112–117
Klein KL, Fuller N, Steigies CT (for the NMDB Team) (2009) The real-time neutron monitor database. http://www.nmdb.eu/
Klimenko VV, Morozova EI, Mikryukova NA, Pisarenko NF, Shafer YG (1982) On efficiency of the coronal transport of protons and electron fluxes through the boundary of unipolar magnetic region. In: Problems of space physics and aeronomy, Bulletin of scientific and technological information. IKFIA, Yakutsk, pp 12–16
Kocharov GE (1978) Opening address at the all-union cosmic ray conference (Yakutsk, 1977). Izv AN SSSR Ser Phys 42(5):898–899
Kocharov GE (1980) Nuclear astrophysics of the Sun, vol 89. Verlag Karl Thiemig-Taschenbucher-Band, Munchen
Kocharov GE (1982) Cosmic radiation bursts and cosmogenic isotopes. The 12th seminar on cosmophysics in Leningrad, L, pp 203–207
Kocharov GE (1983) Solar gamma-quanta and neutrons. Izv AN SSSR Ser Phys 47(9):1716–1737
Kocharov GE (1988) Nuclear processes in the solar atmosphere and the particle- acceleration problem. Sov Sci Rev Sect E Astrophys Space Phys Rev 6(4):313–424
Kocharov GE (1991a) Cosmic rays in the past. Nucl Phys B (Proc Suppl) 22B:153–164

Kocharov GE (1991b) Cosmogenic nuclei, solar neutrinos, neutrons and gamma-rays. In: Proceedings of 22nd international cosmic ray conference, invited and rapporteur papers, vol 5. Dublin, pp 344–347

Kocharov LG (1997) Adiabatic deceleration of solar energetic particles as deduced from Monte Carlo simulations of the focused transport. In: Proceeding of 25th international cosmic ray conference, vol 1. Durban, pp 241–244

Kocharov GE, Kovaltsov GA, Usoskin IG (1990) Solar thermonuclear and flare neutrinos. In: Kocharov GE (ed) Neutrino astrophysics. Physical and Technical Institute, Leningrad, pp 5–44 (in Russian)

Kocharov GE, Kovaltsov GA, Usoskin IG (1991) Solar flare neutrinos. In: Proceeding of 22nd international cosmic ray conference, vol 3. Dublin, pp 752–755

Kocharov GE, Kocharov LG, Kovaltsov GA et al (in all 8 authors) (1993) Evidence for extended neutron and gamma-ray generation during two solar flares. In: Proceeding of 23rd international cosmic ray conference, vol 3. Calgary, pp 107–110

Kocharov LG, Kovaltsov GA, Torsti J, Usoskin IG, Zirin H (1995a) Correlative investigations of the 1990 May 24 solar flare. In: Proceeding of 24th international cosmic ray conference, vol 4. Rome, pp 159–162

Kocharov LG, Kovaltsov GA, Torsti J, Vainio R (1995b) Search for neutron decay protons from the 1990 May 24 solar flare. In: Proceeding of 24th international cosmic ray conference, vol 4. Rome, pp 163–166

Kocharov LG, Kovaltsov GA, Kocharov GE, Chuikin EI, Usoskin IG, Shea MA, Smart DF, Melnikov VF, Podstrigach TS, Annstrong TP, Zhin H (1994) Electromagnetic and corpuscular emission from the solar flare of 1991 June 15: continuous acceleration of relativistic particles. Solar Phys 150:267–283

Kocharov L, Kovaltsov GA, Torsti J, Ostryakov VM (2000) Evaluation of solar energetic Fe charge states: effect of proton-impact ionization. Astron Astrophys 357:716–724

Kocharov L, Kovaltsov G, Torsti J, Usoskin I, Zirin H, Anttila A, Vainio R (1996a) The 1990 May 24 solar flare and cosmic ray event. In: Ramaty R, Mandzhavidze N, Hua X-M (eds) High energy solar physics, vol 374, AIP conference proceedings. AIP, New York, pp 246–255

Kocharov LG, Torsti J, Vainio R, Kovaltsov GA, Usoskin IG (1996b) A joint analysis of high-energy neutrons and neutron-decay protons. Solar Phys 169(1):181–207

Kocharov LG, Torsti J, Vainio R, Kovaltsov GA (1996c) Propagation of solar cosmic rays: diffusion versus focused diffusion. Solar Phys 165(1):205–208

Kocharov LG, Torsti J, Tang F, Zirin H, Kovaltsov GA, Usoskin IG (1997) Impact of magnetic environment on the generation of high-energy neutrons at the Sun. Solar Phys 172(2):271–278

Kocharov LG, Torsti J, Teittinen M, Laitinen T (1999) Post-impulsive phase acceleration of >10 MeV protons. In: Proceeding of 26th international cosmic ray conference, vol 6. Salt Lake City, pp 236–239

Kodama M, Murakami K, Wada M, Tanaka H (1977) Possibility of coronal propagation of relativistic solar protons. In: Proceeding of 15th international cosmic ray conference, vol 5. Plovdiv, pp 94–97

Kodama M, Kohno T, Kanzawa H (1992) Stratospheric sudden cooling after solar proton event over Syowa station, Antarctic. J Geomagn Geoelectr 44:361–368

Kohno T (1991) Solar proton intensity magnitudes of large flares on fall 1989 observed at synchronous orbit. In: Proceeding of 22nd international cosmic ray conference, vol 3. Dublin, pp 125–128

Kolomeets EV, Sevostyanov VN (1987) The propagation of solar cosmic rays in a high- velocity solar wind stream. Geomagn Aeron 27(1):10–14 (English version)

Kolomeets EV, Sevostyanov VN (1988) Solar cosmic rays drift effects in interplanetary magnetic field. Geomagn Aeron 28(3):360–367

Kolomeets EV, Sevostyanov VN, Lyubimov GP (1988) Motion of solar cosmic rays in the presence of complex structures in the interplanetary magnetic field. Geomagn Aeron 28 (5):830–832

Kolomeets EV, Sevostyanov VN, Chebakova EA, Seleznev KB (1991) Study of generation and propagation of high-energy solar cosmic rays in the event of September 29, 1989 by the data of worldwide network of neutron monitors. Izv AN Russia Ser Phys 55(10):1897–1900

Kolomeets EV, Pokudina TV, Chebakova EA (1993) Upper cutoff of energy spectrum of solar cosmic rays. Izv AN Russia Ser Phys 57(7):19–22

Kolomeets EV, Chebakova EA, Kolomeets LE (1995) High energy particle generation during the solar flares. In: Proceeding of 24th international cosmic ray conference, vol 4. Rome, pp 240–243

Kolomeets EV, Barsukov OA, Aitmukhambetov AA (1997) Radiation hazard for airplane and other aircraft. TOO "SANKOM", Alma-Ata, p 264

Koning AJ, Hilaire S, Duijvestijn M (2008) Nuclear data for science and technology. In: Bersillon O et al (ed) EDP Sciences, p 211

Korchak AA (1978) On the problem of power-law spectrum of particles accelerated in solar flares. Solar Phys 56:223–234

Korchak AA (1979) On the concept of "injection energy" in treating of particle acceleration in solar flares. Izv AN SSSR Phys Ser 43(4):724–729

Korn GA, Korn TM (1968) Mathematical handbook for scientists and engineers, Second, enlarged and revised edition, Sections 18.2-6, 18.4-5, McGrow-Hill Book Co, New York

Kota J, Jokipii JR (1997) Energy changes of particles moving along field lines. In: Proceeding of 25th international cosmic ray conference, vol 1. Durban, pp 213–216

Kotov YuD, Bogovalov SV, Ustinov PL, Yoshimori M (1995) Temporal evolution of X- and γ -rays during a flare on 27 October 1991. In: Proceeding of 24th international cosmic ray conference, vol 4. Rome, pp 106–109

Kovalenko VA (1994) Manifestations of fast global changes of solar magnetic fields in the heliospheric properties. In: Solar dynamic phenomena and solar wind consequences, 3rd SOHO workshop, Program and Abstracts, Estes Park, 26–29 Sep 1994

Kovaltsov GA (1981) Production of neutrinos in solar flares. Izv AN SSSR Ser Phys 45 (7):1151–1154

Kovaltsov GA, Usoskin IG (2014) Occurrence probability of large solar energetic particle events: assessment from data on cosmogenic radionuclides in lunar rocks. Solar Phys 289:211–220. doi:10.1007/s11207-013-0333-5

Kovaltsov GA, Efimov YE, Kocharov LG (1993) Ground-level energetic solar particle event on 24 May 1990: possibility of direct solar neutron detection. Solar Phys 144:195–198

Kovaltsov GA, Kocharov LG, Usoskin IG, Kananen H, Tanskanen P (1995) Two injections of high energy neutrons during the May 24, 1990 solar flare. In: Proceeding of 24th international cosmic ray conference, vol 4. Rome, pp 155–158

Kozlov VI, Krymsky GF, Kuzmin AI, Chirkov NI, Filippov AT (1974) Dynamical characteristics of short-term oscillations of cosmic rays. Izv (Bull) AN SSSR Phys Ser 38(9):1908–1911

Kozlovsky B, Murphy RJ, Share GH (2004) Positron-emitter production in solar flares from ^{3}He reactions. Astrophys J 604:892–899

Krasil'nikov DD, Kuz'min AI, Shafer YUG (1955) Cosmic ray intensity bursts. In: Variatsii intensivnosti kosmicheskikh luchei, (Cosmic ray intensity variations). Akad. Nauk SSSR, Moscow, pp 41–47

Kreinin LB, Grigorieva GM (1979) Solar cells in the conditions of impact of cosmic radiation, vol 13. VINITI, Moscow, p 128

Krimigis SM (1965) Interplanetary diffusion model for the time behaviour of intensity in a solar cosmic ray event. J Geophys Res 70(13):2943–2960

Krivonosov YI (2000) Manual writing of S.I. Vavilov to Stalin. Usp Fiz Nauk 170(9):1021–1024

Kropotkin AP (1996) Relativistic electron transport processes associated with magnetospheric substorms. Radiat Meas 26(3):343–345

Kryakunova ON, Chertok IM (1998) On the origin of solar protons of extremely high energies: signatures of the post-eruptive acceleration in the corona. Bull Acad Sci Russia Phys Ser 62 (12):1922–1925 (in Russian)

Krymsky GF (1977) A regular mechanism for the acceleration of charged particles on the front of a shock wave. Sov Phys Dokl 22:327–328

Krymsky GF, Kuzmin AI, Krivoshapkin PA, Samsonov IS, Skripin GV, Transky IA, Chirkov NP (1981). In: Shafer GV (ed) Cosmic rays and solar wind, Nauka, Novosibirsk, p 224 (in Russian)

Krymsky GF, Grigor'ev VG, Starodubtsev SA (2008) New method for estimating the absolute flux and energy spectrum of solar cosmic rays according to the neutron monitor data. Pis'ma Zh Eksp Teor Fiz 88(7):483–485

Krymsky GF, Kuzmin AI, Krivoshapkin PA, Starodubtsev SA, Transky IA, Filippov AT (1990) The exceptional cosmic ray burst of September 29, 1989, as indicated by data from the Yakutsk Array Complex. Trans (Doklady) USSR Acad Sci 314:20–22

Krymsky GF, Grigor'ev VG, Starodubtsev SA (2008) New method for estimating the absolute flux and energy spectrum of solar cosmic rays according to the neutron monitor data. Pis'ma Zh Eksp Teor Fiz 88(7):483–485

Kucera TA, Dennis BR, Schwartz RA, Shaw D (1997) Evidence for a cutoff in the frequency distribution of solar flares. Astrophys J 475:338–347

Kudela K (1990) A search for solar neutron response in neutron monitor data. Astrophys J Suppl 73:297–301

Kudela K, Langer R (2008) Ground Level Events recorded at Lomnicky Stit neutron monitor. In: Proceedings of 30th international cosmic ray conference, vol 1. Mérida, México, pp 205–208

Kudela K, Shea MA, Smart DF, Gentile LC (1993) Relativistic solar particle events recorded by the Lomnicky Stit neutron monitor. In: Proceeding 23rd international cosmic ray conference, vol 3. Calgary, pp 71–74

Kudo S, Wada M, Tanskanen P, Kodama M (1987) Local time and cutoff rigidity dependencies of storm time increase associated with geomagnetic storms. J Geophys Res 92:4719

Kulsrud RM, Cesarsky CJ (1971) The effectiveness of instabilities for the confinement of high-energy cosmic rays in the galactic disk. Astrophys Lett 8(4):189–191

Kumar P, Quataert EJ, Bahcall JN (1996) Observational searches for solar g-modes: some theoretical considerations. Astrophys J Lett 482(2):L83–L85

Kunches JM, Zwickl RD (1997) Delayed-onset solar energetic particle events. In: Heckman G, Marubashi K, Shea MA, Smart DF, Thompson R (eds) Solar-terrestrial prediction-V, proceeding of a workshop at Hitachi, RWC Tokyo, Hiraiso Solar Terrestrial Research Center, Communications Research Laboratory, Hitachinaka, Ibaraki, pp 453–456, 23–27 Jan 1996

Kunow H, Wibberenz G, Green G, Müller-Mellin R, Kallenrode MB (1990). In: Schwenn R, Marsch E (eds) Physics of the inner heliosphere. Springer-Verlag, Berlin, p 243

Kurganov IG, Ostryakov VM (1992) Time trend of the decay protons for the solar events of June 21, 1980, June 3, 1982, and April 24, 1984. Geomagn Aeron 32(3):434–437 (in English)

Kurt VG (1989) Electrons and X-ray emission from solar flares. In: Somov BV, Fomichev VV (eds) Physics of solar plasma. Nauka, IZMIRAN, Moscow, pp 127–162

Kurt VG (1990) Electrons and X-ray emission from solar flares. In: Priest ER, Krishan V (eds) Basic plasma processes on the sun, IAU Symp. No.142, Kluwer Academic Publishers, pp 409–413

Kurt VG, Logachev YI, Pisarenko NF (1976) On the scattering mechanism of non- relativistic electrons in the interplanetary medium. Kosm Issled 14(3):378–382

Kurt VG, Logachev YuI, Stolpovsky VG, Daibog EI (1981) Energetic solar particle spectra according to Venera-11, 12 and Prognoz-5, 6 observations. In: Proceeding of 17th international cosmic ray conference, vol 3. Paris, pp 69–72

Kurt V, Akimov VV, Leikov NG (1995) What can be learnt from H-α, microwave and high energy gamma emission of the March 26, 1991 solar flare? In: Proceeding of 24th international cosmic ray conference, vol 4. Rome, pp 18–21

Kurt VG, Yushkov BY, Belov AV (2010a) On the onset of ground-based registration of solar high-energy protons. Astron Rep 36(7):520–530

Kurt VG, Yushkov BY, Kudela K, Galkin VI (2010b) High-energy gamma-ray emission of solar flares as an indicator of acceleration of high-energy protons. Cosmic Res 48(1):70–79

Kuzhevskij BM (1985) Nuclear processes in solar atmosphere and solar cosmic radiation. Energoatomizdat, Moscow (in Russian)

Kuzhevskij BM, Kuznetsov SN, Troitskaia EV (1998) Development of the solar flare plasma density investigation method based on the 2.2 MeV gamma-line time profile analysis. Adv Space Res 22:1141–1147

Kuzhevskij BM, Miroshnichenko LI, Troitskaia EV (2005a) Gamma-ray radiation with energy of 2.223 MeV and the density distribution in the solar atmosphere during flares. Astron Rep 49 (7):566–577

Kuzhevskij BM, Gan WQ, Miroshnichenko LI (2005b) The role of nuclei-nuclei interactions in the production of gamma-ray lines in solar flares. Chin J Astron Astrophys 5(3):295–301

Kuzmin AI, Filippov AT, Chirkov NP (1983) Large-scale disturbances of solar wind and cosmic ray acceleration in interplanetary space. Izv AN SSSR Phys Ser 47:1703–1706

Kuznetsov SN, Kurt VG (1991) Flare-generated protons of different energies. In: Schmieder B, Priest ER (eds) Flares 22 workshop, dynamics of solar flares. Observatoire de Paris, Paris, pp 63–64

Kuznetsov SN, Kurt VG, Yushkov BY, Kudela K, Galkin VI (2011) Gamma-ray and high-energy-neutron measurements on *CORONAS-F* during the solar flare of 28 October 2003. Solar Phys 268:175–193

Labitzke K, van Loon H (1990) Associations between the 11-year solar cycle, quasi- biennial oscillations and the atmosphere: a summary of recent work. Phil Trans Roy Soc London A330:557–560

Laird CM, Zeller EJ, Dreschhoff GAM, Armstrong TP (1987) Solar and galactic cosmic ray production of atmospheric nitrate. In: Proceeding of 20th international cosmic ray conference, vol 4. Moscow, pp 296–298

Lange I, Forbush SE (1942) Further note on the effect on cosmic-ray intensity of the magnetic storm of March 1, 1942. Terr Magn Atm Electr 47(4):331–334

Lanzerotti LJ (1969) Solar proton radiation damage of solar cells at synchronous altitudes. J Spacecr Rocket 6:1086

Lanzerotti LJ, Maurer DW, Sauer HH, Zwickl RD (1991) Large solar proton events and geosynchronous communication spacecraft solar arrays. J Spacecr Rocket 28(5):614–616

Lanzerotti LJ, Medford LV, MacLennan CG, Thompson DJ (1995) Studies of large-scale Earth potentials across oceanic distances. AT&T Techn J 74(3):73–84

Lario D, Aran A, Decker RB (2008) Major solar energetic particle events of solar cycles 22 and 23: intensities above the streaming limit. Space Weather 6:S12001. doi:10.1029/2008SW000403

Lario D, Aran A, Decker RB (2009) Major solar energetic particle events of solar cycles 22 and 23: intensities close to the streaming limit. Solar Phys 260:407–421. doi:10.1007/s11207-009-9463-1

Lario D, Marsden RG, Sanderson TR et al (2000) Energetic proton observations at 1 AU and 5 AU. I: January–September 1997. J Geophys Res 105:18235–18250

LaRosa TN, Moore RL (1993) A mechanism for bulk energization in the impulsive phase of solar flare: MHD turbulent cascade. Astrophys J 418:912

Lau Y-T, Northrop TG, Finn JM (1993) Long-term containment of energetic particles in coronal loops. Astrophys J 414:908–915

Lean JL (1990) Evolution of the 155 day periodicity in sunspot areas during solar cycles 12 to 21. Astrophys J 363(2):718–727

Lee MA (1991) Solar particle production. In: Proceedings of the 22nd international cosmic ray conference, vol. 5. Dublin, pp 293–312

Lee MA, Ryan JM (1986) Time-dependent coronal shock acceleration of energetic solar flare particles. Astrophys J 303:829–842

Leikov NG, Akimov VV, Volzhenskaya VA et al (in all 17 authors) (1993) Spectral characteristics of high energy gamma ray solar flares. Astron Astrophys Suppl. 97:345–348

Lepping RP, Burlaga LF, Tsurutani BT, Ogilvie KV, Lazarus AJ, Evans DS, Klein LW (1991) The interaction of a very large interplanetary magnetic cloud with the magnetosphere and with cosmic rays. J Geophys Res 96(A6):9425–9438

Leske RA, Cummings JR, Mewaldt RA, Stone EC, von Rosenvinge TT (1995a) Measurements of the ionic charge states of solar energetic particles at 15–17 MeV/nucleon. In: Proceeding of 24th international cosmic ray conference, vol 4. Rome, pp 461–464

Leske RA, Cummings JR, Mewaldt RA, Stone EC, von Rosenvinge TT (1995b) Measurements of the ionic charge states of solar energetic particles using the geomagnetic field. Astrophys J Lett 452:L149–L152

Leske RA, Cummings JR, Mewaldt RA, Stone EC, von Rosenvinge TT (1996) Measurements of the ionic charge states of solar energetic particles at 15–17 MeV/nucleon using the geomagnetic field. In: Ramaty R, Mandzhavidze N, Hua X-M (eds) High energy solar physics, vol 374, AIP conference proceedings. AIP, New York, pp 86–95

Letaw JR, Silberberg R, Tsao CH (1987) Radiation hazard on space missions. Nature 330 (6150):709–710

Lett J, Atwell W, Golightly M (1990) Radiation hazard to humans in deep space: a summary with a special reference to large solar particle events. In: Thompson RJ, Cole DG, Wilkinson PJ, Shea MA, Smart DF, Heckman GH (eds) Solar-terrestrial predictions: proceeding of a workshop at Leura, NOAA ERL, Boulder, Co., vol 1. pp 140–153, 16–20 Oct 1989

Li C, Firoz KA, Sun LP, Miroshnichenko LI (2013) Electron and proton acceleration during the first GLE event of solar cycle 24. Astrophys J 770:34 (11 pp). doi:10.1088/0004-637X/770/1/34

Li P, Hurley K (1995) On the beaming of solar flare energetic photons. In: Proceeding of 24th international cosmic ray conference, vol 4. Rome, pp 110–113

Lin RP (1970) The emission and propagation of ~40 keV solar flare electrons. Solar Phys 12:266–303

Lin RP (1980) Energetic particles in space. Solar Phys 67:393–399

Lin RP, Johns-Krull CM (1996) Inferring the accelerated electron spectrum in solar flares. In: Ramaty R, Mandzhavidze N, Hua X-M (eds) High energy solar physics, vol 374, AIP conference proceedings. AIP, New York, pp 320–328

Lin RP, Mewaldt RA, van Hollebeke MAI (1982) The energy spectrum of 20 keV -20 MeV electrons accelerated in large solar flares. Astrophys J 253:949–962

Lingenfelter RE, Hudson HS (1980) Solar particle fluxes and the ancient Sun. In: Pepin RO, Eddy JA, Merril RB (eds) Proceeding conference ancient sun. Pergamon Press, New York, pp 69–79

Little CG, Leinbach H (1959) The riometer – a device for the continuous measurements of ionospheric absorption. Proc Ins Radio Eng 47(2):315–320

Litvinenko YE (1996a) Particle acceleration in reconnecting current sheet with a nonzero magnetic field. Astrophys J 462:997–1004

Litvinenko YE (1996b) A new model for the distribution of flare energies. Solar Phys 167:321–331

Litvinenko YE (1998) Dimensional analysis of the flare distribution problem. Solar Phys 180:393–396

Litvinenko YE, Somov BV (1993) Particle acceleration in reconnecting current sheets. Solar Phys 146(1):127–133

Litvinenko YE, Somov BV (1995) Relativistic acceleration of protons in reconnecting current sheets. Solar Phys 158:317–330

Lockwood JA, Webber WR, Hsieh L (1974) Solar flare proton rigidity spectra deduced from cosmic ray neutron monitor observations. J Geophys Res 79(28):4149–4185

Lockwood JA, Debrunner H, Flückiger EO, Grädel H (1990a) Proton energy spectra at the Sun in the solar cosmic-ray events on 1978 May 7 and 1984 February 16. Astrophys J 355:287–294

Lockwood JA, Debrunner H, Flückiger EO (1990b) Indications for diffusive shock acceleration of protons in selected solar cosmic ray events. J Geophys Res 95(A4):4187–4201

Lockwood JA, Debrunner H, Ryan JM (1997) The relationship between solar flare gamma-ray emission and neutron production. Solar Phys 173(1):151–176

Logachev YuI, Bazilevskaya GA, Vashenyuk EV, Daibog EI, Ishkov VN, Miroshnichenko LI, Nazarova MN, Petrenko IE, Stupishin AG, Surova GM, Yakovchuk OS (2014) SINP MSU, Catalogue of SEP events of 1997–2009. http://kosmofizika.ru/katalog/katalog.htm

Loginov VF, Sazonov BI (1978) Cosmic rays and tropospheric circulation. Phys Solariterr. (9):85–92

Lovell JL, Duldig ML, Humble JE (1998) An extended analysis of the September 1989 cosmic ray ground level enhancement. J Geophys Res 103(A10):23733–23742

Lu EW, Hamilton RJ (1991) Avalanches and the distribution of solar flares. Astrophys J Lett 380: L89–L92

Lu EW, Hamilton RJ, McTiernan J, Bromund K (1993) Solar flares and avalanches in driven dissipative systems. Astrophys J 412:841–851

Lubimov GP (1988) Reflection model for movement of solar cosmic rays in loop traps. Astronomichesky Circ (1531):19–20

Luhn A, Klecker B, Hovestadt D, Möbius E (1987) The mean ionic charge of silicon in ^{3}He-rich solar flares. Astrophys J 317:951–955

Maeda K, Heath DF, Aruga T (1984) North–south asymmetries of solar particle events in upper atmospheric ozone. Planet Space Sci 32(7):857–870

Makhmutov VS (1983) Investigation of energy spectra of solar cosmic rays at the energies above 100 MeV in the 21st cycle of solar activity. Candidate dissertation, Moscow State University, Institute of Nuclear Physics, Moscow, pp 179

Maksimov VP, Nefedyev VP (1991) The observation of a "negative burst" with high spatial resolution. Solar Phys 136:335–342

Malitson HH, Webber WR (1963) A summary of solar cosmic ray events. In: McDonald FB (ed) Solar proton manual. NASA, TR-R-169, Washington, D.C, pp 1–17

Mandzhavidze N (1994) Solar particles and processes. In: Leahy DA, Hicks RB, Venkatesan D (eds) Proceeding of 23rd international cosmic ray conference: invited, rapporteur and highlight papers, World Scientific Pub. Co. Pte. Ltd., Singapore, pp 157–184

Mandzhavidze N, Ramaty R (1992) Gamma rays from pion decay: evidence for long- term trapping of particles in solar flares. Astrophys J Lett 396:L111–L114

Mandzhavidze N, Ramaty R (1993) Particle acceleration in solar flares. Nucl Phys B Proc Suppl 33:141–160

Mandzhavidze N, Ramaty R (1996) Determination of element abundances in solar atmosphere and solar flare accelerated particles using gamma ray spectroscopy. Bull Am Astron Soc 28:858 (abstract)

Mandzhavidze N, Ramaty R, Akimov VV, Leikov NG (1993) Pion decay and nuclear line emission from the 1991 June 15 flare. In: Proceeding of 23rd international cosmic ray conference, vol 3. Calgary, pp 119–122

Mandzhavidze N, Ramaty R, Bertsch DL, Schneid EJ (1996) Pion decay and nuclear line emissions from the 1991 June 11 flare. In: Ramaty R, Mandzhavidze N, Hua X-M (eds) High energy solar physics, vol 374, AIP conference proceedings. AIP, New York, pp 225–236

Mandzhavidze N, Ramaty R, Kozlovsky B (1997) Solar atmospheric and solar flare accelerated helium abundances from gamma-ray spectroscopy. Astrophys J Lett 489:L99–L102

Marsden RG, Sanderson TR, Tranquille C, Wenzel K-P (1987) ISEE 3 observations of low-energy proton bi-directional events and their relations to isolated interplanetary structures. J Geophys Res 92(A10):11009–11019

Martens PCH (1988) The generation of proton beams in two-ribbon flares. Astrophys J Lett 330: L131–L133

Martens PCH, Kuin NPM (1989) A circuit model for filament eruption and two-ribbon flares. Solar Phys 12:263–302

Martens PCH, Young A (1990) Neutral beams in two-ribbon flares and in the geomagnetic tail. Astrophys J Suppl 73(2):333–342

Martinic NJ, Reguerín A, Palenque E, Taquichiri MA, Wada M, Inoue A, Takahashi K (1985) Search for solar neutrons using NM-64 equipment. In: Proceeding of 19th international cosmic ray conference, vol 4. La Jolla, pp 138–141

Mason GM, Gloeckler G, Hovestadt D (1984) Temporal variations of nucleonic abundances in solar flare energetic particle events. II. Evidence for large-scale shock acceleration. Astrophys J 280:902–916

Mason GM, Mazur JE, Looper MD, Mewaldt RA (1995a) Charge state measurements of solar energetic particles observed with SAMPEX. Astrophys J 452:901–911

Mason GM, Mazur JE, Looper DM, Mewaldt RA (1995b) Charge state measurements of solar energetic particles observed with SAMPEX. In: Proceeding of 24th international cosmic ray conference, vol 4. Rome, p 474 (abstract)

Mason GM, Mazur JE, Dwyer JR (2002) A new heavy ion abundance enrichment pattern in ^{3}He-rich solar particle events. Astrophys J 565:L51–L54

Ma Sung LS, van Hollebeke MAI, McDonald FB (1975) Propagation characteristics of solar flare particles. In: Proceeding of 14th international cosmic ray conference, vol 5. München, pp 1767–1772

Mathews T, Lanzerotti LJ (1973) Detection of relativistic solar particles before the Hα maximum of a solar flare. Nature 241(5388):335–338

Matsubara Y, Muraki Y, Sako T, Tajima N, Kakimoto F, Ogio S, Shirasaki Y, Murakami K, Kaneko T, Yoshii H, Martinic N, Miranda P, Ticona R, Velarde A (1997a) Detection efficiency of the Bolivian solar neutron detector. In: Proceeding of 25th international cosmic ray conference, vol 1. Durban, pp 61–64

Matsubara Y, Muraki Y, Sakakibara S, Koi T, Sako T, Okada A, Murata T, Imaida I, Tsuchiya H, Shibata S, Munakata Y, Tatsuoka R, Sakai H, Wakasa T, Nonaka T, Ohnishi T, Hatanaka K, Miyashida A, Nakagiri M, Okita K, Mizumoto Y (1997b) A new solar neutron telescope in Hawaii. In: Proceeding of 25th international cosmic ray conference, vol 1. Durban, pp 37–40

Matsubara Y, Muraki Y, Sako T, Watanabe K, Masuda S, Sakai T, Shibata S, Fl¨uckiger E, B¨utikofer R, Chilingarian A, Hovsepyan G, Tan YH, Yuda T, Ohnishi M, Tsuchiya H, Katayose Y, Ogasawara R, Mizumoto Y, Nakagiri M, Miyashita A, Velarde A, Ticona R, Martinic N (2005) Search for solar neutrons associated with proton flares in solar cycle 23. In: Proceeding of 29th international cosmic ray conference, vol 1. Pune, pp 17–20

Matthaeus WH, Goldstein ML (1981) Determination of magnetic helicity in the solar wind and implications for cosmic ray propagation. In: Proceeding of 17th international cosmic ray conference, vol 3. Paris, pp 291–293

Matthaeus WH, Goldstein ML (1982) Measurement of the rugged invariants of magnetohydrodynamic turbulence in the solar wind. J Geophys Res 87:6011–6028

Mavromichalaki H, Papaioannou A, Plainaki C, Sarlanis C, Souvatzoglou G, Gerontidou M, Papailiou M, Eroshenko E, Belov A, Yanke V, Flückiger EO, Bütikofer R, Parisi M, Storini M, Klein K-L, Fuller N, Steigies CT, Rother OM, Heber B, Wimmer-Schweingruber RF, Kudela K, Strharsky I, Langer R, Usoskin I, Ibragimov A, Chilingaryan A, Hovsepyan G, Reymers A, Yeghikyan A, Kryakunova O, Dryn E, Nikolayevskiy N, Dorman L, Pustil'nik L (2011) Applications and usage of the real-time Neutron Monitor Database. Adv Space Res 47:2210–2222

Mavromichalaki H, Souvatzoglou G, Sarlanis C, Mariatos G, Papaioannou A, Belov A, Eroshenko E, Yanke V, For the NMDB team (2010) Implementation of the ground level enhancement alert software at NMDB database. New Astronomy 15:744–748

Mazur JE, Mason GM, Klecker B (1995) Heavy ion acceleration beyond 10 MeV per nucleon in impulsive solar flares. Astrophys J Lett 448:L53–L56

Mazur JE, Mason GM, Looper MD, Leske RA, Mewaldt RA (1999) Charge states of solar energetic particles using the geomagnetic cutoff technique: SAMPEX measurements in the 6 November 1997 solar particle event. Geophys Res Lett 26:173–176

McConnel M, Bennett K, Bloemen H et al (in all 11 authors) (1997) COMPTEL all-sky imaging at 2.2 MeV. In: Proceeding of 25th international cosmic ray conference, vol 3. Durban, pp 93–96

McCracken KG (1962a) The cosmic-ray flare effect. 1. Some new methods of analysis. J Geophys Res 67(2):423–434

McCracken KG (1962b) The cosmic-ray flare effect. 2. The flare effects of May 4, November 12 and November 15, 1960. J Geophys Res 67(2):435–446

McCracken KG (1962c) The cosmic-ray flare effect. 3. Deductions regarding the interplenatary magnetic field. J Geophys Res 67(2):447

McCracken KG, Dreschhoff GAM, Zeller EJ, Smart DF, Shea MA (2001) Solar cosmic ray events for the period 1561–1994. 1. Identification in polar ice, 1561–1950. J Geophys Res 106 (A10):21585–21598

McCracken KG, Moraal H, Stoker PH (2008) Investigation of the multiple-component structure of the 20 January 2005 cosmic ray ground level enhancement. J Geophys Res 113:A12101. doi:10.1029/2007JA012829

McCracken KG, Moraal H, Shea MA (2012) The high-energy impulsive Ground-Level Enhancements. Astrophys J 761(101):12. doi:10.1088/0004-637X/761/2/101

McDonald FB (ed) (1963) Solar proton manual, NASA TR R-169. National Aeronautics and Space Administration, Washington, D.C

McGuire RE, von Rosenvinge TT, McDonald FB (1981) A survey of solar proton and alpha differential spectra between 1 and >400 MeV/nucleon. In: Proceeding of 17th international cosmic ray conference, vol 3. Paris, pp 65–68

McGuire RE, Goswami JN, Jha J, Lal L, Reedy RC (1983) Solar flare particle fluences during solar cycles 19, 20, and 21. In: Proceeding of 18th international cosmic ray conference, vol 4. Bangalore, pp 66–69

McGuire RE, von Rosenvinge TT, McDonald FB (1986) The composition of solar energetic particles. Astrophys J 301:938–961

McPeters RD, Jackman CH (1985) The response of ozone to solar proton events during solar cycle 21: The observations. J Geophys Res 90(D5):7945

Melendez-Venancio R, Mendoza B, Miroshnichenko LI, Perez-Enriquez R (1998) Distribution functions of solar proton events. In: Medina J (ed) Rayos cosmicos 98 (Proceeding of 16th European cosmic ray symposium, 20–24 July 1998, Departamento de Fisica, Universidad de Alcala), pp 149–152

Melrose DB (1990) Particle beams in the solar atmosphere. Solar Phys 130(1):3–18

Mewaldt RA, Looper MD, Cohen CMS et al (in all 9 authors) (2005a) Solar-particle energy spectra during the large events of October-November 2003 and January 2005. In: Acharya BS, Gupta SK, Jagadeesan P et al (eds) Proceeding of 29th international cosmic ray conference, Tata Institute of Fundamental Research, Pune, Mumbai, vol 1. pp 111–114

Mewaldt RA, Cohen CMS, Labrador AW et al (in all 10 authors) (2005b) Proton, helium, and electron spectra during the large solar particle events of October-November 2003. J Geophys Res. doi:10.1029/2005JA011038110, A09S18, No.9

Mewaldt RA, Cohen CMS, Haggerty DK, Mason GM, Looper ML, von Rosenvinge TT, Wiedenbeck ME (2007) Radiation risks from large solar energetic particle events. In: Turbulence and nonlinear processes in astrophysical plasmas; 6th annual international astrophysics conference. AIP conference proceedings, vol 932. pp 277–282

Meyer J-P (1985) The baseline composition of solar energetic particles. Astrophys J Suppl 57:151–171

Meyer J-P (1993) Elemental abundances in active regions, flares and interplanetary medium. Adv Space Res 13(9):377–390

Meyer J-P (1996) Accelerated particle composition in impulsive events: clues to the conditions of acceleration. In: Ramaty R, Mandzhavidze N, Hua X-M (eds) High energy solar physics, vol 374, AIP conference proceedings. AIP, New York, pp 461–478

Meyer P, Parker EN, Simpson JA (1956) Solar cosmic rays of February 1956 and their propagation through interplanetary space. Phys Rev 104(3):768–783

Migulin VV, Miroshnichenko LI, Obridko VN (1987) Solar-terrestrial physics:problems and perspectives. Vestn Akad Nauk SSSR. (10):83–89

Miller JA (1991) Magnetohydrodynamic turbulence dissipation and stochastic proton acceleration in solar flares. Astrophys J 376:342–354
Miller JA (1995) Much ado about nothing. EOS Trans AGU 76(41):401–404
Miller JA, Ramaty R (1987) Ion and relativistic electron acceleration by Alfvén and whistler turbulence in solar flares. Solar Phys 113(2):195–201
Miller JA, Ramaty R (1992) Stochastic acceleration in impulsive solar flares. In: Zank GP, Gaisser TK (eds) Particle acceleration in cosmic plasmas, vol 264, AIP conference proceedings. AIP, New York, p 223
Miller JA, Reames DV (1996) Heavy ion acceleration by cascading Alfvén waves in impulsive solar flares. In: Ramaty R, Mandzhavidze N, Hua X-M (eds) High energy solar physics, vol 374, AIP conference proceedings. AIP, New York, pp 450–460
Miller JA, Reames DV (1997) Cascading Alfvén wave acceleration in solar flares. In: Proceeding of 25th international cosmic ray conference, vol 1. Durban, pp 141–144
Miller JA, Roberts DA (1995) Stochastic proton acceleration by cascading Alfvén waves in impulsive solar flares. Astrophys J 452:912–932
Miller JA, Viñas AF (1993) Ion acceleration and abundance enhancements by electron beam instabilities in impulsive solar flares. Astrophys J 412:386–400
Miller JA, Guessoum N, Ramaty R (1990) Stochastic Fermi acceleration in solar flares. Astrophys J 361:701–708
Miller JA, LaRosa TN, Moore RL (1996) Stochastic electron acceleration by cascading fast mode waves in impulsive solar flares. Astrophys J 461(1):445–464
Miller JA, Cargill PJ, Emslie AG, Holman GD, Dennis BR, LaRosa TN, Winglee RM, Benka SG, Tsuneta S (1997) Critical issues for understanding particle acceleration in impulsive solar flares. J Geophys Res 102(A7):14631–14659
Miroshnichenko LI (1970) On the absolute fluxes of particles accelerated at the Sun on February 23, 1956. Geomagn Aeron 10(5):898–900
Miroshnichenko LI (1971) On integral generation multiplicities for the spectrum calculation of solar cosmic rays. Geomagn Aeron 11(6):1083–1086
Miroshnichenko LI (1977) Dynamic model of spectrum formation for solar cosmic rays. In: Proceeding of 15th international cosmic ray conference, vol 5. Plovdiv, pp 214–219
Miroshnichenko LI (1979) Analysis of solar cosmic rays spectra at the source. In: Proceeding of 16th international cosmic ray conference, vol 5. Kyoto, pp 16–21
Miroshnichenko LI (1980) Investigations of solar cosmic rays (a review). Geomagn Aeron 20 (3):369–383
Miroshnichenko LI (1981a) Spectrum formation and energetics of solar cosmic rays. Izv AN SSSR Phys Ser 45(4):588–591
Miroshnichenko LI (1981b) Peculiarities of acceleration and energetics of solar cosmic rays. In: Proceeding of 17th international cosmic ray conference, vol 3. Paris, pp 415–418
Miroshnichenko LI (1983a) On estimates of energetics of solar cosmic rays. In: Problems of physics of solar flares. IZMIRAN, Moscow, pp 120–123 (in Russian)
Miroshnichenko LI (1983b) On the spectrum dynamics of solar cosmic rays. In: Proceeding of 18th international cosmic ray conference, vol 4. Bangalore, pp 16–19
Miroshnichenko LI (1983c) Sources of methodical errors in estimating spectrum of solar cosmic rays. In: Cosmic rays, vol 22. Radio i Svyaz, Moscow, pp 69–78
Miroshnichenko LI (1984) On estimates of the energy density of solar cosmic rays near the Earth. Geomagn Aeron 24(6):871–874
Miroshnichenko LI (1986) The development of diagnostics and prediction methods of solar proton events. In: Simon P, Shea MA, Heckman GR (eds) Solar-terrestrial predictions: proceeding of a workshop at Meudon, NOAA, Boulder, Co., pp 244–262, 18–22 June 1984
Miroshnichenko LI (1987) Dynamics and energetics of accelerated particles in solar flares. In: Shcherbina-Samoilova IS (ed) Solar flares, vol 34. VINITI, Moscow, pp 238–277
Miroshnichenko LI (1990) Dynamics and prediction of radiation characteristics of solar cosmic rays, Doctoral dissertation. IZMIRAN, Moscow, p 326

Miroshnichenko LI (1992a) Generation and transport of solar cosmic rays (a review). Geomagn Aeron 32(6):755–769 (in English)
Miroshnichenko LI (1992b) Cyclic variations and sporadic fluctuations of solar cosmic radiation. Biophysics 37(3):364–377
Miroshnichenko LI (1993) Acceleration parameters and particle spectrum dynamics in solar flares. In: Proceeding of 23rd international cosmic ray conference, vol 3. Calgary, pp 25–28
Miroshnichenko LI (1994) On the ultimate capabilities of particle accelerators on the Sun. Geomagn Aeron 34(4):29–37
Miroshnichenko LI (1995) On the threshold effect of proton acceleration in solar flares. Solar Phys 156(1):119–129
Miroshnichenko LI (1996) Empirical model for the upper limit spectrum for solar cosmic rays at the Earth's orbit. Radiat Meas 26(3):421–425
Miroshnichenko LI (1997) Energetic particles in extended coronal structures: a two-source acceleration scenario. J Moscow Phys Soc 7:17–30
Miroshnichenko LI (2001) Solar cosmic rays. Kluwer Academic Publishers, Dordrecht/Boston, p 492
Miroshnichenko LI (2003a) Radiation hazard in space. Kluwer Academic Publishers, Dordrecht/Boston, p 238
Miroshnichenko LI (2003b) Multiple acceleration at the Sun: new approach to separation of the sources. In: Proceedings ISCS 2003 symposium, "Solar variability as an input to the earth's Environment", Tatranska Lomnica, Slovakia, 23–28 June 2003, ESA SP-535, Sep 2003, pp 625–630
Miroshnichenko LI (2005) Radiation field formation and monitoring beyond LEO. Adv Space Res 36(9):1742–1748
Miroshnichenko LI (2008) Solar cosmic rays in the system of solar-terrestrial relations (review). J Atm Solar-Terr Phys 70:450–466
Miroshnichenko LI (2011) Fizika Solntsa i solnechnozemnykh svyazei. In: Panasyuk MI (ed) Physics of the sun and solar–terrestrial relations. Universitetskaya Kniga, Moscow
Miroshnichenko LI, Gan WQ (2012) Particle acceleration and gamma-rays in solar flares: recent observations and new modeling. Adv Space Res 50:736–756. doi:10.1016/j.asr.2012.04.024
Miroshnichenko LI, Karpov SN (2004) Cosmophysical factors and registration of rare events at the Baksan Underground Scintillation Telescope. Geomagn Aeron 44(5):554–558
Miroshnichenko LI, Nymmik RA (2014) Extreme fluxes in solar energetic particle events: methodological and physical limitations. Radiat Meas 61:6–15
Miroshnichenko LI, Perez-Peraza JA (2008) Astrophysical aspects in the studies of solar cosmic rays (invited review). Int J Modern Phys A (IJMPA) 23(1):1–141
Miroshnichenko LI, Petrov VM (1985) Dynamics of radiation conditions in space. Energoatomizdat, Moscow, p 152 (in Russian)
Miroshnichenko LI, Sorokin MO (1985) Numerical solution of inverse problem for the reconstruction of source spectrum of solar cosmic rays. Geomagn Aeron 25(4):534–540
Miroshnichenko LI, Sorokin MO (1986) Reconstruction of some characteristics of solar cosmic rays in the source based on observations near the Earth. Geomagn Aeron 26(4):535–540
Miroshnichenko LI, Sorokin MO (1987a) Energy spectrum of the solar proton event of February 16, 1984. Geomagn Aeron 27(6):893–899
Miroshnichenko LI, Sorokin MO (1987b) Solution of the inverse problem for determining solar cosmic ray parameters near the source. In: Proceeding of 20th international cosmic ray conference, vol 3. Moscow, pp 117–120
Miroshnichenko LI, Sorokin MO (1989) Temporal and spectral characteristics of particles near the Sun for the proton events of December 7–8, 1982 and November 19, 1949. Geomagn Aeron 29 (2):309–311
Miroshnichenko LI, Petrov VM, Tibanov AP (1973) Determination and interpretation of solar proton spectra near the Earth and in the source. Izv AN SSSR Phys Ser 37(6):1174–1180

Miroshnichenko LI, Nesterov VM, Petrov VM, Tibanov AP (1974) Prediction of solar cosmic ray spectra based on the data of recent measurements. Kosm Issled (Space Res) 12(6):892–898

Miroshnichenko LI, Bengin VV, Petrov VM (1986) The dynamics and prediction of radiation characteristics of solar cosmic rays. In: Simon P, Shea MA, Heckman GR (eds) Solar-terrestrial predictions: proceeding of a workshop at Meudon, NOAA, Boulder, Co., pp 228–243, 18–22 June 1984

Miroshnichenko LI, Perez-Peraza J, Alvarez-Madrigal M, Sorokin MO, Vashenyuk EV, Gallegos-Cruz A (1990) Two relativistic solar proton components in some SPEs. In: Proceeding of 21st international cosmic ray conference, vol 5. Adelaide, pp 5–8

Miroshnichenko LI, Perez-Peraza J, Vashenyuk EV, Rodriguez-Frias MD, del Peral L, Gallegos-Cruz A (1995a) On the formation of relativistic particle beams in extended coronal structures: I. Evidences for two separate sources of solar cosmic rays. In: Proceeding of 24th international cosmic ray conference, vol 4. Rome, pp 34–37

Miroshnichenko LI, Perez-Peraza J, Vashenyuk EV, Rodriguez-Frias MD, del Peral L, Gallegos-Cruz A (1995b) On the formation of relativistic particle beams in extended coronal structures: II. Two source model for solar cosmic rays. In: Proceeding 24th international cosmic ray conference, vol 4. Rome, pp 38–42

Miroshnichenko LI, Rodriguez-Frias MD, del Peral L, Perez-Peraza J, Vashenyuk EV (1995c) Absolute proton fluxes from the Sun at rigidity above 1 GV by ground-bases data. In: Proceeding of 24th international cosmic ray conference, vol 4. Rome, pp 54–57

Miroshnichenko LI, Perez-Peraza J, Vashenyuk EV, Rodriguez-Frias MD, del Peral L, Gallegos-Cruz A (1996) On the formation of relativistic particle fluxes in extended coronal structures. In: Ramaty R, Mandzhavidze N, Hua X-M (eds) High energy solar physics, vol 374, AIP conference proceedings. AIP, New York, pp 140–149

Miroshnichenko LI, Sorokin MO, de Koning CA (1997) Relativistic protons near the Sun on September 29, 1989. In: Proceeding of 25th international cosmic ray conference, vol 1. Durban, pp 165–168

Miroshnichenko LI, Sorokin MO, de Koning CA (1998) Relativistic protons near the Sun on September 29 and October 22, 1989: Pitch angle distributions and ejection functions. Geomagn Aeron 38(2):22–29

Miroshnichenko LI, Mendoza B, Perez-Enriquez R (1999) Energy spectra of accelerated solar protons from different sources: I. Reconstruction and properties of the source spectrum. Solar Phys 186(1-2):381–400

Miroshnichenko LI, de Koning CA, Perez-Enriquez R (2000) Large solar event of September 29, 1989: ten years after. Space Sci Rev 91(3–4):615–715

Miroshnichenko LI, Mendoza B, Perez-Enriquez R (2001) Size distributions of the >10 MeV solar proton evens. Solar Phys 193:151–171

Miroshnichenko LI, Perez-Peraza JA, Velasco-Herrera VM, Zapotitla J, Vashenyuk EV (2012) Oscillations of galactic cosmic rays and solar indices before the arrival of relativistic solar protons. Geomagn Aeron 52:547–560

Miroshnichenko LI, Vashenyuk EV, Perez-Peraza JA (2013) Solar cosmic rays: 70 years of ground-based observations. Geomagn Aeron 53(5):541–560

Miyake F, Nagaya K, Masuda K, Nakamura T (2012) A signature of cosmic-ray increase in AD 774–775 from tree rings in Japan. Nature 486:240–242. doi:10.1038/nature11123

Miyasaka H, Makishima K, Takahashi E et al (2005) The solar event on 20 January 2005 observed with the Tibet YBJ Neutron Monitor Observatory. In: Proceedings of the 29th international cosmic ray conference, vol. 1. Pune, pp 241–244

Moghaddam-Taaheri E, Vlahos L, Rowland HL, Papadopoulos K (1985) Runaway tails in magnetized plasmas. Phys Fluids 28(11):3356–3364

Morfill G, Richter AK, Scholer M (1979) Average properties of cosmic ray diffusion in solar wind streams. J Geophys Res 84(A4):1505–1513

Morishita I, Sakakibara S, Fujii Z, Muraki Y, Koi K, Takahashi K (1995) Anisotropic space distribution of solar particles of the GLE observed on 24 May 1990. In: Proceeding of 24th international cosmic ray conference, vol 4. Rome, pp 220–223

Morozova EI, Likin OB, Pisarenko NF (1977) Propagation of charged particles generated in the flare of August 7, 1972. Geomagn Aeron 17(5):811–819

Morozova EI, Pisarenko NF, Mikryukova NA, Klimenko VV, Timofeev VE, Shafer YuG (1985) Peculiarities of propagation of charged particles in solar corona. In: Proceeding of 19th international cosmic ray conference, vol 4. La Jolla, pp 313–316

Moses D, Dröge W, Meyer P, Evenson P (1989) Characteristics of energetic solar flare electron spectra. Astrophys J 346:523–530

Mullan DJ (1983) Release of solar cosmic rays from the corona: Rayleigh-Taylor instability and reconnection. Astrophys J 269:765–778

Mullan DJ, Kent DW (1979) Can solar flares be the cause of ancient catastrophes? In: Proceeding of 16th international cosmic ray conference, vol 5. Kyoto, pp 323–326

Mullan DJ, Schatten KH (1979) Motion of solar cosmic rays in the coronal magnetic field. Solar Phys 63:153–177

Mullan DJ, Waldron WL (1986) Ionic charge states of solar energetic particles: effects of flare X-rays. Astrophys J Lett 308:L21–L25

Müller-Mellin R, Röhrs K, Wibberenz G (1986) Super-events in the inner solar system and their relation to the solar cycle. In: Marsden RG, Reidel D (eds) The sun and the heliosphere in three dimensions, Hingham, p 349

Muraki Y, Shibata S (1996) Solar neutrons on May 24, 1990. In: Ramaty R, Mandzhavidze N, Hua X-M (eds) High energy solar physics, vol 374, AIP conference proceedings. AIP, New York, pp 256–264

Muraki Y, Murakami K, Shibata S, Yamada T, Miyazaki M, Takahashi T, Mitsui K, Sakai T, Yamaguchi K (1991a) On isolated neutron events of February 22 and May 3, 1991. In: Proceeding of 22nd international cosmic ray conference, vol 3. Dublin, pp 45–48

Muraki Y, Murakami K, Shibata S, Sakakibara S, Yamada T, Miyazaki M, Takahashi T, Mitsui K, Sakai T (1991b) Solar neutrons associated with the large solar flares of June 1991. In: Proceeding of 22nd international cosmic ray conference, vol 3. Dublin, pp 49–52

Muraki Y, Murakami K, Miyazaki M et al (1992) Observation of solar neutrons associated with the large flare on 1991 June 4. Astrophys J Lett 400:L75–L78

Muraki Y, Takahashi T, Matsubara Y, Shibata S, Sakakibara S, Yamada T, Murakami K, Mitsui K, Sakai T (1993) The properties of Norikura Solar Neutron Telescope. In: Proceeding of 23rd international cosmic ray conference, vol 3. Calgary, pp 171–174

Muraki Y, Sakakibara S, Shibata S, Takahashi K (1995a) Comments on second and third enhancements of solar particles on June 4, 1991 flare. In: Proceeding of 24th international cosmic ray conference, vol 4. Rome, pp 179–182

Muraki Y, Sakakibara S, Shibata S, Murakami K, Takahashi K, Pyle KR, Sakai T, Mitsui K, Shea MA, Smart DF (1995b) Simultaneous observations of solar neutrons on June 6, 1991 at Mt. Haleakala and Mt. Norikura. In: Proceeding of 24th international cosmic ray conference, vol 4. Rome, pp 175–178

Muraki Y, Matsubara Y, Sakakibara S, Yamada T, Koi T, Sako T, Nishiyama T, Murata T, Matsuoka T, Kitamura A, Imaida I, Tsuchiya H, Munakata K, Yasue S, Kato C, Sakurai I, Yamaguchi I, Shibata S, Munakata Y, Sakai T, Mitsui K (1997) The 64 m^2 Solar Neutron Telescope at Norikura. In: Proceeding of 25th international cosmic ray conference, vol 1. Durban, pp 53–56

Murphy RJ, Share GH (2005) What gamma-ray de-excitation lines reveal about solar flares. Adv Space Res 35:1825–1832

Murphy RJ, Share GH, Letaw JR, Forrest DJ (1990) Nuclear line spectroscopy of the 1981 April 27 solar flare. Astrophys J 358:298–312

Murphy RJ, Ramaty R, Kozlovsky B, Reames DV (1991) Solar abundances from gamma-ray spectroscopy: comparisons with energetic particle, photospheric, and coronal abundances. Astrophys J 371:793–803

Murphy RJ, Share GH, Grove JE, Johnson WN, Kinzer RL, Kurfess JD, Strickman MS, Jung DV (1996) Abundance study of the 4 June 1991 solar flare using CGRO/OSSE spectral data. In: Ramaty R, Mandzhavidze N, Hua X-M (eds) High energy solar physics, vol 374, AIP conference proceedings. AIP, New York, pp 184–193

Murphy RJ, Share GH, Grove JE, Johnson WN, Kinzer RL, Kurfess JD, Strickman MS, Jung DV (1997) Accelerated particle composition and energetics and ambient abundances from gamma-ray spectroscopy of the 1991 June 4 solar flare. Astrophys J 490:883–900

Murphy RJ, Share GH, Hua X-M (1998) Surprisingly intense neutron emission from a flare behind the limb of the Sun. Eos Trans AGU Spring Meet Suppl 79(17):S279 (abstract)

Murphy RJ, Share GH, Hua X-M, Lin RP, Smith DM, Schwartz RA (2003) Physical implications of RHESSI neutron-capture line measurements. Astrophys J Lett 595:L93–L96

Murphy RJ, Share GH, Skibo JG, Kozlovsky B (2005) The physics of positron annihilation in the solar atmosphere. Astrophys J Suppl 161:495–519

Murphy RJ, Kozlovsky B, Share GH et al (in all 5 authors) (2007) Using gamma-ray and neutron emission to determine solar flare accelerated particle spectra and compositions within the flare magnetic loop. Astrophys J Suppl. 168:167–194

Murphy RJ, Kozlovsky B, Kiener J, Share GH (2009) Nuclear gamma-ray de-excitation lines and continuum from accelerated-particle interactions in solar flares. Astrophys J Suppl 183:142–155

Murray SS, Stone EC, Vogt RE (1971) Interplanetary deceleration of solar cosmic rays. Phys Rev Lett 26(11):663–666

Musatenko SI (1980) Radioemission of near-terrestrial space as a result of influence of solar flares on ionosphere and magnetosphere of the Earth. Geomagn Aeron 20(5):884–888

Musatenko SI, Kurochka LN (1993) Statistical analysis of USW radioemission of near-terrestrial space registered during solar X-ray flares and their precursors. Preprint of Kiev State University, p 79

Nagashima K, Sakakibara S, Murakami K, Morishita I (1989) Response and yield functions of neutron monitor, galactic cosmic-ray spectrum and its solar modulation, derived from all the available world-wide surveys. Nuovo Cimento C 12C(2):173–209

Nagashima K, Sakakibara S, Morishita I (1991) Quiescence of GLE-producible solar proton eruptions during the transition phase of heliomagnetic polarity reversal near the solar-activity-maximum period. In: Proceeding of 22nd international cosmic ray conference, vol 3. Dublin, pp 29–32

Nazarova MN, Pereyaslova NK, Petrenko IE (1992) Solar protons in 22nd cycle of solar activity. Geomagn Aeron 32(5):23–28

Nemzek RJ, Belian RD, Cayton TE, Reeves GD (1994) The October 22, 1989, solar cosmic ray event measured at geosynchronous orbit. J Geophys Res 99(A3):4221–4226

Ng CK, Reames DV (1994) Focused interplanetary transport of ~1 MeV solar energetic protons through self-generated Alfven waves. Astrophys J 424:1032–1048

Ng CK, Reames DV (1995) Interplanetary transport coefficient in an extended quasi- linear theory. In: Proceeding of 24th international cosmic ray conference, vol 4. Rome, pp 253–256

NOAA SESC (2014) Solar proton events affecting the earth environment: January 1976 – present, preliminary list. NOAA, NGDC, Boulder. A current listing can be found at: http://www.swpc.noaa.gov/ftpdir/indices/SPE.txt

Nolte JT, Roelof EC (1973) Large-scale structure of the interplanetary medium. Solar Phys 33 (241–257):483–504

Normand E, Stapor WJ (1990) Variation in proton-induced upsets rates from large solar flares using an improved SEU model. IEEE Trans Nucl Sci 37(6):1947–1952

Nymmik RA (1996) Models describing solar cosmic ray events. Radiat Meas 26(3):421–425

Nymmik RA (1999a) SEP event distribution function as inferred from spaceborne measurements and lunar rock isotopic data. In: Proceeding of 26th international cosmic ray conference, vol 6. Salt Lake City, pp 268–271

Nymmik RA (1999b) Relationships among solar activity, SEP occurrence frequency, and solar energetic particle event distribution function. In: Proceeding 26th international cosmic ray conference, vol 6. Salt Lake City, pp 280–283

Nymmik RA (1999c) Probabilistic model for fluences and peak fluxes of solar energetic particles. Radiat Meas 30:287–296

Nymmik RA (2006) Initial conditions for radiation analysis: models of galactic cosmic rays and solar particle events. Adv Space Res 38:1182–1190

Nymmik RA (2007a) Improved environment radiation models. Adv Space Res 40(3):313–320

Nymmik RA (2007b) To the problems on the regularities of solar energetic particle events occurrence. Adv Space Res 40(3):321–325

Nymmik RA (2011) Some problems with developing a Standard for determining solar energetic particle fluxes. Adv Space Res 47:622–628

Oetliker M, Klecker B, Hovestadt D, Scholer M, Blake JB, Looper M, Mewaldt RA (1995) Charge states of heavy solar energetic particles: observations with the HILT sensor on SAMPEX. In: Proceeding of 24th international cosmic ray conference, vol 4. Rome, pp 470–473

Otaola JA, Valdes-Galicia JF (1995) A study of the magnetic helicity during eight Helios-observed proton events. Solar Phys 160(1):41–52

Pallavicini R, Serio S, Vaiana GS (1977) A survey of soft X-ray limb flare images: the relation between their structure in the corona and other physical parameters. Astrophys J 216:108–122

Palmer ID (1982) Transport coefficients of low-energy cosmic rays in interplanetary space. Rev Geophys Space Phys 20:335–351

Panasyuk MI, Novikov LS (eds) (2007) Cosmos model. SINP MSU, Moscow

Papaioannou A, Souvatzoglou G, Paschalis P, Gerontidou M, Mavromichalaki H (2014) The first ground-level enhancement of solar cycle 24 on 17 May 2012 and its real-time detection. Solar Phys 289:423–436. doi:10.1007/s11207-013-0336-2

Parker EN (1963) Interplanetary dynamical processes. Interscience Publishers, New York/London, p 161

Parker EN (1965) The passage of energetic particles through interplanetary space. Planet Space Sci 13(1):9–50

Pchelkin VV, Vashenyuk EV, Ostapenko AA, Maltsev YP (2000) Relativistic SCR in the event of 7–8 December 1982. Factor of magnetospheric disturbance at the analysis based on neutron monitor data. Geomagn Aeron 40(5):39–44

Pelaez F, Niel M, Vedrenne G, Vilmer N, Lebrun F, Laurent P, Paul J, Salotti L, Sunyaev R, Churazov E, Gilfanov M, Khavenson N, Chulkov I, Novikov B, Kuznetsov A, Dyachkov A (1991) Detection of MeV gamma rays during 5 solar flares observed with the anticoincidence shield of the SIGMA telescope. In: Proceeding of 22nd international cosmic ray conference, vol 3. Dublin, pp 89–92

Pelaez F, Mandrou P, Niel M, Mena B, Vilmer N, Trottet G, Lebrun F, Paul J, Terekhov O, Sunyaev R, Churazov E, Gilfanov M, Denisov D, Kuznetsov A, Dyachkov A, Khavenson N (1992) Solar hard X-ray and gamma-ray bursts observed by the SIGMA anti-coincidence shield aboard GRANAT. Solar Phys 140(1):121–138

Pereyaslova NK, Nazarova MN, Mikirova NA (1983) On relation between the solar cosmic ray characteristics and the coronal magnetic field structure. Izv AN SSSR Phys Ser 47(9):1766–1770

Pereyaslova NK, Nazarova MN, Petrenko IE (1996) An empirical model of the radiation characteristics of solar proton events in near-Earth space. Radiat Meas 26(3):451–454

Perez-Enriquez R (1985) On the role of energetic particles in solar flares. Solar Phys 97 (1):131–144

Perez-Enriquez R, Miroshnichenko LI (1999) Frequency distributions of solar gamma- ray events related and not related with SPEs in 1980–1995. Solar Phys 188(1):169–185

Pérez-Enríquez R, Guerrero M, Miroshnichenko LI, Rodríguez-Taboada RE, Méndez-Berhondo A, Zlobec P (2000) Multiple particle acceleration at the Sun during large extension and long duration gamma ray events. In: Ramaty R, Mandzhavidze N (eds) High energy solar physics – anticipating HESSI, ASP conference Series, vol 206. pp 445–450
Perez-Peraza J (1986) Coronal transport of solar flare particles. Space Sci Rev 44:91–138
Perez-Peraza J, Alvarez-Madrigal M (1990) Evolution of effective charge of accelerated ions. In: Proceeding of 21st international cosmic ray conference, vol 5. Adelaide, pp 382–385
Perez-Peraza J, Gallegos-Cruz A (1993) The role of fluctuational acceleration in the generation of solar particles. Geofis Int 33(2):311–331
Perez-Peraza J, Gallegos-Cruz A (1994) Weightiness of the dispersive rate in stochastic acceleration processes. Astrophys J Suppl Ser 90(2):669–682
Perez-Peraza J, Galvez M, Lara AR (1977) Energy spectra of flare particles from an impulsive process. In: Proceeding of 15th international cosmic ray conference, vol 5. Plovdiv, pp 23–28
Perez-Peraza J, Alvarez-Madrigal M, Gallegos-Cruz A (1989) Electron-capture spectroscopy for identification of source ionization states and acceleration processes of solar cosmic rays. Adv Space Res 9(12):97–101
Perez-Peraza J, Gallegos-Cruz A, Vashenyuk EV, Miroshnichenko LI (1992) Spectrum of accelerated particles in solar proton events with prompt component. Geomagn Aeron 32(2):1–12
Perez-Peraza JA, Gallegos-Cruz A, Miroshnichenko LI, Vashenyuk EV (2008) Relativistic proton production at the Sun in the January 20, 2005 event. Adv Space Res 41(6):947–954
Perez-Peraza JA, Gallegos-Cruz A, Vashenyuk EV, Miroshnichenko LI, Balabin YV (2009) Impulsive, stochastic, and shock wave acceleration of relativistic protons in large solar events of 1989 September 29, 2000 July 14, 2003 October 28, and 2005 January 20. Astrophys J 695 (2):865–873
Perez-Peraza J, Velasco-Herrera V, Zapotitla J, Miroshnichenko LI, Vashenyuk EV, Libin IYa (2011) Classification of GLEs as a function of their spectral content for prognostic goals. In: Proceeding of 32nd international cosmic ray conference, vol 10. Beijing, pp 149–152
Peristykh AN, Damon PE (1999) Multiple evidence of intense solar proton events during solar cycle 13. In: Proceeding of 26th international cosmic ray conference, vol 6. Salt Lake City, pp 264–267
Pesses ME, Klecker B, Gloeckler G, Hovestadt D (1981) Observations of interplanetary energetic charged particles from gamma-ray line solar flares. In: Proceeding of 17th international cosmic ray conference, vol 3. Paris, pp 36–39
Petrov VM, Bengin VV, Andronov EA, Logachev YuI, Podorolsky AN (1987) Solar cosmic ray angular distributions y 9 July, 1985 solar proton event. In: Proceeding of 20th international cosmic ray conference, vol 3. Moscow, pp 155–158
Pfotzer G (1958) On the separation of direct and indirect fractions of solar cosmic radiation on February 23, 1956 and on difference in steepness of momentum spectrum of these two components. Nuovo Cimento Suppl. 8, Series 10, (2):180–187
Pinsky LS, Osborne WZ, Bailey JV, Benson RE, Thompson LF (1974) Light flashes observed by astronauts on Apollo 11 through Apollo 17. Nature 183:957–959
Pisarenko NF (1992) Radiation situation determining the possibility of a manned flight to Mars and back. Adv Space Res 12(2/3):435–439
Plainaki C, Mavromichalaki H, Belov A, Eroshenko E, Andriopoulou M, Yanke V (2010) A new version of the neutron monitor based anisotropic GLE model: application to GLE60. Solar Phys 264:239–254. doi:10.1007/s11207-010-9576-6
Podgorny AI, Podgorny IM (1990) An electromagnetic model of solar flares. In: Russel CF, Priest ER, Lee LC (eds) Geophysical Monograph 58. Physics of magnetic flux ropes, AGU, pp 373–377
Pogorely DK (2001) On possible origin of light flashes in the cosmonaut's eyes. In: Petrukhin AA, Khokonov MKh (eds) Second Baksan Youth School of Experimental and Theoretical Physics. Kabardino-Balkaria, pp 186–191, 18–24 Apr 2001

Poirier J, D'Andrea C (2002) Ground level muons in coincidence with the solar flare of 15 April 2001. J Geophys Res 107(NA11):1376

Pomerantz MA, Duggal SP (1974) The Sun and cosmic rays. Rev Geophys Space Phys 12 (3):343–361

Priest ER (1982) Solar magnetohydrodynamics, vol 21, Geophysics and astrophysics monographs. D.Reidel Publishing Company, Dordrecht

Priest E, Forbes T (2000) Magnetic field reconnection (MHD theory and applications). Cambridge University Press, Cambridge/New York, p 520

Prince TA, Forrest DJ, Chupp EL et al (in all 5 authors) (1983) The time history of 2.22 MeV line emission in solar flares. In: Proceedings of 18th international cosmic ray conference, vol 4. Bangalore, pp 79–82

Pudovkin MI, Babushkina SV (1992a) Influence of solar flares and disturbances of the interplanetary medium on atmospheric circulation. J Atm Terr Phys 54(7/8):841–846

Pudovkin MI, Babushkina SV (1992b) Atmospheric transparency variations associated with geomagnetic disturbances. J Atm Terr Phys 54(9):1135–1138

Pudovkin MI, Raspopov OM (1992) The mechanism of action of solar activity on the state of the lower atmosphere and meteorological parameters (a review). Geomagn Aeron 32(5):1–22

Pudovkin MI, Vinogradova NY, Veretenenko SV (1997) Variations of atmospheric transparency during the bursts of solar protons. Geomagn Aeron 37(2):124–126

Pugacheva GI, Gusev AA, Martin IM, Turtelli A Jr, Pinto HS, Zullo J Jr, Bezerra PC (1995) The influence of geomagnetic disturbances on the meteorological parameters in the BMA Region. In: Proceeding of 24th international cosmic ray conference, vol 4. Rome, pp 1110–1113

Pumar M, Rodriguez-Frias MD, del Peral L, Bronchalo E, Cid C, Medina J, Rodriguez-Pacheco J, Sequeiros J (1997) Computational study of particle energization in solar flares. In: Proceeding of 25th international cosmic ray conference, vol 1. Durban, pp 125–128

Pyle KR, Simpson JA (1991) Observation of a direct solar neutron event on 22 March 1991 with the Haleakala, Hawaii, neutron monitor. In: Proceeding of 22nd international cosmic ray conference, vol 3. Dublin, pp 53–56

Pyle KR, Shea MA, Smart DF (1991) Solar flare generated neutrons observed by neutron monitors on 24 May 1990. In: Proceeding of 22nd international cosmic ray confernce, vol 3. Dublin, pp 57–60

Quack M, Kallenrode MB, von König M, Künzi K, Burrows J, Heber B, Wolff E (2001) Ground level events and consequences for stratospheric chemistry. In: Proceeding of 27th international cosmic ray conference, vol 10. Hamburg, pp 4023–4026

Ramaty R, Lingenfelter RE (1995) Astrophysical gamma-ray emission lines. In: Williams RE, Livio M (eds) Analysis of emission lines. Cambridge Univ. Press, Cambridge, pp 180–186

Ramaty R, Mandzhavidze N (1994a) Theoretical models for high energy solar flare emissions. In: Ryan JM, Vestrand WT (eds) High energy solar phenomena – a new era of spacecraft measurements, vol 294, AIP conference proceedings. AIP, New York, pp 26–44

Ramaty R, Mandzhavidze N (1994b) On the origin of long lasting gamma ray emission from solar flares. In: Proceeding of Kofu symposium, NRO Report No.360, pp 275–278, July 1994

Ramaty R, Mandzhavidze N (1996) Implications of solar flare charged particle, gamma ray and neutron observations: rapporteur paper II for the high energy solar physics workshop. In: Ramaty R, Mandzhavidze N, Hua X-M (eds) High energy solar physics, vol 374. AIP Conference Proceedings, AIP, New York, pp 533–543

Ramaty R, Mandzhavidze N (1998) Solar He abundance from gamma ray spectroscopy. Eos Trans AGU Spring Meet Suppl 79(17):S279 (abstract)

Ramaty R, Murphy RJ (1987) Nuclear processes and accelerated particles in solar flares. Space Sci Rev 45:213–268

Ramaty R, Lingenfelter RE, Kozlovsky B, Reeves H (1997) Light elements and cosmic rays in the early Galaxy. Astrophys J 488:730–748

Ramaty R, Mandzhavidze N, Kozlovsky B (1996) Solar atmospheric abundances from gamma ray spectroscopy. In: Ramaty R, Mandzhavidze N, Hua X-M (eds) High energy solar physics, vol 374. AIP Conference Proceedings, AIP, New York, pp 172–183
Ramaty R, Kozlovsky B, Lingenfelter RE (1975) Solar gamma rays. Space Sci Rev 18(3):341–388
Ramaty R, Kozlovsky B, Lingenfelter R (1979) Nuclear gamma rays from energetic particle interactions. Astrophys J Suppl 40:487
Ramaty R, Miller JA, Hua X-M, Lingenfelter RE (1990) Ion and relativistic electron transport in solar flares. Astrophys J Suppl 73:199–207
Ramaty R, Mandzhavidze N, Kozlovsky B, Skibo JG (1993) Acceleration in solar flares: interacting particles versus interplanetary particles. Adv Space Res 13(9):275–284
Ramaty R, Mandzhavidze N, Kozlovsky B, Murphy RJ (1995) Solar atmospheric abundances and energy content in flare-accelerated ions from gamma-ray spectroscopy. Astrophys J Lett 455: L193–L196
Rank G, Bennet K, Bloemen H, Debrunner H, Lockwood J, McConnel M, Ryan J, Schönfelder V, Suleiman R (1996) Extended γ-ray emission in solar flares. In: Ramaty R, Mandzhavidze N, Hua X-M (eds) High energy solar physics, vol 374, AIP conference proceedings. AIP, New York, pp 219–224
Rank G, Debrunner H, Kocharov L, Kovaltsov G, Lockwood J, McConnel M, Nieminen P, Ryan J, Schönefelder V (1997a) The solar flare on 15 June 1991. In: Proceeding of 25th international cosmic ray conference, vol 1. Durban, pp 1–4
Rank G, Debrunner H, Lockwood J, McConnel M, Ryan J, Schönefelder V (1997b) Extended gamma-ray emission of the solar flares in June 1991. In: Proceeding of 25th international cosmic ray conference, vol 1. Durban, pp 5–8
Rao UR (1976) High-energy cosmic ray observations during August 1972. Space Sci Rev 19 (4/5):533–577
Ray KP, Mullen EG (1991) The effects of the March 1991 storm on microelectronics performance. EOS Trans 72:377
Reagan JB, Meyerott RE, Evans JE, Imhof WL, Joiner RG (1983) The effects of energetic particle precipitation on the atmospheric electric circuit. J Geophys Res 88:3869–3878
Reames DV (1987) Solar particle composition, spectra and frequency of occurrence. In: High-energy radiation background in space: conference, Sanibel Island, Fla., vol 198. AIP, New York, 3–5 Nov 1987, pp 84–86
Reames DV (1989) Wave generation in the transport of particles from large solar flares. Astrophys J Lett 342:L51–L54
Reames DV (1990a) Acceleration of energetic solar particles by shock waves from large solar flares. Astrophys J Lett 358:L63–L67
Reames DV (1990b) Energetic particles from impulsive solar flares. Astrophys J Suppl 73:235–251
Reames DV (1993) Non-thermal particles in the interplanetary medium. Adv Space Res 13 (9):331–339
Reames DV (1994) Acceleration of energetic particles which accompany coronal mass ejections. In: Poland A (ed) Solar dynamic phenomena and solar wind consequences, Third SOHO Workshop, Estes Park, Colorado, ESA SP- 373, ESTEC, Noordwijk, 26–29 Sep 1994, pp 107–116
Reames DV (1995a) Coronal abundances determined from energetic particles. Adv Space Res 15 (7):41–51
Reames DV (1995b) Solar energetic particles: a paradigm shift. Rev Geophys Suppl 33:585–589
Reames DV (1995c) The dark side of the solar flare myth. EOS Trans 76(41):405–406
Reames DV (1996) Energetic particles from solar flares and coronal mass ejections. In: Ramaty R, Mandzhavidze N, Hua X-M (eds) High energy solar physics, vol 374, AIP conference proceedings. AIP, New York, pp 35–44
Reames DV (1999) Particle acceleration at the Sun and in the heliosphere. Space Sci Rev 90:413–491

Reames DV (2013) The two sources of solar energetic particles. Space Sci Rev 175:53–92. doi:10.1007/s11214-013-9958-9

Reames DV, Ng CK (1998) Streaming-limited intensities of solar energetic particles. Astrophys J 504:1002–1005

Reames DV, Ng CK (2004) Heavy element abundances in solar energetic particle events. Astrophys J 610(510–522):2004

Reames DV, Cane HV, von Rosenvinge TT (1990) Energetic particle abundances in solar electron events. Astrophys J 357:259–280

Reames DV, Kallenrode M-B, Stone RG (1991) Multispacecraft observations of solar ^{3}He-rich events. Astrophys J 380:287–292

Reames DV, Richardson IG, Wenzel K-P (1992) Energy spectra of ions from impulsive flares. Astrophys J 387:715–725

Reames DV, Meyer J-P, von Rosenvinge TT (1994) Energetic particle abundances in impulsive solar flares. Astrophys J Suppl 90:649–667

Reedy RC (1977) Solar proton fluxes since 1956. In: Proceeding of 8th Lunar science conference, pp 825–839

Reedy RC (1980) Lunar radionuclide records of average solar cosmic ray fluxes over the last ten million years. In: Pepin RO, Eddy JA, Merril RB (eds) Proceeding conference ancient sun. Pergamon Press, New York, pp 365–386

Reedy RC (1996) Constraints on solar particle events from comparison of recent events and million-year averages. In: Balasubramanian KS, Keil SL, Smartt RN (eds) Solar drivers of interplanetary and terrestrial disturbances, vol 95, Conference series. Astronomical Society of the Pacific, San Francisco, pp 429–436

Reeves GD, Cayton TE, Gary SP, Belian RD (1992) The great solar energetic particle events of 1989 observed from geosynchronous orbit. J Geophys Res 97(A5):6219–6226

Reid GC (1964) A diffusive model for initial phase of a solar proton event. J Geophys Res 69:2659–2667

Reid GC (1991) Solar total irradiance variations and the global sea surface temperature record. J Geophys Res 96:2835–2846

Reid GC (1996) Comment on solar flare debate. EOS Trans AGU 77(8):78

Reid GC, Isaksen ISA, Holzer TE, Crutzen PJ (1976) Influence of ancient solar proton events on the evolution of life. Nature 259(5540):177–179

Reid GC, Solomon S, Garcia RR (1991) Response of the middle atmosphere to the solar proton events in August-December 1989. Geophys Res Lett 18(6):1019–1022

Reinhard R, Wibberenz G (1973) Coronal transport of solar flare protons. In: Proceedings of 13th international cosmic ray conference, vol 2. Denver, pp 1372–1377

Reinhard R, Wibberenz G (1974) Propagation of flare protons in solar atmosphere. Solar Phys 36 (2):473–494

Richardson IG, Reames DV (1993) Bidirectional ~ 1 MeV/amu ion intervals in 1973–1991 observed by the Goddard Space Flight Center instruments on IMP-8 and ISEE-3/ICE. Astrophys J Suppl 85(2):411–432

Richardson IG, Cane HV, von Rosenvinge TT (1991) Prompt arrival of solar energetic particles from far eastern events: the role of large-scale interplanetary magnetic field structure. J Geophys Res 96(A3):7853–7860

Richardson JD, Belcher JW, Paularena KI, Lazarus AJ (1995) Radial evolution of the solar wind from IMP 8 to Voyager 2. Geophys Res Lett 22:325–328

Rieger E (1996) Spectral evolution of an intense gamma-ray line flare. In: Ramaty R, Mandzhavidze N, Hua X-M (eds) High energy solar physics, vol 374, AIP conference proceedings. AIP, New York, pp 194–199

Rieger E, Share GH, Forrest DJ, Kanbach G, Reppin C, Chupp EL (1984) A 154-day periodicity in the occurrence of hard solar flares? Nature 312(5995):623–625

Rieger E, Forrest DJ, Bazilevskaya G, Chupp EL, KanbachG, Reppin C, Share GH (1987) Gamma ray emission from a solar flare observed also as a Ground Level Event. In: Proceedings of 20th international cosmic ray conference, vol 3. Moscow, pp 65–68

Roble RG (1985) On solar-terrestrial relationships in atmospheric electricity. J Geophys Res 90 (D4):6000–6012

Rodríguez-Gasén R, Aran A, Sanahuja B, Jacobs C, Poedts S (2014) Variation of proton flux profiles with the observer's latitude in simulated gradual SEP events. Solar Phys 289:1745–1762. doi:10.1007/s

Roelof EC (1969) Propagation of solar cosmic rays in the interplanetary magnetic field. In: Ogelman H, Wayland JR (eds) Lectures in high energy astrophysics. NASA Spec. Publ., SP 199, pp 111–135

Roldughin VK, Vashenyuk EV (1994) Change of atmosphere transparency under the action of solar cosmic rays. Geomagn Aeron 34(2):155–158

Rosner R, Vaiana GS (1978) Cosmic flare transients: constraints upon models for energy storage and release derived from the event frequency distribution. Astrophys J 222:1104–1108

Roth I, Temerin M (1995) 3He and heavy ion enhancements in impulsive solar flares due to EMIC waves. In: Proceedings of 24th international cosmic ray conference, vol 4. Rome, pp 118–121

Roth I, Temerin M (1997) Enrichment of ^{3}He and heavy ions in impulsive solar flares. Astrophys J 477:940–957

Roussev II, Sokolov IV, Forbes TG, Gomboshi TI, Lee MA, Sakai JI (2004) A numerical model of a coronal mass ejection: shock development with applications for the acceleration of GeV protons. Astrophys J 605:L73–L76

Ruffolo D (1995) Effect of adiabatic deceleration on the focused transport of solar cosmic rays. Astrophys J 442:861–874

Ruffolo D (1997a) Constraints on coronal transport and acceleration times from charge states of interplanetary ions from gradual solar flare events. In: Proceedings of 25th international cosmic ray conference, vol 1. Durban, pp 173–176

Ruffolo D (1997b) Interacting and escaping solar energetic particles. In: Proceedings of 25th international cosmic ray conference, invited, rapporteur, and highlight papers, vol 8. Durban, pp 109–134

Ruffolo D, Khumlumlert T (1995) Propagation of coherent pulses of solar cosmic rays. In: Proceedings of 24th international cosmic ray conference, vol 4. Rome, pp 277–280

Ruffolo D, Youngdee W, Dröge W (1997) A deconvolution technique to determine injection profiles and spectra of solar cosmic rays. In: Proceedings of 25th international cosmic ray conference, vol 1. Durban, pp 97–100

Rust DM (1982) Solar flares, proton showers, and the Space Shuttle. Science 216(4549):939–946

Ryan JM, Lee MA (1991) On the transport of solar energetic particles in a coronal loop. Astrophys J 368:316–324

Ryan JM, McConnel MM (1996) COMPTEL solar flare measurements. In: Ramaty R, Mandzhavidze N, Hua X-M (eds) High energy solar physics, vol 374, AIP conference proceedings. AIP, New York, pp 200–209

Ryan JM, Lockwood JA, Debrunner H (2000) Solar energetic particles. Space Sci Rev 93:35–53

Sakai J-I, de Jager C (1996) Solar flares and collisions between current-carrying loops. Space Sci Rev 77(1/2):1–192

Sakai J-I, Ohsawa Y (1987) Particle acceleration by magnetic reconnection and shocks during current loop coalescence in solar flares. Space Sci Rev 46(1/2):113–198

Sakai J, Zhao J, Nishikawa KI (1995a) Triggering of solar flares and particle acceleration. In: Proceedings of 24th international cosmic ray conference, vol 4. Rome, p 13 (abstract).

Sakai T, Kato M, Muraki Y (1995b) Solar neutron decay protons. In: Proceedings of 24th international cosmic ray conference, vol 4. Rome, pp 143–146

Sako T, Watanabe K, Muraki Y, Matsubara Y, Tsujihara H, Yamashita M, Sakai T, Shibata S, Valdés-Galicia JF, González LX, Hurtado A, Musalem O, Miranda P, Martinic N, Ticona R, Velarde A, Kakimoto F, Ogio S, Tsunesada Y, Tokuno H, Tanaka YT, Yoshikawa I,

Terasawa T, Saito Y, Mukai T, Gros M (2006) Long-lived emission of relativistic solar neutrons in comparison with the electron originated radiation. Astrophys J Lett 651:L69–L72
Sako T, Watanabe K, Valdes-Galicia JF, Gonzalez LX, Hurtado A, Musalem O, Miranda PP, Martinic N, Ticona R, Velarde A, Muraki Y, Matsubara Y, Sakai T, Shibata S, Kakimoto F, Tsunesada Y, Tokuno H, Ogio S (2008) Emission profile of solar neutrons obtained from the ground based observations for the 7 September 2005 event. In: Proceedings of 30th international cosmic ray conference, vol 6. Merida, pp 25–28
Sakurai K (1974) Physics of solar cosmic rays. Tokyo University Press, Tokyo, p 428
Sakurai K (1979) The nature of the solar activity and its relation to solar cosmic-ray production. Astrophys Space Sci 63(2):369–378
Sakurai K (1989) High-energy phenomena on the sun: introductory review. Space Sci Rev 51:1–9
Sarabhai V, Duggal SP, Razdan H, Sastry FSG (1956) A solar flare type increase in cosmic rays at low latitudes. Proc Ind Acad Sci 43:309–318
Sauer HH (1993a) GOES observations of energetic protons to E>685 MeV: Description and data comparison. In: Proceedings of 23rd international cosmic ray conference, vol 3. Calgary, pp 250–253
Sauer HH (1993b) GOES observations of energetic protons E>685 MeV: ground-level events from October 1983 to July 1992. In: Proceedings of 23rd international cosmic ray conference, vol 3. Calgary, pp 254–257
Savun OI, Sladkova AI (1976) Measurement of absorbed radiation dose from solar flare on August 4, 1972 in free space. Kosm Issled 14(1):135–140
Schatten KH, Mullan DJ (1977) Fast azimuthal transport of solar cosmic rays via a coronal magnetic bottle. J Geophys Res 82:5609–5620
Schatzman E (1966) Elementary processes and acceleration mechanisms. In: High energy astrophysics, summer school of the HOUCHES, vol 2. Gordon and Breach, pp 229
Schatzman E (1967) Stellar and solar flares. Solar Phys 1(3/4):411–419
Schindler SM, Kearney PD (1973) Evidence for very high-energy solar particle production. In: Proceedings of the 13th international cosmic ray conference, vol. 2. Denver, pp 1554–1559
Schlickeiser R (1989a) Cosmic-ray transport and acceleration: I. Derivation of the kinetic equation and application to cosmic rays in static cold media. Astrophys J 336(1):243–263
Schlickeiser R (1989b) Cosmic-ray transport and acceleration: II. Cosmic rays in moving cold media with application to diffusive shock wave acceleration. Astrophys J 336(1):264–293
Schlickeiser R, Miller JA (1997) On the mean free path discrepancy of solar cosmic rays. In: Proceedings of 25th international cosmic ray conference, vol 1 Durban, pp 237–240
Schmidt JM, Droge W (1997) Cosmic ray pitch angle diffusion coefficients in the hot and disperse plasma. In: Proceedings of 25th international cosmic ray conference, vol 1. Durban, pp 217–220
Schmieder B, Heinzel P, Wiik JE, Lemen J, Anwar B, Kotrc P, Hiei E (1995) Relation between cool and hot post-flare loops of 26 June 1992 derived from optical and X ray (SXT-Yohkoh) observations. Solar Phys 156(2):337–361
Scholer M (1988) Acceleration of energetic particles in solar flares. In: Havnes O et al (eds) Activity in cool star envelopes. Kluwer Academic Publishers, Dordrecht, pp 195–210
Schrijver CJ, Hudson HS, Murphy RJ et al (in all 5 authors) (2006) Gamma rays and the evolving, compact structures of the 2003 October 28 X17 flare. Astrophys J 650:1184–1192
Schrijver CJ, Beer J, Baltensperger U, Cliver E, Gudel M, Hudson H, McCracken KG, Osten R, Peter T, Soderblom D, Usoskin I, Wolff EW (2012) Estimating the frequency of extremely energetic solar events, based on solar, stellar, lunar, and terrestrial records. J Geophys Res 117, A08103. doi:10.1029/2012JA017706
Schuurmans CJE (1982) Effects of solar flares in atmospheric circulation. In: Solar- terrestrial relationships, weather and climate. Mir, Moscow, pp 129–144
Shakhovskaya NI (1989) Stellar flare statistics - physical consequences. Solar Phys 121(2):375–386

Share GH, Murphy RJ (1995) Gamma-ray measurements of flare-to-flare variations in ambient solar abundances. Astrophys J 452:933–943

Share GH, Murphy RJ (1997) Intensity and directionality of flare-associated α-particles at the Sun. Astrophys J 485(1):409–418

Share GH, Murphy RJ (1998) Accelerated and ambient He abundances from gamma- ray line measurements of flares. Eos Trans AGU Spring Meet Suppl 79(17):S279 (abstract)

Share GH, Murphy RJ (2000) Gamma ray spectroscopy in the pre-HESSI era. In: Ramaty R, Mandzhavidze N (eds) High energy solar physics – anticipating HESSI, vol 206. ASPacific Conference Series, San Francisco, pp 377–386

Share GH, Murphy RJ, Skibo JG (1996) Gamma-ray line measurements and ambient solar abundances. In: Ramaty R, Mandzhavidze N, Hua X-M (eds) High energy solar physics, vol 374, AIP conference proceedings. AIP, New York, pp 162–171

Share GH, Murphy RJ, Smith DM, Schwartz RA, Lin RP (2004) *RHESSI* e^+- e^- annihilation radiation observations: Implications for conditions in the flaring solar chromosphere. Astrophys J Lett 615:L169–L172

Share GH, Murphy RJ, Tylka AJ et al (in all 6 authors) (2011) Physics of solar neutron production: Questionable detection of neutrons from the 31 December 2007 flare. J Geophys Res 116: A03102. doi:10.1029/2010JA015930

Shea MA (1990) Solar cosmic rays, 1960–1989. In: Proceedings of 21st international cosmic ray conference, invited papers, highlight papers, Miscellaneous, vol 12. Adelaide, pp 196–204

Shea MA, Smart DF (1975) Asymptotic directions and vertical cutoff rigidities for selected cosmic ray stations as calculated using the international geomagnetic reference field model for epoch 1975.0, AFCRL-TR-75-0247, 5 May 1975. Airforce Cambridge Research Laboratories, Hanscom AFB

Shea MA, Smart DF (1982) Possible evidence for a rigidity-dependent release of relativistic protons from the solar corona. Space Sci Rev 32(1/2):251–271

Shea MA, Smart DF (1990a) A summary of major solar proton events. Solar Phys 127(2):297–320

Shea MA, Smart DF (1990b) Statistical trends (or lack thereof) in solar proton events during the last three solar cycles. In: Thompson RJ, Cole DG, Wilkinson PJ, Shea MA, Smart DF, Heckman GH (eds) Solar-terrestrial predictions: proceedings of a workshop at Leura, Australia, October 16–20, 1989, vol 1 NOAA ERL, Boulder, pp 586–591

Shea MA, Smart DF (1992) Recent and historical solar proton events. Radiocarbon 34(2):255–262

Shea MA, Smart DF (1993a) Solar proton events: History, statistics and predictions. In: Hrushka J, Shea MA, Smart DF, Heckman G (eds) Solar-terrestrial prediction-IV. Proceedings of a workshop at Ottawa, Canada, May 18–22, 1992, vol 2. U.S. Department of Commerce, NOAA, SEL, Boulder, Co., pp 48–70

Shea MA, Smart DF (1993b) History of energetic solar protons for the past three solar cycles including cycle 22 update. In: Swenberg CE et al (eds) Biological effects of solar and galactic cosmic radiation, Part B. Plenum Press, New York, pp 37–71

Shea MA, Smart DF (1994) Significant proton events of solar cycle 22 and a comparison with events of previous solar cycles. Adv Space Res 14(10):631–638

Shea MA, Smart DF (1996a) Solar-interplanetary phenomena in March 1991 and related geophysical and technological consequences. Radiat Meas 26(3):427–432

Shea MA, Smart DF (1996b) Unusual intensity-time profiles of ground-level solar proton events. In: Ramaty R, Mandzhavidze N, Hua X-M (eds) High energy solar physics, vol 374, AIP conference proceedings. AIP, New York, pp 131–139

Shea MA, Smart DF (1997a) Dual acceleration and/or release of relativistic solar cosmic rays. In: Proceedings of 25th international cosmic ray conference, vol 1. Durban, pp 129–132

Shea MA, Smart DF (1997b) Overview of solar-induced disturbances on the terrestrial environment and the effects on technological systems. In: Heckman G, Marubashi K, Shea MA, Smart DF, Thompson R (eds) Solar-terrestrial prediction-V, proceedings of a workshop at Hitachi, Japan, January 23–27, 1996. RWC Tokyo, Hiraiso Solar Terrestrial Research Center, Communications Research Laboratory, Hitachinaka, pp 441–48

Shea MA, Smart DF (1997c) The >10 MeV solar proton event peak flux distribution. In: Heckman G, Marubashi K, Shea MA, Smart DF, Thompson R (eds) Solar-terrestrial prediction-V, proceedings of a workshop at Hitachi, Japan, January 23–27, 1996. RWC Tokyo, Hiraiso Solar Terrestrial Research Center, Communications Research Laboratory, Hitachinaka, pp 449–452

Shea MA, Smart DF (2012) Space weather and the ground-level solar proton events of the 23rd solar cycle. Space Sci Rev 171:161–188. doi:10.1007/s11214-012-9923-z

Shea MA, Smart DF, Adams JH, Chennette D, Feynman J, Hamilton DC, Heckman G, Konradi A, Lee MA, Nachtwey DS, Roelof EC (1988) Toward a descriptive model of solar particles in the heliosphere; Interplanetary particle environment, JPL Publication 88–28

Shea MA, Smart DF, Wilson MD, Flückiger EO (1991a) Possible ground-level measurements of solar neutron decay protons during the 19 October 1989 solar cosmic ray event. Geophys Res Lett 18(5):829–832

Shea MA, Smart DF, Pyle KR (1991b) Direct solar neutrons detected by neutron monitors on May 24, 1990. Geophys Res Lett 18(9):1655–1658

Shea MA, Smart DF, Allen JH, Wilkinson DC (1992) Spacecraft problems in association with episodes of intense solar activity and related terrestrial phenomena during March 1991. IEEE Trans Nucl Sci 39:1754–1760

Shea MA, Smart DF, Swinson DB, Wilson MD (1993a) High energy cosmic ray modulation in March-June 1991. In: Proceedings of 23rd international cosmic ray conference, vol 3. Calgary, pp 735–738

Shea MA, Smart DF, Dreschhoff GAM, Zeller EJ (1993b) The flux and fluence of major solar proton events and their record in Antarctic snow. In: Proceedings of 23rd international cosmic ray conference, vol 3. Calgary, pp 846–849

Shea MA, Cramp JL, Duldig ML, Smart DF, Humble JE, Fenton AG, Fenton KB (1995a) Comparison of ground level enhancements of 15 November 1960 and 22 October 1989. In: Proceedings of 24th international cosmic ray conference, vol 4. Rome, pp 208–211

Shea MA, Smart DF, Flückiger EO (1995b) A comment on the detection of solar proton events at the Earth with respect to the heliospheric current sheet. In: Proceedings of 24th international cosmic ray conference, vol 4. Rome, pp 309–312

Shea MA, Smart DF, McCracken KG, Dreschhoff GAM, Spence HE (2006) Solar proton events for 450 years: the Carrington event in perspective. Adv Space Res 38:232–238. doi:10.1016/j.asr.2005.02.100

Shibata S (1994) Propagation of solar neutrons through the atmosphere of the Earth. J Geophys Res 99(A4):6651–6665

Shibata S, Murakami K, Muraki Y, Miyazaki M, Takahashi T, Sakai T, Martinic NJ, Capdeville JN (1991) The detection efficiency of a new solar neutron telescope. In: Proceedings of 22nd international cosmic ray conference, vol 3. Dublin, pp 788–791

Shibata S, Murakami K, Muraki Y (1993) Solar neutron events observed in cycle 22. In: Proceedings of 23rd international cosmic ray conference, vol 3. Calgary, pp 95–98

Shibata S, Munakata Y, Tatsuoka R, Muraki Y, Matsubara Y, Sakakibara S, Koi T, Sako T, Okada A, Murata T, Imaida I, Tsuchiya H, Ishida Y, Sakai H, Wakasa T, Nonaka T, Ohnishi T, Hatanaka K (1997) Calibration of neutron monitor using an accelerator. In: Proceedings of 25th international cosmic ray conference, vol 1. Durban, pp 45–48

Shimazaki T (1984) The photochemical time constants of minor constituents and their families in the middle atmosphere. J Atm Terr Phys 46:173–191

Shumilov OI, Raspopov OM, Kasatkina EA, Kokin GA, Chizhov AF, Vashenyuk EV, Trushkin AG, Shcherbin SY, Arefyeva RG (1991) Depletion of the ozone total content inside the polar cap after the solar flares. Dokl (Trans) AN SSSR Geofiz 318:576–580

Shumilov OI, Henriksen K, Raspopov OM, Kasatkina EA (1992a) Arctic ozone abundance and solar proton events. Geophys Res Lett 19:1647–1650

Shumilov OI, Kasatkina EA, Raspopov OM, Henriksen K (1992b) Influence of solar and galactic cosmic rays on polar ozone abundance. In: Proceedings of 19th annual European meeting on atmospheric studies by optical methods, IRF Sci. Report 209. Kiruna, pp 23–28

Shumilov OI, Vashenyuk EV, Henriksen K (1993) Quasi-drift effects of high-energy solar cosmic rays in the magnetosphere. J Geophys Res 98(A10):17423–17427

Shumilov OI, Kasatkina EA, Henriksen K, Raspopov OM (1995) Ozone "miniholes" initiated by energetic solar protons. J Atm Terr Phys 57(6):665–671

Shumilov OI, Kasatkina EA, Raspopov OM, Henriksen K (1996a) High-latitude ozone "miniholes" and solar protons. Geomagn Aeron 36(6):15–21

Shumilov OI, Kasatkina EA, Henriksen K, Vashenyuk EV (1996b) Enhancement of stratospheric aerosols after solar proton event. Ann Geophys 14:1119–1123

Shurshakov VA, Panova NA, Petrov VM, Bazilevskaya GA, Makhmutov VS, Struminsky AB, Dachev TsP (1993) Observations of energetic particles aboard MIR space station and in the stratosphere during solar proton events. In: Proceedings of 23rd international cosmic ray conference, vol 3. Calgary, pp 440–443

Sierks H, Elendt I, Dröge W, Mann G, Aurass H, Kunow H, Müller-Mellin R, Sequeiros J, McKenna-Lawlor S (1997) Electron spectra and associated type II and type III radio bursts of the July 9, 1996 solar event. In: Proceedings of 25th international cosmic ray conference, vol 1. Durban, pp 297–300

Silberberg R, Tsao CH (1979) Biological hazards associated with cosmic-ray and solar flare exposure. In: Proceedings of 16th international cosmic ray conference, vol 5. Kyoto, pp 317–322

Silberberg R, Tsao CH, Adams JH Jr, Letaw JR (1987) Radiation hazard in space. Aerosp Amer 25 (10):38–41

Simnett GM (1985) The fate of sunward streaming protons associated with coronal mass ejections. Astron Astrophys 145:139–143

Simnett GM (1986) A dominant role for protons at the onset of solar flares. Solar Phys 106(1):165–183

Simnett GM (1991) Energetic particle production in flares. Phil Trans Roy Soc London A336 (1643):439–450

Simnett GM (1995) Protons in flares. Space Sci Rev 73:387–432

Simnett GM, Haines MG (1990) On the production of hard X-rays in solar flares. Solar Phys 130 (1/2):253–263

Simnett GM, Roelof EC (2005) Timing of the relativistic proton acceleration responsible for the GLE on 20 January, 2005. In: Proceedings of the 29th international cosmic ray conference, vol 1. Pune, pp 101–104

Simonsen LC, Nealy JE, Sauer HH, Townsend LW (1991) Solar flare protection for manned lunar missions: analysis of the October 1989 proton flare events. SAE technical paper series, No.911351, pp 1–11

Simpson JA (1948) The latitude dependence of neutron densities in the atmosphere as a function of altitude. Phys Rev 73:1389

Simpson JA (1957) Cosmic-radiation neutron intensity monitor. In: Annals of the IGY, vol 4. Pergamon, London, pp 351–373

Simpson JA (1990) Astrophysical phenomena discovered by cosmic rays and solar flare Ground Level Events: the early years. In: Proceedings of 21st international cosmic ray conference, invited papers, highlight papers, Miscellaneous, vol 12. Adelaide, pp 187–195

Sittler ECJ, Guhathakurta M (1999) Semi-empirical two-dimensional magnetohydrodynamic model of the solar corona and interplanetary medium. Astrophys J 523:812–826

Sladkova AI (1996) A solar proton event database. Radiat Meas 26(3):447–449

Sladkova AI, Bazilevskaya GA, Ishkov VN et al (in all 8 authors) (1998) In: Logachev YuI (ed) Catalogue of solar proton events 1987–1996. Moscow University Press, Moscow, 246 pp

Smart DF (1996) Solar cosmic rays. In: N, Lamanna E (eds) Proceedings of 24th international cosmic ray conference, invited, rapporteur and highlight papers, vol 19c. Rome, Il Nuovo Cimento C (5), pp 765–775

Smart DF, Shea MA (1971) Solar proton event classification system. Solar Phys 16(2):484–487

Smart DF, Shea MA (1979) PPS76 – a computerized "event mode" solar proton forecasting technique. In: Donnelly RF (ed) Solar-terrestrial prediction proceedings, vol 1. Washington, D.C., pp 406–427

Smart DF, Shea MA (1984) Cosmic rays exposure factors for Shuttle altitudes derived from calculated cut-off rigidities. Adv Space Res 4(10):161–164

Smart DF, Shea MA (1989a) PPS-87: a new event oriented solar proton prediction model. Adv Space Res 9(10):(10)281–(10)284

Smart DF, Shea MA (1989b) Solar proton events during the past three solar cycles. J Spacecraft Rockets 26(6):403–415

Smart DF, Shea MA (1990a) The concept of using the Deep River and Kerguelen neutron monitors as "flagship" stations for ground-level solar cosmic ray events. In: Proceedings of 21st international cosmic ray conference, vol 5. Adelaide, pp 144–147

Smart DF, Shea MA (1990b) Probable pitch angle distribution and spectra of the 23 February 1956 solar cosmic ray event. In: Proceedings of 21st international cosmic ray conference, vol 5. Adelaide, pp 257–260

Smart DF, Shea MA (1991) A comparison of the magnitude of the 29 September 1989 high energy event with solar cycle 17, 18 and 19 events. In: Proceedings of 22nd international cosmic ray conference, vol 3. Dublin, pp 101–104

Smart DF, Shea MA (1993) Predicting and modeling solar flare generated proton fluxes in the inner heliosphere. In: Swenberg CE et al (eds) Biological effects and physics of solar and galactic cosmic radiation, Part B. Plenum Press, New York, pp 101–117

Smart DF, Shea MA (1994) Geomagnetic cutoffs: a review for space dosimetry applications. Adv Space Res 14(10):787–796

Smart DF, Shea MA (1996) The longitudinal distribution of solar flares associated with solar proton events at the Earth. Adv Space Res 17(2):(2)113–(2)116

Smart DF, Shea MA (1997) The >10 MeV peak flux distribution. In: Heckman G et al (eds) Solar-terrestrial predictions-V. RWC Tokyo, Hiraiso Solar Terrestrial Research Center, Communications Research Laboratory, Hitachinaka, pp 449–452

Smart DF, Shea MA (1997a) World grid of cosmic ray vertical cutoff rigidities for Epoch 1990.0. In: Proceedings of the 25th international cosmic ray conference, vol 2. pp 401–404

Smart DF, Shea MA (1997b) Calculated cosmic ray cutoff rigidities at 450 km for Epoch 1990.0. In: Proceedings of the 25th international cosmic ray conference, vol 2. pp 397–400

Smart DF, Shea MA (1998) Preliminary analysis of GLE of 6 November 1997. Eos Trans AGU Spring Meet Suppl 79(17):S268 (abstract)

Smart DF, Shea MA, Flückiger EO (1987a) Unusual aspects of the ground-level cosmic ray event of 7/8 December 1982. In: Proceedings of 20th international cosmic ray conference, vol 3. Moscow, pp 135–138

Smart DF, Shea MA, Gentile LC (1987b) Vertical cutoff rigidities calculated using the estimated 1985 geomagnetic field coefficients. In: Proceedings of 20th international cosmic ray conference, vol 4. Moscow, pp 204–207

Smart DF, Shea MA, Gentile LC (1993a) The relativistic solar proton event on 11 June 1991. In: Proceedings of 23rd international cosmic ray conference, vol 3. Calgary, pp 55–58

Smart DF, Shea MA, Gentile LC (1993b) The relativistic solar proton event on 15 June 1991. In: Proceedings of 23rd international cosmic ray conference, vol 3. Calgary, pp 59–62

Smart DF, Shea MA, McCracken KG (2006) The Carrington event: possible solar proton intensity–time profile. Adv Space Res 38:215–225

Smart DF, Shea MA, Dreschhoff GAM, McCracken KG (2008a) Solar proton fluence for 31 solar cycles derived from nitrate enhancements in polar ice. In: Caballero R, D'Olivo JC, Medina-Tanco G, Nellen L, Sánchez FA, Valdés-Galicia JF (eds) Proceedings of 30th international

cosmic ray conference, vol 1. Universidad Nacional Autónoma de México, Mexico City, pp 725–728
Smart DF, Shea MA, Dreschhoff GAM, McCracken KG (2008b) Solar proton fluence for 31 solar cycles derived from nitrate enhancements in polar ice. In: Proceedings of 30th international cosmic ray conference, vol 1. Merida, Mexico, pp 718–721
Smart DF, Shea MA, Flückiger EO, Debrunner H, Humble JE (1990) Were solar neutrons observed by neutron monitors on 1984 April 25? Astrophys J Suppl 73:269–272
Smart DF, Shea MA, Wilson MD, Gentile LC (1991) Solar cosmic rays on 29 September 1989: An analysis using the world-wide network of cosmic ray stations. In: Proceedings of 22nd international cosmic ray conference, vol 3. Dublin, pp 97–100
Smart DF, Shea MA, Dreschhoff GAM, McCracken KG (2008a) Solar proton fluence for 31 solar cycles derived from nitrate enhancements in polar ice. In: Caballero R, D'Olivo JC, Medina-Tanco G, Nellen L, Sánchez FA, Valdés-Galicia JF (eds) Proceedings of 30th international cosmic ray conference, vol 1. Universidad Nacional Autónoma de México, Mexico City, pp 725–728
Smart DF, Shea MA, O'Brien K (1995) Comment on the minimum mass path of a unidirectional solar neutron particle beam through the Earth atmosphere. In: Proceedings of 24th international cosmic ray conference, vol 4. Rome, pp 171–174
Smith CW, Bieber JW (1993) Detection of steady magnetic helicity in low-frequency IMF turbulence. In: Proceedings of the 23rd international cosmic ray conference, vol 3. Calgary, pp 493–496
Smith EJ, Tsurutani BT, Rosenberg RL (1978) Observations of the interplanetary sector structure up to heliographic latitudes of 16°: pioneer 11. J Geophys Res 83(2):717–724
Smith DM, Share GH, Murphy RJ, Schwartz RA, Shin AY, Lin RP (2003) High-resolution spectroscopy of gamma-ray lines from the X-class solar flare of 2002 July 23. Astrophys J Lett 595:L81–L84
Solar-Geophysical Data (1989a) Prompt reports 542, part I. National Geophysical Data Center, Boulder, p 18
Solar-Geophysical Data (1989b) Prompt reports 543, part I. National Geophysical Data Center, Boulder, pp 158–160
Solar-Geophysical Data (1990) Prompt reports 547, part II. National Geophysical Data Center, Boulder, pp 112–115
Somov BV (1986) Non-neutral current sheets and solar flare energetics. Astron Astrophys 163:210–218
Somov BV (1987) Solar flares. In: Shcherbina-Samoilova IS (ed) Solar flares, vol 34. VINITI, Moscow, pp 78–135
Somov BV (1992) Physical processes in solar flares. Kluwer Academic Publ, Dordrecht/Boston/London, 248 pp
Somov BV (1991) Magnetically driven coronal transients. Adv Space Res 11(1):179–185
Somov BV (1994) Fundamentals of cosmic electrodynamics. Kluwer Academic Publ, Dordrecht, Boston, London, 364 pp
Somov BV (1996) Reconnection and acceleration to high energies in flares. In: Ramaty R, Mandzhavidze N, Hua X-M (eds) High energy solar physics, vol 374, AIP conference proceedings. AIP, New York, pp 493–497
Somov BV (2012) Plasma astrophysics: 1. Fundamentals and practice. 2. Reconnection and flares. Springer, New York
Somov BV, Oreshina AV (2011) Magnetic reconnection and acceleration of particles on the sun. Bull Russian Acad Sci Phys 75(6):735–737
Speiser TW (1965) Particle trajectories in model current sheets. J Geophys Res 70(17):4219–4226
Spicer DS (1979) A possible observational test to distinguish between slow and transient buildup prior to solar flares. Solar Phys 64(1):121–128
Spruit HC (1974) A model of the solar convection zone. Solar Phys 34:277–290
Starkov GV, Roldugin VK (1994) Relationship between atmospheric transparency variations and geomagnetic activity. Geomagn Aeron 34(4):559–562 (in English)

Steele CDC, Priest ER (1989) The eruption of a prominence and coronal mass ejection which drive reconnection. Solar Phys 119:157–195

Steigies CT, Heber B, Wimmer-Schweingruber RF, Kudela K, Strharsky I, Langer R, Usoskin I, Ibragimov A, Fluckiger EO, Butikofer R, Eroshenko E, Belov A, Yanke V, Klein K-L, Fuller N, Mavromichalaki H, Papaioannou A, Sarlanis C, Souvatzoglou G, Plainaki C, Gerontidou M, Papailiou M-C, Mariatos G, Chilingaryan A, Hovsepyan G, Reymers A, Parisi M, Kryakunova O, Tsepakina I, Nikolayevskiy N, Dorman L, Pustil'nik L, Garcia-Poblacion O (2009a) NMDB collaboration. In: Proceedings of 31st international cosmic ray conference, vol 1. Łodz, (CD-ROM)

Steigies CT, Heber B, Wimmer-Schweingruber RE et al (NMDB collaboration, in all 33 authors) (2009b) NMDB Collaboration. In: Proceedings of 31st international cosmic ray conference. Lodz, CD- ROM

Steinacker J, Meyer J, Steinacker A, Reames DV (1997) The helium valley: comparison of impulsive solar flare ion abundances and gyroresonant acceleration with oblique turbulence in a hot multi-ion plasma. Astrophys J 476:403–427

Steinolfson RS, Mullan DJ (1980) Magnetohydrodynamic shock propagation in the vicinity of a magnetic neutral sheet. Astrophys J 241(3):1186–1194

Stephenson JAE, Scourfield MWJ (1992) Ozone depletion over the polar caps caused by solar protons. Geophys Res Lett 12(24):2425–2428

Stoker PH (1995) Relativistic solar proton events. Space Sci Rev 73(3/4):327–385

Stoker PH, Makgamate S (1990) Spectra of relativistic solar proton ground-level events recorded at Sanae, Antarctica. Astrophys J Suppl 73:263–268

Stoker PH, Heydenreich HEM, Bezuidenhout CN (1993) Ground level solar flare proton events recorded by the Sanae neutron monitor since 1966. In: Proceedings of 23rd international cosmic ray conference, vol 3. Calgary, pp 43–46

Stoker PH, Bieber JW, Evenson P (1995) The phenomenology of the precursor increase on the 19 October 1989 Ground Level Event. In: Proceedings of 24th international cosmic ray conference, vol 4. Rome, pp 224–227

Stolpovsky VG, Erdös G, Kahler SW, Daibog EI, Logachev YuI (1995) Solar energetic electrons observed by Phobos-2. In: Proceedings of 24th international cosmic ray conference, vol 4. Rome, pp 301–304

Stolpovsky VG, Daibog EI, Svertilov SI, Kahler SW, Erdös G (1997) Influence of coronal mass ejections on the time behaviour of energetic electrons in SEP events. In: Proceedings of 25th international cosmic ray conference, vol 1. Durban, pp 189–192

Stozhkov YuI (1978) Global magnetic field of the Sun and effects in cosmic rays. In: Kocharov GE (ed) IX Leningrad international seminar on cosmophysics. Leningrad, pp 39–52

Stozhkov YuI, Svirzhevsky NS, Bazilevskaya GA, Kvashnin AN, Makhmutov VS, Svirzhevskaya AK (2009) Long-term (50 years) measurements of cosmic ray fluxes in the atmosphere. Adv Space Res 44:1124–1137

Struminsky A, Belov A (1997) Neutron monitor sensitivity to primary protons below 3 GV derived from data of Ground Level Events. In: Proceedings of 25th international cosmic ray conference, vol 1. Durban, pp 201–204

Struminsky A, Zimovets I (2009) On estimates of first solar proton arrival. In: Kiraly P, Kudela K, Stehlik M, Wolfendale AW (eds) Proceedings of 21st ECRS. VIENALA s.r.o., Kosice. pp 237–241

Struminsky A, Matsuoka M, Takahashi K (1994) Evidence of additional production of high energy neutrons during the solar flare on 1991 June 4. Astrophys J 429:400–405

Sturrock PA (1974) In: Newkirk G Jr (ed) Coronal disturbances. IAU Symposium No.57. pp 437

Svestka Z (1976) Solar flares. Reidel, Dordrecht, p 400

Svestka Z (1995) On "The solar flare myth" postulated by Gosling. Solar Phys 160:53–56

Svestka Z (1996) Speeds of rising post-flare structures. Solar Phys 169(2):403–412

Svestka Z, Farnik F, Hudson HS, Uchida Y, Hick P, Lemen JR (1995) Large-scale coronal phenomena in Yohkoh SXT images. Solar Phys 161(2):331–363

Swinson DB, Shea MA (1990) The September 29, 1989, ground level event observed at high rigidity. Geophys Res Lett 17(8):1073–1075

Syrovatsky SI (1976) Characteristics of the current sheet and heat trigger of solar flares. Pis'ma v Astron Zh 2(1):35–38 (in Russian)

Syrovatsky SI (1981) Pinch sheets and reconnection in astrophysics. Ann Rev Astron Astrophys 19:163–229

Takahashi K (1989) Detection of solar neutrons by ground-based neutron monitor. Space Sci Rev 51:123–146

Takahashi K, Sakamoto E, Matsuoka M, Nishi K, Yamada Y, Shimoda S, Shikata T, Wada M, Kusunose M, Yoshimori M, Kondo I (1991) Observation of solar neutrons by Mt. Norikura neutron monitor during a period of solar cycle 22. In: Proceedings of 22nd international cosmic ray conference, vol 3. Dublin, pp 37–40

Takahashi K, Wada M, Sakamoto E, Matsuoka M, Munakata K, Kohno I (1990) Observation of high energy solar particles on September 29, 1989, by neutron monitors with high time resolution. Proc Jpn Acad Ser B 66:10–14

Tatischeff V, Kozlovsky B, Kiener J, Murphy RJ (2006) Delayed X- and gamma-ray line emission from solar flare radioactivity. Astrophys J Suppl 165:606–616

Teltsov MV, Tverskaya LV (1992) Radiation dose measurements on board the station MIR during solar proton flares in September-October 1989. In: Fischer S, Vandas M (eds) Proceedings 1st SOLTIP symposium, Liblice, Czechoslovakia, 30 September–5 October 1991, vol 2. Astronomical Institute CSAV, Prague, pp 233–237

Temerin M, Roth I (1992) The production of ^{3}He and heavy ion enrichments in ^{3}He -rich flares by electromagnetic hydrogen cyclotron waves. Astrophys J Lett 391:L105–L108

Temerin M, Roth I (1996) Selective enrichment of energetic ions in impulsive solar flares. In: Ramaty R, Mandzhavidze N, Hua X-M (eds) High energy solar physics, vol 3, AIP conference proceedings. AIP, New York, pp 435–444

Terekhov OV, Kuzmin AG, Sunyaev RA, Tkachenko AY, Denisenko DV, Barat C, Talon R, Vedrenne G (1996) Observations of high energy solar flares by PHEBUS experiment of GRANAT Observatory. Astron Lett (Russia) 22(3):362–370

Thomas BC, Arkenberg KR, Brock R, Snyder BR II, Melott AL (2013) Terrestrial effects of possible astrophysical sources of an AD 774–775 increase in carbon-14 production. Geophys Res Lett 40:1237–1240. doi:10.1002/grl.50222 (accepted)

Thompson DJ, Maclennan CG, Lanzerotti LJ (1995) Propagation of solar oscillations through the interplanetary medium. Nature 376:139–144

Thompson DJ, Lanzerotti LJ, Maclennan CG (1998) Periodicities in interplanetary particles and fields, and relationships to solar modes. Eos Trans AGU Spring Meet Suppl 79(17):S285 (abstract)

Tikhonov AN, Arsenin VY (1979) Incorrect problem solution methods. Nauka, Moscow, p 195

Timashkov DA, Balabin YV, Borog VV, Kompaniets KG, Petrukhin AA, Room DA, Vashenyuk EV, Shutenko VV, Yashin II (2008) Ground-Level Enhancement of December 13, 2006 in muon hodoscopes data. In: Caballero R, D'Olivo JC, Medina-Tanco G, Nellen L, Sánchez FA, Valdés-Galicia JF (eds) Proceedings of 30th international cosmic ray conference, vol 1. Universidad Nacional Autónoma de México, Mexico City, pp 209–212

Timofeev GA, Getselev IV, Kontor NN, Khein LA, Likin OB, Morozova EI, Moszhukhina AR, Pisarenko NF, Polyanskaya MG, Reizman SY (1996) Charged particles fluences and doses on the "INTERBALL" series of spacecraft trajectories. Radiat Meas 26(3):517–519

Tinsley BA, Deen GW (1991) Apparent tropospheric response to MeV-GeV particle flux variations: a connection via electrofreezing of supercold water in high-level clouds? J Geophys Res 96(D12):22283–22296

Tinsley BA, Brown GM, Scherrer PH (1989) Solar variability influences on weather and climate: possible connections through cosmic ray fluxes and storm intensification. J Geophys Res 94 (D12):14783–14792

Tonwar S, on behalf of the L3 Collaboration (2001) Variations in cosmic ray intensity observed with the L3+Cosmics shower array detectors and the intense solar flare on 14 July 2000. In: Proceedings of the 27th international cosmic ray conference, vol 8. Hamburg, pp 3375–3378

Toptygin IN (1973) Direct and inverse problems of propagation of cosmic rays in interplanetary space. Geomagn Aeron 13(2):212–218

Toptygin IN (1985) Cosmic rays in interplanetary magnetic fields. D. Reidel Publ. Co., Dordrecht, p 375

Torsti J, Eronen T, Mahonen M, Riihonen E, Schultz CG, Kudela K, Kananen H (1991) Search of peculiarities in the flux profiles of GLEs in 1989. In: Proceedings of 22nd international cosmic ray conference vol 3. Dublin, pp 141–144

Torsti J, Eronen T, Mahonen M, Riihonen E, Schultz CG, Kudela K, Kananen H (1992) Estimation of high energy solar particle transport parameters during the GLEs in 1989. In: Solar wind seven. Proceedings of 3rd COSPAR Colloq., Goslar, 16–20 Sept. 1991. Oxford, pp 545–548

Torsti J, Halen S, Vainio R (1994) Delayed energetic proton events at 1 AU, 3rd SOHO workshop, solar dynamic phenomena and solar wind consequences. Estes Park, 26–29 Sept 1994. Program and Abstracts

Torsti J, Kocharov LG, Vainio R, Anttila A, Kovaltsov GA (1996) The 1990 May 24 solar cosmic-ray event. Solar Phys 166(1):135–158

Townsend LW, Nealy JE, Wilson JW, Atwell W (1989) Large solar flare radiation shielding requirements for manned interplanetary missions. J Spacecr Rocket 26(2):126–128

Townsend LW, Zapp EN, Stephens DL Jr, Hoff JL (2003) Carrington flare of 1859 as a prototypical worst-case solar energetic particle event. IEEE Trans Nucl Sci 50(6):2307–2309

Townsend LW, Stephens DL Jr, Hoff JL, Zapp EN, Moussa HM, Miller TM, Campbell CE, Nichols TF (2006) The Carrington event: possible doses to crews in space from a comparable event. Adv Space Res 38:226–231

Tranquille C, Daly EJ (1992) An evaluation of solar-proton event models for ESA missions. ESA J 16(3):275–297

Treumann RA, Dubouloz N, Pottelette R (1995) Stochastic acceleration of electrons in localized wave fields. In: Proceedings of 24th international cosmic ray conference vol 4. Rome, pp 9–12

Tripathi RK, Cucinotta FA, Wilson JW (1996) Accurate universal parameterization of absorption cross sections. Nucl Instrum Method Phys Res B 117(4):347–349

Troitskaia EV, Gan WQ, Kuzhevskij BM, Miroshnichenko LI (2007) Solar plasma density and spectrum of accelerated particles derived from the 2.223 MeV line of a solar flare. Solar Phys 242:87–99

Troitskaia EV, Arkhangelskaja IV, Miroshnichenko LI, Arkhangelsky AI (2009) Study of the 28 October 2003 and 20 January 2005 solar flares by means of 2.223 MeV gamma- emission line. Adv Space Res 43(4):547–552

Troitskaia EV, Arkhangelskaja IV, Gan WQ (2011) Study of 20 January 2005 solar flare area by certain gamma-ray lines. In: Proceedings of 32nd international cosmic ray conference. Beijing, SH1.1, ID0702 (CD-ROM)

Trottet G, Barat C, Ramaty R, Vilmer N, Dezalay JP, Kuznetsov A, Mandzhavidze N, Sunayaev R, Talon R, Terekhov O (1996) Thin target γ-ray line production during the 1991 June 1 flare. In: Ramaty R, Mandzhavidze N, Hua X-M (eds) High energy solar physics, vol 374. AIP Conference Proceedings, AIP, New York, pp 153–161

Trottet G, Vilmer N, Barat C, Dezalay J-P, Talon R, Sunyaev R, Kuznetsov A, Terekhov O (1993) Temporal and spectral characteristics of the June 11, 1991 gamma- ray flare. Astron Astrophys Suppl 97:337–339

Tsyganenko NA (1989) A magnetospheric magnetic field model with a warped tail current sheet. Planet Space Sci 37:5–20

Tsyganenko NA, Usmanov AV (1982) Determination of the magnetospheric current sheet system parameters and development of experimental magnetic field models based on data from IMP and HEOS satellites. Planet Space Sci 30(10):985–998

Turner RE, Diamond PS, Baker JC (1998) Solar event forecast tool. Eos Trans AGU Spring Meet Suppl 79(S268) (abstract)

Tverskaya LV, Teltsov MV, Shumshurov VI (1991) Measurement of radiation doze on board the Mir station during solar proton events in September-October 1989. Geomagn Aeron 31 (5):928–930

Tverskoi BA (1967a) On the theory of statistical Fermi acceleration. Sov Phys JETP 52(2):483–497

Tverskoi BA (1967b) On the theory of turbulent acceleration of charged particles in plasma. Sov Phys JETP 53(4(10)):1417–1429

Tverskoi BA (1968) Dynamics of the earth's radiation belts. Nauka, Moscow, p 224

Tverskoi BA (1981) Propagation modes of energetic charged particles in the heliosphere. Adv Space Res 1(3):5–14

Tylka A, Dietrich W (2009) A new and comprehensive analysis of proton spectra in ground-level enhanced (GLE) solar particle events. In: Proceedings 31th international cosmic ray conference. Lod/z

Tylka AJ, Boberg PR, Adams JH Jr, Beahm LP, Dietrich WF, Kleis T (1995) The mean ionic charge state of solar energetic Fe ions above 200 MeV per nucleon. Astrophys J Lett 444: L109–L113

Tylka AJ, Dietrich WF, Boberg PR (1997a) Observations of very high energy solar heavy ions from IMP-8. In: Proceedings of 25th international cosmic ray conference, vol 1. Durban, pp 101–104

Tylka AJ, Adams JH, Boberg PR, Brownstein B, Dietrich WF, Flückiger EO, Petersen EL, Shea MA, Smart DF, Smith EC (1997b) CREME96: a revision of the cosmic ray effects on microelectronics code. IEEE Trans Nucl Sci 44(6):2150–2160

Tylka AJ, Cohen CMS, Dietrich WF, Lee MA, Maclennan CG, Mewaldt RA, Ng CK, Reames DV (2005) Shock geometry, seed population, and the origin of variable elemental composition at high energies in large gradual solar particle events. Astrophys J 625:474–495

Usoskin IG, Kovaltsov GA (1997) Deduction of the neutron flux from solar flares by the ground-based neutron monitors. Geomagn Aeron 37(3):154–157

Usoskin IG, Kovaltsov GA (2012) Occurrence of extreme solar particle events: assessment from historical proxy data. Astrophys J 757:92 [DOI], [ADS], [arXiv:1207.5932]

Usoskin IG, Kovaltsov GA, Kocharov LG, Kananen H, Tanskanen P (1995) High energy neutrons from the June 15, 1991 solar flare as detected by the neutron monitors. In: Proceedings of 24th international cosmic ray conference, vol 4. Rome, pp 151–154

Usoskin IG, Kovaltsov LG, Kananen H, Tanskanen P (1997) The world neutron monitor network as a tool for the study of solar neutrons. Ann Geophys 15:375–386

Usoskin IG, Kovaltsov GA, Mironova IA, Tylka AJ, Dietrich WF (2011) Ionization effect of solar particle GLE events in low and middle atmosphere, *Atmos* Chem Phys 11:1979–1988. doi:10.5194/acp-11-1979-2011. www.atmos-chem-phys.net/11/1979/2011/

Usoskin IG, Solanki SK, Kovaltsov GA, Beer J, Kromer B (2006) Solar proton events in cosmogenic isotope data. Geophys Res Lett 33, L08107. doi:10.1029/2006GL026059

Usoskin IG, Kromer B, Ludlow F, Beer J, Friedrich M, Kovaltsov GA, Solanki SK, Wacker L (2013) The AD775 cosmic event revisited: the Sun is to blame. Astron Astrophys 552, L3, 4 p, http://dx.doi.org/10.1051/0004-6361/201321080

Ustinova GK, Lavrukhina AK (1987) The features of modulation mechanism over a long-time scale. In: Proceedings of 20th international cosmic ray conference, vol 4. Moscow, pp 307–310

Vainio R, Torsti J, Kocharov LG, Anttila A (1995a) GOES response to energetic protons of different origin. In: Proceedings 24th international cosmic ray conference, vol 4. Rome, pp 131–134

Vainio R, Torsti J, Kocharov LG, Anttila A, Kovaltsov GA (1995b) Injection and transport of accelerated particles as deduced from GOES and neutron monitor data. In: Proceedings of 24th international cosmic ray conference, vol 4. Rome, pp 321–324

Vainio R, Desorgher L, Heynderickx D et al (in all 12 authors) (2009) Dynamics of the Earth's particle radiation environment. Space Sci Rev 147:187–231. doi:10.1007/s11214-009-9496-7

Valdes-Galicia JF (1991) Transport of energetic particles in the interplanetary medium. Nucl Phys B (Proc Suppl) 22B:46

Valdes-Galicia JF (1993) Energetic particle transport coefficients in the heliosphere. Space Sci Rev 62(1/2):67–93

Valdes-Galicia JF, Alexander P (1997) Is it possible to find evidence of waves generated by solar energetic protons? Solar Phys 176(2):327–354

Valdes-Galicia JF, Wibberenz G, Quenby JJ, Moussas X, Green G, Neubauer FM (1988) Pitch-angle scattering of solar particles: comparison of "particle" and "field" approach. I. Strong scattering. Solar Phys 117(1):135–156

Valdes-Galicia JF, Alexander PM, Otaola JA (1995) Solar energetic protons self- generated waves in the interplanetary magnetic field (A search for evidences). In: Proceedings of 24th international cosmic ray conference, vol 4. Rome, pp 273–276

Vampola AL (1994) Analysis of environmentally induced spacecraft anomalies. J Spacecr Rocket 31(2):154–159

Van Allen JA, Winckler JR (1957) Spectrum of low-rigidity cosmic rays during the solar flare of February 23, 1956. Phys Rev 106:1072–1073

Van Hollebeke MAI, Ma Sung LS, McDonald FB (1975) The variation of solar proton energy spectra and size distribution with heliolongitude. Solar Phys 41(1):189–223

Van Hollebeke MAI, McDonald FB, Meyer J-P (1990) Solar energetic particle observations of the 1982 June 3 and 1980 June 21 gamma-ray/neutron events. Astrophys J Suppl 73:285–296

Vandas M, Fischer S, Geranios A (1995) On energetic particle flux variations inside a magnetic clouds. J Geophys Res 100(A??): 23507–23513

Vandas M, Fischer S, Geranios A, Dryer M, Smith Z, Detman T (1996) Magnetic traps in the interplanetary medium associated with magnetic clouds. J Geophys Res 101(A10):21589–21596

Vashenyuk EV (2000) Some regularities in the occurrence rate of the events with relativistic solar cosmic rays. Astronomichesky Vestn 34(2):173–176

Vashenyuk EV, Miroshnichenko LI (1998) Characteristics of generation and transport of solar protons during the event of September 29, 1989. Geomagn Aeron 38(2):129–135

Vashenyuk EV, Pchelkin VV (1998) The GLE of September 29, 1989 study by computations and experimental data analysis. In: Medina J (ed) Rayos comicos-98 (proceedings of 16th European cosmic ray symposium). Alcala University Press, Alcalá de Henares, pp 145–148

Vashenyuk EV, Miroshnichenko LI, Sorokin MO, Perez-Peraza J, Gallegos-Cruz A (1993) Search for peculiarities of proton events of 22nd solar cycle based on ground observation data. Geomagn Aeron 33(5):1–10

Vashenyuk EV, Miroshnichenko LI, Sorokin MO, Perez-Peraza J, Gallegos- Cruz A (1994) Large ground level events in solar cycle 22 and some peculiarities of relativistic proton acceleration. Adv Space Res 14(10):711–716

Vashenyuk EV, Fischer S, Vandas M, Shea MA, Smart DF (1995) Peculiarities of anisotropy during Ground Level Events of the 22nd cycle of solar activity. In: Proceedings of 24th international cosmic ray conference, vol 4. Rome, pp 317–320

Vashenyuk EV, Miroshnichenko LI, Perez-Peraza J, Kananen H, Tanskanen P (1997) Generation and propagation characteristics of relativistic protons during the GLE of September 29, 1989. In: Proceedings of 25th international cosmic ray conference, vol 1. Durban, pp 161–164

Vashenyuk EV, Miroshnichenko LI, Gvozdevsky BB (2000) Proton energy spectrum and source parameters of the September 29, 1989 event. Nuovo Cim 23(3):285–291

Vashenyuk EV, Pchelkin VV, Miroshnichenko LI (2001) Flux and spectrum dynamics of relativistic protons inferred from modeling ground-level increases in the event of September 29, 1989. Izv RAN Phys Ser 65(3):350–352
Vashenyuk EV, Balabin YV, Miroshnichenko LI (2008) Relativistic solar protons in the GLE of 23 February 1956: New study. Adv Space Res 41(6):926–935
Vashenyuk EV, Balabin YuV, Gvozdevsky BB (2009) Characteristics of relativistic solar cosmic rays from GLE modeling studies. In: Proceedings of 31st international cosmic ray conference, Łodz. Paper 1304
Vashenyuk EV, Balabin YV, Gvozdevsky BB (2011) Features of relativistic solar proton spectra derived from GLE modeling. Astrophys Space Sci Trans 7:459–463
Veretenenko SV, Pudovkin MI (1993) Effects of cosmic ray variations in the circulation of lower atmosphere. Geomagn Aeron 33(6):35–40
Veretenenko SV, Pudovkin MI (1994) The galactic cosmic ray Forbush decrease effects on total cloudiness variations. Geomagn Aeron 34(4):463–468 (in English)
Vernazza JE, Avrett EH, Loeser R (1981) Structure of the solar chromosphere. III – Models of the EUV brightness components of the quiet Sun. Astrophys J Suppl 45:635–725
Vestrand WT, Forrest DJ (1993) Evidence of a spatially extended component of gamma rays from solar flares. Astrophys J Lett 409:L69–L72
Vestrand WT, Share GH, Murphy RJ, Forrest DJ, Rieger E, Chupp EL, Kanbach G (1999) The solar maximum mission atlas of gamma-ray flares. Astrophys J Suppl 120:409–467
Vitinsky YI, Ol AI, Sazonov BI (1976) The sun and atmosphere of the earth. Gidrometeoizdat, Leningrad, 351 pp
Vlahos L (1989) Particle acceleration in solar flares. Solar Phys 121(1/2):431–447
Vlahos L (1994) Theory of fragmented energy release in the sun. Space Sci Rev 68:39–50
Vlahos L, Machado ME, Ramaty R et al (in all 28 authors) (1989) Chapter 2: Particle acceleration. In: Kundu M, Woodgate B, Schmahl EJ (eds) Energetic phenomena on the sun. Astrophysics and space science library. Kluwer Academic Publishers, Dordrecht/Boston/London, pp 131–224
Völk HJ (1975) Cosmic ray propagation in interplanetary space. Rev Geophys Space Phys 13 (4):547–566
Völk HJ (1981) Acceleration of cosmic rays in the medium with due regard for the losses and diffused sources of gamma-radiation. Izv AN SSSR Phys Ser 45(7):1122–1136
Volodichev NN, Savenko IA (1981) High-energy solar protons. In: Proceedings 17th international cosmic ray conference, vol 3. Paris, pp 45–48
von Rosenvinge TT, Ramaty R, Reames DV (1981) Interplanetary particle observations associated with solar flare gamma-ray line emission. In: Proceedings 17th international cosmic ray conference, vol 3. Paris, pp 28–31
von Rosenvinge TT, Reames DV (1983) The delayed energetic particle event of June 6, 1979. In: Proceedings of the 18th international cosmic ray conference, vol 4. Bangalore, pp 148–151
Wang R (2009) Did the 2000 July 14 solar flare accelerate protons to $\geq$40 GeV? Astropart Phys 31 (2):149–155
Wang HT, Ramaty R (1974) Neutron propagation and 2.2 MeV gamma-ray line production in the solar atmosphere. Solar Phys 36(1):129–137
Wang Y-M, Sheeley NR Jr (1994) Global evolution of interplanetary sector structure, coronal holes, and solar wind streams during 1976–1993: Stackplot displays based on solar magnetic observations. J Geophys Res 99(A4):6597–6608
Wanner W, Wibberenz G (1993) A study of the propagation of solar energetic particles in the inner heliosphere. J Geophys Res 98(A3):3513–3528
Watanabe K, Muraki Y, Matsubara Y et al (in all 31 authors) (2005) Solar neutron events associated with large solar flares in solar cycle 23. In: Proceedings of 29th international cosmic ray conference, vol 1. Pune, pp 37–40

Watari S, Detman T, Joselyn JA (1996) A large scale arcade along the inversion line, observed on May 19, 1992 by Yohkoh, and enhancement of interplanetary energetic particles. Solar Phys 169(1):167–179

Wdowczyk J, Wolfendale AW (1977) Cosmic rays and ancient catastrophes. Nature 268 (5620):510–512

Webb GM (1981) Time-dependent Green functions of the cosmic rays equation of transport. Astrophys Space Phys 80(2):323–335

Webber WR (1963) Solar-flare proton data. Nucleonics 21(8):153–157

Webber WR (1964) A review of solar cosmic ray events. In: Hess WN (ed) AAS-NASA symposium on the physics of solar flares. NASA, Washington, D.C., pp 215–255

Webber WR (1967) Sunspot number and solar cosmic ray predictions for cycle 20 (1965-1975) with preliminary estimates for cycle 21. The Boeing Co. report D2-113522-1

Webber WR, Roelof EC, McDonald FB, Teegarden BJ, Trainor J (1975) *Pioneer 10* measurements of the charge and energy spectrum of solar cosmic rays during 1972 August. Astrophys J 199:482–493

Webber WR, Higbie PR, McCracken KG (2007) Production of the cosmogenic isotopes ^{3}H, ^{7}Be, ^{10}Be, and ^{36}Cl in the Earth's atmosphere by solar and galactic cosmic rays. J Geophys Res 112, A10106

Weeks LH, Cuikay RS, Corbin JR (1972) Ozone measurements in the mesosphere during the solar proton event of November 2, 1969. J Atm Sci 29(6):1138

Wheatland MS, Glukhov S (1998) Flare frequency distributions based on a master equation. Astrophys J 494:858–863

Wheatland MS, Sturrock PA (1996) Avalanche models of solar flares and the distribution of active regions. Astrophys J 471:1044–1048

Wibberenz G (1998) Transient effects and disturbed conditions: observations and simple models. Space Sci Rev 83:310–329

Wibberenz G, Kesckemety K, Kunow H, Somogyi A, Iwers B, Logachev YI, Stolpovsky VG (1989) Coronal and interplanetary transport of solar energetic protons and electrons. Solar Phys 124(2):353–392

Wibberenz G, Cane H, Richardson IG (1997) Two-step Forbush decreases in the inner solar system and their relevance for models of transient disturbances. In: Proceedings of 25th international cosmic ray conference, vol 1. Durban, pp 397–400

Wiik JE, Schmieder B, Heinzel P, Roudier T (1996) Post-flare loops of 26 June 1992. Solar Phys 166:89–106

Wild JP, Smerd SF, Weiss AA (1963) Solar bursts. Annu Rev Astron Astrophys 1:291–366

Wilkinson DC (1992) GOES space environment monitor, format description for 1- and 5-minute averaged data, GOES data distribution disk

Wilkinson DC (1994) National Oceanic and Atmospheric Administration's spacecraft anomaly database and examples of solar activity affecting spacecraft. J Spacecr Rocket 31(2):160–166

Wilkinson DC, Daughtridge SC, Stone JL, Sauer HH, Darling P (1991) TDRS-1 single event upsets and the effect of the space environment. IEEE Trans Nucl Sci 38(6):1708–1712

Wilkinson DC, Shea MA, Smart DF (2000) A case history of solar and galactic space weather effects on the geosynchronous communications Satellite TDRS- 1. Adv Space Res 26:27–30

Wilson BG, Mathews T, Johnson RH (1967) Intercomparison of neutron monitors during solar-flare increases. Phys Rev Lett 18(16):675–676

Wilson JW, Cucinotta FA, Shinn JL, Simonsen LC, Dubbed RR, Jordan WR, Jones TD, Chang CK, Kim MY (1999) Shielding from solar particle event exposures in deep space. Radiat Meas 30(3):361–382

Winckler JR, Bhavsar PD, Peterson L (1961) The time variations of cosmic rays during July 1959 at Minneapolis. J Geophys Res 66:995–1022

Wolff EW, Bigler M, Curran MAJ, Dibb JE, Frey MM, Legrand M, McConnell JR (2012) The Carrington event not observed in most ice core nitrate records. Geophys Res Lett 39:L08503. [DOI], [ADS]

Woodgate BE, Robinson RD, Carpenter KG, Maran SP, Shore SN (1992) Detection of a proton beam during the impulsive phase of a stellar flare. Astrophys J Lett 397:L95–L98

Xapsos MA, Summers GP, Burke EA (1998) Probability model for peak fluxes of solar proton events. IEEE Trans Nucl Sci 45(6):2948–2953

Yanke VG (1980) On the theory of geophysical effects of the secondary cosmic radiation. Candidate dissertation, IZMIRAN, Moscow, 198 pp

Yoshimori M, Morimoto K, Suga K, Matsuda T, Kawabata K (1995a) Downward beaming of ions and electrons in the 1991 November 15 flare. In: Proceedings of 24th international cosmic ray conference, vol 4. Rome, pp 98–101

Yoshimori M, Morimoto K, Suga K, Matsuda T, Kawabata K (1995b) Spectral characteristics of solar gamma-ray flares. In: Proceedings of 24th international cosmic ray conference, vol 4. Rome, pp 102–105

Yoshimori M, Saita N, Ostryakov VM, Kartavykh Yu.Yu (1997) Burst-like electron acceleration associated with plasma heating in a solar flare ofn 22 April, 1994. In: Proceedings of the 25th international cosmic ray conference, vol 1. Durban, pp 269–272

Yoshimori M, Suga K, Shiozawa A (2000) *Yokoh* observations of solar gamma-ray flare on November 6, 1997. Adv Space Res 25(8):1801–1804

Young CA, Ryan JM (1997) Neutron transport in the solar atmosphere. In: Proceedings of 25th international cosmic ray conference, vol 1. Durban, pp 21–24

Young CA, Ryan JM (1998) Low energy neutrons and the 2.2 MeV gamma-ray line. Eos Trans AGU Spring Meet Suppl 79(17):S258 (abstract)

Yucker WR (1972) Solar cosmic ray hazard to interplanetary and Earth-orbital space travel, NASA report-TMX-2440

Yukhimuk AK (1982) Plasma phenomena in geophysics. Naukova Dumka, Kiev, 157 pp

Zadorozhny AN, Kiktenko VN, Kokin GA, Raspopov OM, Shumilov OI, Tuchkov GA, Tyasto MI, Chizhov AF, Shtyrkov OV, Kasatkina EA, Vashenyuk EV (1992) The reaction of the middle atmosphere on the solar proton events in October 1989. Geomagn Aeron 32(2):32–40

Zadorozhny AN, Tyutin AA, Bragin OA, Kiktenko VN (1994) Recent measurements of middle atmospheric electric fields and related parameters. J Atm Terr Phys 56:321–325

Zank GP, Rice WKM, Wu CC (2000) Particle acceleration and coronal mass ejection driven shocks: a theoretical model. J Geophys Res 105A:25079–25095

Zank GP, Li G, Verkhoglyadova O (2007) Particle acceleration at interplanetary shocks. Space Sci Rev 130:255–272. doi:10.1007/s11214-007-9214-2

Zeller EJ, Dreschhoff GAM (1995) Anomalous nitrate concentrations in polar ice cores – do they result from solar particle injections in the polar atmosphere? Geophys Res Lett 22:2521–2524

Zeller EJ, Dreschhoff GAM, Laird CM (1986) Nitrate flux on the Ross Ice Shelf, Antarctic, and its relation to solar cosmic rays. Geophys Res Lett 13(12):1264–1267

Zhao X, Hoeksema JT (1994a) A coronal magnetic field model with horizontal volume and sheet currents. Sol Phys 151:91–105

Zhao X, Hoeksema JT (1994b) The effect of coronal mass ejections on the heliospheric current sheet. In: Poland A (ed) Solar dynamic phenomena and solar wind consequences. 3rd SOHO workshop, 26–29 Sept 1994, Estes Park, ESA SP-373, pp 321–324

Zharkova VV (2012) Electron and proton kinetics and dynamics in flaring atmosphere. Wiley-VCH, Weinheim

Zharkova VV, Kosovichev AG (1998) Seismic response to solar flares observed SOHO/MDI. In: Structure and dynamics of the interior of the sun and sun-like stars. Proceedings of the SOHO 6/GONG 98 Workshop, Boston, 1–4 June 1998, ESA SP-418, pp 661–665

Zharkova VV, Arzner K, Benz AO, Browning P, Dauphin C, Emslie AG, Fletcher L, Kontar EP, Mann G, Onofri M, Petrosian V, Turkmani R, Vilmer N, Vlahos L (2011) Recent advances in understanding particle acceleration processes in solar flares. Space Sci Review 159:357–420. doi:10.1007/s11214-011-9803-y

Zherebtsov GA, Kovalenko VA, Molodykh SI (1997) Heliospheric characteristics during fast global variations of solar magnetic fields. J Geophys Res 102(A2):2137–2145

Zhu FR, Wang RG, Tang YQ et al (2007) Observation of the 2005 January 20 GLE event using Yangbajing solar neutron telescope and neutron monitor. High Energy Phys Nucl 31(4):341–344

Zil MV, Kolomensky AV, Petrov VM (1986) Reduction of the dose from solar cosmic rays by geomagnetic field. Kosm Issled 24(6):944–947

Zil MV, Mitrikas VG, Petrov VM, Okhlopkov VP, Okhlopkova LS, Charakhchyan TN (1987) Kosm Issled 25(2):325–328

Zil MV, Mitrikas VG, Petrov VM (1988) Analysis of quasiperiodic variations and the technique of long-term prediction of the occurrence rate of solar proton events, Moscow, Deposited in VINITI (No.1238-V88), 33 pp

Zirin H (1994) Solar storminess. Sky Telesc 88(5):9

Zusmanovich AG, Shvartsman YE (1989) Solar cosmic rays of high energies. Geomagn Aeron 29 (3):353–358

Zwickl RD (1992) GOES energetic particle correction algorithm, GOES data distribution disk

Index

L. Miroshnichenko, *Solar Cosmic Rays*, Astrophysics and Space Science Library 405, DOI 10.1007/978-3-319-09429-8

MIX
Papier aus verantwortungsvollen Quellen
Paper from responsible sources
FSC® C105338

FSC
www.fsc.org

If you have any concerns about our products,
you can contact us on
ProductSafety@springernature.com

In case Publisher is established outside the EU,
the EU authorized representative is:
Springer Nature Customer Service Center GmbH
Europaplatz 3, 69115 Heidelberg, Germany

Printed by Libri Plureos GmbH
in Hamburg, Germany